HANDBOOK OF PLASTICS ANALYSIS

PLASTICS ENGINEERING

Founding Editor

Donald E. Hudgin

Professor
Clemson University
Clemson, South Carolina

Additional Volumes in Preparation

HANDBOOK OF PLASTICS ANALYSIS

edited by
Hubert Lobo
DatapointLabs
Itchaca, New York, U.S.A.

Jose V. Bonilla
ISP Chemicals
Calvert City, Kentucky

MARCEL DEKKER, INC.   NEW YORK · BASEL

Library of Congress Cataloging-in-Publication Data
A catalog record for this book is available from the Library of Congress.

ISBN: 0-8247-0708-7

This book is printed on acid-free paper.

Headquarters
Marcel Dekker, Inc., 270 Madison Avenue, New York, NY 10016, U.S.A.
tel: 212-696-9000; fax: 212-685-4540

Distribution and Customer Service
Marcel Dekker, Inc., Cimarron Road, Monticello, New York 12701, U.S.A.
tel: 800-228-1160; fax: 845-796-1772

Eastern Hemisphere Distribution
Marcel Dekker AG, Hutgasse 4, Postfach 812, CH-4001 Basel, Switzerland
tel: 41-61-260-6300; fax: 41-61-260-6333

World Wide Web
http://www.dekker.com

The publisher offers discounts on this book when ordered in bulk quantities. For more information, write to Special Sales/Professional Marketing at the headquarters address above.

Preface

Plastics are one of the enabling technologies of the 20th century. They are among the most complex engineering materials being used in the world today, with amazing properties that have revolutionized the way in which products are manufactured. They are used in almost every walk of life, ranging from the mundane to high-end applications in which no other material could serve as a replacement. In each of these applications, it has been crucially important to understand their behavior through various parts of their product life, from manufacture to utilization and eventually their reclaim or disposal. The tools and techniques used to develop this understanding are referred to as plastics analysis.

Plastics analysis can be broadly grouped into two main categories. *Physical analysis* refers to the evaluation of the physical behavior of the material. Properties such as strength, thermal behavior, and flow properties fall into this category, as do failure and morphological characteristics. *Chemical analysis* seeks to evaluate the compositional characteristics of the polymer. The combination of these two broad approaches has been used successfully to correlate the behavior of plastics to their composition.

A wide variety of modern tools are available to the plastics analyst. As the range of available tools continue to expand, it is necessary for the analyst to keep abreast of all these technologies so as to be able to apply the most appropriate technique to the solution of a particular problem. This handbook seeks to highlight the most prominent tools in use by providing information on these diverse techniques and their application, and provides guidelines on the analysis and interpretation of results. It also provides a ready source of detailed references to readers interested in a more complete understanding of the subject matter.

In order to maintain a practical focus, the book concentrates on an approach that is more phenomenological than theoretical. While not going into detailed derivations, the book sets forth the basic governing equations where necessary to provide a good theoretical understanding of the techniques. Through the use of case studies and illustrations, the reader will be aided in the understanding of possible outcomes of each analysis technique.

A number of plastics analysis techniques are currently standardized to national and international norms. The book lists these norms in the form of reference tables and provides brief descriptions where necessary.

The chapters contain:

Introduction of the technique and a brief scientific basis; governing equations if applicable

Illustrations of test instruments along with schematics to aid in the understanding of the techniques

Detailed descriptions, including measurement method(s), highlighting differences in technique(s), if relevant, including merits and deficiencies of the technique

Images of typical outcomes of the analysis

A listing of applicable national and international standards

Applications with typical case studies and corresponding results; these are intended to aid in the analysis and interpretation of results from the analysis

Discussions

Conclusion, including information on the latest advances in the field (noncommercial) so as to provide an indication of future potential of the technology

References and additional reading

The handbook will serve as a concise reference to practitioners in the industry, providing technical information about plastics analysis and descriptions of the technology used to perform the measurements. It is aimed at laboratory personnel who need to have a working knowledge of plastics analysis techniques and would like to keep abreast of the latest developments in the field. These will include laboratory managers, supervisors, and engineers.

The book will also serve as a basic reference to research engineers and scientists who may be looking for techniques to solve problems or investigate behavioral phenomena.

Hubert Lobo
Jose V. Bonilla

Contents

Contributors

Jose V. Bonilla   ISP Chemicals, Calvert City, Kentucky, USA

Anita J. Brandolini   William Paterson University of New Jersey, Wayne, New Jersey, USA

John Coates   Coates Consulting, Newtown, Connecticut, USA

Gary J. Fallick   Waters Corporation, Milford, Massachusetts, USA

Robert Falcone   ICI Fluoropolymers, Bayonne, New Jersey, USA

Kenneth J. Fielder   PerkinElmer Instruments LLC, Shelton, Connecticut, USA

Galina Georgieva   GE Medical Systems, Waukesha, Wisconsin, USA

Scott Kinzy   ICI Fluoropolymers, Bayonne, New Jersey, USA

John Lemmon   GE Corporate Research, Schenectady, New York, USA

Hubert Lobo   DatapointLabs, Ithaca, New York, USA

Kevin P. Menard   PerkinElmer Thermal Laboratory, Materials Science Department, University of North Texas, Denton, Texas, USA

Rick Nielson   Waters Corporation, Milford, Massachusetts, USA

Burke Nelson   Goettfert USA, Rock Hill, South Carolina, USA

Koichi Nishikida   Thermo Electron Corporation, Madison, Wisconsin, USA

Philip Plantz   Microtrac, Inc., Largo, Florida, USA

Andrew W. Salamon   PerkinElmer Instruments LLC, Shelton, Connecticut, USA

1

General Introduction to Plastics Analysis

Hubert Lobo

DatapointLabs, Ithaca, New York, USA

Jose V. Bonilla

ISP Chemicals, Calvert City, Kentucky, USA

INTRODUCTION

Plastics have undoubtedly been the wonder materials of the last century. They have fundamentally revolutionized the manner in which we conceptualize and implement new products. They are ubiquitous in today's environment, appearing in ways that range from mundane to high-tech, from indispensable to completely wasteful. The manner in which these materials have been used has shaped our impressions of plastics as necessary evils as well as miracle materials.

A lot of the negative impressions have come from the fact that the world was not prepared for materials of this complexity. In many ways, our inability to understand plastics affected the manner in which we used or misused them. Before the arrival of plastics, most materials were relatively simple, or if complex, natural. In both cases, either by the application of existing science or by long historical knowledge of their use, it was possible to use these materials in an effective manner. In the case of plastics, the converse occurred. The arrival of plastics heralded the onset of one of the most comprehensive periods of discovery in material science. The very nature

1

of plastics has demanded significant advances in our ability to understand polymers, analyze their composition, and characterize this behavior. These technologies, collectively termed plastics analysis techniques, have come a long way in helping us design novel materials for a better tomorrow.

Our impression of plastics as miracle materials has also stemmed from an incomplete understanding of their complexity. Plastics analysis has allowed us to apply the rigorous application of scientific technique to deepen our knowledge of their behavior, replacing our previous wonder with a deep-rooted scientific understanding that permits us to truly appreciate the capabilities of these materials. Indeed, it is this knowledge that will shape our use of these truly amazing materials in the new millennium.

The introduction of plastics requires us to apply a wide range of techniques to understand their behavior. Plastics exhibit complex molecular characteristics. From our understanding of molecular structure, we are now able to attempt to correlate behavior to structure. This ability, however, is still, far from an exact science. The complex manner in which polymer molecules interact still prevents us from developing strong structure–property relationship theories. Indeed, the disconnection between our understanding of molecular-level behavior and macro-level characteristics is one of the sharpest dividing lines between classical material science and polymer science. In the case of metals, it has been relatively easy to apply our atomic-level understanding of the material and its microstructure to its behavior. In sharp contrast, polymer molecules vary widely in molecular weight. The manner in which polymer molecules interact with each other depends to a great extent on the pendant groups that are attached to the chain. Most frustrating of all, even though we are able to understand these aspects, this still does not permit us to apply our knowledge to understand macro-level behavior. This has forced the simultaneous development of both chemical analysis and physical analysis to grapple with the problem.

Chemical analysis techniques permit us to analyze molecular composition and molecular weight to allow us to characterize plastics precisely. Physical methods allow us to look at the behavior of plastics in response to a variety of influences such as temperature, pressure, and time. This understanding helps us to say how the plastics will behave in their lifetime. Plastics analysis may include identification and chemical composition, thermal properties, mechanical properties, physical properties, electrical properties, and optical properties, among others. Chemical analysis may include material identification and characterization by techniques including FTIR, NMR, GC, GC/MS, HPLC, and GPC. Thermal analysis does provide information such as melting point, glass transition, flash point, heat deflection temperature, melt flow rate, and Vicat softening point. Mechanical properties, on the other hand, provide critical information such as tensile properties, flexural strength, impact strength, hardness, compressive strength, modulus, and fatigue. Recent advances in microscopy also permit us to peek into molecular structure. The atomic force microscope, for example, has truly revolutionized our ability to probe molecular behavior, promising us a startling new understanding of plastics.

The past few years have been characterized by a large growth in development of new polymers and composite materials. Modern research and developments of high-technology materials have also driven the development of new analytical equipment and analytical technologies. As materials become more and more sophisticated and complex, so also must become the analytical techniques required for materials testing and materials characterization.

Innovative, accurate, easy-to-use, performance, and reliability are the requirements that describe instrumentation needed in modern laboratories involved in materials research or materials manufacturing facilities. Instrumentation of this type is needed in order to deliver outstanding performance to downstream customers. Modern instruments, hardware, and software products are designed to support such requirements.

This book covers some of the most significant techniques used in modern analytical technology to characterize plastic and composite materials. A short general introduction to some of them is provided here to the topics covered in more detail in later chapters. A general introduction is also given to other techniques that are not covered in extensive detail in this book but that are of significant use in characterization of certain critical properties of plastic materials.

Much of plastics testing is done by methodology developed and validated in-house by analytical scientists to meet specific needs. There are also a large number of official testing methods developed by agencies such as the American Society for Testing Materials (ASTM). A reference list of ASTM methods used for analysis of plastic and plastic-related materials is included as an appendix at the end of this book.

There are many applications of plastics analysis. It is useful to examine these applications using a life-cycle viewpoint. In each area, one is then able to see the importance of these vital techniques to the understanding and proper application of plastic materials in our lives and the environment.

PLASTICS AND PRODUCT DESIGN

Historically, product developers have had a great deal of difficulty working with plastics. They simply did not behave in the conventional manner. None of the conventional rules of behavior applied either in the manner in which the product was made or in the manner in which it behaved once it was produced. There were serious contradictions to conventional design philo-sophies—for example, the concept that making it thicker did not necessarily make it stronger, or that once it was made, it did not necessarily want to stay that way. Factors such as stress relaxation and residual stresses played havoc with first-generation plastic products, giving plastics the bad reputa-tion that years of continuous improvement have not been able to erase. A lot of improvement can be attributed to our current understanding of physical behavior.

Plastics exhibit large nonlinear stress–strain behavior. Large recoverable deformation is, in fact, one of the defining characteristics of a plastic. This characteristic has been used unfairly to portray plastics as weak materials because they lack the stiffness and strength of metals. In fact, it means that a plastic part can undergo a large amount of recoverable deformation before it breaks. This characteristic has been exploited, for example, in the replace-ment of the metal automotive bumper with TPO. The stress–strain relation-ship of a plastic shows a continuously decreasing stiffness with larger strain. This is because, in contrast to metals, plastics do not undergo complete instantaneous recovery upon unloading. When the plastic is unloaded, the response is viscoelastic, with partial instantaneous "elastic" response and a component of "viscous" time-based recovery. The introduction of the dimen-sion of time is therefore a serious complicating factor, presenting essentially a fourth dimension to be considered in design. Viscoelasticity, creep, and stress relaxation are some of the measures used to characterize this behavior.

PLASTICS AND MANUFACTURING

In conjunction with the injection molding process, plastics have revolutio-nized the manufacturing process over the past 20 years. Major economies in assembly were achieved by molding in features and incorporating subcom-ponents into a single part. For a very long time, however, injection molding was perceived to be a "black art" because of the extreme difficulty of making good parts. It took significant research in areas of polymer rheology, ther-mal properties, and mechanical behavior to develop the scientific under-standing that guides modern injection molding. Other plastics processing methods have also benefited from these advances. Characterizations of vis-coelastic behavior and extensional rheology guide modern blow molding and thermoforming, transforming this industry into a producer of high-quality engineering products.

PLASTICS AND THE ENVIRONMENT

Just as steel rusts and wood rots, plastics degrade in the environment. This is in stark contrast to the picture of the 1960s, which presented plastic as both the wonder material that would never die and the terrible scourge that would be present on the earth unchanged long after people had disappeared. Neither of these extreme positions helps us understand plastics. Instead, the past 40 years have helped us develop a healthy perspective on the true characteristics of plastics. We now see that plastics are not as unaffected by the environment as we had hoped, or feared, and that the properties and appearance of these materials do change adversely with time. This has spawned widespread effort to develop an understanding of environmental effects on plastics. Today, detailed analyses are conducted by exposing plastics to a variety of environments.

Lastly, the product life cycle of plastic does not end in the landfill. Materials recovery and recycling is becoming an important means to reuse rather than waste these valuable materials. As milk jugs are converted to automobile fuel tanks and old soda bottles to fleece sweaters, a brand-new area of plastics analysis arises, challenging us to find new ways to understand the factors that affect the recycling: issues of cross-contamination and material degradation and their effects on product performance; material identification and differentiation techniques to reduce cross-contamination; process monitoring systems that permit us to produce good products with a feed stream that is not very consistent.

SEPARATION METHODS

Gas Chromatography

Gas chromatography (GC) instruments may be equipped with various detectors to accomplish different analytical tasks. Flame ionization and thermal conductivity detectors are the most widely used detectors for routine analyses, nitrogen-phosphorus detectors are used for the trace analysis of nitrogen-containing compounds, and electron-capture detectors are used for halogen-containing compounds. GCs may also be equipped with peripheral accessories such as autosamplers, purge and trap systems, headspace samplers, or pyrolyzer probes for special needs in sample introduction.

High-Speed Gas Chromatography

In recent years, the need for rapid GC analysis has led to the development of gas chromatographs with fast separation times. Several recent technologies have been developed to decrease analysis times in GC, such as using short, narrow-bore columns and optimizing flow rates, temperature rates, and sample focusing parameters. Two major requirements for successful fast GC are fast data acquisition rates and fast detector response.

High-speed GC offers several benefits:

Quicker results for timely decisions about sample or product fate
Faster sample turnaround times
Lower operating costs per sample analysis
Ability to handle more samples with fewer pieces of equipment

Three major types of commercially available systems provide high-speed GC.

High-Speed GC Using a Standard Instrument

High-speed GC analysis is now possible using recently developed GC equipment that allows rapid heating of the GC oven and precise control of the carrier-gas pressure. Although this is not yet a widely used approach in chromatography, it has already been demonstrated that this new technique can reduce analysis time by a factor of 5 or better, compared to conventional GC analyses. One of the many remarkable features of modern GC instruments is their ability to perform fast GC without special modifications or expensive accessories. These GCs offer the capability to carry out fast GC without the need for cryofocusing or thermal desorption devices that may limit the flexibility or performance of the instrument. Properly configured for fast GC, these systems can perform all types of analyses using existing detectors, injectors, and flow controllers. Minimal system requirements include electronic pneumatics control (EPC), off-the-shelf capillary columns, split/splitless or on-column inlets, standard detectors optimized for capillary columns, and a fast acquisition data system. At any time, users can switch from fast GC back to the original method without major difficulties, or optimize new methods to meet new analytical demands.

Flash GC

Flash temperature programming is a new technique for rapidly heating capillary GC columns. The technique utilizes resistive heating of a small-bore metal sheath that contains the GC column. This technology is based on the flash GC system, an innovative chromatographic system that accomplishes in 1–2 min what takes a conventional GC from 30 min to an hour or more. The flash- GC can be over 20 times faster than conventional GC, and can be more sensitive and far more versatile. Development of the flash GC began in the 1980s. The first application was for the detection of explosives. The basic concepts, the proof of principle, and the initial designs were initially classified by the U.S. Ggovernment. The patents were declassified in the early 1990s, and have been issued to produce this equipment for public use. The flash GC provides great speed and flexibility to the analyst for the characterization of a great variety of chemical compounds. This technology has been proven to work in conjunction with mass spectrometry. There is also an upgrade kit available to easily convert a conventional GC into a flash GC. It is called the EZ Flash. This kit converts a conventional GC to a flash GC, providing the benefits from the speed and accuracy of flash CC technology with minimal investment. EZ Flash columns mount inside the oven of a conventional GC, replacing the existing column. The system offers column heating rates up to 20°C per second and a temperature control range of ambient to 400°C.

Cryofocusing Technology

Cryofocusing technology permits high-speed GC on already-existing GC equipment. Some GC manufacturers offer fast GC technology that enhances conventional capillary gas chromatographs with a novel sample inlet system to allow very rapid separations to take place. These are ideal for use in a wide range of applications including plastic materials, industrial chemicals, and environmental applications. Using a unique cryofocusing inlet system, samples can be preconcentrated and subsequently desorbed onto the analytical column in a very narrow band, thus eliminating the band broadening that occurs with conventional inlet systems. The instruments can also be interfaced to other automated sample introduction devices such as autosamplers, purge and trap, headspace, etc. Fast GC is extremely sensitive, capable of sub-parts-per-billion detection. This approach allows high-speed separations to be achieved using short lengths of conventional, 0.25-mm columns that provide increased sample handling capacity over microbore columns. A high-speed inlet system can deliver injection bandwidths of 5 ms. The main components include restrictor columns, solenoid valves, a source of hydrogen or helium carrier (CG), vacuum pump, detector, and split injector for sample introduction. Short lengths of 0.25-mm capillary columns are typically used at high linear velocities. This provides the maximum rate of plate production as opposed to number of plates. The cold trap consists of a metal tube with inert coating that is cooled using liquid nitrogen and rapidly heated using a capacitive discharge power supply.

Gas Chromatography/Mass Spectrometry

Mass spectrometry (MS) instruments are normally equipped with electron impact (EI) and chemical ionization (positive and negative CI) sources, and a solid probe. This type of equipment is used for product identification, and analysis of a variety of components in various polymer products and raw materials or any unknown compounds in laboratory samples.

As an analytical technique, mass spectrometry offers special advantages over other techniques that derive from its properties as both a highly specific and universal detection method. It can be more sensitive than other analytical techniques. It can also be made highly specific for the analysis of a target component. This results in a high signal-to-noise ratio for an analyte, due to a reduction in the detection of unrelated background interference. The MS instrument can also be operated as a universal detector. Molecular fingerprint information can be derived from the structurally dependent fragmentation patterns produced and used to identify unknown compounds or to confirm their presence.

All modern mass spectrometers have the following essential components:

1. Sample introduction or inlet system
2. Ion source and ion focusing system
3. Mass analyzer
4. Vacuum system(s)
5. Ion detection and signal amplification system
6. Data handling system

Inlet System

A sample inlet system is used to introduce the sample into the ionization chamber. The inlet system can range from a very simple solid probe to a highly sophisticated liquid interface that permits the introduction of non-volatile molecules or high-molecular-weight polymeric materials. Inlets may also be static (one sample at a time) or dynamic (allowing for continuous analysis). In recent years the use of dynamic inlets in MS has greatly expanded, due to the enormous numbers of samples being generated by the applications MS can be used with.

There are several alternative methods of introducing a sample into a mass spectrometer. Samples to be analyzed may come in a variety of different forms: solids, liquids and gases, either as single compounds or as mixtures. Each sample type brings its own unique handling problems. The type of inlet system used depends on the physicochemical properties of the sample and its thermal stability. Thermally stable gas and liquid samples with a high vapor pressure at room temperature are usually admitted to the MS via the static solid probe inlet or through a dynamic gas chromatography interface. Gas chromatography is commonly used to introduce medium- to high-volatility compounds. Often, other ancillary techniques are used in turn to introduce different types of samples into the GC. These ancillary techniques include headspace, purge and trap, thermal desorption, cryogenic concentrators, and thermal pyrolysis devices. Special applications using these devices are described in more detail in the chapter on GC. GC is most commonly used in conjunction with either electron impact, or chemical ionization MS ion sources.

Compounds with little or no volatility, or thermally labile compounds, are normally introduced through a dynamic interface employing liquid chromatography (LC). The type of LC interface used is dependent on the nature of the process used to ionize the analyte and includes techniques such as particle beam, electrospray, and APCI. Static probe inlets such as Maldi, field desorption, desorption chemical ionization (DCI), and FAB can also be used. The details of the individual interfaces will be presented in the section on hyphenated techniques.

Ion Source

Ionization methods in mass spectrometry are divided into gasphase ionization techniques and methods that form ions from the condensed phase, either inside or outside of the MS. All ion sources are desired to produce ions without mass discrimination from the sample and to transport them into the mass analyzer. Ideally, ions should be produced with high efficiency (ion yield) and transported to the mass analyzer with no loss (high transport efficiency).

Electron Impact Ionization

Electron impact (EI) ionization is the most commonly used ionization method. Electrons are produced from the cathode of a resistively heated filament located perpendicular to the incoming gas stream and collide with the sample molecules to produce a molecular ion. The source normally operates with an electron energy of 70 eV, the optimum ionizing potential. This provides sufficient energy to cause ionization and the characteristic fragmentation of sample molecules. Some compounds do not produce a molecular ion in EI source. This is a disadvantage of this ionization mode, as is the low ionization efficiency (typically less than 0.01% of all neutrals admitted are ionized). The typical MS employing EI has an ion extracting and focusing system that operates at high vacuum, resulting in high transport efficiency for the ions that are formed. As a result, modern MS instruments have detection limits in the mid-femtogram (10^{-12} g) range in full-scan mode. Perfluoroalkanes are often used as calibration compounds in EI because they provide ions at known masses corresponding to the loss of CF, CF_2 and CF_3 groups.

EI is widely used in MS for volatile compounds that are insensitive to heat and as a result are generally low in molecular weight (< 1000 Da). The spectra, usually containing many fragment ions, are useful for structural characterization and identification. Small impurities in the sample are easy to detect if chromatographic separation is employed. EI is therefore compatible with GC and with LC interfaces such as moving belt and particle beam. Large libraries of spectra obtained using EI have been compiled across a wide range of manufacturers' instruments and are available to aid in the identification of unknown compounds. Mass analyzers used with EI include the quadruple, ion trap, magnetic sector, time-of-flight, and Fourier transform MS systems.

Chemical Ionization

Chemical ionization (CI) results from ion–molecule reactions between sample molecules and reagent gas ions. The reagent gas is produced by admitting a suitable gas (H_2, methane, isobutane, ammonia, etc.) into an EI source modified to allow the buildup of a small amount of gas pressure. Gas molecules are ionized by the electron beam in the source (generally operated at 100 eV or higher). Subsequent reactions between the initial reagent gas ions and additional reagent gas neutrals produce complex reagent gas plasma. When a reagent gas ion encounters a sample molecule, ion–molecule reactions occur by adduct ion formation or charge transfer.

Essentially all the spectral information obtained are related to the molecular weight of the compound (in the quasi-molecular ion) with little or no fragmentation. This process is also referred to as soft ionization. This is a result of the low excess internal energies remaining after adduct formation. The rules governing how likely an adduct is to form depend primarily (in positive ion mode, PCI) on gas phase acid–base chemistry between the reagent gas ions and the neutral analyte molecules. In general, the reagent gas ions must always be more acidic than the target neutral molecule to be ionized. The use of PCI is desirable when an analysis of a mixture of compounds is needed and the list of possible components is limited. The use of negative-ion CI (NCI) is generally limited to molecules containing electronegative substituents such as halogens, nitro- functional groups, and oxygen-containing functional groups (sugars, etc.). To extend the utility of this technique, an analyte lacking a suitably electronegative functional group may be chemically derivatized with an agent containing these electronegative functional groups (e.g., fluorine atoms or nitrobenzyl groups). Such derivatization is generally performed on the target analyte after isolation and before mass spectrometric analysis. The ionization mechanisms most prevalent in NCI are either resonance or dissociative electron capture, in which low-energy electrons created in the plasma are captured by an analyte with an electronegative substituent. The sensitivity of NCI analyses can be two to three orders of magnitude greater than that of PCI- or EI-based analyses. Little fragmentation occurs during NCI, and this mode of ionization is generally employed for quantitative analyses of trace amounts of compounds of known structure, often in conjunction with the use of stable isotope-labeled internal standards. The mass analyzers (as well as the mass ranges) used with CI are the same as those used for EI. Samples to be analyzed can be admitted into the source via a GC inlet or by a probe inlet technique such as DCI. In DCI a sample in solution is deposited on a rhenium filament and the solvent is allowed to evaporate. Then the probe is inserted into the vacuum system and the filament is rapidly heated so that the sample molecules rapidly desorb into the gas phase before any pyrolysis occurs.

High-Pressure Liquid Chromatography

High-press (HPLC) is a separation technique employed for the analysis of low- to medium-molecular-weight compounds, typically under 2000 Da. The technique is particularly effective for the separation of multicomponent samples containing nonvolatile, ionic, isomeric, and thermally labile components. Major applications include the determination of residual monomers, additives, and solvents in polymers. HPLCs are normally equipped with UV detectors, diode-array detectors, or other appropriate detectors depending on the nature of the analyte of interest. Options to perform precolumn or postcolumn derivatization for samples that may need introduction of special functionalities for detection are also available.

Additional detectors available for HPLC analysis include fluorescence detectors, high-sensitivity diode-array detectors, refractive index detectors, and electrochemical detectors.

Liquid Chromatography/Mass Spectrometry

A mass spectrometer coupled to a high-performance liquid chromatograph is a powerful analytical system that provides structural chemical information. The particle beam interface was for a long time a popular interface, providing spectra essentially similar to conventional mass spectra. Recently, the most popular type of interface between MS and HPLC systems has been the atmospheric pressure inlet (API), which accepts either electrospray or atmospheric-pressure chemical ionization sources. This instrument is capable of providing structural elucidation as well as very specific detection and quantitation. The API inlet and sources are designed for medium- to high-polarity analyses. This instrument is used for profiling of components in plastic materials, raw materials, polymer additives and residuals, and reaction monitoring for analytes amenable to liquid chromatographic applications.

Capillary Electrophoresis

Capillary electrophoresis systems are used for chromatographic separation of charged species based on mobility in an applied electric field. UV-visible absorbance and fluorescence detectors are utilized for specific and sensitive quantitation of different analytes. Applications include the determination of inorganic and organic components or additives in plastic formulations, determination of inorganic anions at parts-per-billion levels, and determination of small organic acids in various polymers.

Ion Chromatography

Ion chromatography is used to perform separation and detection of inorganic anions and cations, organic acids, and electroactive organic compounds. This instrument may be configured with a variety of detectors such as conductivity, pulsed electrochemical, and UV-visible detectors. Applications include the determination of anions in electronic-grade materials, residual organic acid in polymers and copolymers, and blend formulations.

Thin-Layer Chromatography

Normal and reverse-phase thin-layer chromatography (TLC) chromatograms are quantitated using a dual-wavelength scanning densitometer. Trace levels of organics in a mixture are separated and detected using UV-visible, reflectance, or fluorescence modes. Hydrazine, for example, can be analyzed for trace components on polymeric materials, determined to a 50-ppb or even lower level via established official (i.e., ASTM, USP) analytical procedures. TLC plates can be scanned unattended and quantitated automatically. The technique has been found to be particularly useful for compositional analysis, since individual component spots can be collected for subsequent spectral analysis or other available analytical techniques.

Size-Exclusion Chromatography (Gel Permeation Chromatography)

Size-exclusion chromatography (SEC) and gel permeation chromatography (GPC) are very important tools in polymer characterization. They can provide the answers to key properties in polymer materials such as starches, cellulosics, nylon, polyethylene, PET, etc. GPC instruments are equipped with columns for aqueous or organic mobile phases. A refractive index detector is a very popular detector for most GPC applications. SEC (or GPC) is used to chromatographically separate polymer molecules by size (or hydrodynamic volume) and, through various calibration methods, determine the molecular-weight averages (molecular weight, viscosity, and average molecular number) and molecular weight distribution of a sample. Data collection and processing is accomplished using sophisticated PC-based software systems. Calibration can be performed with direct (narrow) standards, broad standards, the universal method, or light-scattering detection. The technique is generally applicable in the molecular weight range of 500 to 5,000,000 amu. Some applications include determination of relative and absolute molecular weights and distributions of polymers and copolymers. Fraction collectors may also be used in conjunction with preparative GPC columns to fractionate various polymers and copolymers for further characterization using other analytical capabilities such as NMR, FTIR, and other spectroscopy instrumentation. In order to obtain maximum data from this type of analysis, a triple detector system can be used which may include a four-capillary differential viscometer, right-angle laser light-scattering detector, and a differential refractometer. This system set up readily provides information about molecular weight, molecular size, intrinsic viscosity, branching, copolymers and aggregation information. A multitude of detectors is available for GPC to provide different information about the polymers.

Low-Angle Laser Light-Scattering Photometry

A photometer can be used to determine the weight-average molecular weight of polymer samples by the static method or in conjunction with gel permeation chromatography. Laser differential refractometers are available to determine the differential refractive index increment (dn/dc) as required by the Debye equation to calculate Rayleigh scattering. Applications of this technique include the determination of weight-average molecular weights of various polymers and copolymers.

Multiangle Laser Light-Scattering Photometry

A multiangle laser light scattering detector is used to determine absolute molecular weight, molecular weight distribution, radius of gyration, and radius distribution of polymeric materials. This technique can be applied to all polymers and copolymers with composition distributions that are not dependent on molecular weight. Indirect information on branching characteristics of polymers can be obtained from the relationship between molecular size and molecular weight.

Viscosity Detector for SEC (or GPC)

A differential viscometer is used to determine weight-average molecular weights of polymers by a universal calibration. The differential viscometer may be used in conjunction with a GPC apparatus to measure the intrinsic viscosity of polymer solutions eluted from the GPC instrument. These data are used togeher with data obtained from polymer standards of known molecular weights and distributions. Absolute molecular weights, coefficients, intrinsic viscosities, and relative degree of branching can be obtained for unknown polymers by the use of this technique.

High-Osmotic-Pressure Chromatography

HPLC equipment dedicated to high-osmotic-pressure chromatography is used for the fractionation of narrow polymer fractions from broad distribution samples. This technique, which employs columns that are packed with a control pore glass of very narrow pore distribution, separates polymers by molecular weight as a function of osmotic pressure. When this approach is coupled with a fraction collector the technique can generate polymer fractions in significant quantities for further study by nuclear magnetic resonance, (NMR), FTIR, or other spectroscopic techniques. This technique can offer superior resolution to the previously mentioned preparative GPC. This technique has been applied to the characterization of both copolymers and homopolymers.

FLOW INJECTION POLYMER ANALYSIS

Flow injection polymer analysis (FIPA) is a new method for routine process and quality control. FIPA is a fast, precise, accurate, and reliable technique for measuring the molecular weight, molecular size, intrinsic viscosity and percent polymer of a process sample without the need for elaborate sample preparation. Proper control of the polymerization process requires fast turnaround time for sequential measurements of the extent of reaction. FIPA measures polymer concentration, molecular weight, and intrinsic viscosity of a solution or emulsion reactor sample in less than 5 min. The typical system consists of a pump, an injector, a FIPA column, and a triple detector array (TDA). The TDA combines a light-scattering detector, viscometer detector, and refractometer in a single integrated unit with fixed interdetector volumes and temperature control. The procedure for sample analysis is simple:

1. The sample is diluted with an approximate amount of mobile-phase solvent.
2. The sample is injected into the instrument.
3. The polymer is separated from the low-molecular constituents (diluent, additives, monomer, oligomers) by the FIPA column. Typical run time is 3–4 min.

4. FIPA software calculates the molecular weight, molecular size, intrinsic viscosity, and percent polymer from the light-scattering, viscometer, and refractometer response.

Precision on molecular weight, molecular size, and intrinsic viscosity is typically less than 0.3% RSD, which permits excellent feedback for control of the polymerization process. The integrated detector array consists of three primary detectors: a light scattering detector that measures molecular weight, a four-capillary differential viscometer that determines molecular density and size, and a differential refractometer that measures concentration.

FIPA systems are highly flexible and may be interfaced to a variety of instruments including UV-visible absorbance, pH, and redox detectors, and mass spectrometers (by the use of a particle beam interface, for example). Specific reaction chemistries can be accomplished through the use of dedicated in-line heaters and reaction coils. These types of systems can be used for the determination of the acid number of polymer solutions, additives content from polymer extracts, and trace analyses of other inorganic or organic components.

SPECTROSCOPIC ANALYSIS

Infrared Spectroscopy

FTIR spectrometry is a widely used technique for materials analysis by quality control technicians, engineers, students, synthetic chemists, and other scientists. Every operator, whether a spectroscopist or not, can get quick, accurate, precise results by this technique. Today's QA/QC labs require instruments that are reliable and rugged, yet sensitive and easy to use. These features are distinctive to the FTIR instruments, which makes it the perfect system for QA/QC applications in the routine analysis of polymers and polymer components.

FTIR is ideal for raw materials testing and acceptance: reducing scrap and increasing product quality has made raw material acceptance testing critical to manufacturing. Manufacturers who use FTIR spectroscopy know the advantages FTIR has over other techniques, such as time-consuming chromatographic separations. Modern FTIR systems normally come with a sealed and desiccated optics with a diversity of sampling accessories, including fiber optics, which lets the analyst do sample testing even outside the controlled environment of the laboratory. Raw material analysis can be performed outside the laboratory for acceptance testing in the warehouse, or right at the loading dock.

Modern FTIR equipment offer a set of correlation charts that can be superimposed over a spectrum to help the analyst interpret the data and also permits software customization to meet any special needs in materials characterization. For spectral library searches, application-specific spectral libraries can be easily built or obtained from commercially available libraries.

A Fourier transform infrared (FTIR) spectrometer equipped with an attenuated total reflectance (ATR) accessory is used for the study of surfaces and coatings. A microscope attachment is useful for identifying particulate impurities. Through the technique of computerized spectral subtraction, many sample mixtures may be identified by FT-IR without prior physical separation and, thus, the technique lends itself to compositional analysis. Typical applications of FT-IR are the study of surfaces of polymers by ATR, identification of samples isolated by thin-layer chromatography or other preparatory chromatographic techniques, identification of impurities in polymer and polymer blends, product characterization and product formulations by spectral subtraction, as well as routine analysis of solid and liquid materials.

Nuclear Magnetic Resonance Spectrometry

Nuclear magnetic resonance (NMR) spectrometers offer spectral capabilities to elucidate polymeric structures. This approach can be used to perform experiments to determine comonomer sequence distributions of polymer products. Furthermore, the NMR can be equipped with pulsed-flied gradient technology (PFG-NMR), which not only allows one to determine self-diffusion coefficients of molecules to better understand complexation mechanisms between a chemical and certain polymers, but also can reduce experimental time for acquiring NMR data. Some NMR instruments can be equipped with a microprobe to be able to detect microgram quantities of samples for analysis. This probe has proven quite useful in GPC/NMR studies on polymers. Examples include both comonomer concentration and sequence distribution for copolymers across their respective molecular-weight distributions and chemical compositions. The GPC interface can also be used on an HPLC, permitting LC-NMR analysis to be performed too. Solid-state accessories also make it possible to study cross-linked polymers by NMR.

Near-Infrared Spectrophotometry

A near-infrared (NIR) spectrophotometer allows analytical utilization of absorption bands in the 1100–2500 nm region. These bands are primarily overtones and combinations of fundamental vibrational modes originating in the mid-IR. The strongest absorbencies in the NIR region are typically generated from $C-H$, $O-H$, and $N-H$ vibrational modes. A liquid sampling module is available that allows for temperature-controlled analysis of samples in high-quality quartz and disposable cuvettes. The instrument is also available with fiber-optic sampling modules that include both a reflectance (for solids, slurries) and an immersion probe (liquids). This instrument is primarily used for research and development of on-line applications of NIR spectroscopy.

A transmission pair analyzer combines the advantages of laboratory NIR analysis in a process instrument. The NIR monochromator is interfaced to the process via fiber-optic bundles and stainless steel transmission probes. This instrument is used primarily for on-line investigations of pilot plant or production plant applications. Typical applications include monomer moisture content, residual solvents, and solids content in polymerization processes, and solvent and solutions compositions in continuous chemical processes.

UV-Visible Spectrophotometry

UV-visible spectrophotometers afford both detection and quantification of inorganic and organic species that absorb electromagnetic radiation in the wavelength range 200–900 nm. Both spectrophotometers are PC-based. This technique has been extensively applied to color determinations and for quantitation of aromatic components in polymer formulations. It has also been used for determination of various additives in polymers and polymer solution studies.

POLYMER SOLUTION PROPERTIES

Solution Viscometry

Solution viscometers are available to determine the relative, specific, inherent, and/or intrinsic viscosities of polymer solutions. Absolute viscosities in centipoises or centistokes may be determined for neat, nonviscous liquids. Recent applications include analysis of viscosity-related properties of solution-based polymers.

Relative Viscosity

Capillary viscometry is used for the determination of relative viscosity of polymer solutions in an appropriate solvent at diluted concentrations. Relative viscosity is dependent on concentration, so the concentration is specified for the solution. Relative viscosity is also dependent on the viscometer used; therefore, the size of the tube used for the measurement

(Cannon Fenske viscometer or Ubbelohde viscometer) is also specified. The relative viscosity of the solution is determined by dividing the flow time of the polymer solution through a capillary viscometer by the flow time of pure solvent through the same viscometer at the same temperature conditions.

Absolute Viscosity

The flow time of a polymer solution through a small-diameter Ubbelohde viscometer or Cannon Fenske viscometer is measured at a specified temperature. The absolute viscosity (cp) equals the product of the flow time, the calibration constant of the viscometer, and the density of the solution in the solvent used. The viscometer should be calibrated with an ASTM oil standard if needed.

The viscosity is calculated as viscosity (cp) = flow time (s) × calibration constant solution density (at measurement temperature).

Automatic Relative Viscometers

The automatic relative viscometer is ideally suited for measuring dilute polymer viscosities. It provides faster analysis and greater precision than is obtainable with conventional glass tube viscometers (Ubbelohde or Cannon-Fenske), which it replaces. The principle of operation is based on measurement of pressure drops due to the continuous forced flow of solvent and sample through two stainless steel capillary tubes placed in series. The pressure drop across each capillary tube obeys Poiseuille's law. The pressure drop is measured by a differential pressure transducer. The sample solution is loaded into a sample loop via a syringe pump and then pushed into one of the two capillaries. A steady-state condition is reached when the sample solution completely fills capillary 2, solvent remaining in capillary 1 at all times. The relative viscosity of the sample solution is determined simply and directly by the ratio of the pressure drops. From the measured relative viscosity, all other solution viscosity measurements can be calculated. Solution viscosities are determined by the viscosity of the sample relative to the reference solvent. The relative viscometer measures the solvent and sample viscosity simultaneously, so errors due to temperature fluctuation and solvent variations are avoided. The main advantages of this approach are:

Significant solvent savings compared to traditional methods
Faster analysis than with conventional glass tubes
Higher accuracy and precision
Availability of various levels of automation
Autosampling, for unattended measurement of a series of samples

Rotational Viscometers

Many product specifications call for the use of a rotational viscometer, a simple device that measures the resistance to turning a cylindrical metal spindle dipped in the liquid polymer solution being tested. The higher the viscosity, the greater the resistance. Although this measurement is simple in principle, great care must be taken not to misuse or misinterpret the results.

Some plastics are mostly Newtonian (viscosity independent of shear rate); for Newtonian materials, viscosity is independent of the spindle speed and spindle type. On the other hand, some polymer materials are mostly non-Newtonian (viscosity dependent not only on shear rate, but also on time). For these materials, viscosity is dependent on spindle type and speed, and it is very important to define the test conditions carefully.

Selection of the proper spindle and speed is critical to obtain accuracy and precision in viscosity. To maintain adequate precision and accuracy, it is desirable to select spindle and speed so that the scale reading is between 50% and 100% full scale to cover the viscosity specification of a product. Readings at the low end of the scale (much less than 50% of full scale) increase the error in the measurement. For materials with very low viscosities, care must be taken in speed selection: too high a speed can lead to turbulence in the liquid and to error in the measurement. Centering of spindle in the sample jar is important when larger spindles are used.

Viscosity is dependent on temperature, and the time required for a sample to reach equilibrium temperature is often the most time-consuming step.

Calibration of a rotational viscometer with a standard should be done regularly and recorded. ASTM D-2196 considers a viscometer calibrated if the viscosity reading is within ±5% of the stated viscosity.

Falling-Ball Viscometers

The falling-ball viscometer is a traditional and simple viscometer for very precise measurements of Newtonian fluids according to ISO12258 and DIN53015 in the temperature range −20 to 120. It is a proven tool for precise viscosity measurements which requires only a stopwatch and no power connection. Viscosity measurements of transparent fluids with Newtonian flow behavior give the highest precision possible. This rheometer is especially suitable for low-viscosity liquids such as polymer solutions, beverages, oils, and diluted solutions. The viscosity is proportional to the time required for a ball to fall through the test liquid contained in a precise and temperature-controlled glass tube.

Refractometry

Temperature-controlled refractometers are available for refractive index measurements of liquid samples and solutions. Such refractometers are stable and precise, and are applicable for both aqueous and volatile organic solutions such as those made with acetone or alcohol. A differential refractometer is used to obtain the exact percent solids of diluted solutions (nominal 1% polymer solutions or lower). In the case of both types of refractometers, the percent solids measurements are made relative to a calibration of the refractive indices of standard solutions of known concentration.

Densitometry

Digital density meters measure density, specific gravity, and other related values (i.e., percent alcohol, BRIX, API degrees) with high precision and short measuring time. A hollow glass tube vibrates at a certain frequency. This frequency changes when the tube is filled with the sample: the higher the mass of the sample, the lower the frequency. This frequency is measured and converted into density. A built-in Peltier thermostat controls the temperature (no water bath is required).

MELT VISCOSITY/RHEOLOGY MEASUREMENTS

Some type of melt viscosity is included in the specification for almost every polymeric or plastic product. This is because viscosity is related to the molecular weight and to the performance of a polymer. Equipment used for rheological measurements range from the simple and ubiquitous melt flow indexer to the precise and quantitative capillary and cone-and-plate rheometers.

Melt Flow Index

The melt flow index test method is used to monitor the quality of plastic materials. The quality of the material is indicated in this test by melt flow rate through a specified die under prescribed conditions of temperature, load, and piston position in the barrel as timed measurement is being made. The melt flow rate through a specified capillary die is inversely proportional to the melt viscosity of the material if the melt flow rate is measured under constant load and temperature. The melt viscosity of the material or MFR is related to the molecular weight of the material if the molecular structures are the same. The extrusion plastometer as specified in

ASTM D-1238 is equipped with a piston rod assembly and weights, removable orifice of $L/D = 4/1$, temperature controller and temperature readout, orifice drill, charging tool, and cylinder cleaning tool.

Capillary and Slit-Die Rheology

Capillary and slit-die rheometers are used to determine the dependency of viscosity on shear rate. Since most molten polymers exhibit non-Newtonian behavior, it is important to be able to characterize this behavior. Measurements are made using a piston-driven cylinder that drives the molten polymer through a die of precise dimensions. The pressure drop across the die is measured, as is the flow rate through the die. Temperature is precisely controlled throughout the measurement. This test yields precise viscosity measurements as a function of temperature and shear rate. However, measurements tend to have artifacts in them, which need to be corrected in order to obtain "true viscosity" using Bagley and Rabinowitsch corrections. Capillary rheometers are also used to determine the effects of slip, a phenomenon in which the velocity of the melt at the capillary wall is nonzero. Slip has important implications for highly filled materials.

Additionally, rheometers may also be equipped with slit dies, which have the advantage of eliminating end pressure loss artifacts, but suffer from the limitation of lower maximum shear rates, and being more difficult to use.

Cone-and-Plate Rheology

Cone-and-plate rheology represents one of the most ideal means of measuring viscosity of molten plastics. The melt is placed between a plate and a cone that are located a precise distance apart. The torque required to impose a constant rate of angular rotation is measured. Cone angles of 3° or less are used, which result in a geometry where the shear rate is constant with radius so that no assumptions about the flow kinematics are needed. The main drawback of the technique is that it is limited to very low shear rates. It remains an ideal method for the determination of the zero shear viscosity, a number that is found to correlate well to the molecular weight of the plastic. The first and second normal stresses can also be measured using this apparatus.

In order to get to higher shear rates, it is customary to change over to a "dynamic" mode, in which the steady rotation is replaced by a sinusoidal oscillation, performed at a constant strain but varying frequencies. Viscosities at frequencies of as high as 500 rad/s are achievable with this technique. Assumption of the Cox-Merz relationship gives a 1:1 relationship between frequency and shear rate. The relationship is found to hold generally for unfilled homopolymers.

Numerous variants of this technique have been created to adapt it for more practical applications. The most popular of these is the parallel-plate apparatus. Here the melt is placed between two parallel disks, 1–2 mm apart. The shear rate varies from the center to the circumference of the disk and corrections need to be performed [1]. The parallel-plate geometry is less sensitive to errors in gap and is also more able to handle filled materials. In contrast, cone-and-plate geometries are not useful in cases where the filler dimension is of the same order of magnitude as the gap. Parallel plates are also recommended in situations where rheology is measured as a function of temperature, where tool thermal expansion would otherwise affect the accuracy of the measurement.

For low-viscosity, low-elasticity melts, a cup-and-cone arrangements is needed in order to contain the material. In cases of highly filled materials, there is a significant concern of slip between the plates and the material. Serrated plates are often used to overcome this problem.

Extensional Viscosity

There is an increasing interest in the study of melt elasticity. Such data are used in the study of die swell in extrusion and stretching in blow molding, factors which can affect finished-part appearance and quality. In general, such characterizations are fairly complex. Maxwell's melt elasticity tester represents one of the simplest approaches, characterizing the elastic recovery of a melt which is subjected to a instantaneous step rotational shear. The Rheotens is an attachment to a capillary rheometer and uses two counter-rotating wheels to draw down a vertical polymer melt strand at constant velocity or at constant acceleration. This instrument provides an easy means to measure the melt tensile strength. It is able to achieve large stretch ratios and high strain rates. However, it does not yield quantitative elongational viscosity data, for a number of reasons. Chief of these is lack of strand temperature control and the fact that the strand cross-sectional area changes continuously between the die exit and the pickup rolls.

The Meissner rheometer utilizes a horizontal sample supported on a gas cushion in a temperature-controlled environment, which is then drawn horizontally by metal belts moving in opposing directions. It yields precise quantitative measures of tensile extensional viscosity to large stretches. It is somewhat difficult to use, however, and it is limited to low strain rates.

There has recently been a great deal of interest in biaxial extensional viscosity. The lubricated squeezing test was proposed by Harrison and Hsu [2] as an elegant means to achieve such measurements. Here, the sample, in the form of a thick disk, is melted and squeezed between lubricated plates. A reasonable range of strain rates and stretch ratios is attainable. The technique and subsequent data analysis are simple. Concerns revolve around the issue of whether the squeezing mode of deformation is truly equivalent to the mode of biaxial tension. Also, it is not always possible to prepare specimens thick enough for the measurement. Bubble inflation has been proposed as an alternative measure of biaxial extensional viscosity. Here heated molten sheet is inflated using air or hot oil. Complexities arise from the measurement of change in area of the bubble. The test is sensitive to variations in thickness and temperature of the sheet, which lead to nonuniform bubble inflation. Variations of the technique have used a containment cage and other means to reduce this effect.

THERMAL ANALYSIS

Thermal analysis represents a broad spectrum of analytical techniques designed to assess the response of materials to thermal stimuli, typically temperature change. Various techniques evaluate changes in enthalpy, specific heat, thermal conductivity and diffusivity, linear and volumetric expansion, mechanical and viscoelastic properties with temperature.

Heat Deflection Temperature and Vicat Softening Point

Heat deflection temperature and the Vicat softening point provide a simple measure of melting transitions. Both tests involve placing a specimen under a stress state in an oil bath that can be heated at a programmed rate. The temperature at which the material undergoes a specified amount of deformation is recorded.

In the heat deflection test, the specimen is loaded in flexure at a specified stress defined by ASTM or ISO standards. Specimens are flexural bars suspended in a three-point bending frame. The specimens are heated at a heating rate of 50°C/h until they deflect by the amount specified in the standard.

In the Vicat test, the samples are placed under a thin cylindrical probe, again at a stress specified in the standards. The apparatus is heated in a similar manner. The temperature at which the probe penetrates the sample is recorded.

Both tests yield a qualitative measure of the maximum operating temperature of a plastic. They are used extensively in quality control and material qualification applications. The data are not used in design.

Differential Scanning Calorimetry

The differential scanning calorimeter (DSC) is perhaps the instrument that has dominated the field of thermal analysis in the past decade. It measures heat flows and temperatures associated with exothermic and endothermic transitions. The ease with which important properties such as transitions, heat capacity, reaction, and crystallization kinetics are characterized have made the DSC widely used in the plastics laboratory. Significant efforts to simplify the technique have put this form of analysis within the reach of most plastics analysts. The DSC can operate in one of two ways: with a power-compensated design in which energy absorbed or evolved by the sample is compensated by adding or subtracting an equivalent amount of electrical energy to a heater located in the sample holder. Or, alternatively, it can operate based on a heat flux design by which it measures the differential heat flow between a sample and an inert reference. Modulated DSC (MDSC) is an extension of conventional DSC in which the material is exposed to a cyclic, rather than linear, heating profile. Deconvolution of the heat flow results obtained provides unique benefits, including improved resolution of closely occurring or overlapped transitions, increased sensitivity for subtle transitions, and separation of reversing and nonreversing thermal phenomena.

DSC is routinely used for investigation, selection, comparison, and end-use performance of materials. It is used in academic, industrial, and government research facilities, as well as quality control and production operations. Material properties that are routinely measured include glass transitions, melting point, freezing point, boiling point, decomposition point, crystallization, phase changes, melting, crystallization, product stability, cure and cure kinetics, and oxidative stability.

Differential Photocalorimetry

A differential photocalorimeter can be interfaced with a thermal analysis system to measure the heat absorbed or released by a sample as the sample and an inert reference are simultaneously exposed to UV radiation of known wavelength and intensity during programmed heating or in an isothermal manner. Important applications include UV curing studies of compositions, formulations, and mechanisms of UV-curable materials and coatings.

Thermogravimetric Analysis

Thermogravimetric analysis (TGA) is the second most commonly used thermal technique. It measures weight changes in a material as a function of temperature or time under a controlled atmosphere. The main uses include measurement of a material's thermal stability and composition. TGA instruments are routinely used in all phases of research, quality control, and production operations.

Thermogravimetric analyzers are used to study thermal stability of materials. The high-resolution mode provides variable heating rate, constant heating rate, step-wise isothermal and linear heating to increase the resolution of any thermal events which are occurring in very close neighborhood. Applications include the thermal stability of polymers, determination of bound and free water in polymer products, and magnetic properties of composite materials.

Thermomechanical Analyzers and Other Dilatometers

Thermomechanical analyzers (TMAs) and other dilatometers are based on the measurement of dimensional changes with temperature. The most common of these is linear thermal expansion measurement using some form of quartz dilatometer.

The TMA is an easy-to-use analytical instrument that measures dimensional changes in a material as a function of temperature or time under a controlled atmosphere. Its main uses in research and QC include accurate determination of coefficient of linear expansion of plastic materials. It also is used to detect transitions in materials (e.g., glass transition temperature, softening and flow information, delamination temperature, creep and stress relaxation, and melting phenomena). A wide variety of measurement modes are available (expansion, penetration, flexure, dilatometry, and tension) for analysis of solids, powders, fibers, and thin film samples.

Key features of the TMA include:

Wide temperature range (−150 to 1000°C)
Easy probe exchange
Programmable force and programs for ramp, step, constant-load, and isostrain conditions

Another means of examining fundamental thermodynamic phenomena is the use of high pressure dilatometry to measure the pressure–volume–temperature dependence of polymers. This results in the development of an equation of state describing the variation of specific volume with temperature and pressure. As with DSC, these curves show thermodynamic as well as kinetic phenomena. The data yield three important pieces of information: the volumetric thermal expansion coefficient, the compressibility, and the pressure dependence of glass, melting, and crystallization transitions. Data are utilized primarily for mold analysis simulations. The most commonly used technique is based on the used of a confining fluid, typically mercury, in which the sample is immersed. The confining fluid provides a hydrostatic state for the application of pressure. Volume change is measured by measuring the deflection of a flexible bellows. Other techniques exist, based on a piston–cylinder arrangement, that work well for fluids and soft solids where the piston is able to transmit pressure in a hydrostatic manner throughout the sample.

Dynamic Mechanical Analysis

A dynamic mechanical analyzer (DMA) is a controlled-stress or controlled-strain instrument that provides information on mechanical properties such as modulus, energy dissipation, and material stability. This versatile instrument is able to perform studies of materials in both the melt and solid states. Melt-state studies have been covered earlier, in the introduction to rheology. DMA is a powerful technique for developing a fundamental understanding of polymer behavior. A vast amount of research has led to the development of a mathematically rigorous science that allows for the characterization of polymers as well as the extrapolation of properties.

There are several modes of operation of DMAs. The most common is the rotational/torsional type of instrument, although a number of linear tensile-compressive type are now available. These may operate in either a constant-strain or a constant-stress mode. In the former, the specimen is always deflected to a defined strain while the stress is measured. Constant-stress machines are the converse. These instruments are preferred for creep-mode type experiments while constant-strain instruments lend themselves better to stress-relaxation studies. The decision of what type of DMA to use for a test then depends not only on the type of behavior under study but also on the mode of deformation. Torsional DMAs provide data in a shear mode, while tensile-compressive machines yield a tensile or compressive mode. With the application of proper fixtures, the linear DMAs are also able to perform flexural and cantilever-type measurements.

The DMA is routinely used for the analysis of a wide variety of polymers, polymer blends, elastomers, and composites. It is also used to study foods, pharmaceuticals, coatings, and various inorganic materials (metals, mineral composites, glasses).

The key features of DMA are

Measures viscoelastic material properties of solid materials as a
function of temperature, time, frequency, stress, and strain
Creep and stress relaxation measurements can be performed
Broad modulus range 10^3–10^{12} Pa with high level of precision

Thermal Conductivity

Thermal conductivity is essential in almost every heat transfer calculation,
from such basic applications as component heat conduction and dissipation
to the complex computer programs used to simulate plastic manufacturing
processes. Plastics fall into the category of materials of intermediate thermal
conductivity (0.15–0.4 W/m-K). They are an order of magnitude more conductive than foams and insulation but about five times less conductive than
ceramics and glass. This puts them in a unique class by themselves, making
this measurement quite important. A number of techniques are available,
arising from two basic philosophies as described below.

In the *guarded hot-plate method*, a sheet of the material is sandwiched
between a metal heater plate and a heat sink plate. The temperature differential across the thickness of the specimen is measured when the system
is at equilibrium. This technique is well suited for measurements of solids,
particularly in cases where orientation exists within the test specimen. It is
unsuitable for measurements of molten polymers. Thermal contact resistance is a problem that must be overcome with such measurements.

The *hot-wire method* is a transient technique in which the temperature
rise of a hot wire immersed in the sample is monitored as function of time. It
is better suited for molten materials, and variants have been designed to
measure both melts and solids.

Thermal Diffusivity

The laser flash method is used for thermal diffusivity measurements. In
this technique, a laser is used to pulse one side of a test specimen uniformly. The temperature rise of the other side is measured using an infrared detector. This transient is then used to calculate the thermal diffusivity.
While the technique is simple in concept, nonidealities such as heat loss
from the front and back surfaces complicate the resulting data analysis, so
that fairly complex models need to be used to extract the thermal diffusivity. These calculations can be easily performed by the computers used to
run the instruments.

Dielectric Analyzer

A dielectric analyzer (DEA) measures the capacitive and conductive properties of materials as a function of temperature, time, and frequency under a controlled atmosphere. It provides high-sensitivity studies of the chemistry, rheology, and molecular mobility of materials. It can offer considerably improved sensitivity to low-energy transitions over what is available from DSC or TMA. A key feature of DEA is its flexibility for analysis of liquids, paste, and powder samples.

MECHANICAL TESTING

Modulus and Stress–Strain Behavior

Stress–strain behavior represents the response of a material to loading. Tests are performed on a universal testing machine (UTM), sometimes referred to as a tensile tester because of the primary mode of deformation used to characterize this form of behavior. Specimens are typically deformed at a constant speed, for reasons explained later. Since the properties vary significantly with temperature, tests may be conducted within an environmental chamber to obtain data at elevated and subambient temperatures. The most common information obtained from these tests are the modulus and tensile strength.

The modulus is the slope of the initial portion of the stress–strain curve. Its measurement is complicated by the fact that, unlike metals, most plastics do not have a linear relationship between stress and strain. This means that the modulus decreases with increasing strain, and the measured value depends on the strain region used as well as whether the slope is taken as a secant, a chord, or a tangent to this region. There is justification for each of these methods, and the choice depends on the application of the data. In order to ensure comparability, standards are used to define criteria for modulus measurement.

The tensile strength is the maximum stress that the material can withstand before failure. This term is well defined for most plastics. It suffers from variability in cases where the material is brittle, where the failure may be dictated by microscopic defects in the test specimen. In contrast, ductile materials exhibit well-defined maxima that result in repeatable tensile-strength measurements.

Stress–strain curves are by far the most important data that are derived from the UTM. These curves show the complete relationship and so permit a more realistic picture of material behavior. Tensile strength and modulus are, then, mere conveniences that permit comparing the properties of plastics—in contrast to metals, for which they can be used to provide a fairly accurate representation of the metal behavior.

Because of viscoelastic effects, the stress–strain relationship of plastics is rate dependent. Performing measurements at higher speeds results in a stiffer response: higher modulus means higher tensile strength. This behavior is important in the understanding of impact situations.

Several modes of deformation have been devised along the lines of the principal modes of deformation seen in the material: tension, compression, shear, and flexure.

The tensile test is the most commonly performed test. Artifacts in measurement are seen because of deformation of material in the grip region. To overcome this problem, dogbone-shaped specimens are used. Truly accurate measurements, however, require the use of extensometers, localized displacement-measuring devices. Extensometers may be contact or noncontact. For design purposes, it is also of interest to measure the lateral contraction of the plastic in response to the applied tensile stress. The ratio of lateral contraction to tensile extension is termed Poisson's ratio.

Flexural tests are commonly performed because of the ease of measurement. The test is considered to have limited scientific merit because the stress state of the sample changes continuously through the thickness of the sample, from a tensile state on one side to compressive on the other. Typical values taken from the test include a modulus and flexural strength.

Compressive tests are relatively uncommon and quite difficult to perform on plastics, mainly because of the difficulty of obtaining the right kind of specimens. Compressive tests typically require thick specimens to prevent buckling. This is difficult to do with most plastics because of the tendency of void and sink formation during molding. Further, the morphology of thick specimens can be dramatically different from that of thin specimens. The use of antibuckling fixtures does help, and the ASTM Standards offer a few options. However, caution must be exercised during compressive measurements to eliminate data that might have been subjected to experimental artifact.

Creep

Creep measurements bring in the dimension of time by characterizing the extension with time of a test specimen subjected to a constant load. Loads are applied in tension, flexure, or compression. Classical creep is extremely time consuming because it involves subjecting the test specimens to the desired load for the period of time of interest, which can be several months or even years. Consequently, creep data are not common. Sometimes, it is possible to apply time–temperature superposition to generate a master curve from data taken at several temperatures. Time–temperature superposition assumes that tests at higher temperatures are equivalent to tests at longer times.

Creep obeys viscoelastic theory at small strains and it is possible to apply predictive methods using data from dynamic mechanical analysis to obtain creep data. Here, remarkable amounts of data can be obtained from a relatively short test.

A further complication of creep is that it is nonlinear in strain, just as the stress–strain relationship is nonlinear for plastics. Since plastics are typically subjected to large deformations during their life, it is unfortunately essential to characterize this phenomenon. This phenomenon is best seen in the classic isochronous stress–strain curve which plots stress–strain relationships at several times, decades apart. These curves are invaluable for design and product performance evaluation.

Fatigue

Fatigue is a very important but unfortunately not well understood characteristic of plastics. Fatigue refers to the failure of a test specimen to repeated cyclic oscillation. Specimens may be fatigued in flexural or tensile modes. The most common test is ASTM D671, in which a specially designed test specimen is subjected to cantilever oscillation at 30 Hz with a constant amplitude of force mode: the specimen sees the same maximum force during the test and will continue to deflect to larger and larger strains as it weakens, until failure occurs. This is in contrast to a constant-deflection fatigue test in which the sample is subject to the same maximum deflection while the resultant force is reduced asymptotically with increasing cycles. The results from the two types of tests may not correlate. In the constant-deflection mode, it is feasible to prepare a cyclic stress–strain curve, which represents the stress–strain relationship of a material that has already been subjected to the cyclic oscillations and is at its ultimate stress level for the applied strain. These data are often useful in design.

Fatigue has important application in the automotive industry, in which components are subject to continuous vibration and, sometimes, rather large oscillations. Under these circumstances, components that use the inappropriate material or are inadequately designed may fail. In order to aid the design process, it is therefore vital to have some measure of the fatigue life of the material.

Fatigue life measurement is complicated by the fact that failure is observed to depend on the temperature as well as the frequency of oscillation. Particularly at high frequencies, considerable temperature rise may occur which cannot be easily dissipated due to the poor conductance of plastics. This makes the material more ductile, therefore tainting the result. Workarounds involve measurement and reporting the surface temperature of the test specimen.

Impact

Impact is a catastrophic event that has become very important for plastics. While measurements of impact performance have always been made in the past, new applications in the automotive, electronic, and consumer applicance industries have placed considerable importance on the quantification of this behavior. Historically, the Izod test has been used routinely to characterize impact. A notched rectangular bar is clamped in a vise and broken by a sharp impact from a hammer attached to a moving pendulum. The test has been widely criticized for being unsuitable for plastics, but it remains the most common test for impact and failure characterization. The Charpy test, widely used in Europe, has seen better acceptance by the scientific community. Here the test specimen, similarly notched, is held in a flexural mode while it is subjected to the impact.

More recently, attention has shifted to the falling-dart, multiaxial impact test, in which a sheet of the material is punched through by a heavy, instrumented, hemispherical striker (tup). The load versus time trace is used to characterize the impact characteristics of the plastic.

Because the impact behavior may vary considerably with temperature, tests are often carried out at other temperatures. Of particular interest is low-temperature tests to characterize the ductile–brittle transition. In many product applications, particularly for the automotive industry, it is important to ensure that external components do not undergo brittle failure at subambient conditions.

WET CHEMISTRY TESTING

Automatic Titration

Automatic titrators are used to perform potentiometric (pH, redox), amperometric, and coulometric titrations on a wide variety of organic and inorganic compounds. Major applications include acid number, hydroxyl number, and un-saturation level determinations in various materials.

Chlorine Analysis

A chlorine analyzer is used for low-level (parts-per-million) determination of chlorine in a variety of materials. This analyzer works by burning samples in an argon/oxygen atmosphere, transferring the resultant chloride to a titration cell, and automatically titrating it with silver ions that are generated coulometrically. With additional options, this analyzer can be used to determine sulfur and total organic halogens down to the parts-per-million level in different matrices.

Electrochemical Analysis

An electrochemical analyzer is used to perform polarography, voltammetry, and coulometry experiments using dropping mercury, noble metal, rotating disk, and carbon electrodes. These techniques are amenable to the trace analysis of electroactive anions, cations, metals, and organics. Applications include the determination of individual aldehydes in polymer solutions by voltammetry and the determination of stabilizers in polymer formulations.

pH/Ion Measurements

A specific-ion meter is capable of direct potentiometric determinations of electroactive species using pH, redox, or various ion-selective electrodes (i.e., chlorine, calcium, nitrate, ammonia/carbon dioxide, copper, and halides). Microprocessor control allows for instrument calculation of analyte levels by known additions, standard additions, or activity. An automatic temperature-compensation feature helps to reduce analytical errors.

By choosing the right electrode, pH meter, and accessories, the analyst can build a dedicated system for his or her application. A system capable of giving accurate, repeatable results time after time is critical, especially for difficult high-viscosity solutions. Measuring parameters as diverse as pH, ISE, conductivity, redox, and dissolved oxygen, the electrochemical meters offer great flexibility to measure different properties in polymeric solutions.

A critical part of any measuring system is the sensor. That is why a sensor with the right properties of reliability, accuracy, and robustness must be carefully chosen for each application. This technique offers a flexible variety of electrodes which are ideal for use for a wide range of electrochemical meters and a variety of products as well.

ELEMENTAL ANALYSIS

Inductively Coupled Plasma-Mass Spectrometry

Inductively coupled plasma-mass spectrometry (ICP-MS) is one of the most sophisticated and versatile analytical techniques available to the analyst. ICP-MS is a rapid and precise analytical technique that provides a high-quality, multielement and isotopic analysis package in a single analysis.

Some of the highlights of the technique are as follows: The detection limit for most elements is in the sub-parts-per-billion (ppb) range. For some elements it may lie in the sub-parts-per-trillion range. The versatility of the ICP-MS technique makes it a multidisciplinary analytical tool. It can be used in environmental sciences, the nuclear and semiconductor industries, materials science, medicine, agriculture, and food and biological sciences.

Clean-room facilities must be used to ensure contamination-free instrument environment as well as clean sample preparation. The instrument is normally equipped with a number of different sample introduction techniques. These accessories enhance the versatility of the ICP-MS in terms of handling samples of various types (biological, geological, etc.), concentrations, chemical matrix, mineralogy, and structure. The sample is introduced into a radiofrequency (RF) induced plasma in the form of a solution, vapor, or solid. The temperature of the plasma may reach up to 6000°C at the center and 8000°C at the periphery of the plasma. The high thermal energy and electron-rich environment of the ICP results in the conversion of most atoms into ions. A quadruple mass spectrometer permits the detection of ions at each mass in rapid sequence, allowing signals of individual isotopes of an element to be scanned. Various analytical packages (trace, rare earth, platinum-group elements, etc.) are currently available. The ICP-MS clean room facility normally requires multiple HEPA filters set into the ceiling or walls with the air return and filtering systems; the lab can be maintained at positive pressure versus outside air. Argon plasma is used to volatilize, atomize, and ionize the sample prior to extraction into the lens stack and quadruple mass spectrometer. Auto sampling capabilities are available for sample throughput. Additionally, a variety of sample introduction devices are available. These include a LaserProbe, an electrothermal vaporization (graphite furnace) unit, ultrasonic nebulizer, hydride generator, and a flow injection unit for the rapid analysis of chemical elements (ultratrace to percent levels) in a variety of matrices including aqueous materials, semiconductors, petrochemicals, plastics, etc.

Inductively Coupled Plasma Atomic Emission Spectrometry

The inductively coupled plasma (ICP) atomic emission spectrometer (AES) is used for the high-sensitivity detection of metals in dissolved samples. Applications include metals analysis of polymers, additives, catalysts, and other components on polymers and plastic formulations as well as advanced composite materials. The operating principle is essentially the same as in ICP-MS, instrument with the main difference being the detector. While the ICP-MS detector is a quadruple mass spectrometer which detects elements by their mass, the ICP-AES uses a detector based on the specific energy frequency emitted by each element in the plasma.

Energy-Dispersive X-Ray Fluorescence

Energy-dispersive X-ray fluorescence technology (ED-XRF) provides one of the simplest, most accurate, and most economical analytical methods for the determination of the chemical composition of many types of materials. It is nondestructive and reliable, requires no or very little sample preparation, and is suitable for solid, liquid, and powdered samples. It can be used for a wide range of elements, from sodium to uranium, and provides detection limits at the sub-parts-per-million level; it can also measure concentrations of up to 100% easily and simultaneously. ED-XRF is highly versatile and is available in a wide range configurations, from small benchtop instruments for special tasks to multipurpose laboratory spectrometers. Typical uses include the analysis of oils and fuel, plastic, rubber and textiles, pharmaceutical products, heat-resistant materials, glass, ceramics and wafers; the determination of coatings on paper, film, polyester, and metals; and the sorting of metal alloys, glass, and plastic according to their constituent materials. The instruments are extremely robust and designed for harsh environments. Application examples include sulfur and calcium in polymer formulations, polymer additives, and other engineering material formulations.

Atomic Absorption Spectrophotometry

The atomic absorption (AA) spectrophotometer approach is a powerful and cost-effective solution to a variety of elemental analyses. It is normally available to meet the performance and flexibility required in many analytical laboratories, including ease of lamp settings, secondary wavelengths, and alternative methods of analysis for multiple elements by flame, furnace, or hydride techniques. Background correction by a unique, in-line deuterium lamp is normally a nice option offered by most equipment manufacturers. The deuterium lamp emits radiation from the far-UV region (< 190 nm through approximately 350 nm), which corresponds precisely with the spectrum of the analyte. Using modulated signals, the absorbance of analyte and background interferences are ratioed, resulting in a clean, unbiased absorbance signal. Additionally, a variable giant pulse (VGP) correction is also offered for background correction. Hollow cathode lamps normally operate at currents of 3–15 mA. If the applied power is raised to several hundred milliamperes, they exhibit a phenomenon called self-reversal. This giant pulse of current changes the nature of the analyte absorption line so it will only measure the background absorbance. Like deuterium lamp correction, the background absorbance is subtracted from the total signal to give the corrected sample reading. The VGP system removes interferences for elements outside the normal D_2-UV region. A continuous-flow hydride accessory provides access to part-per-trillion detectability for hydride metals in a semiautomated, continuous -flow system. Low-level determinations of As, Se, Sb, Bi, Te, Sn, and Ge are easily performed. The cold vapor mercury model achieves trace-level analysis.

An AA spectrometer is also available with a graphite furnace and vapor generation accessories for the trace analysis of lead, antimony, arsenic, and mercury at parts-per-billion levels. AA is used for quantitative analysis of these metals in polymers as well as finished formulations. It has been used to determine the elemental composition of catalysts and plastic additives, polymer formulations, and composite materials. Samples may be rapidly acid digested prior to analysis using a microwave oven or similar techniques. Microwave furnaces are also available for dry ashing.

Flame Photometry

Thermal excitation of sodium, lithium, and potassium in a low-temperature flame gives atomic line emission at sufficient detection levels of 8 ppt sodium, 5 ppt lithium, and 15 ppt potassium. The optical range of this instrument is 589 to 766. The system is designed for high throughput in the visible range with sufficient resolution to discriminate against flame background radiation. The ASTM D1428 (20–200 ppb) and ASTM D2791 (0.1–20 ppb) procedures are simplified and freed from previous instrumental limitations associated with light leaks and electronic stability.

Kjeldahl Analysis

A Kjeldahl analyzer is used to determine ammonia (by a digestionless method), or total organic nitrogen by acid digestion followed by steam distillation and potentiometric titration. This unit performs all steps automatically and calculates the %N in the sample. The technique is suitable for liquids and powders that can be catalytically converted to ammonium ion. This method is complementary to the combustion method, which is much faster and has become the method of choice for analysis of nitrogen content.

Carbon, Hydrogen, Nitrogen, Sulfur, and Oxygen Analysis

The CHNS and O analyzer offers speed, precision, and ease of use for the compositional analysis of solid or liquid materials. This analyzer is based on the classical organic analysis Pregl-Dumas technique and provides important information to the scientist regarding product compositional chemistry. This analyzer offers the user multiple analysis options, including carbon, hydrogen, nitrogen, sulfur, and oxygen (CHN or CHNS and O_2 analysis). The user may choose one or more options to suit the needs of the laboratory. A fully configured system is capable of CHN, CHNS, and oxygen analysis. Alternatively, the user could start with the CHN option and simply add the CHNS and or oxygen analysis capability as the need arises.

Nitrogen Analysis

The nitrogen analyzer measures the protein content, determining the nitrogen percentage in a wide variety of materials by an advanced combustion process based on the original Dumas modified method. This technique is widely used in the polymer industry as well as in the food and agricultural industries. The nitrogen analyzer can be utilized instead of the time-consuming Kjeldahl method to analyze polymer materials, additives, raw materials, etc.

Halogen Analysis

Rapid and precise measurements of total organic halogens are available from many equipment providers. A microcoulometric and activated-carbon technology is officially certified by the EPA, ASTM, and ISO. This type of equipment is easy to use and very flexible for the analysis of different sample types.

MOISTURE ANALYSIS

One of the most useful tests for various plastics and additives is a moisture test. Often the moisture content is a critical measurement for plastics products. Lack of moisture control can cost a business significantly in out-of-specification product, downtime, and raw materials. Alternatively, effective moisture control can save a company thousands of dollars.

Karl Fischer Titration

A Karl Fischer titrator biamperometrically determines the level of water in organic and inorganic matrices. A volumetric type is used to analyze samples with water contents in the range of 0.1–100% employing a pyridine-free reagent. This unit may be equipped with a vapor generation oven that is used for solid samples or samples containing interfering substances (e.g., olefinic species). The sample is dissolved in anhydrous methanol or other appropriate solvent and titrated with a Karl Fischer reagent. Water reacts with the titrant in the following manner:

$$2RN + H_2O + I_2 + (RNH)SO_3CH_3 \rightarrow (RNH)SO_4CH_3 + 2(RNH)I$$

The endpoint (excess I_2) is determined by monitoring the change in current between two polarized platinum microelectrodes.

The Karl-Fischer method is widely accepted as a significant process for moisture measurement and has earned a permanent place in the modern laboratory.

Some types of Karl Fischer instruments provide special capability of moisture analysis by utilizing a ramping oven feature that can differentiate surface from bound moisture in a single analysis. Optimizing moisture methods is simplified by computer software available to fully document sample analysis data treatment.

KF titration is available in coulometric or volumetric techniques: volumetric titration is normally used for a broad range in moisture concentration (0–100%), while the coulometric techniques is normally used for small water concentrations (ppm levels to low percent levels).

On a heat transfer option for moisture analysis by KF, the sample is heated at a temperature greater than 100°C in an attached oven for a specified amount of time under a flow of dry nitrogen. The water evolved is carried by the flowing gas and concentrated by bubbling into the Karl Fischer titration vessel containing methanol. Karl Fischer reagent is used, and water reacts with the titrant as indicated above.

Moisture by Microwave Analysis

Microwave analysis of moisture is also known as LOD (loss on drying), drying test, or total volatile solids. For many years, the standard oven test was used for moisture determination. Although very accurate, the test could take many hours to complete, depending on the sample. The time lapse rendered it practically useless for anything except a final quality control check. The new microwave moisture/solids analyzers were introduced about 20 years ago. They offer rapid, accurate moisture/solids analysis in a fraction of the time it took to run a standard oven-drying test.

Moisture by Regular Convection Oven

The convection oven method is used to determine the amount of water (volatile matter) in a polymeric material to constant weight. The method is used to determine compliance with a loss on drying (LOD) test for regulatory agencies or customer requirements. The method provides accurate results for LOD values ranging from 0 to 15%. The sample to be analyzed is heated at 100 to 105°C for 2–4 h, which may be sufficient time to reach constant weight for the samples. The sample is then cooled in a desiccator and weighed to determine the weight loss.

ANALYSIS OF PARTICLE SIZE

Particle-Size Distribution Measurement

A particle size analyzer determines the particle size distribution of powders either dry or dispersed in solvent by laser light scattering based on the Fraunhofer scattering theory. This type of equipment has an optical bench whose combined dynamic range is nominally 0.7–2000 µm. The instrument calculates mean diameters and distribution data. An interfaced computer generates sample histograms. This technique has been applied to the study of particle size and particle size distributions for polymer powders and polymer suspensions in a variety of solvents.

A second type of instrument determines particle size distribution of either dry powders or liquid dispersions by using Mie laser light-scattering theory. Mie theory, unlike Fraunhofer theory, allows consideration of particle refractive index, required for reliable results on particles < 10 µm. Capabilities may range from 0.05 to 900 µm for wet analysis and 0.5 to 900 µm for dry samples. This type of instrument is particularly useful for obtaining information on very fine particle polymers.

A third type of analyzer uses a modified form of photon correlation spectroscopy to generate particle-size distributions in the range of 0.005–6.5 µm, which is useful for the measurement of microemulsions and micelles in polymers or polymer suspensions.

Optical Microscopy

An optical microscope is used for the optical examination of solids and liquids up to 1000× magnification. Cross-polarizers are available to enhance the contrast between analyte and background as well as to study crystalline forms. Video or photographic cameras are also available for the generation of photomicrographs. Various staining techniques can be employed for the identification of minerals and other components. A dedicated computerized image analyzer allows for particle-size distribution measurement as well.

Sieving

Several instruments are available to perform semiautomated particle sizing by sieving. Modes of agitation include shaking, forced air, and sonic irradiation. The first two types are suited for large sample sizes (10 g or more), and the third type is suited for smaller samples size (< 1 g). These instruments may be interfaced to a balance with software for collection of data by a PC. Particle size analysis by these approaches may require the use of rifflers to generate representative powder sample if the product is not uniform to begin with.

OTHER ANALYTICAL TECHNIQUES

Scanning Electron Microscopy/Energy Dispersive X-Ray Fluorescence Spectrometry

A digital scanning electron microscope (SEM) with an electron source is used to obtain high-resolution images of solid samples. This instrument is capable of magnification up to 100,000×. Special sample preparation techniques such as polishing and etching can be employed for examination of the internal structure of particle additives in composite materials. Stereoimaging and particle-size range measurements can also be performed. Qualitative elemental information can be obtained via a backscattered electron detector. An energy-dispersive X-ray analyzer may be interfaced to the SEM system. This allows for qualitative and semiquantitative determination of elements (atomic number > 6) in powders and thin films at discrete locations or for mapping of compositional distribution of chemical compositions. The equipment software normally contains image analysis capability that permits determination of quantitative particle size distribution.

Helium Pycnometry

A helium pycnometer is used to measure the true density of powder materials. The automated pycnometers run tests and report data from consecutive runs that are within a user-specified tolerance. This instrument can be used to perform particle density measurements on various composite and polymeric materials.

Ash Content (Also Known as Loss on Ignition or Residue on Ignition)

Whether burning off organic material for filler content, moisture determination, or as a preparatory step for further analysis, ashing is a useful process which can provide valuable information for material use or processing. Traditional electric muffle furnaces can take hours to ash a sample; a microwave-muffled furnace can do it in minutes. Several hours are required to bring a traditional furnace to temperature; the microwave muffle furnaces heat quickly up to 1200C. The microwave muffle furnace is an excellent tool for ashing of samples and can reduce ashing time significantly compared to conventional ovens, up to five times faster than conventional muffle furnaces, and eliminates the need for fume hoods, hot plates, or Bunsen burners.

Color Measurements

A large diversity of color measurement equipment from different manufacturers is available to the analyst, offering great flexibility and capabilities. Each system is designed to collect specific color data for specific applications. Although color measurement devices can be technologically sophisticated, they are normally easy to use and dependable. Systems are engineered for durability and to withstand the rigors of day-to-day use. They are also designed to accommodate sample handling devices that make color measurement fast and efficient. Color measurement systems offer a variety of sample measurement precisions and versatility for a variety of products including liquids, solids (powders or flake), smooth or textured. In addition to routine quality control, these instruments are also well suited for research and process control applications. All the spectrophotometers can be used with color-matching software to create color formulas and adjust color in the product. Sensors are available with sample measuring areas from several inches in diameter for measuring coarse or textured products as well as for measuring small areas or amounts (e.g., small powders). To accommodate this wide range of samples, numerous sample holding and positioning devices are available. For measuring the color of opaque or transparent samples, different sensor configurations are available and can measure apparent color changes due to both the sample color and surface shine or texture. A sphere sensor minimizes the surface effect and only measures differences due to sample color. Sphere sensors also have the ability to measure the transmitted color of clear samples. A personal computer offers high versatility in data processing and reporting. This enhanced capability is due to the power and flexibility of the various PC software packages available. Color formulation software packages create the colors the analyst or the customer wants.

Turbidity

Turbidity is a critical quality parameter in many applications of polymers and polymer solutions. Turbidimeters measure turbidity in either ratio or nonratio mode. Light from a tungsten-filament lamp is focused and passed through the sample. The $90°$ scatter detector receives light scattered by particles. Transmitted and forward-scatter detectors receive light that passes through the sample. A back-scatter detector measures light scattered back toward the light source. Benefits of the ratio mode include long-term calibration stability, wide measurement range with excellent linearity, and the ability to measure turbidity in the presence of color. A ratio turbidimeter is used to perform solution turbidity measurements by the principle of nephelometry. Results are reported in nephelometric turbidity units (NTUs). This type of instrument is also capable of APHA color measurements.

Molecular Modeling

Computer systems for molecular modeling are available from various sources. These systems are designed for visualization and properties estimation of large polymeric materials. These systems allow one to identify and visualize low-energy conformations, and to observe the dynamics of a molecule at a certain temperature. Electronic densities and related properties, such as charge distribution, nucleophilic, electrophilic, and free-radical susceptibility, can be calculated. Reactions between large molecules can be modeled, and transition-state energies can be estimated. IR and UV-visible spectra and other properties can also be calculated. The molecular simulations systems are designed for molecular modeling of polymers. In addition, estimations of many physical properties of polymers can be projected using modeling techniques. Thermodynamic properties of molecules in solution as well as binding energies can also be estimated. Recent applications include studies on complexation between polymeric materials and correlations between molecular and physical properties under different environmental conditions.

REFERENCES

1. P. Carreau, D. De Kee, and R. Chhabra, *Rheology of Polymeric Systems.* Hanser, 1997.
2. I. Harrison, T-C. Hsu, *An Introduction to Elongational Flow and the Multifunctional Axial Rheometer System (MARS)*, 1991.

2

Capillary Rheometry

Burke Nelson

Goettfert USA, Rock Hill, South Carolina, USA

INTRODUCTION

Rheology is the study of the deformation of materials. This includes the elastic deformation of solids such as metals as well as the viscous behavior of fluids such as water or oil. There is a wide range of materials that exhibit both a viscous and an elastic response to an applied force, and polymers fall into this group. One of the best ways to determine the viscous nature of these materials is with a capillary rheometer.

A capillary rheometer operates by measuring the force required to extrude a material through a small channel at a known flow rate. High temperatures and forces are required to work with polymers; capillary rheometers are robust instruments that are capable of reaching the temperatures and shear rates typical of a wide range of processing equipment, from compression molders to extruders to injection molders. This makes them the instrument of choice for evaluating the suitability of a new polymer for a specific process. Other types of rheometers, such as cone-and-plate-type rotational rheometers, are designed to study materials at very low deformation rates and cannot reach the shear rates necessary to directly evaluate the processability of polymers. Capillary rheometers are also very useful in quality control applications, for process troubleshooting, and process flow modeling.

43

The basic output of a capillary rheometer is the viscosity curve (or flow curve), the viscosity of a test sample expressed as a function of shear rate at a single test temperature. Many things influence the shape of the viscosity curve. Viscosity data can be used to study the effects of temperature, additives, and/or fillers on material processability, and to determine optimum process conditions. Viscosity may also be correlated with other, more difficult to measure properties such as molecular-weight distribution.

This chapter starts with a presentation of the basic equations used to calculate viscosity, the limitations of their assumptions, and the procedures used to correct for these problems. A basic capillary rheometer is described, and then a detailed description of the steps used to set up and run an experiment is presented. Methods of interpreting the data and some procedures to gather data other than viscosity curves are also discussed.

THEORY

Rheology studies the relationship between force and deformation in a material. To investigate this phenomenon we must be able to measure both force and deformation quantitatively. Steady simple shear is the simplest mode of deforming a fluid. It allows simple definitions of stress, strain, and strain rate, and a simple measurement of viscosity. With this as a basis, we will then examine the pressure flow used in capillary rheometers.

Before starting, the following Greek symbols are used in the field of rheology and throughout this chapter:

γ (gamma): shear strain
$\dot{\gamma}$ (gamma dot): shear rate
σ (sigma): shear stress
η (eta): viscosity

Steady Shear Viscosity

The most important mode of deformation during flow is shear. Shearing deformation results when the material is moved in a direction parallel to the surface as shown in Fig. 1. This can be visualized as layers of material sliding across one another, similar to shearing a deck of cards.

The amount of deformation is measured by the shear strain as defined below:

$$\gamma = \frac{x}{h} = \frac{X}{H} \tag{1}$$

The shear strain is normalized by the height of the sample and is zero when the sample is undeformed. Each element in the sample sees the same amount

Undeformed   Sheared

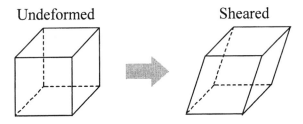

Figure 1 Shear strain of a cube.

of strain regardless of the size of the element (see Fig. 2). Strain is dimensionless and is often reported as a percentage.

Fluids flowing in a capillary rheometer see a shearing mode of deformation. Since fluids are always in motion and cannot hold a shape, a static measure of the strain is not useful; a continuous deformation rate must be determined. Differentiating the shear strain with respect to time gives the shear strain rate or "shear rate":

$$\dot{\gamma} = \frac{d\gamma}{dt} = \frac{1}{H}\frac{dX}{dt} = \frac{V}{H} \tag{2}$$

which is a measure of how fast the shearing process is occurring. The units of shear rate are reciprocal seconds (s^{-1}), with parameters V and H as defined in Fig. 3.

Once the deformation is determined, the force causing the deformation must be measured. A stress, σ, is defined as a force F applied to an area A:

$$\sigma = \frac{F}{A} \tag{3}$$

A is the area of the sample over which the force is applied; in Fig. 1 it corresponds to the area of the top of the cube and in Fig. 3 it corresponds

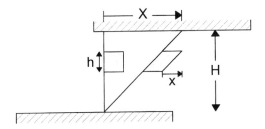

Figure 2 Shear strain—the top plate is moved to the right by distance X.

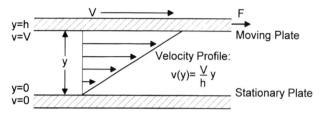

Figure 3 Simple steady shear of a fluid between two infinite plates. The fluid is stationary at the fixed plate and moves at velocity V at the moving plate.

to the area of the sample that wets the plates. A shear stress acts parallel to the surface to which it is applied, and the units of stress are force per unit area, or Pa (N/m²) in the SI system.

Figure 3 illustrates a steady simple shear flow between infinitely large plates (hence no edge effects, leakage, etc., which would be experienced in a real-world experiment). The top plate moves with a constant velocity V and shears a fluid sample of height H. The shear rate is constant across the gap so that all material sees the same deformation. The force F is that which is required to keep the plate moving at constant speed V. The stress and shear rate are determined from the force and the plate speed and geometry using Eqs. (2) and (3). The viscosity, defined as the fluid's resistance to flow, is determined from the ratio of the force and the deformation rate:

$$\eta = \frac{\sigma}{\dot{\gamma}} \tag{4}$$

The units for viscosity are Pa-s. A Newtonian fluid is a fluid that follows Newton's law of viscosity, which states that the viscosity is independent of shear rate. This implies a linear relation between force and plate speed; doubling the force on the plate will double its velocity.

Newtonian Equations for Round-Hole Capillary Flow

The flow in Fig. 3 is called a *drag flow*; the top plate is dragging the material across the stationary plate to create the velocity profile that is shearing the fluid. In contrast, flow in a capillary rheometer is *pressure-driven flow*. All of the wall area inside the capillary is stationary so that the material has zero velocity at the walls and a maximum velocity along the centerline. Calculating the shear rate in a capillary is not as straightforward as with steady simple shear. Each fluid element still sees steady simple shear, but the shear rate is no longer constant; it varies across the radius of the die. It runs from zero at the centerline to a maximum at the wall. Since the shear rate is not constant, a point must be chosen where the shear rate and stress can both be determined. The forces that generate stress interact with the material at the wall, and at this point the shear rate is a maximum. Thus this is where the viscosity is calculated.

Assuming Newtonian fluid behavior (constant viscosity), the no-slip assumption ($v = 0$ at the die wall), and fully developed flow, a parabolic velocity distribution can be derived for the material flowing down the capillary:

$$v(r) = 2v^* \left[1 - \left(\frac{r}{R} \right)^2 \right] \tag{5}$$

where v^* is the average velocity of the fluid and R is the radius of the capillary. This average velocity can be defined in terms of the cross section of the die and the volumetric flow rate of the fluid, Q, as follows:

$$v^* = \frac{Q}{\pi R^2} \tag{6}$$

The shear rate profile can now be calculated by differentiating the velocity profile. The shear rate at the wall ($r = R$) is determined and expressed as a function of the volumetric flow rate:

$$\dot{\gamma}_{ap} = \frac{dv}{dr}\bigg|_{r=R} = \frac{4Q}{\pi R^3} \tag{7}$$

The meaning of the subscript "ap" is discussed in the section on the Rabinowitch correction.

Once the shear rate is established, the stress required to drive the flow must be determined. A force balance between the driving pressure across the cross section of the capillary and the shear stress experienced along the wetted surface inside the capillary links the measured pressure drop across the die (ΔP) to the stress at the wall (σ_W):

$$\Delta P \cdot \pi R^2 = \sigma_W \cdot 2\pi R L \tag{8}$$

where L is the length of the die. This can be rearranged to give

$$\sigma_W = \frac{\Delta P \cdot R}{2L} \tag{9}$$

Pressure is measured in units of pascals (Pa, the same unit as stress); other common usages are MPa (10^6 Pa) and bar (10^5 Pa). Using eqs. (7) and (9), the Newtonian viscosity can then be determined from the flow rate and the pressure drop:

$$\eta_{ap} = \frac{\sigma_W}{\dot{\gamma}_{ap}} = \frac{\pi \cdot \Delta P \cdot R^4}{8 \cdot Q \cdot L} \tag{10}$$

Again, the units for viscosity are Pa-s (Pascal-seconds).

Slit Die Capillaries

In addition in round-hole capillaries, it is possible to use a slit die to perform viscosity measurements. Theoretically such a die would have no side walls (be infinitely wide), but in practice it has been found that wall effects are negligible if the slit die is at least 10 times wider than it is high. The equations for slit flow are similar to those for round-hole capillaries:

$$\dot{\gamma}_{ap} = \frac{6Q}{H^2 W} \tag{11}$$

$$\sigma_W = \frac{\Delta P \cdot H}{L} \tag{12}$$

$$\eta_{ap} = \frac{\sigma_W}{\dot{\gamma}_{ap}} = \frac{\Delta P \cdot H^3 \cdot W}{12 \cdot Q \cdot L} \tag{13}$$

where H is the height and W is the width of the slit.

The main advantage of a slit die is that, since it has flat walls, the pressure transducers can be mounted directly in the die. This eliminates the entry and exit effects discussed in the section on the Bagley correction below. A slit die's disadvantages tend to be of a practical nature. While slit dies can usually be made wide enough to avoid edge effects, they are much more difficult to flush clean with a new sample, especially near the edges. They are usually multipiece dies that require much more effort to disassemble and clean. Round-hole dies are more convenient and are the die of choice unless a special test procedure specifically requires a slit die. (This might happen if a tapelike extrudate is desired for further testing, or if three or more pressure transducers are to be mounted along the die to study the pressure profile along the die.)

Rabinowitch Correction

The steady simple shear flow in Fig. 3 is fully controllable, because the velocity profile of the flow depends only on the geometry and motion of the plates, regardless of the rheological properties of the fluid. Fully developed flow in a capillary tube is only partially controllable. The material properties can affect the velocity profile, which must be known in order to calculate the shear rate that the sample is experiencing. If a Newtonian fluid (constant viscosity at all shear rates) is assumed, then a parabolic flow profile can be calculated as in Fig. 4 with $n = 1$. Shear rates and viscosities calculated using this assumption are termed "apparent" values, with the subscript "ap." Apparent values are correct for Newtonian fluids. Most materials, including most polymers and polymer solutions, are not Newtonian fluids because their viscosity is not constant with changing shear rate.

Figure 5 shows some of the basic curve shapes and names for materials whose viscosity varies with shear rate. Most polymers are pseudo-plastic, or shear thinning; as the shear rate increases, their viscosity decreases. Figure 6 shows a typical flow curve for a molten polymer. It can be divided into three regions: the Newtonian plateau at low shear, the power law region at high shear, and the transition zone in between. At very low shear rates (often $\ll 0.01$ s^{-1} for polymer melts), η is independent of γ; this constant value is called the zero-shear viscosity (η_0). As the shear rate is raised, the structure of the molecular chains begins to break down and the viscosity is reduced. At higher shear rates, this can be modeled with the power law model,

$$\tau = k\dot{\gamma}^n \tag{14}$$

which has parameters n and k. When combined with Eq. (4), this gives

$$\eta = k\dot{\gamma}^{n-1} \tag{15}$$

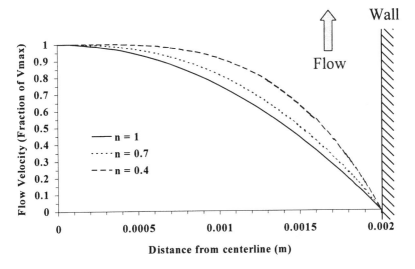

Figure 4 Velocity profiles as a function of the power law exponent.

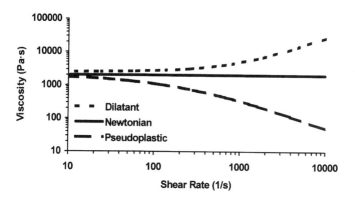

Figure 5   Viscosity curves varying with shear rate.

Because viscosity is nonconstant, it must be measured across a range of shear rates to fully characterize a material. The equations used to calculate the apparent viscosity are not exactly correct due to the shear-thinning phenomenon. A correction has been developed (by Rabinowitch [1]) that uses the slope of the $\sigma_{ap}(\gamma_{ap})$ curve,

$$n = \frac{d\log(\sigma_{ap})}{d\log(\dot{\gamma}_{ap})} \tag{16}$$

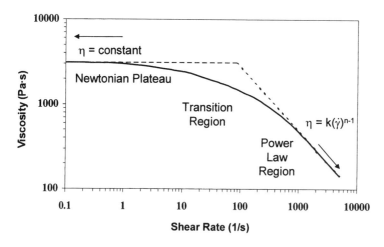

Figure 6   Regions of a typical flow curve for a shear-thinning polymer.

to correct for the non-constant viscosity. For a Newtonian fluid, $n = 1$. The smaller the value of n, the more shear thinning is occurring and the larger the correction that is required. The true wall shear rate is calculated using

$$\dot{\gamma}_w = \dot{\gamma}_{ap}\left(\frac{3n+1}{4n}\right) \tag{17}$$

with n calculated at the point of the measurement to be corrected. Note that multiple measurements must be available so that the calculation of a slope is possible; it is not possible to correct a single point measurement. Most modern capillary rheometers perform this correction automatically in their software, and then use the true shear rate values in eq. 4 to calculate the true viscosity.

Another approach that is available involves calculating the viscosity at a point different from the wall. The Schümmer correction [2] involves determining where the actual velocity profile intersects with the Newtonian profile; at this point,

$$\eta(\dot{\gamma}^*) = \eta_{ap}(\dot{\gamma}_{ap}) \tag{18}$$

i.e., the apparent viscosity equals the true viscosity at the Schümmer-corrected shear rate. If the Schümmer shear rate is

$$\dot{\gamma}^* = x\dot{\gamma}_{ap} \tag{19}$$

then for a power law fluid,

$$x = \left(\frac{3n+1}{4n}\right)^{n/n-1} \tag{20}$$

The value of x varies by only a small amount over a wide range of values of n, so that a representative value of x^* may be chosen for a material with very little loss in accuracy. This method can therefore be used to correct the shear rate of single point measurements if an approximate value of n is known for the tested material. Schümmer suggests that if a constant value of $x = 0.83$ is used for all materials, the error in the viscosity will be 3% or less.

Bagley Correction

Most capillary rheometers use round-hole dies with the pressure (or force) measured above the die in the barrel. The melt flow undergoes a large contraction as it moves from the barrel into the die and an expansion as it leaves the die to the atmosphere. These entrance and exit effects add an extra pressure drop (P_e) to the measured pressure (P_m) in addition to the pressure drop due to flow through the die (ΔP). This will result in a viscosity measurement that is erroneously high due to the P_e component.

Bagley [3] developed a method to determine P_e so that capillary data could be corrected. The method requires multiple measurements using dies of the same diameter and varying lengths. By keeping the diameter the same, the entrance and exit effects will be the same for each die; the total measured pressure P_m can be determined as a function of die length for each shear rate. Extrapolating the pressure back to a die with zero length allows the determination of P_e. In other words, whatever pressure still remains after ΔP is removed is the pressure due to entrance and exit effects.

The following steps are taken for a Bagley correction.

1. Determine the apparent flow curve for the sample using at least two different die lengths (three or more improves accuracy). The same apparent shear rate points must be used for all tests.
2. Construct a Bagley plot from this data (see Fig. 7). The measured pressures are plotted against the L/D of the die for each apparent shear rate. (*Note*: By convention, the Bagley plot

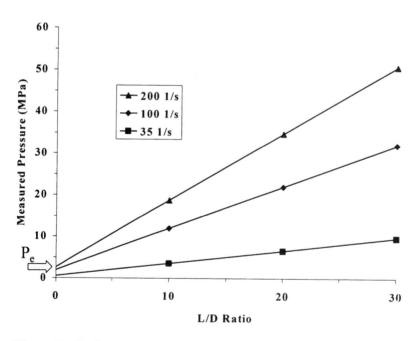

Figure 7 Bagley correction performed for a three-point curve run three times with 30/1-, 20/1-, and 10/1-mm dies. For each shear rate, P_e is determined by extrapolating back to $L/D = 0$.

uses the L/D ratio instead of just the die length, but since D is held constant the results are the same using L or L/D).

3. For each shear rate, perform a linear extrapolation back to zero. The value of pressure at $L/D = 0$ is P_e, the extra pressure due to entrance and exit effects.

4. To apply the correction, subtract P_e from the measured pressure P_m to get the correct pressure for each data point. Use the corrected pressure to calculate the stress and viscosity for that point. Once each curve (from the different dies) has been corrected, the curves can be averaged and Rabinowitch corrected to get a final true viscosity curve.

The entrance and exit effects are often quite small relative to the measured pressure, especially if longer dies are used that increase the magnitude of ΔP compared to P_e. A study of the effect of both the Bagley and Rabinowitch corrections [4] demonstrated that the corrections become more substantial with shorter dies and higher shear rates As a rule of thumb, dies with an L/D ratio of 20 or greater often have P_e values that are 5% or less of P_m, in which case the effort of performing the Bagley correction is probably not necessary. However, there are many materials (especially with strong elastic components) that have appreciable P_e values even with long dies, so it is wise to check an unknown material at least once to determine the magnitude of the Bagley correction before deciding if it is necessary.

CAPILLARY RHEOMETERS

A capillary rheometer measures the viscous properties of a fluid by determining the pressure required to cause it to flow through a small cylindrical tube or rectangular slit (the capillary) at a set flow rate. Many rheometers control the flow rate (by setting the piston speed) and measure the pressure drop across the capillary; some gas- or weight-driven models fix the pressure and measure the flow. The American Society for Testing and Materials has issued a standard entitled "Standard Test Method for Determination of Polymeric Materials by Means of a Capillary Rheometer," designated D3835–96 [5]. This method covers the measurement of the viscosity of materials at temperatures and shear rates found in common plastics processes. Issues such as temperature control and calibration, rheometer and die specifications, common temperature ranges, procedures, data analysis, and reproducibility are addressed.

Figure 8 shows a schematic of a typical capillary rheometer. The molten polymer is held in a cylindrical reservoir and heated to the desired temperature. A motor-driven piston is used to force the material down the barrel

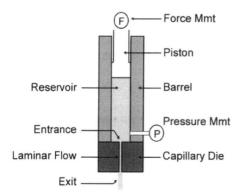

Figure 8   Schematic of a capillary rheometer.

and through the die, with the flow rate determined by the speed of the piston. Once the pressure has reached equilibrium it is noted, and the piston speed may be changed to measure the material at a different rate. The important parameters in this measurement to be controlled and/or measured are the temperature, pressure (or force on the piston), the material flow rate (calculated from the piston speed and the barrel cross section), and the die geometry.

The geometry of the capillary rheometer, both the barrel and the die, must be known exactly. Typical barrels range in size from 9 to 15 mm in diameter. Larger barrels hold more material and achieve higher flow rates for the same piston speed but require longer holding times for the samples to reach temperature equilibrium before a test can begin. Barrel, piston, and die are made of very hard metals to resist wear over time, but tolerances must be checked for wear on a regular basis. The piston must seal well with the barrel to prevent backflow of the material. Often a double-land or O-ring design is used.

The capillary dies must be very smooth, accurately machined, and hard. They are often made of tungsten carbide. Accuracy is especially critical. In a round-hole capillary, for example, the shear rate is a function of the radius cubed [see Eq. (7)]. A 1% error in the radius will result in a 3% error in the shear rate determination. Some dies have entry angles machined into their entrance to ease the abruptness of the transition from barrel to die and reduce the entrance pressure losses (see Bagley correction). Typical dies have diameters in the range of 0.5 to 2 mm, but specialty dies are available in almost any practical dimension. Depending on the piston speed and die selection, a wide range of shear rates can be achieved on capillary rheometers (see Fig. 9).

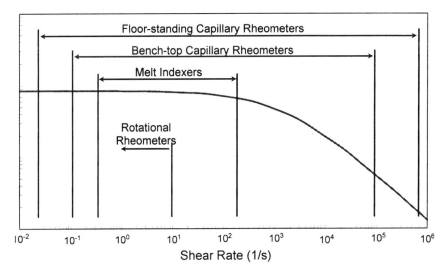

Figure 9 Shear rate ranges of common rheometers.

Once the flow rate (and thus the shear rate) is determined from the piston speed and the geometry, the pressure drop across the die must be determined. This is measured with a pressure transducer in the barrel near the die or a force transducer situated on top of the piston. Each method has pros and cons, so choosing the best method depends on a particular situation (Table 1).

Pressure transducers are often preferred when high-accuracy measurements on a broad range of materials are required. Force transducers see use in quality control labs, where ease of operation and cleaning are important, or when maximum corrosion resistance is desired. Further discussions will assume that a pressure measurement system rather than a force transducer is used. (To calculate stress, the force measurement is converted into a barrel pressure by dividing by the cross-sectional area of the barrel, so no generality is lost.)

Often the pressure measurement is the limiting factor determining the maximum shear rate range of a single experiment. Pressure and force transducers typically have 0.5% of full scale total accuracy (linearity, repeatability, and histerysis). This means that a measurement registering 1% of full scale has a relative accuracy of ±50%! Manufacturers often recommend measuring pressures between 10% and 100% of full scale for this reason. This is a range of only a single decade, while most rheometers can accurately control piston speeds over a range of four decades or more. New transducers

Table 1 Pros and Cons of Measurement with Pressure and Force
Transducers

Pressure measurements	Pressure is measured right at the die entrance
	Transducers are less expensive and easier to change, allowing a better match range to material viscosity
	But a pressure hole error is generated (usually negligible since material flows slowly in the barrel)
Force measurements	Easier to clean (no pressure tap to collect material)
	Transducer will not come in contact with corrosive or very-high-temperature samples
	But piston friction can be significant and material flow along barrel as well as through die is measure; the measurement is dependent on the amount of material in the barrel

are being released with better accuracy as high as ±0.1%, a significant
improvement but not enough to remove the restriction completely.

One of the most important criteria for any instrument is repeatability,
not only with the same instrument running the same sample, but between
different machines with different manufacturers in different laboratories. As
well as machine design and setup, repeatability depends on the actual test
procedure used and the consistency of sample conditioning. ASTM D3835
includes the results of a round-robin test in which a number of different labs
measured flow curves on a number of materials. It was found that, within a
single lab, the standard deviation in the viscosity results was typically less
than 2%. Between the labs, considerably higher variability was seen; most
measurements had between 3% and 8% standard deviation. An important
observation is that the standard deviations tended to be much higher for the
lower shear rate points. This is because stress measurements taken at the low
end of the pressure/force transducer ranges, as mentioned above, have much
higher relative errors.

MELT INDEXERS

The melt index is an industry standard test used to assess the processability
of a polymer melt. A schematic of a melt indexer is shown in Fig. 10. It is
essentially a stress-controlled capillary rheometer using a weight-driven piston to force material through a round-hole capillary die. A melt index is
obtained by measuring the amount of time required for a specific volume
of material to be extruded from the die, with the results presented in units of
g/10 min.

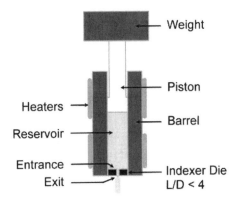

Figure 10   Schematic of a melt indexer.

ASTM Method 1238, "Standard Test Method for Flow Rates of Thermoplastics by Extrusion Plastometer" [6], completely defines the melt index, both the instrument and the experimental procedure. Temperatures and instrument configurations for common materials are given along with recommended calibration procedures and round-robin results. This method must be strictly followed if consistent melt index results are to be obtained. A melt index measurement is performed at low stress levels compared to capillary rheometers measurements (typical weights are in the range of 1.2–21.6 kg) and is sensitive to variations in operating procedure, instrument cleanliness (between tests), and operator bias (items such as packing and sample preparation).

The melt index itself is not a material property like temperature or viscosity, but a machine dependent index. This is because the die is short relative to its length (8 mm long and 2.1 mm in diameter), with an $L/D < 4$. Typical L/D values for capillary rheometers are greater than 15. As a result, the melt index results are heavily influenced by entrance and exit effects at the die (see the section on the Bagley correction) that cannot be exactly duplicated on anything other than another melt indexer. While it is possible to calculate a viscosity from the flow data generated by the melt indexer, the entrance and exit effects also negatively influence its accuracy.

TECHNIQUES

Setup Parameters

Before running tests on a polymer sample, a number of parameters must be chosen for the instrument. These include the test temperature, shear rate ranges, and the die and pressure transducer selections. This section discusses these considerations.

Temperature

Temperature is one of the most important parameters to be specified, as viscosity is a strong function of temperature. The temperature must be high enough to melt the sample completely and low enough that the sample will not degrade during the test. Once a temperature is chosen, a thermal degradation study should be performed to determine the maximum residence time of the material in the rheometer.

Meaningful test temperatures can often be found in two places: an ASTM specification or a processing temperature. A standard testing temperature suggested by ASTM (either in D3835 or in the melt index specification, D1238) is often used if available, because it has been chosen by experts to provide a good molten sample with a reasonable dwell time in the rheometer. Also, it will likely be chosen by other people, allowing comparison of results with other laboratories and literature sources. The ASTM temperature is often not, however, the temperature at which the material is processed. Using the process temperature will give the most relevant results when the test is performed as part of a troubleshooting effort or to screen candidate samples for a process.

A pretest heating period must be specified to allow a material to melt and come to temperature before a test is begun. Polymers are notoriously poor conductors of heat and require time for the material in the center of the barrel to come to temperature. The amount of time required depends on the temperature and the barrel size: higher temperatures and larger barrel sizes require longer times for temperature stabilization. If too long a period is specified, the duration of the test may be extended to the point that polymer degradation occurs. Typical times for barrels from 9.5 to 12 mm in diameter in the 150–300°C range are 4–7 min.

Shear Range and Die Diameter Selection

Before selecting a die or transducer, the desired shear rate range for the flow curve must be determined. The process for which the material is intended usually determines this. Some typical ranges are listed in Fig. 11.

A typical test range for an average extrusion process might be from 20 to 2000 s^{-1}, whereas testing for an injection molding process might be more relevant across a range of 200–20,000 s^{-1}. Keeping in mind that pressure and force transducers have a maximum usable range of two decades, it is usually a good idea to also limit the shear rate range to two decades in a single test, even if the rheometer is capable of more. Once you are familiar

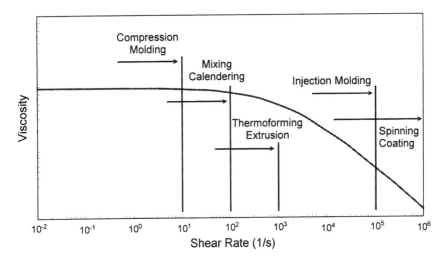

Figure 11 Shear rate ranges for common polymer processes. Compare these ranges with the rheometer capabilities in Fig. 9 to see why capillary rheometers are more suited to test polymer processability.

with a material, you may find that you can add extra points at either the high or low end of the range.

Once the shear rate range is determined, the appropriate die can be selected. The radius of the die, R, is selected first; it should be such that the piston speeds required to generate the selected shear range fall into the mid-range (if possible) of the piston speed range of the instrument. Using Eq. (7) to relate shear rate and volumetric flow rate, and the cross-sectional area of the barrel piston (πR_B^2, where R_B is the rheometer barrel radius), the piston speed is calculated as

$$v = \frac{\dot{\gamma} R^3}{4 R_B^2} \tag{21}$$

The piston speed is a linear function of the shear rate that is strongly affected by the geometry of both the barrel and die. Usually the barrel geometry is fixed and the die geometry is variable; reducing the radius of the die by one-half will increase the shear rate through the die by eight times for the same piston speed. For an instrument with $R_B = 6$mm and $R = 0.5$ mm (1-mm diameter), the 20–2,000 s^{-1} range mentioned above would correspond to piston speeds of 0.0174–1.74 mm/s, easily achievable by most rheometers. On the same instrument, a die with $R = 0.25$ mm would be more suitable for testing an injection molding-grade material.

Die Length and Pressure Transducer Selection

The length of the die controls the magnitude of the pressure drop the sample will generate and determines the range of the pressure transducer that will be required. In many cases, a rheometer has only a single transducer available and so the die length must be properly selected to assure test results in a range suitable for the transducer. The pressure drop is linear with the die length; doubling the length doubles the pressure drop at the same flow rate.

In considering die length, a second criterion is the L/D (length/diameter) ratio of the die. In order to assure fully developed flow, an $L/D >$ 15 is recommended, with values in the range of 20–30 being common. If a die is too short, then entrance and end effects will form a large percentage of the pressure drop and the viscosity measurement will be incorrect (see the Bagley correction). If a die is too long, the pressure drop may be too great for the rheometer to handle, or pressure/density effects may begin to become significant in the viscosity measurement.

With a material that is completely unfamiliar, one or two trial tests may be required before the best transducer selection can be determined. Usually the rheometer is equipped with a transducer that covers the high end of its capabilities. (Most capillary rheometers can generate pressures up to 1000 bar; most floor-standing units can achieve close to 2000 bar.) Obviously, the first test should use a high-range transducer to avoid possibly overpressuring a more sensitive transducer. Most transducers will take 25–50% overload before damage occurs—check the unit's specifications. If more sensitive transducers are available for your unit, the maximum pressure generated in your first experiment will tell you which transducer to choose for subsequent tests. Note that pressures below 1 bar are extremely difficult to measure even with sensitive transducers.

Determining the Number of Shear Rate Points per Curve

The number of points that can be achieved in a single test is highly dependent on the piston speed required to generate the desired shear rates. The higher the piston speed, the more material is extruded for the given point. Generally, only a few high-speed points (where the test piston is traveling more than ˜1 mm/s) can be obtained with a single sample. Rheometers with longer barrels will hold more melt and are able to generate more points. Some experimentation is usually necessary to before a final shear rate configuration can be achieved.

The speed with which a point can be taken depends on how fast the pressure stabilizes after the piston reaches the correct speed. The transient nature of the pressure response in a capillary rheometer has been discussed and modeled by Dealy and Hatzikiriakos [7]. The pressure will stabilize quickly at high extrusion rates but can take 4–5 min at low rates. This affects both the overall length of the test and the amount of material expended for a given point.

Various algorithms exist to determine automatically when the pressure is stable. One procedure is outlined in an appendix to ASTM D3835. Usually, the operator sets a tolerance (a percentage change in the pressure) over a specified period of time. When the pressure changes less than the tolerance over the comparison interval, then the pressure is considered steady and the viscosity point is taken. If the pressure is still changing, the routine waits for another comparison interval to pass and then checks again.

If the tolerance is made small and the comparison interval long, the pressure will have to be very level before it is considered stable. If the tolerance is relaxed or the comparison interval shortened, then the viscosity point will be taken quickly, while the pressure is still slowly changing. There is a trade-off between speed and accuracy: viscosity points taken more quickly may not be as accurate. Generally, one chooses a reasonable pressure tolerance and a long comparison interval to start (for many materials, a 1% tolerance over a 10–20-s interval is a good starting point). Gradually reducing the length of the interval until the viscosity measurement starts to change allows the determination of the optimum parameters, which in turn determines the number of points that can be obtained in a single test (7–10 points is reasonable in most cases).

Shear Rate Order

The order in which the shear rate points are run during a test influences the dynamics of the experiment. Running in decreasing order (from high to low rates) as opposed to increasing order will affect the speed of the measurement. When low shear measurements are performed at the end of an experiment rather than the beginning, there is much less material in the rheometer barrel. Therefore there will be less material that must equilibrate before a stable pressure reading is generated; faster stabilization means that the viscosity points can be taken faster.

Unfortunately, the amount of sample saved by taking points faster at low shear rates is offset by extra sample required by the first points at high shear rates, so in most cases the number of viscosity points obtainable with a single barrel of material is about the same. However, running decreasing shear rates can reduce the overall length of an experiment by 30–50% compared to increasing shear rates and in most cases is the preferred mode of operation. One must be careful, however, with unknown materials: starting at a high piston speed can damage an undersized pressure or force transducer.

Sometimes the shear rate order is randomized to ensure that the viscosity measurement is independent of the order in which the points are taken. This causes no problems but, with the piston speed and the pressure swinging up and down, more time will be needed for stabilization. It may not be possible to get as many points as when the shear rates are ordered.

Running a Test

Running a test involves three steps: preparing the sample, preparing and loading the rheometer, and running the test.

Sample Preparation

Preparation of the sample may involve drying the sample; moisture absorbed from the air can cause some polymers (e.g., polyesters and polycarbonates) to degrade chemically when melted. These samples must be dried in a heated convection oven or vacuum oven to drive off moisture before testing. Since the presence of moisture dramatically affects the rheological properties of the material, it is very important that the drying protocol be followed consistently to ensure repeatable test results.

To develop a drying protocol, test materials that have been in the oven for varying lengths of time. The shortest length of time that gives consistent results should be considered the drying time. In general, higher temperatures (especially over 100°C, if the material does not begin to melt or degrade) and the application of a vacuum will reduce the amount of drying time required. Typical times range from 2 to 6 h.

When loading moisture-sensitive materials into the rheometer, speed is usually the best defense against the material picking up any moisture. Ensure that the drying oven is close to the rheometer and that any sample removed from the oven is immediately loaded or discarded. Some rheometers have a nitrogen system available that blows dry nitrogen into the barrel to keep a blanket over the sample as it is loaded.

Preparing the Rheometer

Preparing the rheometer involves bringing it to the correct temperature, installing the selected die and pressure transducers, and ensuring that it is clean. When installing cold transducers or dies into a hot instrument, the items should be inserted and tightened only loosely until they have had a chance to reach the rheometer temperature. Final tightening should be performed with items at the test temperature.

One of the most common causes of sample variation is an improperly cleaned rheometer. Making sure the barrel walls and piston are clean will free up piston travel and ensure that degraded material will not contaminate new material as it flows through the die. Swabbing the barrel with cotton patches and using a rotating brass brush are typical methods of cleaning out the barrel. The barrel wall is often not very dirty, since the close tolerance between the piston and barrel keeps it scraped clean. However, piston travel usually ends a predetermined distance from the die, leaving a disk of material on the top of the die that needs to be removed. Often the cotton patches are sufficient to remove this material. If in doubt, remove the die after cleaning the barrel to see how much material is left behind.

If a pressure tap is present, it must be cleaned at least on a daily basis. Often the material trapped inside the tap degrades very slowly, since it is not exposed to oxygen, and the material does not leave the tap to contaminate the new sample. As long as the material does not carbonize and block the opening into the barrel, it often does not need to be cleaned after each experiment. (Be careful when operating with materials at high temperatures, 300°C or above: the high heat is more likely to carbonize the sample. It is usually better to clean the tap more often in this case.)

The die itself must be purged of the old material before a new test can begin. Often when loading and prepressuring the sample a small amount will be purged through the die; this will suffice to clean the die if the tested materials are similar or if the new material is more viscous than the old. To further aid purging, a high to low shear rate profile causes a large amount of material to flow through the die at the start of the test. By the time the pressure has leveled out for the first point, the die is well purged.

When a lower-viscosity material is used for purging, a half or full barrel of material might be required to remove the old sample before the new sample can be tested. (*Note*: If the sample being tested is very different in viscosity from the previous sample, then a "blank" test should be run, in which the die is purged with an entire barrel of new material to ensure that the previous sample is completely removed.)

If a muffle furnace or sand bath is available, the old sample can be burned out of the die (usually at ~ 500–$550°C$) to assure that only the new sample will be present during the test. This is time consuming, however, so often it is done only at the end of the day to prevent buildup from occurring in the die. Not all materials can be successfully burned out of a die; materials that carbonize (such as rubbers and some high-temperature engineering plastics) can permanently plug a die if burning is used to try to remove them. They should be purged with another material (such as a polyolefin) before cleaning.

Loading the Rheometer

Samples for testing come in many forms, including pellets, powder, cut strips, pourable liquids, or molded slugs. The material is loaded into the barrel with the aid of a tamping rod. After pouring in a small amount of material, use the tamping rod to press it to the bottom and compact it to remove as much air as possible. Repeat this step roughly 6–10 times, until the barrel is full. Liquids may be poured slowly down one wall of the barrel to avoid air pockets. Semiliquid pastes may have to be loaded with a syringe having a tube attached that is long enough to reach the bottom of the barrel (considerable effort may be involved to get the material to flow down the tube). Every effort must be taken to avoid the introduction of air bubbles, as they will interfere with the measurement.

Once the sample is in place, the piston should be inserted and slowly driven down on top of the sample until a small flow is extruded from the die. This will compress the sample and drive out air bubbles. Usually, by the time the piston is inserted into the barrel the material at the bottom is melted enough to flow, though care should be taken not to overpressure the system. Pushing a small amount of material through the die will help to purge any old material left from previous testing. A pressure spike indicates that the material has filled the pressure tap and no blockage is present.

At this point, the melt timer is started and the instrument holds the sample to allow it to come to temperature. Once the allotted time is over, the piston begins to move and the first data point is begun. Figure 12 shows a typical pressure trace from a nine-point viscosity curve for polyethylene.

Possible Problems

Once operators gain some experience, they will find a large number of indicators telling them if there is something wrong with their experiment. The first check is to compare your data with previous runs on similar samples to make sure it is reasonable. The viscosity–shear rate function should be a smooth curve similar to the typical curve shown in Fig. 6. If the points are very noisy, check to make sure that the pressure readings are at or above 10% of the full scale of the pressure transducer. Often the low-shear-rate points are not, but some noise in the low-pressure measurements might have to be accepted to get a pressure transducer large enough to measure the high-rate points.

If the data are different than expected, check to make sure the pressure/force transducer that was used was calibrated properly, that the correct die was inserted in the rheometer, and that the instrument's temperature controllers accurately reflect the temperature in the rheometer barrel. Most laboratories keep a well-characterized "lab standard" material available that can be run on an instrument to make sure it is giving the correct results.

When working with an unknown material, double check your results by running the sample at least twice to ensure that your results are repeatable. Exactly what else might occur depends heavily on the material being tested,

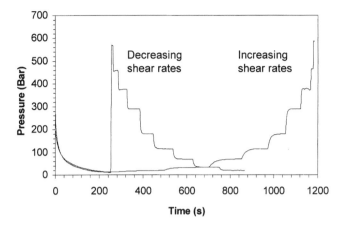

Figure 12 Two typical pressure traces from a polyethylene flow curve at 190°C, one with an increasing shear rate profile and one with decreasing rates. Each time the pressure levels off, the instrument saves the point and changes the piston speed to move to the next shear rate. The resulting flow curve is similar to Fig. 14.

but the following list gives some situations that might require special handling.

1. *No pressure reading.* If no pressure is indicated across the range of shear rates for the experiment, check for a blocked pressure tap. Other possible causes are a faulty transducer or a transducer that is not sensitive enough to register the pressures that are generated.

2. *Viscosity curve is shifted.* There are a number of items that can cause the viscosity curve to shift up or down. Check to see that the proper calibration values have been entered for the pressure/force transducer. Check the die to ensure that it is not partially blocked, causing greater than normal pressure drops. Lastly, check for a shift in either the barrel or die temperature. Burned-out heater bands or maladjusted temperature controllers will cause large shifts in experimental results.

3. *Thermal degradation.* Examine the polymer extrudate for discoloration or other signs of degradation. The residence time of the material in the rheometer may need to be shortened if significant degradation occurs. A thermal degradation test, described later in this chapter, can be performed to determine the maximum residence time in the rheometer.

4. *Chemical degradation.* Look for the formation of bubbles, as this often exhibits itself as off-gassing; listen for hissing and popping as the sample is extruded. The bubbles are often evident in the extruded strand.

Reducing the residence time and/or lowering the test temperature can help with this problem. Improper drying of the material can also cause this problem.

5. *Melt fracture and distortion.* The phenomena occur in materials at higher stresses and can be an indication of slip/stick in the die. The extrudate can display a wide variety of distortions from simple sharkskin (a very rough surface on the strand) to twists and kinks in the strand (gross melt distortion). These distortions usually occur at the exit of the die and do not affect the viscosity curve as long as the material does not slip.

6. *Slip/Slip-stick.* Slip occurs when the material "slips" through the die. The start of melt slip usually manifests as an abrupt change in slope in the stress-versus-shear rate curve; the curve suddenly levels off and there is little or no increase in stress as the shear rate increases. The stress on the material has become high enough that it slips along the wall (the mechanism of this failure to adhere to the wall is still being debated). Once the material begins to slip, the equations used to calculate shear rate become invalid: the no-slip wall boundary condition (material velocity is zero at the wall) is no longer true. Viscosity measurements are not possible without using a multi-experiment procedure [8] to determine the wall speed of the material. Since slip occurs once the material reaches a critical stress (σ_c), increasing the test temperature (and thus reducing the melt viscosity) will allow testing at higher flow rates until the σ_c is reached again. σ_c is a strong function of the type of wall material

Slip-stick occurs when the material is on the verge of slipping. This manifests as a sawtooth-shaped oscillation in the pressure signal. The pressure builds to the point that the material slips. As it slips, the pressure in the barrel is relieved until the material begins to stick and the pressure builds again [9].

7. *Shear heating.* Shear heating occurs at high shear rates due to viscous dissipation in the material. Some of the flow energy is converted into thermal energy. It is difficult to detect and monitor shear heating because it is extremely difficult to get an accurate measurement of the temperature of the melt strand exiting the die. The thermal generation term varies according to

$$E_{\text{Thermal}} \sim \eta \dot{\gamma}^2 \tag{22}$$

and can raise the temperature significantly (i.e., 10–20°C or more) when testing high viscosity materials at high rates. Since viscosity is a strong function of temperature, erroneously low viscosity measurements will occur.

8. *Compressibility/density effects.* Compressibility and density effects will not affect most materials at typical testing pressures (less than 100 MPa). Polymeric materials are considered incompressible at normal pressures and temperatures, and the equations used to calculate shear rates and

viscosity assume this. It is also assumed that viscosity is not a function of the material density. Capillary rheometers, however, can generate such high pressures (when testing very tough materials or if a die with a very large L/D ratio is used) that the density of a material can change appreciably as it flows through a die. The material sees the barrel pressure as it enters the die and atmospheric pressure when it exits, a pressure drop that can reach up to 2000 times on a powerful rheometer! The material will swell as the pressure decreases along the die, increasing the shear rate. Also, viscosity is actually a weak function of density; this can become significant for some materials when large density changes occur. One way to check for compressibility effects is to perform a Bagley correction as discussed earlier in the chapter. Curved lines on a Bagley plot rather than straight lines can indicate the possibility of pressure effects in the viscosity. Note that when performing high-pressure, high-rate measurements, shear heating and pressure effects can both occur, complicating data interpretation.

DATA INTERPRETATION

The main uses for capillary rheometers fall into two broad categories, quality control functions and material characterization for research, design, or trouble shooting applications. The use of models for data presentation and interpretation is also discussed.

Material Characterization

The main purpose of a capillary rheometer is to generate a viscosity–shear rate curve over as wide a range of shear rates as possible. The effect of temperature on the viscosity can also be determined by running curves across a range of temperatures. The selection of specific dies and pressure transducers tailor the shear rate range to match a desired process window. This information can be used in many ways.

1. *Material selection.* Flow curves can be used to screen new materials for a process by ensuring that their viscosity is suitable at process shear rates.
2. *Polymer development.* The effect of additives, recipe changes, or adding or changing fillers in a material can be determined to see if the material is still in (or has now achieved) a desired process window.
3. *Effect of processing.* Virgin material is tested and compared with material that has been run through a process. Differences in the viscosity curves can indicate if any significant property degradation has occurred.

4. *Process troubleshooting.* The viscosity curve of a troublesome material can be run and compared against previous materials that have performed well. If the viscosity is the problem, determining the viscosity–temperature relation for the material by running curves at several temperatures will determine if a simple temperature adjustment to the process will bring the viscosity back to an appropriate value. A thermal stability test can be used to determine the maximum residence time of a material in a process before its properties begin to degrade.
5. *Process modeling/equipment design.* A viscosity function is necessary for computer simulations of polymer flow through extrusion dies, injection molds, or almost any type of equipment.
6. The viscosity function can be correlated with other, more difficult to measure material properties such as the molecular-weight distribution [10–12].

The last item above deserves to be expanded upon, since one of the most common questions asked is "What does this viscosity curve say about the molecular-weight distribution of this material?" The answer is "quite a bit," but much of the information is qualitative in nature rather than quantitative. It is often used to compare and/or rank samples rather than generate absolute numbers defining the distribution.

As shown in Fig. 6, the viscosity at low shear rate approaches a constant value called the zero-shear value (η_0). In almost all cases η_0 is independent of the shape of the molecular-weight distribution and depends on the weight-average molecular weight (M_w) only. Above a critical value of M_w (M_c), there is a very strong correlation between η_0 and M_w.

$$\eta_0 \sim M_W^{3.4} \tag{23}$$

M_c varies by material, but almost all commercial polymers have $M_w > M_c$.

Capillary rheometers cannot measure accurately at shear rates low enough to determine η_0 directly. For many polymers, this determination would require measurements in the 0.01–0.001 s^{-1} range. Correlations are sometimes made using the lowest reproducible shear rate possible instead of η_0, or models (such as the Carreau or Yasuda model) are used to extrapolate back to η_0. When comparing viscosity curves that extend to sufficiently low shear rates (where the curve begins to flatten out), it is possible to make the statement that material A has a higher M_w than material B because its low-shear viscosity is higher.

The second item of information that can be gleaned from the flow curve has to do with the breadth of the molecular-weight distribution. Broader distributions tend to have an earlier onset of shear thinning and a more gradual transition into the power law region. Narrow distributions have a relatively sharp transition and achieve a steeper slope in the power law region. This relation is illustrated graphically in Fig. 13. It is much more difficult to get quantitative correlations for this kind of information. Attempts made to correlate viscosity curve slopes or "shape factors" from the transition region with parameters such as polydispersity, a measure of the breadth of a log-normal molecular-weight distribution, have met with limited success for specific materials [13].

Data Presentation

The viscosity curve generated by a capillary rheometer is a series of viscosity–shear rate points. Depending on how the data are to be used, it is often convenient to fit a model to the data. This reduces the amount of data to the parameters of the chosen model and allows easy interpolation of viscosity values between the measured points. It is also the preferred form for use in computer modeling of process flows. Usually a viscosity curve for this purpose is fitted to a model so that the computer program can deal with an equation rather than a series of data points. This way the form of the

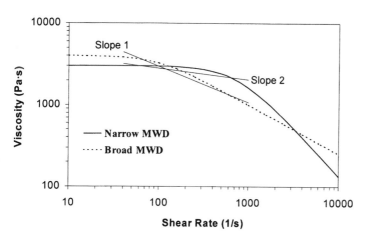

Figure 13 The shape dependence of flow curves on the molecular-weight distribution. Viscosity curve tangents taken from the transition region can quantify the shape of the curve.

equation can be programmed ahead of time and the actual viscosity data are entered as the parameters for the model.

There are a number of models in common use, a few of which will be listed here. The most common two-parameter model is the power law model (parameters n and k), whose form is given in Eq. (15). This model corresponds to a straight line on a log-log plot, and fits viscosity data well at high shear rates. It is the simplest model to account for shear thinning. In logarithmic form, the equation for the model becomes

$$\log \eta = \log k + (n-1)\log \dot{\gamma} \tag{24}$$

If $\log(\eta)$ is plotted against $\log(\dot{\gamma})$ on linear axes, then the slope of the curve is $n-1$ and the intercept is $\log k$.

Two more complicated models are the three-parameter Carreau model [14],

$$\eta = \eta_0 \left(1 + (\lambda\dot{\gamma})^2\right)^{-p} \tag{25}$$

and the four-parameter Yasuda model [15],

$$\eta = \eta_0 [1 + (\lambda\dot{\gamma})^a]^{n-1/a} \tag{26}$$

These models are useful because their mathematical structure mimics the shear-thinning curve expected for a viscosity function. This gives more reliable extrapolation beyond the data used to determine the model parameters. A four-parameter polynomial, for example, does not have a predefined shape; it may fit the data well but gives unpredictable extrapolation results. Also, the parameters in these equations have physical significance, which aids in data interpretation. The η_0 parameter contains the model's estimate of the zero-shear viscosity, while the n and p parameters correspond to the power law behavior seen at high shear rates. Also, the transition region of the viscosity curve occurs in the shear rate range around $\dot{\gamma} = 1/\lambda$. The Yasuda equation contains an extra shape factor, a, which allows it to better fit the transition region. Figure 14 compares the fit of these models with the viscosity data from a nylon sample. The disadvantage of these models is that the parameters are more difficult to determine: nonlinear optimization routines must be used. Fortunately, most modern instruments have software with curve-fitting capabilities already included.

Quality Control

Quality control laboratories use the capillary rheometer to monitor incoming and outgoing material. In general, the emphasis is to search for any differences between the tested material and previously established standards and norms. The test procedures are repetitive and the general nature of the

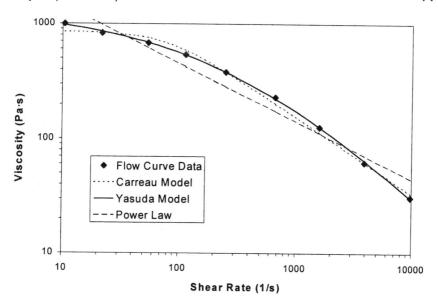

Figure 14 A comparison of three common viscosity models. Increasing the number of parameters improves the fit of the model to the data.

results is known beforehand. Often only a single, process-relevant shear rate is selected for monitoring. The sample passes the test if the measured viscosity falls within a specified range of the target viscosity. Multipoint measurements, whether a two-point, high-/low-shear rate test, or an entire 3- to 10-point curve, provide a more thorough evaluation of a polymer. In this case, each point must fall within specifications for a material to pass.

When performing quality control measurements a major goal is to optimize the sensitivity and repeatability of the results for a specific test [16]. Especially for a single point measurement, the pressure or force transducer can be selected so that the measurement falls at the upper end of the transducer's range to ensure the best accuracy. The die can be selected so that the desired shear rate is generated at a reasonable piston speed. This ensures that the test finishes in a reasonable amount of time, and that it will not run out of material before finishing. Often Bagley and Rabinowitch corrections are not made for these tests, as it is consistency (and convenience) rather than absolute accuracy that is desired. This introduces a small instrument dependency to the results, but this is usually neglected since many company labs are standardized on a single brand of rheometer, and the data are not for outside consumption.

Many laboratories use a melt indexer to generate a single-point property measurement for quality control. A capillary rheometer offers a number of advantages over a melt indexer. First, the capillary rheometer measures the viscosity rather than a machine-dependent index. Viscosity results can be compared to tests from other types of rheometers, including on-line process rheometers that measure directly from the process. Reproducibility tends to be better as well: operating at higher pressures and flow rates often makes the measurement less sensitive to operator bias in the sample preparation and loading procedures.

The most important advantage that the rheometer holds over the indexer, however, is the shear rate at which the test occurs. The melt index measurement occurs at shear rates that fall into the lower ranges of a capillary rheometer's capabilities (see Fig. 9). In most cases, these shear rates are also considerably below the shear rate range of the process for which the polymer is intended (see Fig. 11). Two materials with differently shaped viscosity curves could easily have the same viscosity at low shear rates and different viscosities at high rates. They would have the same melt index, pass QC inspection, and behave completely differently on the production floor. The ability of the rheometer to test at shear rates pertinent to the process allows the development of more relevant quality control test procedures.

OTHER MEASUREMENTS

While the basic function of the capillary rheometer is to produce viscosity–shear rate curves, there are a number of procedures that can be performed which extend the range of the viscosity curve. There are also tests that measure completely different material properties.

Time–Temperature Superposition

The viscosity of polymeric materials is a very strong function of temperature. A temperature difference of $10°C$ can make a 30–50% difference in the viscosity of many polyolefins, and as much as 200% for some engineering polymers such as polyvinyl chloride (PVC). Time–temperature superposition determines the effect of temperature by shifting viscosity curves measured at different temperatures onto a single, temperature-independent master curve. The shift factor is

$$a_T = \frac{\eta_0(T)}{\eta_0(T_0)} \tag{27}$$

where T is the actual temperature of the viscosity test and T_0 is the reference temperature of the master curve. Once the shift factor is determined for each curve, plotting $\eta(\gamma)/a_T$ verses $a_T\gamma$ produces the master curve. The master curve will cover a larger shear rate range than the individual curves because the curves are shifted along the shear-rate axis.

Once the shift factor has been determined for a large enough number of curves, the shift factors themselves may be fitted to a model. This allows the determination of shift factors and master curves at arbitrary temperatures. For viscosity curves taken within 100°C of a material's glass transition temperature (T_g), the WLF (Williams-Landel-Ferry) equation [17] is used:

$$\log a_T = \frac{A(T - T_0)}{B + (T - T_0)} \tag{28}$$

The parameters A and B are the model parameters used to fit the shift data. When the temperatures are more than 100C above T_g, an Arrhenius equation,

$$a_T = \exp\left[\frac{E_A}{R}\left(\frac{1}{T} - \frac{1}{T_0}\right)\right] \tag{29}$$

is useful, where R is the universal gas constant and E_A, an activation energy, is the parameter adjusted to fit the data.

Thermal Stability

One important item of information that can be determined with a capillary rheometer is the thermal stability of a material, or how long it can be maintained at a specific temperature before it begins to degrade. This information is critical for process designers and can be used to test the performance of stabilization-additive packages. This simple test is involves loading the rheometer barrel with a sample and running intermittent viscosity tests over a period of time, each test requiring only a small portion of the material in the barrel.

The rheometer is set up at the desired temperature and a single (or usually not more than two) shear rate at which to perform the test is chosen. Usually a low shear rate is chosen for two reasons: low shear viscosity is more sensitive to the effects of degradation (the main effect is usually a reduction in the average molecular weight) and the piston speed should be low enough to allow for many tests on the same barrel of material. Make sure a proper pressure transducer is selected, one that is sensitive enough to track the low-shear measurements accurately. Perform the viscosity measurement at specified intervals until the barrel is empty or a sufficient time period is covered.

A constant viscosity measurement implies that there is no significant degradation of the material. If the viscosity drops off over time, degradation is occurring and a maximum recommended residence time recommendation can be made based on when the viscosity reduction becomes significant. A sample test result is shown in Fig. 15.

Melt Density

Determining the melt density of a material with a capillary rheometer is simple. It involves only extruding and weighing a known volume of sample. Most rheometers show piston displacement values, but even with instruments that do not it is possible to measure piston travel with a ruler. Perform the measurement as follows:

1. Load a sample into the rheometer carefully, to avoid air bubbles, and allow it to come to temperature.
2. Purge some material through the die to ensure that no air is in the system.
3. Note the initial position of the piston and use a brass scraper or similar tool to remove any drool from the die.
4. Extrude and collect a sample of the material. When the piston stops, wait until the barrel pressure drops before using the scraper to ensure that all extruded sample is collected.
5. Note the final position of the piston and determine how far it traveled. Multiply the distance traveled by the cross-sectional area of the piston (πR^2) to get the volume of the material extruded.

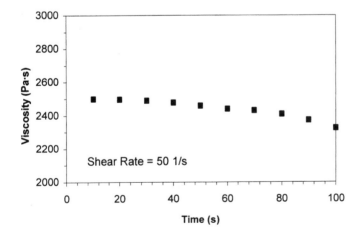

Figure 15   Sample stability test performed on polyethylene at 190°C.

6. Weigh the extruded sample.
7. Divide the weight by the volume to get the density.

Typical values fall between 0.7 and 1.2 g/cm^3.

Common Accessories

A number of common accessories are available to further enhance the capabilities of a capillary rheometer. The rheometer is designed to determine viscous data and minimize the effects of other properties. These other instruments work in conjunction with the rheometer to produce other types of measurements.

Die Swell Measurement

A die swell measurement of a strand extruded from a capillary die is used to measure the elastic properties of a material. As a polymer flows down a capillary tube, its molecules become stretched out and aligned in the direction of the flow. Once the material leaves the confines of the die, the molecules recoil and draw back up to themselves, causing the extrudate to swell beyond the size of the opening in the die. The amount of swell displayed by a polymer is a characteristic of its elastic nature.

Die swell is measured by determining the diameter of the extruded strand after it has exited the capillary die. Usually a noncontact laser measuring system is used to get accurate strand diameter measurements. Die swell is a strong function of shear rate, so most systems are configured to get a die swell measurement each time a viscosity point is taken by the rheometer. Die swell data are usually presented as a ratio of the cross-sectional area of the die to the cross-sectional area of the strand:

$$\text{Die swell} = \frac{A_{\text{Strand}}}{A_{\text{Die}}} = \left(\frac{D_{\text{Strand}}}{D_{\text{Die}}}\right)^2 \tag{30}$$

Melt Tensile Tests

In polymer processes such as film blowing, fiber spinning, and blow molding, the extensional properties and melt strength of a material are crucial pieces of information. Melt tensile testers are devices which pull on a melt strand as it is extruded from the capillary die and measure the force necessary to generate a desired draw ratio or melt extension. These instruments are capable of measuring very small forces and can generate the extensional strain rates found in many processes.

The melt strand is exposed to the atmosphere and is not at a constant temperature, complicating efforts to determine extensional viscosity directly. However, the data from these tests are very useful for qualitative testing of materials to rank their suitability for a process, troubleshoot, or perform quality control. Much work has been done recently to expand the usefulness of this measurement, including techniques that can generate temperature- and process-independent master curves [18].

CONCLUSION

The capillary rheometer is a versatile and robust instrument. Easily inter-changeable transducers and dies give the instrument the flexibility to study a wide range of materials over a wide range of shear rates, especially in the high-shear-rate ranges. It can handle tough materials at the temperatures, pressures, and flow rates typically found in high-performance plastics pro-cesses. The viscosity curve that the instrument produces finds many practical applications in quality control, process design, and troubleshooting, as well as in the study of the material properties themselves.

Aside from viscosity curves, the capillary rheometer can be used to determine other material properties. The effects of time and temperature on processability and chemical stability can be studied, and other properties such as the melt density can be measured. Elastic data can be collected with accessories such as a die swell measurement system, and extensibility mea-surements can be performed with a melt tensile tester. The capillary rhe-ometer is the instrument of choice for any practically oriented polymer laboratory.

REFERENCES

1. Rabinowitch, B. Z., *Phys. Chem. 145*:1, 1929.
2. Schummer, P., and Worthoff, R. H., *Chem. Eng. Sci.*, *33*:759, 1978.
3. Bagley, E. B. *J. Appl. Phys. 28*:624, 1957.
4. Watson, S. J., and Miller, R. L. *SPE Annu. Tech. Conf. Proc.*, *42*:1100, 1996.
5. *ASTM Annual Book of Standards*, Volume 08.02, Designation D3835-96, American Society for Testing and Materials, West Conshohocken, PA, 1999, pp. 466–476.
6. *ASTM Annual Book of Standards*. Volume 08.01, Designation D1238-95, American Society for Testing and Materials, West Conshohocken, PA, 1999, pp. 255–263.
7. Hatzikiriakos, S., and Dealy, J. M. *Polymer Eng. Sci.*, *34* (6):493, 1994.
8. Mooney, M., *Trans. Soc. Rheol. 2*:210, 1931.
9. Hatzikiriakos, S., and Dealy, J. M., *J. Rheol. 36*(5):845, 1992.
10. Liu, Y.-M., Shaw, M. T., Tuminello, W. H., and Bu, L., *SPE Annu. Tech. Conf. Proc.*, *42*:1074, 1996.

11. Ross, C., Malloy, R., and Chen, S., *SPE Annu. Tech. Conf. Proc., 36*:243, 1990.
12. Tzoganakis, C., Vlachopoulos, J., and Hamielec, A. E., *Polymer Eng. Sci. 29*(6):390, 1989.
13. Goettfert, A., Reher, E.-O., and Schulze, V., *SPE Annu Tech. Conf. Proc. 41*:1166, 1995.
14. Carreau, P. J. Ph.D. thesis, University of Wisconsin, 1969.
15. Yasuda, K. Y., Armstrong, R. C., and Cohen, R. E., *Rheol. Acta*, 20:163, 1981.
16. Bafna, S. S., and Ford, W. L. *SPE Annu. Tech. Conf. Proc.*, 40:1193, 1994.
17. Ferry, J. D., *Viscoelastic Properties of Polymers*, 3rd ed., Wiley, New York, 1980, p. 274.
18. Wagner, M. H., Schulze, V., and Goettfert, A., *Polymer Eng. Sci.*, *36*(7):925, 1996.

3

Practical Uses of Differential Scanning Calorimetry for Plastics

Andrew W. Salamon and Kenneth J. Fielder
PerkinElmer Instruments LLC, Shelton, Connecticut, USA

INTRODUCTION

Although differential scanning calorimetry (DSC) is used in many different industries, its application and use in the plastics industry is widely accepted. It is used to characterize materials for melting points, softening points, and other material and material-reaction characteristics such as specific heat, percent crystallinity, and reaction kinetics. This chapter addresses the practical uses of DSC in the plastics industry, focusing on the most common tests and experiments. Advanced analysis will be mentioned briefly, but not reviewed in detail. For the best results and the most reproducible data, consider all of the suggestions about operational variables discussed, and you will experience successful and reliable thermal analysis.

DEFINITIONS OF THERMAL ANALYSIS TERMS

Differential scanning calorimetry is an analytical technique that measures the heat flow to or from a sample specimen as it is subjected to a controlled temperature program in a controlled atmosphere. In other words, a DSC

79

heats or cools a material and energy is either absorbed or released as the material experiences physical or chemical changes.

Heat is a form of energy. Heat is not temperature. It is measured in joules. This is an important factor when doing DSC work, because the material is heated or cooled and, in some DSCs, the energy is measured directly. Often the process engineer will want to know the amount of heat required to melt a material or how much heat is given off by a material as it cools in the mold.

A *joule* (J) is the amount of work done when the force of 1 newton (1 N) acts through the distance of 1 m. Although the joule is mechanical in nature, it is a suitable form for heat energy. Heat flow is measured in joules per second.

Temperature is the degree of heat measured on a definite scale. The centigrade scale is internationally accepted.

A watt (W) is the power expended when 1 J of work is done in 1 s of time. Heat flow is measured in watts or milliwatts. A *milliwatt* (mW) is 1000th of a watt.

A *calorie* is the amount of heat required to raise 1 g of water by 1°C.

Heat flow is measured in milliwatts. Heat flow is the flow of energy measured by the DSC as it heats or cools the sample specimen.

Enthalpy, H, is the "heat content" of a material. The absolute enthalpy of a material cannot be measured directly; however, a change in enthalpy, ΔH, can be measured directly by DSC. The change in enthalpy of a material is either a gain (endothermic, such as melting) or a loss (exothermic, such as curing or recrystallization). This information is very helpful for process engineering.

Specific heat or *heat capacity* is the amount of heat required to raise the unit mass of a material 1° in temperature. *Note:* Specific heat and heat capacity should be assumed to be equal for engineering purposes.

$$C = \frac{Q}{m \, \Delta T}$$

where

C = specific heat

m = mass of materia

Q = heat added

ΔT = change in temperature

Heat of fusion is the amount of heat per unit mass needed to change a substance from a solid to a liquid at its melting point. A melting transition is an endothermic reaction, which is represented as a peak on a DSC thermal curve. The heat of fusion is the amount of energy added to the material divided by its mass. It is determined from a thermal curve by calculating the area under the melting peak. This is important information for process engineering as a plastics processing system is designed.

$$H_f = \frac{Q}{m}$$

where

H_f = heat of fusion
m = mass of material
Q = heat added

Mass is often incorrectly interchanged with weight because the mass of a material is determined by using a balance and weighing the sample. Most thermal software does this by using the sample weight in the above equations. As a practical matter, remember that weighing the mass of the sample will directly affect the DSC's specific heat measurement and heat of fusion measurement. It is best to use a microbalance (accurate to six decimal places of a gram) for weighing.

APPLICATIONS

Glass Transition Temperature

The glass transition temperature is the softening region of a plastic. It actually refers to the amorphous region of semicrystalline materials or to amorphous materials; it is the molecular motion of the material as it is heated that makes the material pliable and no longer brittle.

The glass transition temperature is represented graphically in the thermal curve as a shift in the baseline. It is a change in the heat capacity of the material as the amorphous regions melt. For the easiest identification of glass transition temperature, use a large sample specimen and quench the material from just above the melting temperature to below the glass transition temperature region. Faster quenching creates larger shifts in the baseline (see Figs. 1 and 2).

The calculation of the glass transition temperature can be done in several different ways. The most common calculation is the half-C_p or half-height of the shifted baseline. The limits are set up on the flat portion of the preshifted and postshifted baseline with the calculation construction lines extrapolating the baselines as shown in Figs. 1–3.

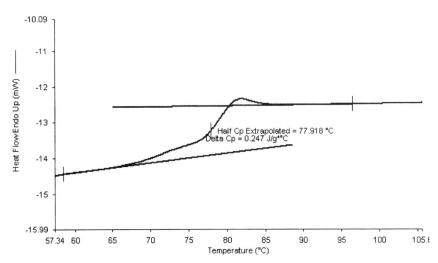

Figure 1   Thermal curve of PET after quenching (cooling at 200°C/min) with a glass transition temperature calculation using the half-height (half-C_p).

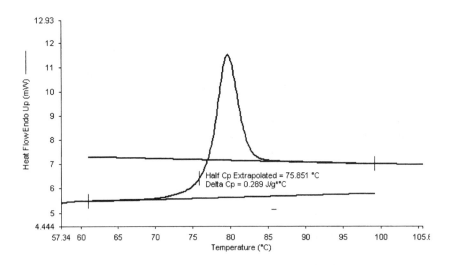

Figure 2   Thermal curve of polystyrene with a glass transition calculation.

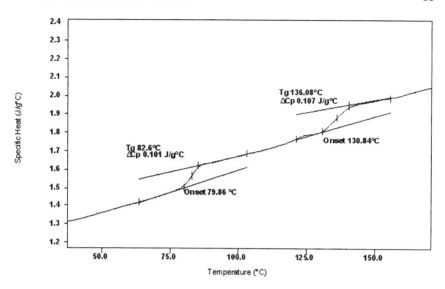

Figure 3 This material exhibits two glass transitions (T_g). The T_g is calculated by establishing the calculation limits and extrapolating the pre-T_g baseline and the post-T_g baseline and calculating the half-height as the T_g. This is the most common representation of the T_g.

Melting Temperature

Melting temperatures of different materials are identified as different portions of the thermal curve. During calibration the melting temperature of the calibration reference materials (usually a pure metal such as indium) is identified as the onset temperature of the melting peak (Fig. 4).

Because plastics process engineering has defined the minimum process temperature of a plastic (its melting temperature) as its peak temperature on a thermal curve, the melting temperature of a plastic is sometimes confused with the melting temperature of a nonplastic. Figure 5 shows the melting peak of polyethylene.

The melting temperature is calculated by setting the calculation limits before and after the peak, keeping them on the flat portion of the baseline. This will give the most reproducible results (Fig. 5).

Enthalpy of Melting

The area under the melting peak represents the amount of energy required to melt the polymer. This is the heat of fusion of the material or how much

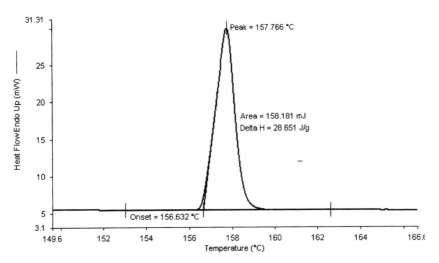

Figure 4 Thermal curve of indium melting with a peak area calculation including the onset temperature and peak temperature.

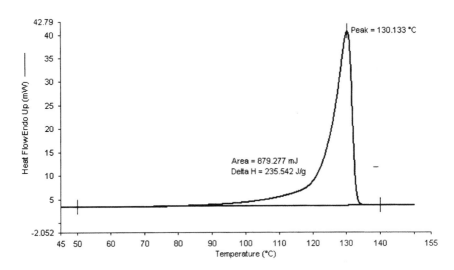

Figure 5 Thermal curve of polyethylene melting with the peak area calculation including the peak temperature.

energy is required to melt the material. The ΔH or change in enthalpy is expressed in normalized terms of joules per gram (J/g). Because this term, has the sample weight as the denominator, accurate sample weighing is important. Melting is an endothermic reaction, that is, it is a reaction that absorbs energy. The calculation of the peak area is very helpful in process development and material characterization. The peak area can also be used to identify the percent crystallinity of the material. This is explained later.

The peak area calculation is used with the limits of the calculation on the flat portion of the baseline before and after the melting peak. Some thermal analysts use the horizontal flat portion of the first derivative curve as a guide to set the peak calculation limits (Figs. 4 and 5).

Curing Studies

Curing of a resin material is an exothermic event, which means that energy is released as the material goes through the transition. This is the opposite of melting, which is an endothermic event. Curing studies may be conducted as DSC scanning experiments or DSC isothermal experiments.

Rates of reaction of both these types of curing experiments are very useful to polymer scientists. There are several different kinetics software packages available to characterize materials. There are scanning kinetics, isothermal kinetics and model-free kinetics calculations. Each has its own unique application or is best suited for certain applications. Figure 6 depicts the thermal curve of an epoxy cure with a second curve superimposed representing the percent area as the reaction advances.

In curing studies, the glass transition temperature will vary depending on the degree of cure of the material. Remember that the glass transition temperature is the softening region of the material. If the cure reaction is more advanced in sample A than in sample B, the glass transition temperature will be higher for sample A than for sample B.

Recrystallization Temperature upon Cooling

Recrystallization studies are useful for process engineers who want to optimize the process by identifying the recrystallization temperature of the material in the mold. A simple recrystallization study is accomplished by running a multiple-step temperature programmed (heat-iso-cool) experiment, in which the material is heated slightly above its melting temperature and held there for a minute or so, then cooled in a controlled fashion at a certain scanning rate such as 20°C/min. The temperature of recrystallization of the material can be identified by using the onset temperature calculation (Figs. 7 and 8).

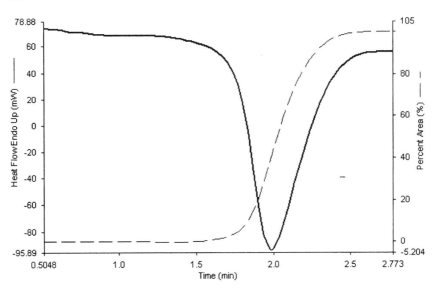

Figure 6 An epoxy cure is an exothermic event as depicted by the heat flow curve. The second curve with the dashed line is the percent area curve, which relates directly to the percent reacted assuming the epoxy is 100% cured.

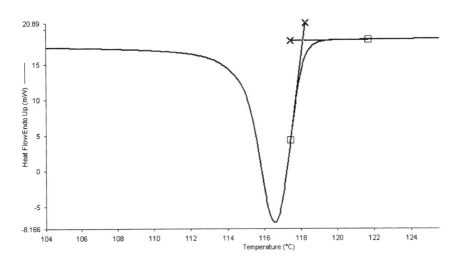

Figure 7 The limits of an onset calculation of this recrystallization-upon-cooling curve are identified and the construction lines are adjusted correctly. Many analysts use the peak of the first derivative curve (not shown) to identify the left limit in this case.

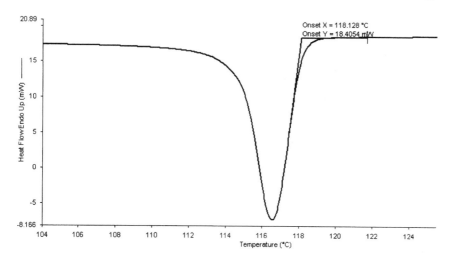

Figure 8 The onset of recrystallization upon cooling is calculated.

Note that for best results when conducting cooling experiments, calibrate the calorimeter upon cooling. This will correct for instrumental effects and sample supercooling phenomena.

Recrystallization Time

Recrystallization time is very useful for the development of processing times for recrystallization upon cooling from the melting temperature. Recrystallization time is measured in an isothermic experiment in which the material is cooled to a predefined temperature and then held there isothermally until the recrystallization is complete. The time it takes to recrystallize is measured. A slight shift in mold temperature often saves time, thus increasing manufacturing throughput. The sensitivity and temperature control of a power-compensation DSC is ideal for this experiment (Fig. 9).

This type of experiment is also used to identify the curing cycle of an epoxy material. Although an epoxy material is usually mixed at room temperature (or at subambient temperature), it is heated to an elevated curing temperature and examined isothermally. At the selected isothermal temperature, the material is examined with regard to the length of time it requires to cure fully.

Oxidative Stability Testing

The investigation into the effectiveness of antioxidant additives in a material is easily done by DSC. This a test that determines the length of time that a

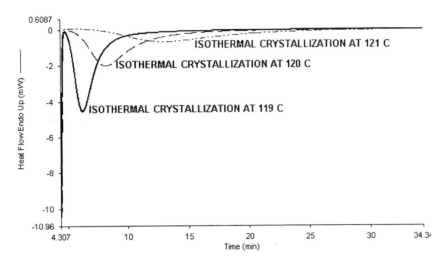

Figure 9   The three curves displayed are the same sample quickly cooled to slightly different temperatures and then held isothermally until fully crystallized. This is an example of the type of resolution power compensation DSC is known for. This information may have been used to optimize a process.

polymer can withstand an oxidative environment at an elevated temperature. This test is regularly done for cable and wire insulation materials, among others.

A brief overview of this test is that the sample is prepared and installed in a calorimeter that has been purged with oxygen. The sample is heated to 200°C and held isothermally at that temperature. The time it takes the material to decompose is measured. Decomposition is indicated by an exothermic downturn of the baseline. The extrapolated onset is calculated to determine the amount of time it took for decomposition. The longer the time, the more effective is the antioxidant (Fig. 10).

There are two very important considerations when conducting this test: (1) very accurate temperature calibration and (2) consistent and accurate purge gas flow rate.

For more information on this test, refer to ASTM Publication STP1326, "Oxidative Behavior of Materials by Thermal Analytical Techniques."

Effects of Additives in Plastics or Treatments of Plastics Can Be Analyzed by DSC

Additives will affect the thermal characteristics of a material. The glass transition temperature (T_g), the melting peak, or the recrystallization

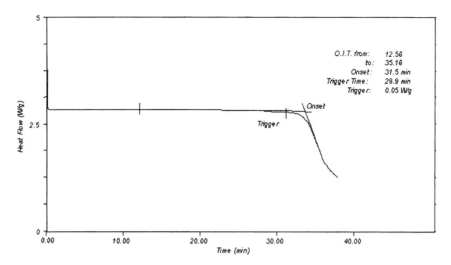

Figure 10 The calculated extrapolated onset temperature determines the oxidative induction time.

point may shift due to the additive's effect on the material. Some general categories of additives are plasticizers, antioxidants, stabilizers, reinforcements, fillers, impact modifiers, cross-linking agents, foaming agents, flame retardants, lubricants, and antistatics.

One of the most common thermal laboratory practices is to analyze a virgin material, then add plasticizer of a known percentage to the untreated material and measure the effects of the additive by comparing the T_g or other thermal characteristics of each specimen. Figure 11 compares untreated and treated polypropylene samples.

Temperature-Dependent Crystallinity

Temperature-dependent crystallinity is a relatively new data calculation of a temperature scan of a material. PerkinElmer Instruments LLC has the exclusive rights to this software package, which was developed by Vincent B. F. Mathot of DSM, The Netherlands, and Gosse van der Plaats of the Netherlands. Note that it does not matter what instrument you use to collect the data, just that it is in ASCII format so that it can be imported into this standalone software package.

Before the temperature-dependent crystallinity calculation was available, the estimate or calculation of the percent crystallinity of a semicrystalline material was more subjective. The melting peak area of a semicrystalline

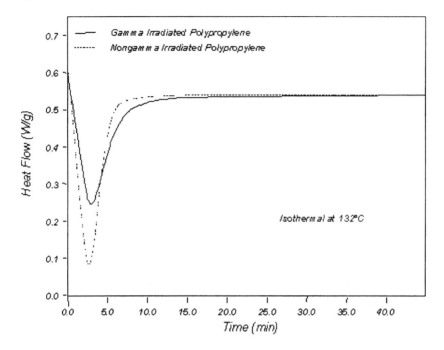

Figure 11 A recrystallization of two polypropylene samples; one is gamma-irradiated and the other is untreated. Notice the differences of endsets (where the exothermic peak returns to the baseline), the peak temperatures, and the areas.

material was compared to the melting peak area of a 100% crystalline structure. A simple ratio was done of the two and the percent crystallinity was assumed. The subjectivity arises from the placement of the calculation limits chosen by the analyst. Many times they are in error or are not reproducible. Temperature-dependent crystallinity calculations eliminate many of these errors and are more reproducible.

Temperature-dependent crystallinity software analyzes the percent crystallinity of a material during a DSC scanning experiment. As the material changes during heating, the calculation may be done on any portion of the thermal curve through the melt. It extrapolates the postmelt baseline down to the lower temperature limit, which is usually in the glass transition region. This extrapolation is a reference line; deviations in the thermal curve are compared to it and to published theoretical crystallinity values. The crystallinity values of the material are then presented in tabular form.

Understanding the percent crystallinity in a material will be helpful in designing products, prolonging product life, or designing new processes (Figs. 12 and 13).

QA/QC Applications

Some mechanical tests of materials are now being correlated to DSC testing. Examples of these are abrasion testing and impact testing. They are now being tested by both mechanical means and by DSC. A material that passes the mechanical test should be characterized by DSC and likewise the failed products being characterized by DSC. The differences in the materials characterization by DSC are noted and used in collaboration with the mechanical test findings. Generally the DSC testing is faster and more reproducible.

Routine QA/QC material testing is now being automated so that the samples are accurately prepared and encapsulated by the thermal analyst, but loaded into an autosampling system for analysis and data calculations with automatic tolerance testing. That is, the software automatically identifies the sample as a passed or failed sample. This type of automation

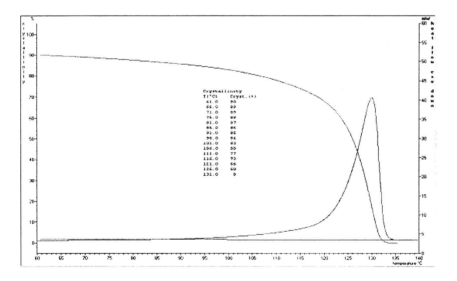

Figure 12 Temperature-dependent crystallinity is determined for most common polymers by using a simple software package that employs user-defined limits for the calculation on the DSC heat flow curve. The resultant temperature crystallinity curve is produced with the percent crystallinity table.

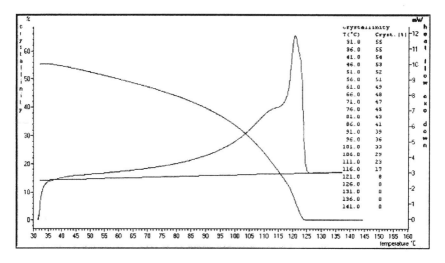

Figure 13 An example of temperature-dependent crystallinity showing the postmelt extrapolated baseline. The table indicates the change in crystallinity as the material is heated. This information may be useful for determining processes and the material's useful working range.

system addresses the needs of business today by providing not only higher throughput, but also a simple handling of samples and data. Figures 11 and 14 show examples of QA/QC comparisons.

A BRIEF HISTORY OF DIFFERENTIAL SCANNING CALORIMETRY

Early civilizations stumbled over the effects of temperature on materials. Since those early stages, there have been significant advances that are worth mentioning.

The current DSC evolved from the first thermal analysis done by Sir Roberts-Austen, who, in 1899, invented differential thermal analysis (DTA). His investigations were of clays and silicate materials. DTA was used as a laboratory setup that measured the temperature difference between a sample specimen and an inert reference material.

Currently there are two DSC designtypes that are used to characterize plastics. They are power-compensation DSC and heat flux DSC. The power-compensation DSC measures the heat flow directly to or from the sample specimen, while the heat flux DSC measures the change in temperature between the sample specimen and an empty reference pan and then uses

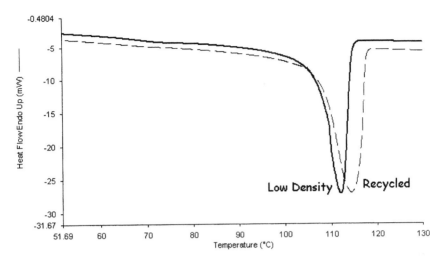

Figure 14   A comparison of two types of polyethylene upon cooling.

algorithms to calculate the heat flow. They both serve the plastics industry well.

Heat flux DSC and power-compensation DSC are used to analyze most polymers. The difference between them, besides the temperature-sensing device, is the speed and accuracy at which they can heat and cool a sample specimen. The power-compensation DSC has the ability to heat and cool at very fast rates (up to 500°C/min), while the heat flux DSC effectively heats or cools at slower rates (up to 30°C/min). Their differences in heating/cooling are due to the design of the calorimeters and the heat transfer characteristics of each. For a heat flux DSC diagram, see Fig. 15.

Because of the fast cooling rates, the power-compensation DSC is typically used for materials that are difficult to analyze or that require fast cooling rates (quenching). Depending on what analysis is, desired, quenching may be required to condition the sample in such a manner that the next heating scan will produce the desired results consistently. Power-compensation DSC has the best temperature control and energy sensitivity of all calorimeters. Figure 16 is a power-compensation DSC diagram.

DSC OPERATING VARIABLES: REPRODUCIBLE EXPERIMENTS AND TESTS

The best way to ensure that material investigations that are reproducible and accurate is to be aware of the operational variables that will affect your analysis. For best results, none of these variables should be overlooked:

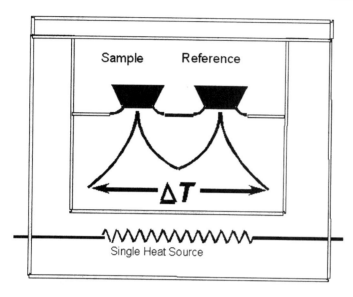

Figure 15 Heat flux DSC utilizes a large, single heating element that heats both sample and reference materials. It should be operated slowly, at about 10°C/min to minimize thermal lag. It utilizes thermocouples to sense the temperature difference between the sample and the reference.

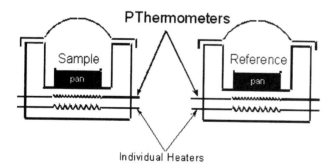

Figure 16 Power-compensation DSC schematic depicting individual platinum heaters and sensors. Power-compensation DSC is typically used for very exact calorimetry work which demands accuracy as well as precision. It is used for fast quenching experiments such as subtle T_g determination as well as melting studies and specific heat studies. It is also used for routine analysis.

Cleanliness of the instrument
Instrument coolant
Calibration parameters
Sample pan selection
Sample size
Sample preparation
Sample encapsulation
Sample atmosphere
Temperature range of interest
Scanning rate

The operating variables of the DSC will affect the results. The point of DSC experiments is to have reproducibility from today's experimental run to next week's experimental run. These operational variables are important; if they are not checked they will lead to nonreproducible results.

Checking Instrument Calibration

The first instrument variable to verify is that the analyzer is calibrated correctly. This is easily done by running an encapsulated calibration reference material and evaluating its melting temperature and the transition energy required to melt that material. A typical reference material used for calibration is indium. The melting temperature of indium is 156.61°C. It is 99.999% pure, which translates into very accurate temperature calibration and energy calibration (ΔH), which is the integrated area under the melting peak (Fig. 4). This test may be done regularly—daily or weekly, or when deemed necessary.

To see if the analyzer is calibrated correctly, compare the results of an indium run (run under the same conditions as the sample—same scan rate, etc.) to the expected or theoretical values of indium. Let's say that your calibration check reveals that the onset temperature of indium is 159.0°C. Is this value out of tolerance? The acceptable tolerance range for any analyzer is chosen by the operator. Indeed, 159.0°C might be acceptable for someone who is doing a very broad review of materials. On the other hand, it might be out of tolerance and totally unacceptable for more precise material characterization work.

Assuming that the instrument calibration check is not acceptable and the DSC requires recalibration, the first thing to do is to clean the analyzer.

Cleaning a DSC

Visual observation of the DSC furnaces, lids, and covers should be done. If you see contamination in the furnaces, remove the covers or swing them out of the way and remove all sample pans from both the sample and reference sides of the calorimeter. Then, with the furnace open to ambient air, heat it to 600°C. This will burn off the organic contaminants. If volatile gases evolve during cleaning, you may want to hold a small exhaust line above the furnace to evacuate the volatile gases safely out of the laboratory.

Cleaning the remaining furnace covers must also be done. If you have a power-compensation DSC, flame the platinum covers and reshape them before installing them into the furnaces. Check the underside of the swing-away cover, and if there are any blemishes or stains from volatiles, clean them away with an appropriate solvent, followed by an isopropyl alcohol rinse, followed by a water rinse, and air dry.

If you have a heat flux DSC, the cover over the sample area should be cleaned in the same manner as the swing-away cover described above. Never use abrasives on the sample area components or furnaces. If there is a persistent stain that heat or solvent cannot remove, then call the instrument manufacturer and follow its recommendations.

Now reassemble the clean DSC and run the indium sample again. Sometimes cleaning the DSC is all that is required to return the instrument to working order. Check the indium onset temperature of melting against the expected values and determine if the DSC still needs recalibration.

If the DSC is still out of calibration, you must recalibrate it.

Getting Ready to Calibrate the DSC

There are certain things that remain to be checked on the instrument.

1. Check the sample purge gas. If you are using bottled gas, now is the time to check whether the tank should be replaced. It is best not to let tanks run to empty. If 100–300 psi is left in the tank, replace it.
2. Check the coolant that you are using and replenish it, if required.

Calibrating the DSC

When calibrating a DSC or any other thermal analysis analyzer, the previous calibration values that are applied in the software should be erased. This is easily done via the software by restoring defaults. Defaulted values are calibration settings that the factory sets as a common value for all DSCs produced. If you don't restore defaults, you will be calibrating incorrectly by installing your calibration factors on top of previous calibration factors. To understand exactly how your calibration software operates, refer to the instrument documentation provided by the manufacturer.

Today's calorimeters have a furnace calibration routine that is separate from the temperature and enthalpy calibration routines. Follow the manufacturer's recommended sequence of which order to perform these routines. The first thing to identify is the temperature range of interest that you will be working in. Let's use the example of −10 to 200°C for your temperature range of interest. Based on this information the furnace calibration should bracket this range. That is, set up the furnace calibration with limits equal to these or slightly beyond these limits.

A general overview of temperature calibration and enthalpy calibration (heat of fusion) is best if DSC operational basics are understood before calibration is begun.

1. Use the same sample pans that you will use when running your samples.
2. Set the sample purge gas so the flow rate, in milliliters per minute, is exactly the same as when you run your samples.
3. Use the same instrument coolant as you will when you run your samples.
4. Use the same scanning rate as you will use when you run your samples.

The instrument should be thermally equilibrated and stable before calibration. This is accomplished by turning on the purge gas, coolant, and instrument 45 min before calibration. The analyzer should be equilibrated at load temperature, which is usually set to 25°C.

The next thing to do is to select two calibration reference materials that melt within the experimental temperature range. It is best to use the purest materials available. Materials that may be are commonly used for this temperature range (−10 to 200°C) are distilled water and indium. (Note that there are many other subambient calibration materials. Water was chosen to explain how to run a liquid sample.) The melting of these materials is well documented and reliable. At room temperature, water is a liquid and indium is a solid. Each sample will have to be prepared slightly differently.

We will make the assumption that the sample pans selected are proven to be best for the samples that will be run. In this case we will use standard aluminum pans. For the liquid sample we will not crimp the sample pan and for the indium sample we will crimp the sample pan.

We will use the indium and water to calibrate temperature, and indium to calibrate energy.

To run the water sample, place an empty uncrimped sample pan on the reference side of the calorimeter. This is always the right side of the analyzer. Using a pipette or an eyedropper, fill another sample pan halfway with distilled water and place it on the sample side of the analyzer. Weighing the water sample is not necessary because we are only using the melting onset temperature, which is unaffected by sample weight. Install the furnace covers and cool the analyzer to −25°C; let it equilibrate for 5 min. The distilled water is now a solid that can be used as a melting standard.

Set up the method as you would for the samples that you will run, using the same scanning rate (usually 10 or 20°C/min) and experiment with temperature limits of −25 and 25°C. This temperature span should be large enough to encompass the melting peak of the distilled water. Begin the run and observe the melting.

After the run is over, calculate the melting temperature of water. This is easily done by calculating the peak temperature and including the onset temperature. In Fig. 4, an indium melt, observe that the limits of the calculation are set on the flat portion of the baseline before and after the melting peak. Use the onset temperature as the melting point of water.

Now return the analyzer to the load temperature (25°C) and remove the sample and reference pans. Preparing indium for a calibration run takes a little more time than preparing the distilled water because you will crimp the sample and weigh it accurately before using it.

Using the same type of standard aluminum pans, crimp one empty pan and lid together and place them in the reference side of the calorimeter. This point is a good time to check the calibration of the weighing balance and calibrate it if required. For the sample side of the calorimeter, we will tare the weight of the sample pan and lid in the balance before encapsulating the indium in the pan.

Indium is supplied as a $\frac{1}{8}$-in. diameter wire. It is soft enough to use a razor blade to slice off a thin slice of the indium. The ideal indium sample will weigh between 6 and 10 mg and lay flat on the bottom of the pan. Take the sliced piece of indium and smear it against a clean work surface with a handle of a forceps or tweezers until it is a thin foil. Then cut a small rectangle of indium from that foil and weigh it. Keep trimming it until it is the desired weight. Place the indium in the sample pan, flat along the pan bottom, and encapsulate it per the manufacturer's instructions. Record the weight and enter it into the method being used. Install the crimped sample in the calorimeter on the left side (sample side). Indium melts at 156.61°C, so set up the method to begin and end approximately 25° below and above this temperature. Once the analyzer contains the sample and reference crimped pans and it is equilibrated at the initial starting temperature of the experiment, begin the run.

Again, calculate the peak area of the indium melt and include the onset temperature in the calculation (Fig. 4). Observe that the limits of the calculation are set on the flat portion of the baseline before and after the melting peak. Use the onset temperature as the melting point of indium and use the normalized peak area (ΔH) as the enthalpy value for indium. Indium is the purest material available for calibration of a DSC and should always be used for the energy or heat flow calibration. Even if indium is not in the temperature range of interest for your analysis, it should still be used for the energy calibration.

Now enter the calculated and expected values for distilled water and indium into the calibration software per the manufacturer's instructions and apply them to the DSC. The analyzer is now properly calibrated and ready to run sample specimens.

Sample Size

For the very best results, the sample weight should be the same from run to run. The sample shape and preparation should be the same. The sample should cover the entire bottom of the sample pan and ideally should weigh more than 3 mg. Of course, the sample weight is dependent on the density of the material and there might be times where you will struggle to reach 2 mg of sample; 3 mg is only a guideline.

Sample Preparation

DSC sample specimens can have different physical forms: thin films, thick films, composite sheets, powders, chunks, pellets, liquids, or highly aerated foams. Sample preparation is important because a poor choice of sample preparation tools can actually change the crystalline structure of the sample. An example of this is the use of a saw, which creates friction during cutting that translates mechanical energy into heat energy, which may change the structure of the polymer.

Thin films are easily prepared for encapsulation. Typically, a cork borer or a clean paper punch is used to punch several sample specimen disks from the larger thin film sheet. These disks should be sized to fit snuggly in the sample pan and cover the bottom of the pan. They should be stacked until the sample specimen mass is more than 3 mg. Other tools that can be used for thin film preparation are scissors or razor blades. If irregularly shaped thin film sample specimens are prepared, then cover the bottom of the sample pan and stack them evenly in a similar fashion as the disks.

Thick films are prepared in the same manner as thin films; razor blades are frequently the first choice as a cutting tool. Razor blades are much better than other mechanical cutting devices, because they do not introduce heat energy into the sample during preparation.

Composite sheet material sample specimens are prepared for the DSC by means of shears, saws, and nibblers. These tools all generate heat that may be detrimental to the sample specimen. Many times, however, there is no other choice. A wise choice is to use cutter coolant during preparation.

Choose a coolant, such as a water-based coolant, that will not react with the sample.

Powders are the simplest to prepare for DSC analysis. Each powder particle is usually the same size in diameter or mesh. Carefully scoop the powder into the sample pan using a spatula. Cover the bottom of the pan and fill the sample pan halfway or enough to reach 3 mg.

Chunks and pellets are samples that are larger than a powder and irregular in size. The best way to prepare irregular chunks is to use a razor blade and slice the chunks into smaller chunks. Smaller chunks are better than larger chunks, as are small slices of a pellet as opposed to a large section of pellet. Uniform heating of the sample is the main concern. A large solid sample can be compared to an iceberg, in that an iceberg melts on the outside and is cold on the inside. The ideal situation is to have small sample specimen pieces, which afford more surface area to the controlled atmosphere, so each piece will heat in a uniform manner. If large chunks cannot be avoided, then use a slower scanning rate so uniform heating of the sample takes place.

Liquids are easily run in the DSC because there are special sample pans designed for liquids. Use a pipette, eyedropper, or a thin glass rod as a dipstick to load the sample into the proper sample pan. Be aware that sometimes liquids will wick up the sides of the sample pan; if this happens, use a smaller sample or choose an alternative sample pan.

Highly aerated foams may be difficult to prepare for the DSC because of their low density. What results is a very large sample that is low in weight. To overcome this, you may have to use the largest-volume pan available and use a piece of platinum mesh as a cover. Use a tweezers to pinch the mesh in place. Avoid crushing or compressing the foam to make it fit into a standard pan or small pan, because the material's characteristics will change as a result of the mechanical energy transferred to the sample during compressing. It is not unusual to have to run a sample as light as 1 mg. A high-quality DSC can analyze foams, but not all DSCs have enough sensitivity. Usually a power-compensation DSC is required.

Sample Pan Selection

A sample specimen is encapsulated in a sample pan to protect the calorimeter from the sample during heating, and the pans provide a convenient way to handle the samples. Most sample pans are crimped or sealed after installing the sample specimen. Crimping and sealing usually enhance the heat transfer properties between the calorimeter heating element and the sample specimen. A crimper is the best choice to seal a pan, rather than tweezers or other devices, because a crimper ensures the same crimp each time. Crimpers eliminate differences between analysts' style and provide a reproducible seal each time. Numerous sample pans and crimping assemblies are available. Sample pan selection is based on sample type and temperature range of interest. The following general categories of the pans and crimpers are discussed in more detail in this section:

Standard sample pans and crimpers
Specialty pans
DPA 7 pans
Volatile sample pans and crimpers
Vapor pressure pans and crimpers
Large-volume stainless steel pans and crimpers
Reusable high-pressure capsules and sealers
Autosampler system pans and universal crimpers

Standard sample pans are made of aluminum, gold, or copper. Aluminum standard sample pans are most commonly used for polymer analysis. They are the least expensive pans and provide for easy crimping, and they have excellent heat transfer properties. A simple, push-type, manual, mechanical crimper is used to seal them. The standard aluminum pans are used in the temperature range of -170 to $600°C$. These pans are mechanically sealed, but not hermetically sealed. Use the gold or copper pans when aluminum is not appropriate for the sample specimen due to a chemical reaction, physical change, or temperature range of interest.

Specialty sample pans and covers are used when standard pans are not suitable. Specialty pans are usually used to study inorganic materials such as metals, ceramics, and soils. They are available in alumina (Al_2O_3), platinum, and graphite. No crimper press is required with specialty pans.

DPA 7 aluminum pans and covers are intended to be used with a photocalorimetric accessory. The covers are actually quartz windows that allow ultraviolet (UV) energy from the accessory light source to transfer to the sample specimen. Photocalorimetry is used to study UV-curing reactions. A crimper is not required with these pans.

Volatile sample pans are used for volatile solid or liquid samples that exert significant vapor pressure in the temperature range of interest.

Aqueous solutions can be scanned up to and through sublimation to observe solute behavior. The heats of fusion of materials which require an enclosed atmosphere (e.g., water vapor evolved in dehydration below $100°C$).

The capacity of these pans is 20 μl, with an internal pressure limit of 2 atm. A simple, rotational-type, manual mechanical crimper is used to seal them. The operating temperature range is from -170 to $600°C$.

Vapor pressure sample pans are similar to volatile sample pans, but they have a hole 50 μm microns in diameter in the center of the cover. These pans and covers are used for more reproducible measurements of boiling points, heats of vaporization, and sublimation temperatures. They use the same crimper as the volatile sample pans. The operating temperature range is from −170 to 600°C.

Large-volume O-ring-sealed stainless steel sample pans and covers are used with samples that vaporize or contain a volatile reaction product in the temperature range of interest. These are used when a large volume or high internal pressure limit is required. These pans have a capacity of 60 μl and an internal pressure limit of 24 atm. They utilize a rubber O-ring to seal and maintain pressure. They are sealed by a manual lever-type press. Curing reactions (phenolic or epoxy resins) and the study of dilute aqueous solutions between −40 and 300°C are some of the applications for these pans. They can also be used with large samples to detect subtle glass transition temperatures.

Reusable high-pressure capsules are used to suppress the endothermic signal resulting from volatilization of sample material or from the volatilization or decomposition of reaction by-products. They are used in any situation where the advantage of a self-generating atmosphere is to be employed. They have a capacity of 30 μl and an internal pressure limit of 150 atm. They are available in titanium or stainless steel and utilize gold-plated seals. These pans screw together with a sealer that has a slip clutch to ensure the same torque each time. The temperature range for these pans is −170 to 750°C.

Autosampler system sample pans are designed to be used with an auto-sampling system. There are several types, made of aluminum or gold. The capacity range is 10, 25, 30, 40, or 50 μl. There are pans that have a pressure limit of 2 atm, and there are also vented pans. They are sealed with a manual lever-type universal crimper that uses an interchangeable die to crimp the pan. The universal press with the appropriate die can crimp all sample pans.

Purge Gas

A purge gas is used in the calorimeter to control the sample environment. The sample environment is controlled so that each experiment is reproducible from day to day. The typical purge gas used with DSC is dry and with a purity of 99.9% pure. The purge rate should be from 20 to 50 ml/min. For general thermal analysis choose a rate between 20 and 30 ml/min and always run the DSC at that purge rate. For best results, install a gas dryer in line with the gas supply. If a gas cylinder is used, it is best not to run it dry but to replace it when 100–300 psi remains, to ensure that no contaminates will enter the calorimeter. Nitrogen is the most commonly used purge gas with polymers in the -120 to $750°C$ region. Helium has a much higher thermal conductivity than nitrogen and should be used when conducting experiments beginning at -170 to $200°C$. Above $200°C$, helium molecules become too excited and do not enhance the analysis. Compressed air or oxygen are also used when performing oxidative stability testing such as OIT or shelf-life testing.

There are some situations in which a combination of gas sequences is required. Then a combination of gases can be programmed into the experimental method and gas switching can be done on a time basis or a temperature basis. Up to four different gases can be switched on and off.

Coolants

Always use a coolant when performing DSC analysis. It does not matter if you are above ambient temperature or not. Use the appropriate coolant to accommodate sample equilibrium in the temperature range of interest. There are commercial chillers that attach to the calorimeter that enable the DSC to operate from -120 to $750°C$. Liquid nitrogen is used to achieve lower temperatures.

Selecting the Pan and Sample Size for the Experiment

Never load a sample into a sample pan and run it in a DSC without first performing a preliminary muffle furnace or hot-plate test. This simple test will determine what is the maximum sample size and best pan for the analysis without damaging the calorimeter. The desired results of this test is to choose a sample pan the sample size that will not spill out of the pan and contaminate or damage the DSC. This is accomplished by selecting the pan you think will be best for the analysis. That is usually the cheapest-priced pan. We will assume that the standard aluminum pan will be used. Then crimp 5-, 10-, and 20-mg samples in separate pans and wrap them in foil, mapping the size and position in the foil and place them in a muffle furnace. Heat the furnace up to the maximum temperature of the experiment and hold it there for a few minutes until the temperature is stable. Cool the furnace and then carefully remove the pans from the furnace and unwrap them.

Observe whether any sample bubbled out of the pans. If some did, use a smaller sample size. If all pans exhibit material flowing out of them, choose a different sample pan style and test three more samples with the new style pans.

Running a Sample Specimen: The Effects of Scanning Rates and Sample Size

There are two variables to consider after you have successfully selected the best sample pan: the maximum sample size and the best scanning rate for your analyses.

A *slow scanning rate* will yield a thermal curve that has good resolution with the baseline and a smaller peak that is sharp. Slower scanning rates are used to separate or resolve components that are close to each other in temperature. Slow scanning rates are 5°C/min or slower.

A *fast scanning rate* yields a thermal curve that has a large broad peak with poor baseline resolution. Faster scanning rates are used for subtle events that need to be magnified to see. Glass transition temperatures are usually scanned at faster rates, such as 20°C/min or faster.

A *small sample size* has an effect on the thermal curve comparable to that of a slower scanning rate. A small sample size will yield a thermal curve that has good resolution with the baseline and a smaller peak that is sharp. Small samples are used to separate or resolve components that are close to one another in temperature.

A *large sample size* has an effect on the thermal curve comparable to that of a faster scanning rate. A large sample yields a thermal curve that has a large broad peak with poor baseline resolution. Larger sample sizes are used for subtle events that need magnification, such as glass transition temperatures.

As an example, to separate two peaks, use a smaller sample at a slower scanning rate. Adversely, there might be a shoulder on a peak which cannot be resolved into two separate peaks.

Running a Sample Specimen: The Effects of Heat History

Thermal history, also known as heat history, describes the memory effect in polymers. In simple terms, polymers are affected by hot or cold temperatures and remember the last effect. This is considered a variable that should be controlled in the laboratory. Thermal history has a strong influence on the material's glass transition, melting, and crystallization temperatures. Temperature changes, both hot and cold temperatures, induce thermal stresses in the material. The extent and implication of thermal stresses to a polymer are usually unknown and uncontrolled. This variable may be of interest for those who want to know more about the sample as it was received in the laboratory or how it reacts to an uncontrolled end-use environment. Usually this is of interest, but because it is uncontrolled it is very difficult to rely on for material comparisons.

Comparison of polymers is best when a controlled thermal history is applied to each sample. To apply a controlled heat history to a polymer, the as-received heat history must be erased. This is simple done by heating the material above the melting temperature and holding the material isothermally at that temperature for several minutes. Usually the material is heated just above the melting peak, by about 10°C or so, and held for 5–10 min. Beware that no decomposition or volatilization occurs during this erasure. Now a known heat history can be applied to the sample by cooling it rapidly or slowly. When comparing samples, each sample should be subjected to the identical heat history. A simple way to accomplish this is to set up the analysis as a multiple-step *heat-iso-cool-iso-heat* experiment.

Important: Always compare first heats with first heats, and second heats with second heats. Never compare first heats with second heats, because the conditioning of each sample (heat histories) is different.

The *first heat and isothermal temperature* is used to examine the "as-received" condition of the sample and to erase the previous unknown heat history. This is done by heating the sample past its melting point and holding it at that temperature isothermally for several minutes to ensure temperature equilibration.

The *controlled cooling* step is the most important step for this analysis. Controlled cooling can be deliberately slow or fast. Controlled cooling is used to give the material a known heat history. Fast controlled cooling at 200°C/min from just above the melting peak to below the glass transition temperature or room temperature is called *quenching*. Quenching is used to trap amorphous regions of a semicrystalline material and to minimize the crystalline ordering that takes place upon cooling. Quenching is usually necessary to investigate glass transition temperatures. Slow controlled cooling at 10 or 20°C/min from just above the melting peak to the initial experimental temperature is called *annealing*. This is used to investigate crystalline melting (Fig. 17).

The *second heating* step should start after thermal equilibrium of the sample/DSC has been achieved. Equilibrium is identified when the ordinate signal (heat flow or temperature) is stable. The second heat is used to compare sample specimens that have been conditioned in the same manner. This heating is used because all samples being compared now have the same controlled heat history. In this step, the sample is heated from below the glass transition temperature through the melting temperature.

Running a Sample: Subtract a Baseline

A baseline run is a temperature scan across the temperature range of interest. It is done under the same experimental conditions as the sample speci-

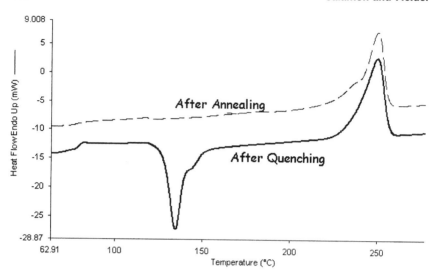

Figure 17 The thermal curves displayed are of the same PET sample sub-jected to different heat histories. These are different segments of the same experiment. Before quenching (cooling at 300°C/min) or annealing (cooling at 10°C/min), the sample was heated above the melting peak to remove the previous heat history and stresses.

men run, but with no sample and reference pans in the furnace or an empty sample pan on each side. This is your choice; both baseline methods are acceptable. Whichever you choose, stick with it. Remember that you always want your experiments to be run exactly the same way each time. It is best to begin and end every run with an isothermal segment of 1–2 min. This ensures that temperature equilibrium is achieved. Run a baseline before sample specimens are run and save the baseline as a separate file. Then input that file into the sample specimen method so that it is automatically subtracted from your sample specimen run, in real time. This will remove any instrumental effects of noise or baseline curvature. This is strongly recommended for all analyses.

Characterizing a Sample: What Type of Experiment Should I Run?

It is best to know as much about the material as possible before starting an experiment:

Of the polymer materials in engineering use, the plastics form the largest group by production volume. It is common to subdivide plastics into thermoplastics and thermosets (or thermosetting resins). Thermoplastics comprise the four most important commodity materials—polyethylene, polypropylene, polystyrene and poly (vinyl chloride)—together with a number of more specialized engineering polymers. The term "thermoplastic" indicates that these materials melt on heating and may be processed by a variety of molding and extrusion techniques. Important thermosets alkyds, amino, and phenolic resins, epoxies, and unsaturated polyesters, and polyurethanes, substances that cannot be melted and re-melted but which set irreversibly. The distinction is important in that production, processing and fabrication techniques for thermoplastics and thermosets differ (Hall, 1989, pp. 1–2).

For *partially crystalline thermoplastics*, analyze the glass transition region, the melting region, recrystallization upon cooling, and isothermal crystallization.

For *amorphous thermoplastics*, analyze the glass transition regions.

For *thermosets*, analyze the glass transition regions and curing studies.

For *elastomers*, analyze the glass transition regions.

Common ASTM Tests and Test Methods for DSC

D2471	Test Method for Gel Time and Peak Exothermic Temperature of Reacting Thermosetting Resins
D5028	Test Method for Curing Properties of Pultrusion Resins by Thermal Analysis
D4816	Test Method for Determining the Specific Heat Capacity of Materials by DSC
D4565	Test Method for Determining the Physical/Environmental Performance Properties of Insulation and Jackets for Telecommunications Wire and Cable
D4591	Test Method for Determining Temperatures and Heats of Transitions of Fluoropolymers by DSC
D3012	Test Method for Thermal Oxidative Stability of Polypropylene Plastics Using a Biaxial Rotator
D4803	Test Method for Predicting Heat Buildup in PVC Building Products
D2117	Test Method for Melting of Semicrystalline Polymers by the Hot Stage Microscopy Method
D3417	Test Method for Heats of Fusion and Crystallization of Polymers by Thermal Analysis
D3418	Test Method for Transition Temperature of Polymers by Thermal Analysis

D3895	Test Method for Oxidative Induction Time (OIT) of Polyolefins by Differential Scanning Calorimetry
D4419	Test Method for Determining the Transition Temperatures of Petroleum Waxes by DSC
E698	Standard Test Method for Arrhenius Kinetic Constants (of thermally unstable materials) Using DSC
E1559	Standard Test Method for Contamination Outgassing Characteristics of Space Craft Materials by DSC
E537	Standard Test Method for Determining the Thermal Stability of Chemicals by DSC
E793	Standard Test Method for Determining the Heat of Crystallization (of solid samples in granular form) by DSC
E1269	Standard Test Method for Specific Heat Capacity by DSC
E1356	Standard Test Method for Glass Transition Temperature by DSC

BIBLIOGRAPHY

Books

Bryce, D. M., *Plastic Injection Molding*, Society of Manufacturing Engineers, Dearborn, MI, 1996.

Hall, C., *Polymer Materials*, 2nd ed., Wiley, New York, 1989.

Hatakeyama, T., and Quinn, F. X., *Thermal Analysis: Fundamentals and Applications to Polymer Science*, Wiley, New York, 1994.

Mathot, V. B. F., *Calorimetry and Thermal Analysis of Polymers*, Hanser, Cincinnati, OH, 1993.

McNaughton, J. L., and Mortimer, C. T., *Differential Scanning Calorimetry*, IRS. Physical Chemistry Series 2, Vol. 10, Butterworths, London, 1975.

Perkin-Elmer Corporation, *Differential Scanning Calorimetry Training Manual*, N020-0607, Perkin-Elmer, 1990.

Riga, A. T., and Patterson, G. H., *Oxidative Behavior of Materials by Thermal Analytical Techniques*, ASTM Publication STP1326, ASTM, Philadelphia, PA, 1997.

Turi, E., *Thermal Characterization of Polymeric Materials* (two vols.), Academic Press, Boston, 1996.

Wendlandt, W. W., *Thermal Methods of Analysis*, Wiley, New York, 1974.

Wunderlich, B., *Thermal Analysis*, Academic Press, Boston, 1990.

Journals

Buzagh, E., and Simon, J., *Journal of Thermal Analysis*, Wiley, New York.
Ferris, R., *Plastics Engineering Magazine*, Society of Plastics Engineers, Brookfield, CT.
Wendlandt, W., *Thermochimica Acta*, Elsevier Science, Amsterdam.
Witzler, S., *Injection Molding Magazine*, Abby Communications, Chatham, NJ.

Articles

"DTA Instrumentation for High-Temperature Materials Characterization," *American Laboratory Magazine*, January 1993.
"A Power Compensation Differential Scanning Calorimeter and Thermal Analysis Software," *American Laboratory Magazine*, January 1996.

Plastics Application Notes from PerkinElmer Instruments LLC

PETAN-39, Gamma-Irradiated Polypropylene Studies Using DSC.
PETAN-50, Determination of Kinetic Parameters for the Crystallization of PET.
PETAN-51, Measurement of the Fictive Temperature of Polystyrene.
PETAN-56, Curing Kinetics of an Epoxy Adhesive Using Isothermal DSC Kinetics.
PETAN-58, Polyethylene: The Effect of Thermal Conditioning on Percent Crystallinity.
PETAN-78, Low Temperature Phase Transitions of Siloxane using Pyris 1 DSC.
PETAN-79, Low Temperature Phase Transitions of ABS using Pyris 1 DSC.
PETAN-82, Heat Capacity of Sapphire using the "Two Curve" Method.
PETAN-83, Measurement of the Temperature Dependent Crystallinity of Polyethylene.

4

Thermogravimetric Analysis of Polymers

Scott Kinzy and Robert Falcone

ICI Fluoropolymers, Bayonne, New Jersey, USA

PRINCIPLES OF OPERATION

Thermogravimetric analysis (TGA) is defined as the study of the change in mass as a function of temperature, time, and/or atmosphere. TGA consists of two essential components, mass and temperature detection.

The mass is measured using a balance. A balance uses the physical principle of leverage: a sample is placed at the end of the arm while placing a reference weight in the opposite end and the weight is recorded as the change in force. In the TGA, the mass detection is based on the principle of substitution weighing introduced by Mettler in 1945 (Fig. 1). In the TGA shown in Fig. 1, the calibration weights and the sample are suspended from a single arm. The calibration weight is substituted as the sample mass substitutes for it. The weighing principle is still leverage.

Most modern TGAs, however, in addition incorporate the principle of electromagnetic balances that have relatively little dependence on vibration (one of the common problems for weight measurements), have high sensitivity, and display little thermal drift. The beam position is monitored by a photodetection scheme. This concept was introduced by Cahn Instruments. As shown in Fig. 2, after taring the sample, the balance is assumed to be in equilibrium. Addition of the sample to the left side of the beam will cause the right side of the beam to be displaced upward. Sufficient current is

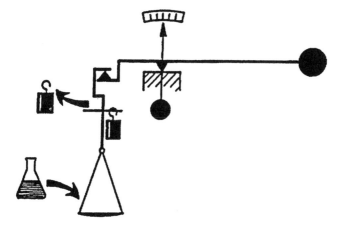

Figure 1   The principle of substitute weighing.

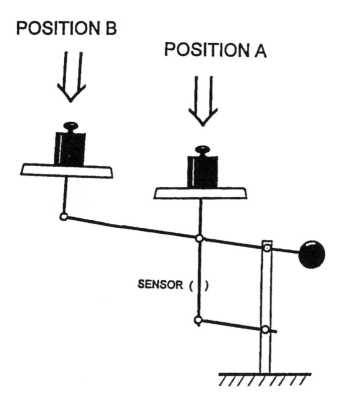

Figure 2   The principle of electromagnetic balance.

supplied to the electric torque motor to restore it to the original beam position. The restoring force, and hence the current, is proportional to the change in mass. Although manufacturers will claim a sensitivity of 0.1 μg, due to the temperature changes experienced during the course of the testing, 1 μg is the more typical sensitivity. This is the equivalent of 0.1% error (1 μg to a 1-mg sample).

Figures 3A and 3B show typical TGA assemblies. The heating is primarily by convection and radiation. This is achieved by placing the sample inside a miniature furnace capable of high heating and cooling. The heating elements of the furnaces are based mostly on resistive heating. Table 1 lists the most commonly used resistive elements. Nichrome and Kanthal are the alloys of choice when operating in the range up to 1000–1200°C. This is the range for polymeric applications. There are actually four types of furnaces used to cover the entire normal working range from −150 to 2800°C. These are −150 to 500°C, 25 to 1000°C, 25 to 1600°C, and 400 to 2800°C. As to be expected, the higher the temperature, the more complex is the design of the furnace and controls, thus making them more expensive. In most conventional TGA instruments, the temperature sensor is located in the vicinity of the sample, thus avoiding disturbing the weigh signal. Usually, the temperature sensor is very close to the furnace wall, or sometimes it can be found outside to avoid accidental movement and exposure of the sensor to corrosive gases.

CALIBRATION TECHNIQUES

All TGA instruments have the capability of being calibrated for mass and temperature. The mass calibration usually consists of placing a standard with a specific traceable NIST (National Institute of Standards, formerly the NBS) weight, ranging between 50 and 100 mg. The scale then measures the weight of the standard and by placing the expected weight it then adjusts the tension in the weighing arm.

The recommended calibration procedure for a TGA is as follows:

Step 1: weight calibration
Step 2: temperature calibration
Step 3: furnace calibration

Step 1: Weight Calibration

Weight calibration is done by comparing the balance readout of a standard weight. Most modern TGA instruments include this procedure in their software.

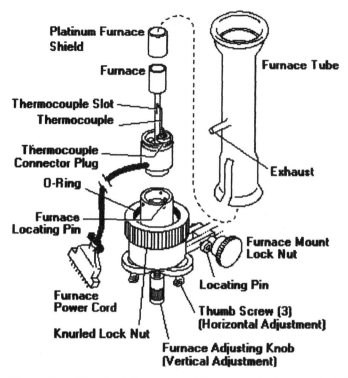

Figure 3A Standard furnace and furnace tube in a typical TGA assembly. (Courtesy of the Perkin Elmer Corporation.)

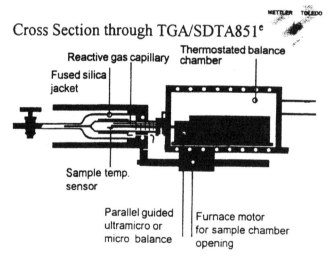

Figure 3B Cross section in a typical TGA assembly. (Courtesy of Mettler Corporation.)

114

Table 1   Maximum Temperature Limits for Furnace Resistance
Elements (approximate values)

Element	Approx. temperature (C°)	Required atmosphere
Nichrome	1000	Oxygen/oxidizing
Chromel A	1100	Oxygen/oxidizing
Tantalum	1330	Nitrogen/inert/vacuum
Kanthal	1350	Oxygen/oxidizing
Platinum	1400	Nitrogen/inert/vacuum/oxygen/oxidizing
Globar	1500	Oxygen/oxidizing
Platinum–10% rhodium	1500	Nitrogen/inert/vacuum/oxygen/oxidizing
Platinum–20% rhodium	1500	Nitrogen/inert/vacuum/oxygen/oxidizing
Kanthal Super	1600	Oxygen/oxidizing
Rhodium	1800	Nitrogen/inert/vacuum/oxygen/oxidizing
Molybdenum	2200	Nitrogen/inert/vacuum/hydrogen
Tungsten	2800	Nitrogen/inert/vacuum/hydrogen

Step 2: Temperature Calibration

The temperature calibration can be done in two ways. One is by measuring the melting points of materials, the other is by measuring the Curie temperature, the temperature at which materials lose their ferromagnetism. This is usually done by placing a sample of ferromagnetic material in the furnace and letting it heat while a magnet is placed outside the furnace. The temperature at which the loss in weight is noticed is the Curie temperature (Fig. 4).

It is advisable to use several materials whose Curie temperatures will encompass the temperature ranges of the intended plastics analysis. Table 2 shows materials and their Curie temperatures.

Step 3: Furnace Calibration

After completing the determination of the Curie temperatures, then proceed with furnace calibration. All instrument manufacturers include this step in their calibration protocols.

Measuring Weight Loss with Temperature Using Calcium Oxalate Monohydrate

The measurement of weight loss with temperature using calcium oxalate monohydrate is performed as follows:

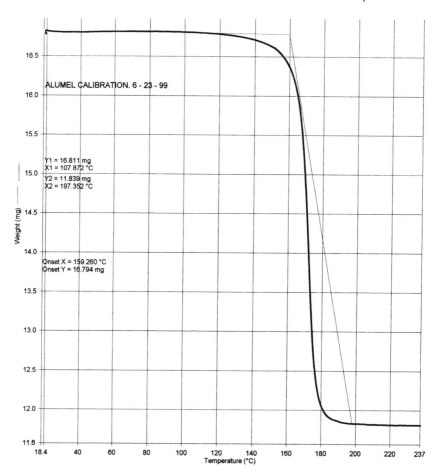

Figure 4 Typical Curie temperature analysis.

$$CaC_2O_4H_2O \longrightarrow CaC_2O_4 + H_2O$$

Loss will occur between 113 and 207°C.

$$CaC_2O_4 \longrightarrow CaCO_3 + CO$$

Loss will occur between 405 and 523°C.

$$CaCO_3 \longrightarrow CaO + CO_2$$

Loss will occur between 626 and 793°C.

Table 2 Typical Curie
Temperatures

Metal	Curie temperature ($C°$)
Alumel	154.2
Nickel	355.2
Perkalloy	596
Iron	780
Hisat-50	1000

The procedure is to place a sample of $CaC_2O_4 \cdot H_2O$ and set the instrument to run between 50 and 850C at a rate of $15°C/min$ (Table 3 and Fig. 5).

APPLICATIONS

Application 1: Determination of the Degradation Temperature

The *degradation temperature* is the temperature at which a plastic reverts to its original component(s). This is also known as the *ceiling temperature*. The importance of the degradation temperature is that it helps to determine the maximum limits at which the plastic will no longer be a plastic.

There are two TGA techniques to determine the degradation temperature of a plastic. The first is called the *isothermal method* and is based on setting the sample on a temperature that is between 50 and 150C above its melting temperature and letting the instrument measure the weight loss with time. This is one technique used to determine physical aging of a sample under extreme conditions. A typical decomposition curve is shown in Fig. 6.

The second technique is based on running the sample under an increasing temperature pattern while measuring the weight loss. This is known as a *nonisothermal degradation test*. This technique is used when the decomposition temperature of the sample is not known. Figure 7 shows the use of this

Table 3 $CaC_2O_4 \cdot H_2O$
Weight Percent Loss

Step number	Weight loss (%)
1	12.33
2	21.88
3	44.00
Overall	61.64

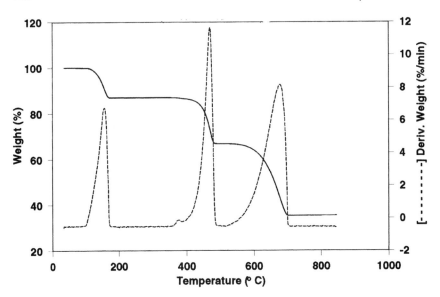

Figure 5 Typical calcium oxalate monohydrate run. (Courtesy of TA Instruments, TS 13.)

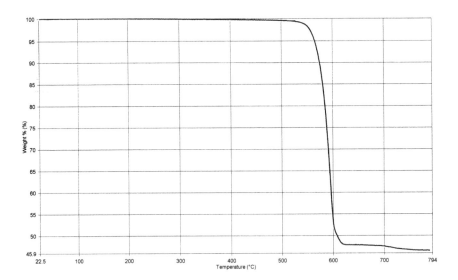

Figure 6 Isothermal decomposition method.

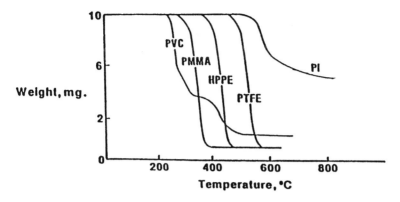

Figure 7 Constant heat rate decomposition method. PVC = polyvinylchloride; PMMA = poly(methyl methacrylate); HPPE; high-pressure polyethylene; PTFE; PTFE = polytetrafluoroethylene; PI = polyimide.

technique for determining the degradation temperatures of some industrial plastics.

The isothermal method should be used when the melting temperature of the sample is known and a better understanding is needed of the degradation of the material. The nonisothermal method should be used when the degradation temperature is not known.

Application 2: Physical Aging

It is often essential to know the lifetime performance of a plastic. The TGA can provide an accelerated testing to predict lifetime performance. A typical example is shown in Fig. 8.

Application 3: Moisture Determination

Some plastics are hygroscopic in nature. This means that they retain/get water from the surroundings. Water, like any other chemical, can chemically deteriorate the structure of the plastic, thus weakening some or all of its mechanical properties. Therefore, in plastics such as nylon the determination of moisture content is very important. Although the water molecule is smaller than all of the plastic molecules, it is very important to choose the temperature(s) for moisture determination, because some plastics may soften at 100°C (212°F), making it very difficult to determine the water content in the plastic. Usually this is done by placing the sample at a temperature below its melting point (40–60°C below) and by using compressed/

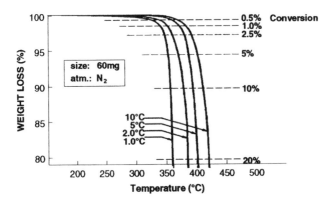

Figure 8 Physical aging method—wire insulation thermal stability. (Courtesy of TA Instruments, TS-125.)

dry nitrogen atmosphere to run the test for a period between 1 and 2 h. Ensure that the calibration procedure has been done before starting this measurement. A typical example is shown in Fig. 9.

Application 4: Reverse Kinetics

Reverse kinetics involves a hyphenated technique. "Hyphenated techniques" are procedures that require the use of more than one instrument or techniques simultaneously. The reason for using these techniques is that the measurement may be lost because of the limitations of one technique used.

Reverse kinetics is a procedure by which, using the degradation method described in application 1 in combination with either a gas chromatograph (GC) or a Fourier transform infrared spectrophotometer (FTIR), you can find the following:

1. The original components of the plastic being degraded (FTIR and TGA)
2. The amount of components of the plastic being degraded (GC and TGA).

One of the limitations of the above techniques is that either the type of components or the amount of each is known but not both. However, there is a third hyphenated technique that combines all of the above. This technique uses the method described in applications in combination with a gas chromatograph and a mass spectrometer (GC-MS) that is also coupled with an infrared spectrometer (FTIR). Most GC-MS units are already equipped with infrared capabilities. The advantage of this technique is that not only

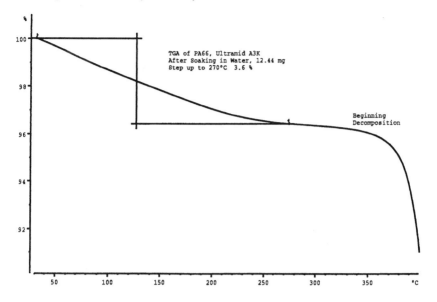

Figure 9   Moisture determination. (Courtesy of Mettler Instruments.)

the quantities of the original components are determined but also the types of components. Figure 10 shows a typical instrument setup to run the TGA/GC-MS technique, Fig. 11 shows the connection between the TGA and GC-MS unit, and Fig. 12 shows typical GC-MS scans done on TGA polymer samples.

Application 5: Determination of Filler/Additive Content

In most industrial applications, plastics do not possess the properties needed, thus the use of additives and fillers. The procedure to determine the content of additives/fillers is to run the sample to and above its decomposition temperature. Most additives will decompose at temperatures below the material's decomposition temperature, while most fillers will still be present after the sample has decomposed. The typical heating rate is between 15 and 30°C/min. See Figs. 13 and 14.

Application 6: Determination of Different Types of Plastics in One Sample

Besides the use of additives and fillers to enhance the mechanical properties of industrial plastics, it is also common to blend different plastics in order to

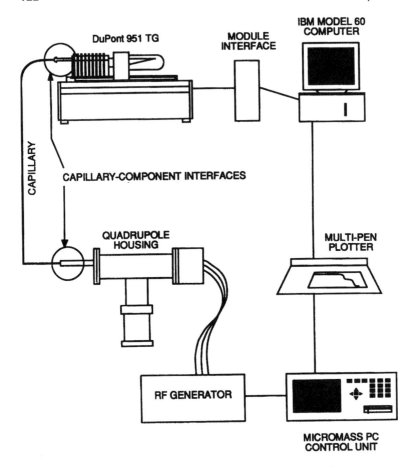

Figure 10   Typical instrument setup to run the TGA/GC-MS technique. (Courtesy of TA Instruments, H-16781.)

create a material whose properties are better than those of the individual components.

The procedure is to run the sample under a heating rate of 20–30°C from around room temperature to 100C above the component having the highest decomposition temperature. With the help of Table 4, the components can be identified. This technique is approximate, and the best way to determine the types of components present in a sample is by using DSC and FTIR testing along with the TGA (Fig. 15).

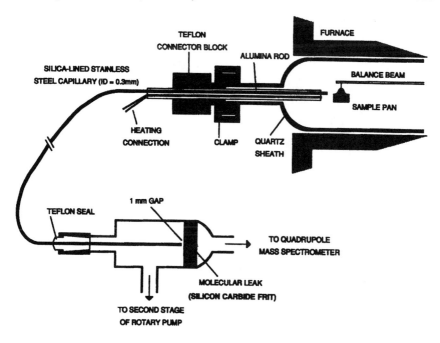

Figure 11 Connection between the TGA and GC-MS unit. (Courtesy of TA Instruments, H-16781.)

Application 7: Modulated Thermal Gravimetric Analysis

Modulated thermal gravimetric analysis (MTGA) is a recently introduced technique that provides the advantage of running a typical TGA while modulating the temperature response (Fig. 16). Kinetic parameters can be determined in a much faster mode. In some plastics, total decomposition could be more than a single weight loss and typical weight loss procedure as indicated in application 1 will not determine the multi-weight loss reactions. Although there are ASTM procedures outlining such methods (E1641 and E1877), these procedures can be time consuming when rates are below 2C°/ min. In such cases, the use of an MTGA technique will enable you to determine the weight loss in less time.

ACKNOWLEDGMENTS

We are grateful to Mr. James Creedon and Dr. Roger Blaine, TA Instruments; Mr. Jon Foreman, Mettler Corp.; Ms. Karen Gillette, Perkin Elmer Instruments; and Mr. Tom Wampler, CDS Analytical.

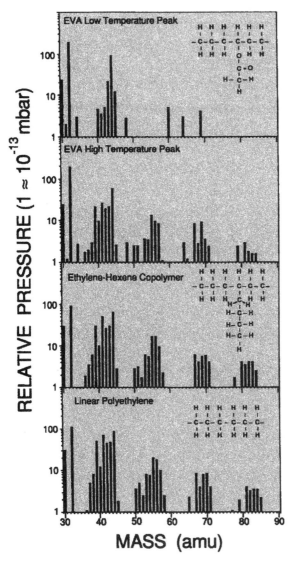

Figure 12 Typical GC-MS scans done on TGA polymer samples. (Courtesy of TA Instruments, H-16781.)

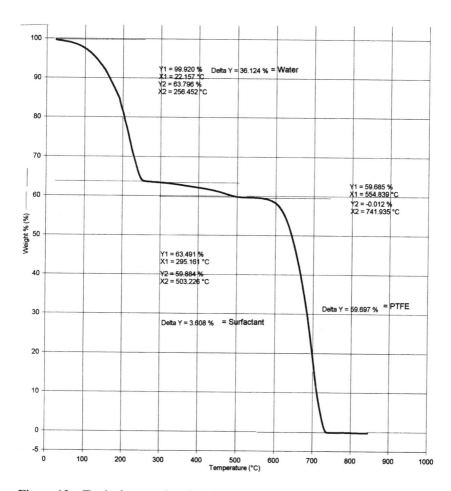

Figure 13 Typical curve showing the presence of additives in PTFE. In this particular case, the additive is a surfactant. (Courtesy of ICI Fluoropolymers.)

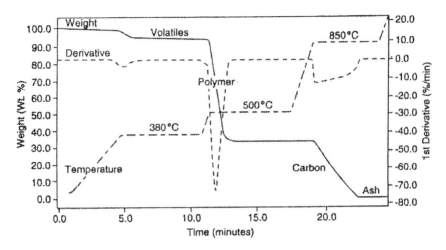

Figure 14 Typical curve showing the presence of a filler in polyethylene. In this particular case, the filler is carbon. (Courtesy of Perkin Elmer Instruments.)

Table 4 TGA Decomposition Temperatures of Common Polymers

Acronym	Polymer	Decomposition temperature (C°)
ABS	Acrylonitrile-butadiene-styrene	375
PMMA	Poly methyl methacrylate	313
PTFE	Polytetrafluoroethylene	525
PVDF	Polyvinylidene fluoride	470
Nylon 6	Nylon 6	400
Nylon 6,6	Nylon 6,6	426
PC	Polycarbonate	473
PBT	Polybutylene terephthalate	386
PET	Polyetheretherketone	575
LDPE	Low-density polyethylene	459
HDPE	High-density polyethylene	469
PPO	Polyphenylene oxide	400
PPS	Polyphenylene sulfide	508
PP	Polypropylene	417
PS	Polystyrene	351
PSO	Polysulfone	510
PVC	Polyvinyl chloride	265

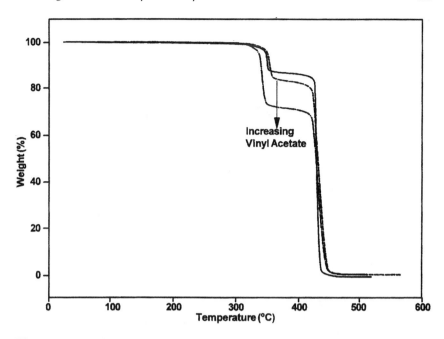

Figure 15   TGA analysis of ethylene vinyl acetate blend. (Courtesy of TA Instruments, TA023.)

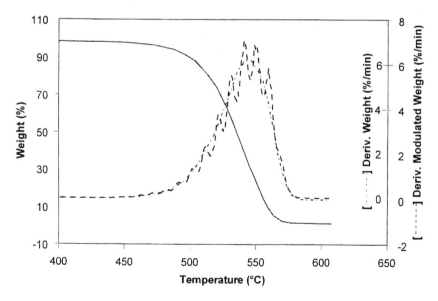

Figure 16   MTGA curve of PTFE. (Courtesy of TA Instruments.)

BIBLIOGRAPHY

General

Bordoloi, B. K, Wright, C .E., and Pearce, E. M., *Laboratory Experiments in Polymer Synthesis and Characterization.* EMMSE, 1982, p. 269.

Mathot, Vincent B. F., *Calorimetry and Thermal Analysis of Polymers*, Hanser, 1993.

Turi, Edith, *Thermal Characterization of Polymeric Materials*, Vols 1 and 2, Academic Press, 1997.

Wendlandt, W. W. M., *Thermal Analysis*, 3rd ed., Wiley, New York, 1986.

Wunderlich, B., *Thermal Analysis*, Academic Press, 1990.

Standard Testing Methods

ASTM E473-85, Terminology Relating to Thermal Analysis.

ASTM E914-83, Evaluating Temperature Scale for Thermogravimetry.

ASTM E1131-93, Standard Test Method for Compositional Analysis by Thermogravimetry.

ASTM E1582-93, Standard Practice for Calibration of Temperature Scale for Thermogravimetry.

ASTM E1641-94, Decomposition Kinetics by Thermogravimetry.

ASTM E1868-97, Standard Test Method for Loss-on-Drying by Thermogravimetry.

ASTM 1877-97, Standard Practice for Calculating Thermal Endurance from Thermogravimetric Decomposition Data.

Hyphenated Methods

Charsley, E. L., Warrington, S. B., Jones, G. K., and McGhie, A. R., TGA-MS Using a Simple Capillary Interface. "*Am. Lab.*, 1990.

Mettler Toledo Technical Notes.

Nicolet Instruments Technical Notes.

Perkin Elmer Technical Notes.

TA Instruments Technical Notes.

5

Thermal Conductivity and Diffusivity of Polymers

Hubert Lobo

DatapointLabs, Ithaca, New York, USA

INTRODUCTION

Plastics fall into the category of materials of intermediate thermal conductivity (0.15–0.4 W/mK). They are an order of magnitude more conductive than foams and insulation but about five times less conductive than ceramics and glass. This puts them in an unique class by themselves. They have not been heavily characterized in the past for many reasons: the measurements have been difficult, with highly variable results, coupled with a lack of demand. In recent years, the need for thermal conductivity data has grown dramatically. This important property is essential in almost every heat transfer calculation, from such basic applications as component heat conduction and dissipation to the complex computer programs used to simulate plastics manufacturing processes. In the latter case, because thermoplastics are typically processed in the melt state, the use of room-temperature-measured thermal conductivity data may result in serious errors, because the melt state thermal conductivity of polymers can differ significantly from solid-state values. Empirical rules have been used in the past to estimate melt state values from the typically available solid-state data. These rules are often suspect because have they been based on avail-

129

able information in the literature, which is scarce and often inaccurate, especially at high temperatures. Additionally, some rules do not take into consideration effects such as those of crystallinity, composition, or matrix–filler interfacial contact, on which thermal conductivity is strongly dependent. Theoretical studies on the thermal conductivity of polymers have been hampered by the inconsistency and paucity of experimental data.

The measurement of thermal conductivity of plastics is a difficult process, and until recently, the availability of such data was quite limited. Conventional guarded hot-plate techniques have traditionally been used to make measurements of thermal conductivity. These remain the reference technique for thermal conductivity measurements, since they do not require calibration against a material of known thermal conductivity.

Guarded heat flow meters are variants on the guarded hot-plate technique, which utilize a heat flow meter to measure the heat flux. These techniques are comparative in nature because a reference material is needed to calibrate the heat flux meter. They utilize smaller samples and are thus able to equilibrate faster, improving the versatility of the guarded hot-plate method.

These steady-state techniques, while successful in measuring thermal conductivity in the solid phase, tend to be laborious, requiring meticulous specimen preparation and long measurement times. They are not conducive to making measurements at nonambient temperatures, and they tend to fail or give poor results for the melt. This is because the sample cannot be properly contained and tends to flow and create voids upon melting. Also, significant difficulties exist with raising the temperature of the typically large apparatus to the melting temperature of the polymer.

More recently, the transient line-source technique has been successfully applied to these measurements. In contrast to the guarded hot-plate method, the sensing apparatus and sample are very small, making it simple to attain any desired temperature. Additionally, the sample can be easily contained, so problems with void formation can be avoided. The technique was initially applied exclusively to measurements of melt-state thermal conductivity, where it was shown to yield highly reproducible data [1]. Later, the technique was extended to enable measurements in the solid state as well. This makes the line-source method a powerful technique for thermal analysis of plastics, since few experimental techniques have the capability to span transitions. The utilization of a single technique for both states permits the creation of consistent data, free from errors that might arise from the use of separate instruments for the melt and solid states. Because of these attractive attributes, the technique has been used extensively for measurement of thermal conductivity of plastics. A large body of data now exists which permits us to develop an understanding of the dependency of thermal conductivity on such variables as temperature, pressure, composition, and degree of cure. Because of its importance to plastics, special attention will be devoted to the line-source technique in this chapter. Further, most of the data that will be presented in the analysis and data interpretation section has been developed using this technique.

Finally, the effects of thermal contact resistance cannot be neglected in the realm of plastics thermal conductivity. This important heat transfer effect, which is due to the presence of an interface between the specimen and the sensor, needs to be compensated for, or eliminated. The success of a thermal conductivity measurement technique often depends on how well it addresses this issue. This issue is discussed in more detail later.

The laser flash method is used for thermal diffusivity measurements. In this technique, a laser is used to uniformly pulse one side of a test specimen. The temperature rise of the other side is measured using an infrared detector. This transient is then used to calculate the thermal diffusivity. While the technique is simple in concept, nonidealities such as heat loss from the front and back surfaces complicate the resulting data analysis, so fairly complex models need to be used to extract the thermal diffusivity. These calculations can be easily performed by the computers used to run the instruments.

STANDARDS AND REFERENCE MATERIALS

Standardization efforts have yielded national and international norms for thermal conductivity measurement techniques. Table 1 summarizes the relevant standards that are currently available.

One of the main drawbacks in the generation of thermal conductivity data on plastics remains the paucity of reference materials. Reference materials are important because they provide an important baseline for the calibration of instruments. Most thermal conductivity instruments require calibration to materials of known thermal conductivity. The National Institute of Standards (NIST) and other standards-setting institutions have characterized a number of metals, ceramics, glasses, and even insulation. These are not suitable as reference materials because they tend to possess a thermal conductivity that is higher or lower than those of plastics. For example, the NIST reference glass materials Pyrex 7740 and Pyroceram 9606 possess a thermal conductivity of 1 W/m-K which is five times greater than that of most plastics. The other reference materials are still more inapplicable. Many fluids, on the other hand, possess thermal conductivities similar to plastics. The transient line-source technique, however, is the only technique that can use such reference materials.

Table 1   ASTM and ISO Standards Currently
Available for Thermal Conductivity Measurements

Technique	Standard
Guarded hot plate	ASTM C177, ISO 3802
Heat flow meter	ASTM E1530
Comparative method	ASTM E1225
Line-source method	ASTM D5930
Laser flash thermal diffusivity	ASTM E1461

GUARDED HOT-PLATE TECHNIQUE

The guarded hot-plate test is based on the principle of heat flow across a large flat sample at steady state, in which the heat flow is unidirectional through the thickness. A temperature differential across the sample provides the driving force for heat transfer, following the Fourier equation:

$$Q = kA \, dT/dx \tag{1}$$

where

Q = heat flow (W)

k = thermal conductivity (W/m-K)

A = cross sectional area (m^2)

T = temperature ($^\circ$C)

X = thickness of sample (m)

In its practical application, the instrument requires two identical samples. A sample is placed on either side of the primary heater as shown in Fig. 1. Auxiliary heaters are then placed on the other sides of the samples. A guard heater surrounds the main heater, minimizing lateral heat flow. During the test, both auxiliary heaters are kept at the same temperature and the primary and guard heaters are held at a higher temperature. An outer guard heater may be used to provide additional insulation at extreme hot or cold temperatures. Measurements can also be performed in alternative gas or vacuum environments. Thermocouples, placed at the interface between the heater plate and the sample, measure the temperature drop across the sample. Specimens are typically round or square sheets, 8 in. or more in diameter and can be as much as 2 in. thick. Larger specimen sizes are suited to inhomogeneous materials for which it is desired to average out variations in morphology of the sample. This is in general unnecessary for

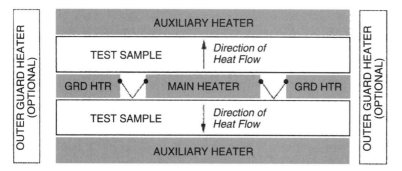

Figure 1 Schematic diagram of the guarded hot-plate technique. (Courtesy of Holometrix Corp.)

plastics. Instruments that utilize smaller specimens have a lower thermal mass and are therefore better suited to nonambient measurements.

During the test, the sample/heater stack is clamped tightly together and the heaters are started. After a while, usually 30 min, the system equilibrates. Longer equilibration times are needed at nonambient temperatures. Steady temperature and voltage readings indicate thermal equilibrium. The heat flow through the sample is equal to the power supplied to the primary heater. Readings are recorded and a test at a new temperature can begin.

Using Eq. (1), thermal conductivity is determined from the temperature drop across each sample, the thickness of each sample, the power applied to the main heater and the area of the main heater plate. Measurements in the range 0.014–2.0 W/m-K are typical. The guarded hot plate remains the reference technique for thermal conductivity measurements, since it does not require calibration against a material of known thermal conductivity. It has a repeatability of ±1% with a stated accuracy of ±4%.

HEAT FLOW METERS

The heat flow variant of the guarded hot-plate technique utilizes a heat flux transducer to measure the heat flow through the sample. ASTM E1530 provides a detailed description of the measurement. ASTM C518 and ISO 8301 also describe this method. The technique offers significant operational simplifications to the hot-plate technique: only a single specimen is needed, and a smaller test apparatus permits the stack to reach steady state faster. As with the hot-plate technique, this method works best with solid and thin sheet materials, but it is possible to use other specimen holders to accommodate pastes, powders, and polymers through the melt.

Heat flux transducers are typically thermopiles, which produce an output proportional to the heat flux. They present an experimental means to measure heat flux rather than calculate it from the voltage applied to the primary heater. However, the heat flux transducer must be calibrated using materials of known thermal conductivity. Calibration establishes a relationship between the voltage signal of the transducer and the heat flow through it. Typical materials used include Pyroceram 9606 and Pyrex 7740 glasses, both of which are NIST standards.

During the test, a sample is held under a compressive load between two polished metal surfaces as shown in Fig. 2. The upper surface is temperature controlled. The lower surface is part of a calibrated heat flux transducer that is attached to a liquid-cooled heat sink maintained at a constant temperature. When city water is used for cooling, the mean sample temperature may vary slightly, as the water temperature changes with the seasons. If this variation is undesirable, a chiller can be connected to the unit. Cryogenic measurements are achieved by circulating cryogenic fluids through the system. In such cases, it is important to use a dry gas environment to prevent ice buildup from affecting the measurement. An axial temperature gradient is established through the stack as heat flows from the upper surface through the test sample to the heat sink. Samples are in the form of 2 in. (5-cm) diameter disks, 0.004 to 0.8 in. (0.1 to 20 mm) thick. Some instruments permit the use of samples as small as 1 in. (2.5 cm) in diameter.

The temperature drop through the sample is determined from temperature sensors in the metal surfaces on either side of the sample. A heat flux transducer mounted on the cold plate measures a voltage proportional to the heat flow through the sample. After reaching thermal equilibrium, typically 30 min for an ambient measurement, the temperature difference across

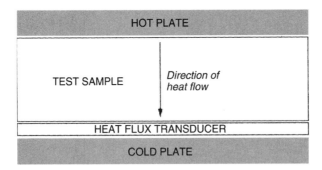

Figure 2 Schematic diagram of the guarded heat flow meter. (Courtesy of Holometrix Corp.)

the sample is measured along with the output from the heat flux transducer. The amount of heat that flows from the hot plate to the cold plate is determined by the thermal conductivity and thickness of the sample according to the Fourier equation:

$$Q/A = \frac{k(T_h - T_c)}{x} \tag{2}$$

where

$$Q/A = \text{heat flux through the sample (W/m}^2)$$
$$k = \text{thermal conductivity (W/m-K)}$$
$$T_h \text{ and } T_c = \text{hot and cold plate temperatures, respectively}$$
$$x = \text{sample thickness (m)}$$

Thermal conductivity is calculated from the calibration data, the sample thickness, and the temperature drop across the sample. The instruments can measure in the range of 0.1 to 10 W/m-K over a temperature range of 40 to 200°C. By changing the circulating fluids, it is possible to perform measurements as low as −100°C and up to 300°C. Repeatability is in the ±2% range, with a stated accuracy of ±5%.

Thermal contact resistance is an issue with the heat flow meter. Some of its effects can be "calibrated out" by applying the same load to the test stack as was used during calibration of the instrument. Further, a heat transfer compound can be applied if needed.

COMPARATIVE METHOD

The comparative instrument measures heat flow based on the known thermal properties of standard reference materials. ASTM E1225 provides a detailed description of the test method. A test specimen is clamped between two identical reference pieces of known thermal conductivity and identical cross section (Fig. 3). A temperature gradient is established in this test stack by controlling the top heater at a higher temperature than the bottom heater, resulting in heat flow through the specimen and references from the top to the bottom. Radial heat flow is minimized by placing a guard around the stack. This active guard closely mirrors the thermal gradient of the test stack with the aid of a top guard heater, a bottom guard heater, and a water-cooled heat sink. The tubular guard is typically split lengthwise and hinged to readily expose the test stack. The annular space between the guard and the stack contains thermal insulation to further reduce heat loss.

At thermal equilibrium, typically after 2 h, heat flow through each reference is calculated from the measured temperature gradients. The comparative method is based on the premise that the heat flow through the

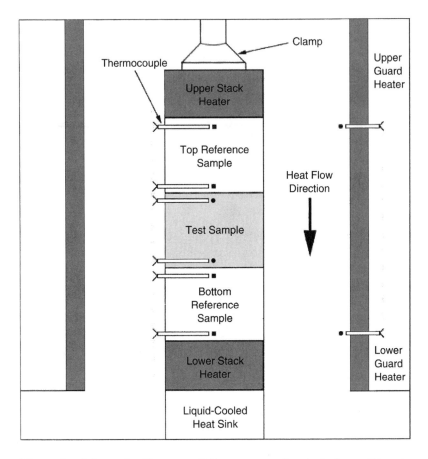

Figure 3 Schematic diagram of the comparative technique. (Courtesy of Holometrix Corp.)

unknown test specimen equals the average heat flow through the two references. From this, and the measured temperature gradient in the test specimen, the thermal conductivity is then calculated. For best results, it is important to match the test specimen to reference materials of similar thermal conductivity. Commonly available references include electrolytic iron, stainless steel, Inconel, nickel, Pyroceram 9606, Pyrex 7740, fused silica, Vespel, graphite, and alumina. Depending on the material, the sample thickness must be properly chosen and the thermal conductivity of the reference materials must be closely matched. For higher-conductivity materials, thicker samples are required. Otherwise, the insufficient thermal gradients will adversely affect the results.

Typical test specimens are flat solid disks, 1 or 2 in. diameter and 0.5 to 1.5 in. thick. Some instruments provide a special sample holder, also referred to as a fluid cup, for testing liquids and pastes. The technique is well suited for evaluating materials in the moderate-to-high conductivity range (0.2–100 W/m-K). At its lower end, therefore, it should be applicable for plastics. The test accuracy lies between ±5% and ±10% over the entire thermal conductivity range with a reproducibility of ±3%. As stated earlier, an important factor affecting overall test accuracy lies in achieving a good match between the test sample and the reference material. Test temperatures in the range of −160 to 500°C can be achieved.

LINE-SOURCE METHOD

The line-source technique is a transient method capable of very fast measurements. A line source of heat is located at the center of the sample being tested as shown in Fig. 4. The whole is at a constant initial temperature. During the course of the measurement, a known amount of heat is produced by the line source, resulting in a heat wave propagating radially into the sample. The rate of heat propagation is related to the thermal diffusivity of the polymer. The temperature rise of the line source varies linearly with the logarithm of time. Starting with the Fourier equation, it is possible to develop a relationship which can be used directly to calculate the thermal conductivity of the sample from the slope of the linear portion of the curve:

$$k = \frac{CQ'}{4\pi \, \text{Slope}} \tag{3}$$

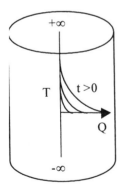

Figure 4 Schematic diagram of the line-source technique.

where

k = thermal conductivity (W/m-K)

Q' = heat input per unit length of line source, (W/m)

C = probe constant

$$\text{Slope} = \frac{T_2 - T_1}{\ln(t_2/t_1)} (\text{K})$$

T_2 = and is the temperature recorded at time t_2

T_1 = and is the temperature recorded at time t_1

Typical transients show an initial nonlinearity due to the heat wave propagating through the finite thermal mass of the probe. This is a region of high conductivity and, hence, low slope. With typical melt-state transients, where the sample has no contact resistance, the transient approaches linearity directly after it overcomes this effect, typically within a few seconds. The slope of interest is the linear region that follows the initial nonlinearity.

In real life, the line source takes the form of a hypodermic sensor probe of finite length and diameter [2]. Typical probes are 50 mm (2 in.) long and about 1.5 mm (1/16 in.) in diameter. They contain a heater element that runs the whole length of the hypodermic. A thermocouple sensor is also located halfway down the length of the probe, to measure the temperature rise associated with the transient. These and other nonlinearities require that the probes be calibrated against a reference material. The resultant probe constant is the ratio of the actual thermal conductivity of the reference material to that measured by the instrument. Silicone fluids have been used for the purpose.

Acquisitions typically range in the 30–60-s timeframe. This is very important in gathering melt-state thermal conductivity because it dramatically reduces the possibility of thermal degradation. Scanning methods have been devised, permitting the automated acquisition of data at different temperatures, so that measurements over a wide range of temperatures are possible. With this, the same sample that was used for the melt-state measurements is used for solid-state measurements, thereby permitting measurements across the melt-to-solid transition.

Plastics tend to shrink significantly upon solidification. This is especially so for the semicrystalline materials, which experience a significant change in specific volume upon crystallization. This can result in gaps developing between the sample and the sensing device, which aggravate the contact resistance problem. To compensate for shrinkage, a simple compression scheme is used to move the line-source probe downward along with the sample as it shrinks.

The test method for the line-source technique is unusual and is therefore described below in more detail. Predried pellets of the polymer are loaded into a cylindrical sample cell, which is maintained above the melting temperature of the material. A typical loading temperature is the lowest processing temperature of the polymer. At this temperature, the pellets melt slowly and are easily compacted to obtain a uniform, air-free sample. The probe, fitted with a dynamic seal, is inserted into the sample and the system is allowed to equilibrate for 2–5 min. Dead weights are then hung to supply a static load upon the probe. Pressures in the range of 300–1000 psi are used. This vibration-free, constant force permits the probe to move with the sample as it shrinks during solidification.

Measurements are made by supplying a known, constant voltage to the probe heater. Durations are set at 45–50 s, beyond the range of effect of the thermal contact resistance factors. Data are acquired by the computer and stored for later analysis. As seen in the theory [1,5], a plot of temperature against log time is a straight line. Thermal conductivity can be calculated from the slope of this line. Since the amount of heat added during the measurement is small, the system returns to equilibrium quickly, and thus reproducibility checks are easily made. The scans can be carried out in both cooling and heating modes to examine hysteresis effects, if any.

When measurements are complete, the probe is removed and cleaned. The sample is purged by opening a plug at the bottom of the cell.

LASER FLASH THERMAL DIFFUSIVITY

Laser flash thermal diffusivity is a versatile technique that permits the measurement of thermal diffusivity of a wide range of materials including plastics. Details of the test method are given in ASTM E1461. A laser fires a pulse at the sample's front surface and an infrared detector measures the temperature rise of the sample's back surface (Fig. 5). Typical lasers used are ND:glass with wavelength 1060 nm and pulse energy 15 J. The duration of the pulse is small compared to the time taken for the temperature of the back surface to reach its peak, typically 400 μs. A typical transient is shown in Fig. 6. Initial theoretical analyses assumed that there was no heat loss from the front and back surfaces of the specimen during the test. This assumption yielded a fairly simple analysis developed by Parker [6]. It is observed that in actual measurements, there is heat loss from both surfaces. This is best seen in Fig. 6, where the dimensionless temperature is observed to fall after reaching its maximum. The consequence of this nonideality is that not all the supplied heat is transmitted through the sample. Heat loss from the front and back surfaces will change the shape of the transient. Several attempts have been made to correct for this problem. Of these the

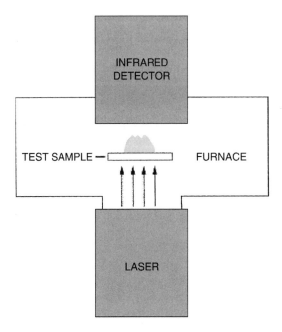

Figure 5 Schematic diagram of the laser flash technique. (Courtesy of Holometrix Corp.)

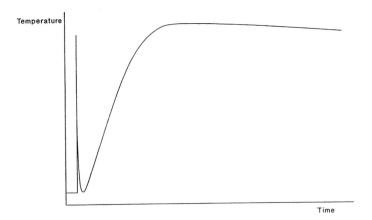

Figure 6 Typical laser flash transient.

method of Koski [7] has been used commercially. The analysis is performed by computer, which uses these analysis routines to match a theoretical curve to the experimental temperature rise curve.

The technique can also measure specific heat, where the infrared detector measures the actual temperature rise of the sample. For this measurement, the response of the infrared detector must be calibrated with a reference sample of known specific heat. The instrument can thus measure thermal diffusivity and specific heat simultaneously. If the density of the sample is known, the technique can also determine thermal conductivity from the equation:

$$k = \alpha \rho C_p \qquad (4)$$

where

α = thermal diffusivity (m^2/s)

ρ = density (kg/m^3)

C_p = specific heat (J/kg-K)

Specimens as small as 0.5 in. (12.7 mm) in diameter, 0.04 to 0.12 in. (1–3 mm) thick, can be used. The small sample size is convenient for expensive materials and those manufactured in thin sheets. Tests are typically performed at ambient; however, the test can also be performed on material at elevated temperatures by placing specimens in a heated chamber. Once the specimen equilibrates at the desired temperature, the test takes a few seconds. Tests are typically conducted in a vacuum.

The technique can measure materials in the thermal diffusivity range of 0.001 to 10 cm^2/s with a repeatability of ±3% and a stated accuracy of ±5%. Laser flash instruments offer the advantages of rapid testing and wide utility over a broad range of material thermal conductivities. Materials suited to laser flash testing include composites, alloys, ceramics, coatings, and plastics. Typical test temperatures range between ambient and moderate temperatures. Wider temperature ranges are possible with more complex instrument configurations.

THERMAL CONTACT RESISTANCE

There are a number of difficulties that may be encountered in measuring thermal conductivity in the solid and melt states. Problems such as thermal contact resistance and shrinkage are intrinsic to the polymer system and appear regardless of the method used for the measurements.

Thermal contact resistance is the resistance to heat transfer across the interfacial boundary between solids. The resistance is observed to depend on the quality of the contact between the surfaces; the presence of gaps results in a significant increase in the resistance to heat transfer. Thermal contact does not affect measurements when the plastic is a melt because of the tenacious contact between the polymer and the sensing device. In the solid state, however, there can develop a contact resistance. With guarded hot-plate techniques, efforts are made to minimize the effect of thermal contact resistance by preparing smooth and parallel surfaces and through the use of thermally conductive pastes. With careful application of these methods, it is possible to reduce but not eliminate the effect of thermal contact resistance.

In the line-source method, contact resistance manifests itself as an additional nonlinearity in the initial portion of the transient. As described earlier, the line-source transient is actually a superposition of two or more transients. The transient possesses an initial nonlinearity due to the heat wave propagating through the finite thermal mass of the probe. With typical melt-state transients, in which the sample has no contact resistance, the transient approaches linearity directly after it overcomes this effect. In solid-state transients, however, there appears a second nonlinearity, in which the heat wave is propagating across the region of thermal contact resistance (Fig. 7). This is a region of high slope, corresponding to a region of high resistance to heat flow. Once the transient overcomes this effect, typically after 10–15 s, it again becomes linear and independent of time. By extending the time of the measurement, it is possible to "progress"

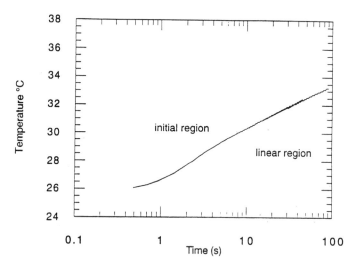

Figure 7 Line-source transient with nonlinear region due to thermal contact resistance.

beyond the region of thermal contact resistance and achieve a state in which the contact resistance does not contribute to the measured transient. Indications of the existence of these phenomena have been pointed out in theoretical analyses by Jaeger and Blackwell, and are reviewed by Tye [8].

It has been observed that the larger the contact resistance, the greater is the time before linearity is attained. It is therefore important to make a long enough measurement to exclude the portion of the transient that shows the effect of the contact resistance.

RESULTS

Thermal conductivity data as a function of temperature are presented for five commercial polymers, three amorphous and two semicrystalline. Most measurements have been carried out over a temperature range from the processing temperature down to room temperature. The polymers selected for these measurements are known to be reasonably stable. In many cases, it has been possible to carry out an entire heating scan to verify the cooling curve or examine hysteresis effects.

Amorphous Materials

Figure 8 shows a plot of thermal conductivity as a function of temperature for polystyrene (Styron 678D, Dow Chemical) obtained using the line-source method [10]. One can observe a point of inflection at the glass transition T_g with the data above T_g, displaying a slight upward trend, leveling off at higher temperatures.

Comparing the data with the literature, it is observed that there is a lot of scatter in the data. The reported melt-state data appear to fall into two groups. The data from Lobo and Newman [10] are almost coincident with those of Wynter [12] and close to those of Underwood and Taylor [13], both of whom used line-source methods. They are about 2% higher than those of Lohe [19] and 8.5% higher than that of Fuller and Fricke [16]. All show the same trend, although the data of Fuller and Fricke display a stronger temperature dependence on thermal conductivity. Kline [17] and Pasquino [15] have reported a much lower melt-state thermal conductivity. Figure 9 presents data for a polymethylmethacrylate (PMMA). In comparing trends, the melt-state data of Shoulberg and Shetter [21] display a decreasing trend, in agreement with Hands [22], but contrary to that of Lohe [19] and Lobo [10], where relatively little temperature dependence above T_g are reported. The polycarbonate (Makrolon CD2000, MOBAY) data of Lobo and Newman [10] presented in Fig. 10 show similar behavior. Almost no temperature dependence is observed in the melt state over the temperature range covered.

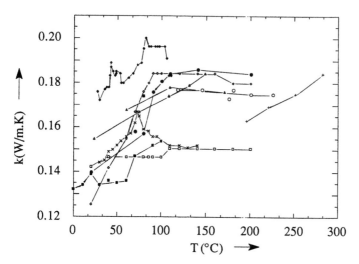

Figure 8 Thermal conductivity of polystyrene. The data of Lobo ●, Kline ■, Fuller +, Hall ▲, Lohe ○, Uberreiter x, Holzmuller ◆, Underwood △ and Wynter ◇ are presented.

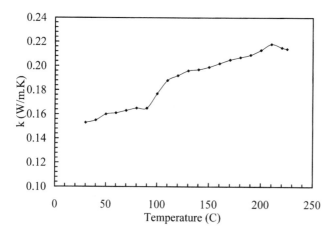

Figure 9 Thermal conductivity of polymethylmethacrylate.

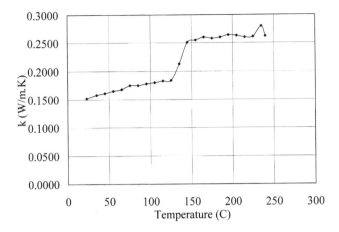

Figure 10 Thermal conductivity of polycarbonate.

All three materials show a sharp transition at their respective glass transition temperatures, with solid-state data being lower than melt-state data. It has been debated whether this transition is real or is an artifact. The transition has been observed by Holzmuller [14], Hattori [24], Kline [17], Lobo [10] and Wynter [12] for polystyrene and by Shulz [25] for glycerol. However, Sandberg et al. [26] suggest that the transition is due to the introduction of thermal contact resistance upon solidification, a factor which is absent in the melt state. In that case, pressurizing the sample during solidification should raise the measured solid-state thermal conductivity by reducing the contact resistance. Such data, however, have not been reported.

Two mechanisms could account for a change in thermal conductivity across the glass transition: free volume and segmental mobility. At the glass transition, there is an increase in free volume that would cause the thermal conductivity to be lowered above the glass transition. At the glass transition, however, the segmental mobility also increases, a mechanism that seeks to raise the thermal conductivity above the glass transition. The line-source data seem to indicate that the segmental mobility mechanism has a greater influence on thermal conductivity than free volume change, which could account for why we observe an increase in conductivity above the glass transition. In any case, the correspondence between the glass transition measured by differential scanning calorimetry (DSC) and by this technique is striking.

Effect of Crystallinity

Data on a polypropylene (#5092, Exxon Chemical) taken using the line-source technique by Lobo and Newman [10] are observed to be relatively constant in the melt state (Fig. 11). The crystallization transition is reproducible and is observed at 130C during the cooling scan. Upon crystallization, the thermal conductivity rises due to the appearance of highly conductive spherulites in the solid phase. Except for the change at this transition, the thermal conductivity remains relatively insensitive to temperature. Upon heating, the melting transition is noted at 170C. DSC runs on this material have shown similar shifts in the transition with heating and cooling scans. Good correlation is reported in the solid state with the data of Tomlinson and Kline [29].

Thermal conductivity of low-density polyethylene (LDPE) is presented in Fig. 12. The solid-state thermal conductivity is observed to decrease with increasing temperature, due mainly to the breakdown of the crystalline structure accompanied by a sharp transition at the melting temperature. Melt-state thermal conductivity is relatively insensitive to temperature. While cooling from the melt, a sharp crystallization transition occurs at 120°C, resulting in an increase in thermal conductivity. Once again, the data show clear differentiation between melting and crystallization temperatures. As before, the transitions are seen to correspond with those observed using DSC. The presence of such a transition is substantiated by most workers [10,22,27,28].

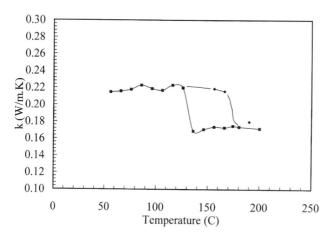

Figure 11   Heating ◆ and cooling ■ thermal conductivity scans for polypropylene.

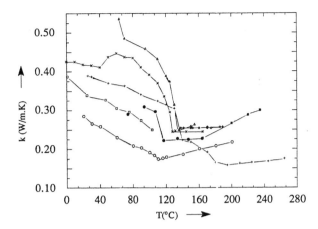

Figure 12 Thermal conductivity of polyethylene. The data of Wynter ○, Underwood □, Lohe ◇, Tomlinson X, Lobo ●, Hansen △, Hands + and Fuller ■ are presented.

Effect of Pressure

Thermal conductivity of polymer melts is seen to increase with pressure. Tests on a number of polymers (Fig. 13) exhibit varying degrees of pressure dependence [10]. The trend is as expected and has been observed by others [19,30]. In general, the increase is gradual. Andersson and Bäckström [31] observed high-pressure transitions for solid polytetrafluoroethylene, but no such transitions were observed for the materials shown in the figure.

Effect of Composition

The presence of fillers will tend to increase the thermal conductivity of polymers. Figure 14 shows the effect of glass fibers on the thermal conductivity of polypropylene as reported by Lobo and Cohen [1]. They noted that the inverse result of mixtures was able to provide a reasonable means to calculate the thermal conductivity:

$$\frac{1}{k} = \frac{\varphi_1}{k_1} + \frac{\varphi_2}{k_2} \tag{5}$$

where

φ = volume fraction

k = thermal conductivity

1 and 2 = continuous and dispersed phases, respectively

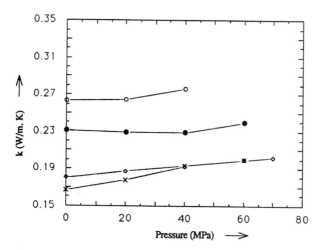

Figure 13 Effect of pressure on thermal conductivity: ○, polycarbonate at 250°C, X; polypropylene at 200°C, ● polyethylene at 160°C (◇) polystyrene at 195°C.

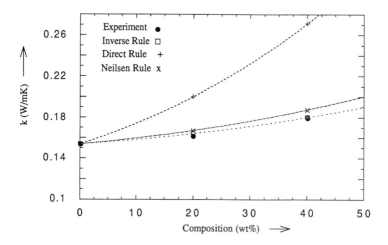

Figure 14 Melt thermal conductivity of polypropylene with glass fiber reinforcement. (From Ref. 1.)

The above equation should be used with caution, however, because it does not account for the quality of interfacial contact between the plastic and the filler system. Poor interfacial contact has the same effect as a thermal contact resistance and can result in a significant lowering in the ability of the highly conducting filler particles to transmit heat to the low-conductivity polymer matrix. What complicates the matter further is that these systems may possess good interfacial contact while the polymer matrix is molten but then become lower in thermal conductivity as interfacial contact resistance develops between the filler and the now-solidified polymer. This can be particularly confusing in the case of some filled semicrystalline polymers, where the appearance of the crystalline phase upon solidification should result in increased thermal conductivity, while the actual value appears to decrease. For this reason, it is considered safer to measure the thermal conductivity of filled materials.

Effect of Chemical Reaction

Thermosetting plastics solidify by chemical reaction between monomer units or by cross-linking between short-chain molecules. The effect of this chemical reaction is to increase the molecular weight with degree of cure. Thermal conductivity measurements made using the line-source technique [32] show an increase in thermal conductivity as the reaction progresses. In this study, measurements were made on an unfilled, slow-curing, low-viscosity liquid bisphenol A epoxy resin system. Isothermal cure data were gathered at three temperatures. It was observed that the thermal conductivity of the uncured resin system was not a strong function of temperature, as evidenced by the close overlap of data at low cure levels (Fig. 15). Conductivity began to increase significantly with the progression of the reaction, leveling off as the reaction ceased. This is in agreement with the observations of Bates [9] and Hansen [28], who note that thermal conductivity is a strong function of molecular weight at low molecular weights but that the effect is less dramatic at high molecular weights. It is interesting to note that the final thermal conductivity of the 70°C isothermal scan was much lower than the 90°C and 110°C isotherms. One possible explanation is that the final degree of cure achieved at 70°C was lower than that which could be achieved at the higher temperature. The thermal conductivity of the cured polymer used in this study was observed to be fairly temperature dependent and is shown in Fig. 16.

THERMAL DIFFUSIVITY

Thermal diffusivity data are still relatively sparse compared to thermal conductivity data. In most cases, this property is calculated using Eq. (4)

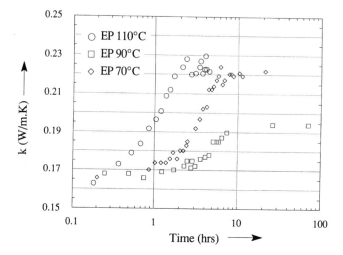

Figure 15 Effect of cure on thermal conductivity under isothermal conditions.

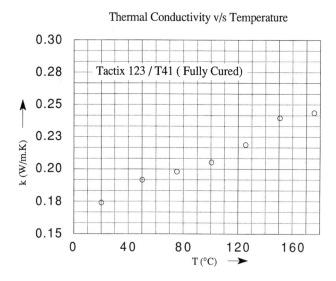

Figure 16 Thermal conductivity of cured epoxy resin.

because the other terms in the formula are much more readily available. It is difficult to use the equation, however, to determine the dependency of thermal diffusivity on parameters such as temperature and composition. This is because density, specific heat, and thermal conductivity are themselves dependent on these parameters. Any error or assumptions made can influence the measured properties. Diffusivity data are compiled for a few polymers in the monograph by Schramm et al. [35]. Additionally, Fig. 17 presents data on a Vespel SP1 (R. Campbell, personal communication, 1999) over the temperature range of 25–300°C. In general, it is observed that thermal diffusivity decreases with increasing temperature. Diffusivity measurements on PTFE indicate an increase in diffusivity with level of crystallinity.

CONCLUSION

Several techniques exist for the measurement of thermal conductivity. The guarded hot-plate method and its variants, the heat flux meter and the comparative instrument, provide an important means to measure thermal conductivity in the solid state. While measurements using the guarded hot-plate may be tedious, absolute determinations may be made using this technique, giving this instrument an important place among the available test methods. These techniques also permit measurements of thermal conductivity of plastics with unique morphological characteristics such as those imparted by processing. Measurements on structured composites are also best performed using these techniques. The line-source technique has been

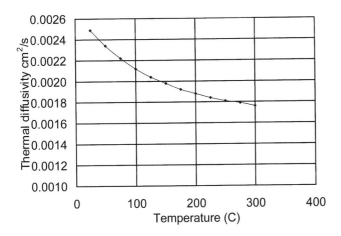

Figure 17 Thermal diffusivity of Vespel SP1.

applied most extensively for measurements on polymers. Its unique ability to generate data in both melt and solid states has enabled the creation of an extensive body of data, from which the following important statements can be made about the thermal conductivity of polymers.

The thermal conductivity of molten polymers is seen to either increase with temperature or show little temperature dependence. In the solid state, the presence of crystallinity will increase the thermal conductivity in a manner proportional to the degree of crystallinity. For semicrystalline materials, an increase in thermal conductivity is observed upon solidification, corresponding to the formation of a crystalline structure as the polymer solidifies. The transition is distinct and corresponds well with that reported using DSC. Hysteretic effects are seen depending on whether measurements are performed in heating or cooling. Thermal conductivity is seen to increase with pressure and with the degree of cure.

Thermal diffusivity data are still not commonly available even though this is the ultimate property used in heat transfer calculations and simulations. The lack of data is attributed to the difficulty in making these measurements over a wide range of test conditions. With the availability of modern refinements to the laser flash method, it is anticipated that this situation will change and that our understanding of this important property will improve.

ACKNOWLEDGMENTS

I thank AC Technology, ANTER Laboratories, and Holometrix Instruments for providing information, figures, and data for this chapter.

REFERENCES

1. H. Lobo and C. Cohen, *Polymer Eng. Sci. 30:65* (1990).
2. H. Lobo and K. K. Wang, U.S. Patent 4,861,167 (1988).
3. F. C. Hooper and F. R. Lepper, *Trans. Am. Soc. Heat. Vent. Eng.* 56:309 (1950).
4. D. D'Eustachio and R. E. Schreiner, *Trans. Am. Soc. Heat. Vent. Eng.* 58:331 (1952).
5. W. M. Underwood and R. B. McTaggart, *Chem. Eng. Prog. Symp. Ser. 56* (30;261 (1960).
6. Parker et al., *J. Appl. Phys. 32*:1679 (1961).
7. Koski, proc. 8th Symposium on Thermophysical Properties, Vol. 2, p. 94 (1981).
8. R. P. Tye, ed., *Thermal Conductivity*, Vol. p. 377.
9. O. K. Bates, *Ind. Eng. Chem. 41* (9): 1966 (1949).
10. H. Lobo and R. Newman, *SPE ANTEC '90 Proceedings*, p. 892, (1990).
11. K. Eiermann, *Kunststoffe 51*:512 (1961).

12. R. C. M. Wynter, M. S. thesis, Mcgill University, Montreal, Canada (1978).
13. W. M. Underwood and J. R. Taylor, *Polymer Eng. Sci. 18*:556 (1978).
14. W. Holzmüller and M. Münx. *Kolloid Z. 159*:25 (1958).
15. A. D. Pasquino and J. Pilsworth, *J. Polymer Sci. B 2*:253 (1964).
16. T. R. Fuller and A. L. Fricke, *J. Appl. Polymer Sci. 15*:1729 (1971).
17. D. E. Kline, *J. Polymer Sci. 50*:441 (1961).
18. K. Uberreiter and S. Nens, *Kolloid Z. 123*:92 (1951).
19. P. Lohe, *Kolloid Z. Z. Polymere 203*:115 (1965).
20. J. A. Hall, W. V. Ceckler, and E. V. Thompson, *J. Appl. Polymer Sci. 33*:2029 (1987).
21. R. H. Shoulberg and J. A. Shetter, *J. Polymer Sci. 6*:532 (1962).
22. D. Hands, *Rubber Chem. Technol. 50*:480 (1977).
23. W. Knappe, *Kunststoffe. 51*:707 (1961).
24. M. Hattori, *Bull. Univ. Osaka Prefect. Ser. A9*:51 (1960).
25. A. K. Shulz, *J. Chim. Phys. 51*:530 (1954).
26. O. Sandberg, P. Anderson, and G. Bäckström, *J. Phys. E 10*:474 (1977).
27. J. N. Tomlinson, D. E. Kline, and J. A. Sauer, *SPE Trans 5:44 (1965)*.
28. D. Hansen and C. C. Ho, *J. Polymer Sci. A3*:659 (1965).
29. J. N. Tomlinson and D. E. Kline, *J. Appl. Polymer Sci 2*:1931, (1967).
30. C. L. Yen, M.S. thesis, National Taiwan Institute of Technology, Taiwan (1988).
31. P. Andersson and G. Bäckström, *Rev. Sci. Instrum. 47*:205 (1976).
32. H. Lobo, *SPE ANTEC Proceedings*, p. 1281 (1991).
33. *K-System User's Manual*, AC Technology, Ithaca, NY, p. 15 (1988).
34. A. R. Challoner and R. W. Powell, *Proc. Roy. Soc. (Lond.) 238A*:90 (1956).
35. R. E. Schramm, A. F. Clark, and R. P. Reed, *A Compilation and Evaluation of Mechanical, Thermal and Electrical Properties of Selected Polymers*, NBS Monograph 132 (1973).

6

Thermomechanical and Dynamic Mechanical Analysis

Kevin P. Menard

*PerkinElmer Thermal Laboratory, Materials Science Department,
University of North Texas, Denton, Texas, USA*

INTRODUCTION

Thermomechanical and dynamic mechanical tests represent what may be the most useful and yet least understood techniques in modern thermal analysis. Three techniques are commonly used to study polymers: (1) thermomechanical analysis (TMA), (2) pressure–volume–temperature (PVT) measurements, and (3) dynamic mechanical analysis (DMA).

TMA is the technique of measuring the dimensional changes in a specimen as a function of time or temperature. One can argue that rheology and traditional mechanical tests should be included in this classification. In its purest form, the changes in a material's dimensions under minimal load are recorded and used as an indicator of the changes in the material's free volume. These data allow the calculation of a material's expansivity or coefficient of thermal expansion as well as detection of transitions in the material. TMA on inorganic glass was the first measurement of the glass transition and it still remains the preferred technique for that measurement in many applications. It is often said to be more sensitive to the glass transition than DSC by an order of magnitude.

Pressure–volume–temperature instruments are designed to probe more deeply into a material's free volume. They can be considered a subset of TMA and are treated here as such. Capable of applying literally tons of force and of reaching high temperatures, they are used to collect information of how pressure and temperature affect the volume of a material and how the transitions in a material shift as a function of pressure.

Dynamic mechanical analysis is the technique of applying a stress or strain to a sample and analyzing the response to obtain phase angle and deformation data. These data allow the calculation of the damping or tan delta (δ) as well as complex modulus and viscosity data. Two approaches are used: (1) forced frequency, in which the signal is applied at a set frequency; and (2) free resonance, in which the material is perturbed and allowed to exhibit free resonance decay. Most DMAs are the former type, while the torsional braid analyzer (TBA) is the latter. In both approaches the technique is very sensitive to the motions of the polymer chains, and they are powerful tools for measuring transitions in polymers. DMA is estimated to be 100 times more sensitive to the glass transition than differential scanning calorimetry (DSC), and it resolves other, more localized transitions not detected by DSC. In addition, the technique allows the rapid scanning of a material's modulus and viscosity as a function of temperature or of frequency.

THEORY AND OPERATION OF THERMOMECHANICAL ANALYSIS

The basis of TMA is the change in the dimensions of a sample as a function of temperature; a simple way of looking at a TMA is as a very sensitive micrometer. TMA is believed to have developed from hardness or penetration tests and was used on polymers in 1948 [1]. Subsequently, it has developed into a powerful tool in the analytical laboratory. TMA measurements record changes caused by changes in the free volume of a polymer [2]. While the latter tends to be preferred by engineers and rheologists, in contrast to chemist and polymer physicists who lean toward the former, both descriptions are equivalent in explaining behavior. Changes in free volume, v^f, can be monitored as a volumetric change in the polymer; by the absorption or release of heat associated with that change; the loss of stiffness; increased flow; or by the change in relaxation time. The free volume of a polymer, v^f, is known to be related to viscoelasticity [3], aging [4], penetration by solvents [5], and impact properties [6]. Defined as the space a molecule has for internal movement, it is shown schematically in Fig. 1.

The glass-transition-temperature, T_g, in a polymer corresponds to the expansion of the free volume, allowing greater chain mobility above this

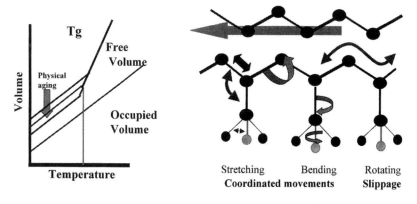

(a) Free Volume   **(b) Crankshaft Model**

Figure 1   Free volume, v^f, in polymers: (a) the relationship of free volume to transitions, and (b) a schematic example of free volume and the crankshaft model. Below the T_g in (a), various paths with different free volumes exist depending on heat history and processing of the polymer, where the path with the least free volume is the most relaxed. (b) shows the various motions of a polymer chain. Unless enough free volume exists, the motions cannot occur. (From Ref. 20a.)

transition (Fig. 2). Seen as an inflection or bend in the thermal expansion curve, this change in the TMA can be seen to cover a range of temperatures, of which the T_g is an indicator calculated by an agreed-upon method (Fig. 3). This fact seems to be forgotten by inexperienced users, who often worry why perfect agreement is not seen in the value of the T_g when comparing different methods. The width of the T_g can be as important an indicator of changes in the material as the actual temperature.

Experimentally, a TMA consists of an analytical train that allows precise measurment of position and can be calibrated against known standards. A temperature control system of a furnace, heat sink, and temperature-measuring device (most commonly a thermocouple) surrounds the samples. Fixtures to hold the sample during the run are normally made out of quartz because of its low coefficient of thermal expansion (CTE), although ceramics and invar steels may also be used. Fixtures are commercially available for expansion, three-point bending or flexure, parallel-plate, and penetration tests (Fig. 4).

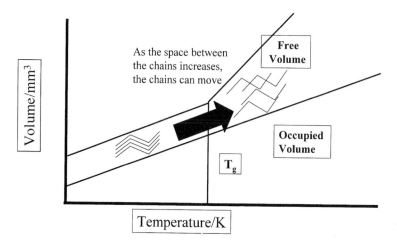

Figure 2 The increase in free volume is caused by increased energy absorbed in the chains and this increased free volume permits the various types of chain movement to occur.

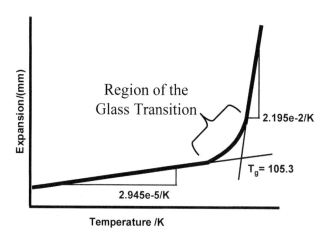

Figure 3 The T_g is the region shown here between the points where the tangents depart from the curve. The T_g, by convention, is taken as the intersection of those two tangents.

(a)

Figure 4 A PerkinElmer TMA 7 is shown in (a) and as a schematic in (b). (Used with the permission of PerkinElmer Instruments LLC, Shelton, Conn.). Various test geometries are shown in (c). Normally these are made in quartz glass to take advantage of its low CTE values.

APPLICATIONS OF THERMOMECHANICAL ANALYSIS

TMA applications are in many ways the simplest of the thermal techniques. We are just measuring the change in a the size or postion of the a sample. However, they are also increasingly important in supplying information needed to design and process everything from chips to food products to engines. Because of the sensitivity of modern TMA, it is often used to measure T_g values that are difficult to obtain by DSC, for example, those of highly cross-linked thermosets.

Expansion and CTE

TMA allows the calculation of the thermal expansivity [7] from the same data set as used to calculate the T_g. Since many materials are used in contact with a dissimilar material in the final product, knowing the rate and amount

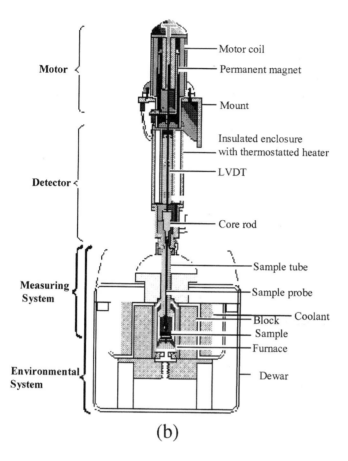

(b)

Figure 4 Continued

of thermal expansion helps design around mismatches that can cause failure of the final product. These data are available only when the T_g is collected by thermal expansion, not by the flexure or penetration method. Different T_g values will be seen for each mode of testing [8] (Fig. 5). The coefficient of thermal expansion (CTE) is calculated by

$$\alpha = \frac{1\Delta L}{L\Delta T}$$

where L is the sample length and ΔT is the temperature range.

Once this value is obtained, it can be compared to that of other materials used in the same product. Large differences in the CTE can lead to

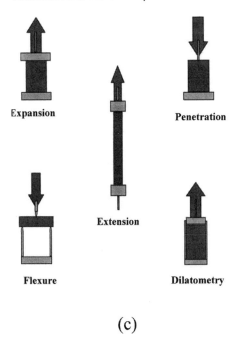

Expansion

Penetration

Extension

Flexure

Dilatometry

(c)

Figure 4   Continued

motors binding, solder joints failing, composites splitting on bond lines, or internal stress buildup.

If the material is heterogeneous or anisotropic , it will have different thermal expansions depending on the direction in which they have been measured. For example, a composite of graphite fibers and epoxy will show three distinct thermal expansions corresponding to the x, y, and z directions. Blends of liquid crystals and polyesters show a significant enough difference between directions that the orientation of the crystals can be determined by TMA [9]. Similarly, oriented fibers and films have different thermal expansivities in the direction of orientation than in the unoriented direction. This is normally addressed by recording the CTE in the x, y, and z directions (Fig. 6a).

Dilatometry and Bulk Measurements

Another approach to anisotropic materials is to measure the bulk expansion of the material using dilatometry (Fig. 6b). The technique itself is fairly old. It was used extensively to study initial rates of reaction for bulk styrene

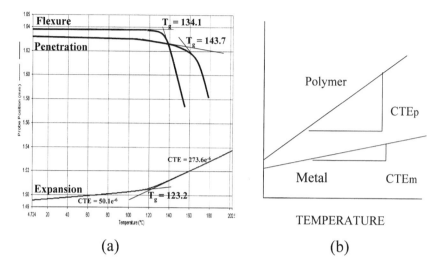

(a)   (b)

Figure 5   Different methods of measuring the T_g in the TMA give different values as shown in (a); an overlay of the penetration, flexure, and expansion runs above. In (b) we see the comparison of a polymer to a mtal CTE run.

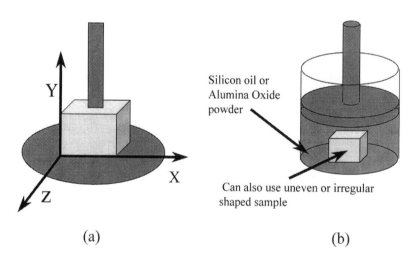

(a)   (b)

Figure 6   Heterogeneous samples require the CTE to be determined in the x, y, and z planes (a) or in bulk to obtain a volumetric expansion in the dilatometer (b).

polymerization in the 1940s [10], an experiment which the author has used in his thermal analysis class on the TMA. By immersing the sample in a fluid (normally silicon oil) or powder (normally Al_2O_3) in the dilatometer, the expansions in all directions are converted to vertical movements, which are measured by the TMA. This technique has enjoyed a renaissance in the last few years because modern TMAs make it easier to perform than previously. It has been particularly useful for studying the contraction of a thermoset during its cure [11,12]. The technique itself is rather simple: a sample is immersed in either a fluid such as silicon oil or buried in alumina oxide in the dilatometry and run through the temperature cycle. If a pure liquid or a monomer is used, the dilatomer is filled with that liquid instead of the silicon oil or alumina oxide.

3.3 *PVT* Relationship Studies

The temperatures of a polymer's melting, glass transition, crystallization, and its solid-state annealing are all known to have pressure dependencies [13]. High-pressure instruments, such as the Gnomix, have been developed to study pressure–volume–temperature (*PVT*) relationships in polymers. Figure 7 is a schematic of the Gnomix instrument. In these experiments, the sample is placed in an incompressible fluid and then the desired pressure is applied. Full details of this technique as well as a collection of *PVT* relationships for a wide range of polymers up to 200 MPa (~30,000 psi) and 400°C have recently been published [14]. These data have been mainly collected isothermally to report the affect of pressure and temperature on the volume of the polymer and to monitor the respective changes in melt and glass transitions. For example, data on polymer liquid crystals have been obtained as a function of the concentration of the liquid crystal (rigid) constituent in a series of copolymers [15].

The measurement of the volume of the polymer above and below the glass transition under high pressure is an attempt to determine the occupied volume of the polymer. As the pressure is increased, the glass transition, which occurs in the free volume, becomes generally greater and broader (Fig. 8) as the material changes between different types of glasses before the T_g is reached. This pressure sensitivity can be useful information for determining processing conditions. While the Gnomix can apply pressures that greatly exceed any high-pressure DSC, the experimental times are long and DSC may be more useful for studies up to ~6.8 MPa (~1000 psi). The agreement between these techniques is quite good [15] and hence high-pressure DSC, within its limits, is often a suitable alternative to a PVT instrument.

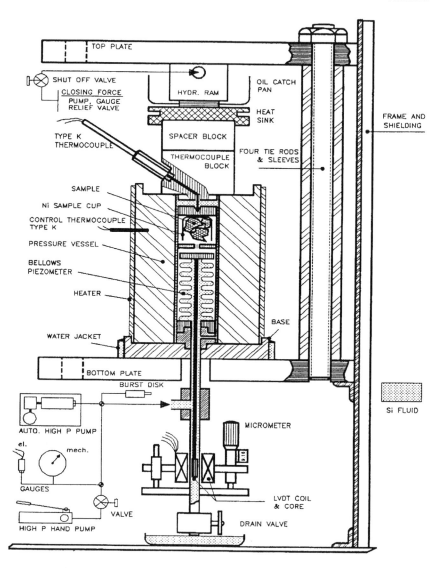

Figure 7 Schematic of the Gnomix, a commercial PVT instrument. (From Ref. 14. Reprinted with permission of the Technomics Publishing Company.)

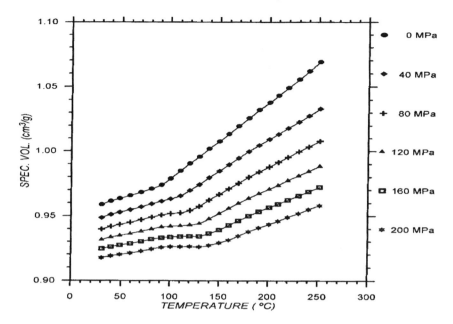

Figure 8 The effect of pressure of the T_g of a polystyrene (M ≈ 1.1e5, D ≈ 1.06). (From Ref. 14. Reprinted with permission of the Technomics Publishing Company.)

Mechanical Tests

A wide variety of tests are performed in the TMA that are adapted from physical tests that were used before the instrument became commonly available [16]. Examples of these were already mentioned above where the flexure and penetration tests for T_g were compared to expansion data (Fig. 5). Other tests may also be modeled or mimicked in the TMA, such as heat distortion (Fig. 9) and softening point [16]. Methods to obtain the modulus [16], compressive viscosity [17], and penetrative viscosity [18] have also been developed. Many of these methods, such as ASTM D648 for example, specify the stress to which the sample needs to be exposed during the run. In ASTM D684, a sample is tested at 66 and 264 psi. TMAs on the market today include software that allows them to generate stress–strain curves and to run creep-recovery experiments [19].

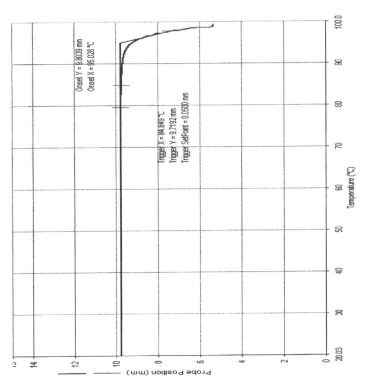

(b)

(a)

Figure 9 Heat distortion tests, which look for sample flexure while the sample is immersed in oil, can be run in the TMA. (a) A TMA 7 with a furnace liner. (b) Results of the heat distortion test runs.

THEORY AND OPERATION OF DYNAMIC MECHANICAL ANALYSIS

Dynamical mechanical analysis is a very important tool in the modern polymer laboratory. Despite that, only a few books have concentrated on the technique [20]. The first attempts to do oscillatory experiments to measure the elasticity of a material were made by Poynting [21] in 1909. Other early work gave methods to apply oscillatory deformations by various means to study metals [22], and many early experimental techniques were reviewed by te Nijenhuis [23] in 1978. Miller's book on polymer properties [24] referred to dynamic measurements in this early discussion of molecular structure and stiffness. Early commercial instruments included the Weissenberg rheogoniometer (~1950) [25] and the Rheovibron (~1958) By the time Ferry wrote *Viscoelastic Properties of Polymers* [26] in 1961, dynamic measurements were an integral part of polymer science, and his is still the best development of the theory available. In 1967, McCrum et al. [27] collected the current information on DMA and dielectric analysis (DEA) into his landmark textbook. About 1966, J. Gillham developed the torsional braid analyzer [28] and started the modern period of DMA. In 1971, J. Starita and C. Macosko [29] built a DMA that also measured normal forces. In the late 1970s, Murayani [30] and Read [31] wrote books on the uses of DMA for material characterization. Several thermal and rheological instrument companies introduced DMAs in the same time period, and currently most thermal and rheological vendors offer some type of DMA. The revolution in computer technology, which has affected all parts of the laboratory, has caused DMAs of all types to become more user friendly and made their use in polymer analysis more of a routine task.

Forced Frequency Analyzers

If a constant load applied to a sample begins to oscillate sinusoidally (Fig. 10a), the sample will deform sinusoidally. This motion will be reproducible if the material is deformed within its linear viscoelastic region. For any one point on the curve, the stress applied is described in Eq. (1):

$$\sigma = \sigma_o \sin \omega t \tag{1}$$

where σ is the stress at time t, σ_o is the maximum stress, ω is the frequency of oscillation, and t is the time. The resulting strain wave shape will depend on how much viscous behavior the sample has as well as how much elastic behavior it has. In addition, the rate of stress can be determined by taking the derivative of the above equation in terms of time:

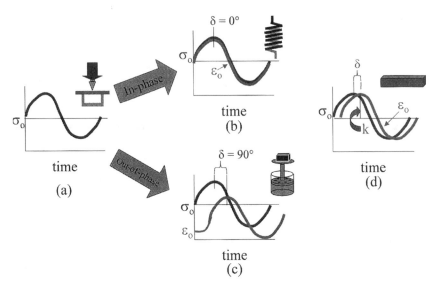

Figure 10   (a) When a sample is subjected to a sinusoidal oscillating stress, it responds in a similar strain wave provided the material stays within its elastic limits. When the material responds to the applied wave perfectly elastically, an in-phase, storage, or elastic response is seen (b), while a viscous response gives an out-of-phase, loss, or viscous response (c). Viscoelastic materials fall in between these two extremes as shown in (d). For the real sample in (d), the phase angle, δ, and the amplitude at peak, k, are the values used for the calculation of modulus, viscosity, damping, and other properties.

$$\frac{d\sigma}{dt} = \omega\sigma_o \cos \omega t \tag{2}$$

The two extremes of the material's behavior, elastic and viscous, provide the limiting extremes that sum to give the strain wave. The behavior can be understood by evaluating each of the two extremes. The material at the springlike or Hookean limit will respond elastically with the oscillating stress. The strain at any time can be written as

$$\varepsilon(t)E\sigma_o \sin(\omega t) \tag{3}$$

where $\varepsilon(t)$ is the strain at any time t, E is the modulus, σ_o is the maximum stress at the peak of the sine wave, and ω is the frequency. Since in the linear region σ and ε are linearly related by E, the relationship is

$$\varepsilon(t)\varepsilon_o \sin(\omega t) \tag{4}$$

where ε_o is the strain at the maximum stress. This curve shown in Fig. 10b has no phase lag (or no time difference from the stress curve) and is called the in-phase portion of the curve.

The viscous limit is expressed as the stress proportional to the strain rate, which is the first derivative of the strain. This is best modeled by a dashpot and for that element, the term for the viscous response in terms of strain rate is described as

$$\varepsilon(t)\frac{\eta d\sigma_o}{dt} = \eta\omega\sigma_o\cos(\omega t) \tag{5}$$

or

$$\varepsilon(t)\eta\omega\sigma_o\sin(+\pi/2) \tag{6}$$

where the terms are as above and η is the viscosity. Substituting terms as above makes the equation

$$\varepsilon(t)\omega\sigma_o\cos(\omega t) = \omega\sigma_o\sin\left(\omega t + \frac{\pi}{2}\right) \tag{7}$$

This curve is shown in Fig. 10c.

Now, take the behavior of the material that lies between these two limits. The difference between the applied stress and the resultant strain is an angle, δ, and this must be added to the equation. So the elastic response at any time can now be written as

$$\varepsilon(t)\varepsilon_o\sin(\omega t + \delta) \tag{8}$$

Using trigonometry this can be rewritten as

$$\varepsilon(t)\varepsilon_o[\sin(\omega t)\cos\delta + \cos(\omega t)\sin\delta] \tag{9}$$

This equation, corresponding to the curve in Fig. 10d, can be separated into the in-phase and out-of-phase strains that corresponds to curves like those in Figs. 10b and 10c, respectively. These are the in- and out-of-phase moduli and are

$$\varepsilon'\varepsilon_o\sin(\delta) \tag{10}$$

$$\varepsilon''\varepsilon_o\cos(\delta) \tag{11}$$

and the vector sum of these two components gives the overall or complex strain on the sample:

$$\varepsilon* = \varepsilon' + i\varepsilon'' \tag{12}$$

Free-Resonance Analyzers

If a suspended sample is allowed to swing freely, it will oscillate like a harp string as the oscillations gradually come to a stop. The naturally occurring damping of the material controls the decay of the oscillations. This produces a wave, shown in Fig. 11, which is a series of sine waves that decrease in amplitude and frequency. Several methods exist to analyze these waves and are covered in the review by Gillham [32]. These methods have also been successfully applied to the recovery portion of a creep-recovery curve, where the sample goes into free resonance upon removal of the creep force [33]. From the decay curve, the period, T, and the logarithmic decrement, Λ, can be calculated. Both manual and digital processing methods have been reported. [32,34] Fuller details may be found in McCrum et al. [34] and Gillham [32]. The decay of the amplitude is evaluated over as many swings as possible to reduce error:

$$\Lambda = \frac{1}{j} \ln\left(\frac{A_n}{A_{(n+j)}}\right) \tag{13}$$

where j is the number of swings and A_n is the amplitude of the nth swing. For one swing, where $j = 1$, the equation becomes

$$\Lambda = \ln\left(\frac{A_n}{A_{(n+j)}}\right) \tag{14}$$

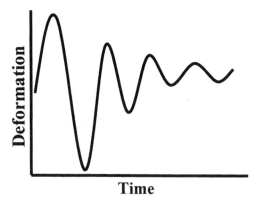

Figure 11 The decay wave from a free-resonance analyzer show the decreasing amplitude of signal with time. (From Ref. 20a.)

If for a low value of Λ where A_n/A_{n+1} is approximately 1, the equation can be rewritten as

$$\Lambda \approx \frac{1}{2}\left(\frac{A_n^2 - A_{n+1}^2}{A_n^2}\right) \tag{15}$$

From this, since the square of the amplitude is proportional to the stored energy, $\Delta W / W_{st}$, and the stored energy can be expressed as $2\pi \tan \delta$, the equation becomes

$$\Lambda \approx \frac{1}{2}\left(\frac{\Delta W}{W_{st}}\right) = \pi \tan \delta \tag{16}$$

which gives us the phase angle, δ. The time of the oscillations, the period, T, can be found using the following equation:

$$T = 2\pi \sqrt{\frac{M}{\Gamma_1}} \sqrt{\frac{1 + \Lambda^2}{4\pi^2}} \tag{17}$$

where Γ_1 is the torque for one cycle and M is the moment of inertia around the central axis. Alternatively, T can be calculated directly from the plotted decay curve as

$$T = \frac{2}{n}(t_n - t_0) \tag{18}$$

where n is the number of cycles and t is time. From this, the shear modulus, G, can be calculated, which for a rod of length L and radius r is

$$G = \left(\frac{4\pi^2 ML}{NT^2}\right)\left(1 + \frac{\Lambda^2}{4\pi^2}\right) - \left(\frac{mgr}{12N}\right) \tag{19}$$

where m is the mass of the sample, g is the gravitational constant, and N is a geometric factor. In the same system, the storage modulus, G', can be calculated as

$$G' = \frac{1}{T^2}\left(\frac{8\pi ML}{r^4}\right) \tag{20}$$

where I is the moment of inertia for the system. Having the storage modulus and the tangent of the phase angle, the remaining dynamic properties can be calculated.

Free-resonance analyzers normally are limited to rod or rectangular samples or materials that can be impregnated onto a braid. This last approach is how the curing studies on epoxy and other resin systems are done in torsion and gives these instruments the name torsional braid analyzers (TBA).

Instrumentation

One of the most important choices made in selecting a DMA is to decide whether to chose stress (force) or strain (displacement) control for applying the deforming load to the sample. Strain-controlled analyzers, whether for simple static testing or for DMA, move the probe a set distance and use a force balance transducer or load cell to measure the stress. These parts are typically located on different shafts. The simplest version of this is a screw-driven tester, in which the sample is pulled one turn. This requires very large motors so the available force always exceeds what is needed. These testers normally have better short-time response for low viscosity materials and can normally perform stress relaxation experiments easily. They also usually can measure normal forces if they are arranged in torsion. A major disadvantage is that their transducers may drift at long times or with low signals.

Stress-controlled analyzers are cheaper to make because there is only one shaft, but they are somewhat trickier to use. Many of the difficulties have been alleviated by software, and many strain-controlled analyzers on the market are really stress-controlled instruments with feedback loops that make them act as if they were strain controlled. In stress control, a set force is applied to the sample. As temperature, time, or frequency varies, the applied force remains the same. This may or may not be the same stress: in extension, for example, the stretching and necking of a sample will change the applied stress seen during the run. However, this constant stress is a more natural situation in many cases and it may be more sensitive to material changes. Good low force control means less likelihood of destroying any structure in the sample. Long relaxation times or long creep studies are more easily preformed on these instruments. Their biggest disadvantage is that their short-time responses are limited by inertia with low-viscosity samples.

Since most DMA experiments are run at very low strains ($\sim$0.5% maximum) to stay well within a polymers' linear region, it has been reported that both types of analyzers give the same results. However, when one gets to the nonlinear region, the difference becomes significant, as stress and strain are no longer linearly related. Stress control can be said to duplicate real-life conditions more accurately, since most applications of polymers involve resisting a load.

DMA analyzers are normally built to apply the stress or strain in one of two ways (Fig. 12). One can apply force in a twisting motion so one is testing the sample in torsion. This type of instrument is the dynamic analog of the constant-shear spinning-disk rheometers. They are used mainly for liquids

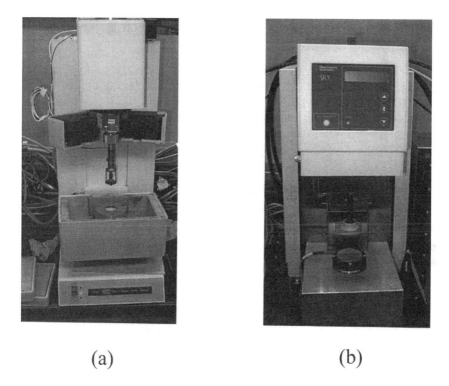

(a) (b)

Figure 12 Torsion versus axial analyzers: The PerkinElmer DMA 7e (a) is am axial analyzer, while the Rheometric Sciences SR-5 (b) is a torsional instrument. Both are controlled stress but can act as strain controlled because of the feedback loop programmed in.

and melts, but solid samples may also tested by twisting a bar of the material. Torsional analyzers normally also permit continuous shear and normal force measurements. Most of these analyzers can also do creep-recovery, stress-relaxation, and stress–strain experiments. Axial analyzers are normally designed for solid and semisolid materials and apply a linear force to the sample. These analyzers are usually associated with flexure, tensile, and compression testing, but they can be adapted to do shear and liquid specimens by proper choice of fixtures. Sometimes the instrument's design makes this inadvisable, however. (For example, working with a very fluid material in a system in which the motor is underneath the sample has the potential for damage to the instrument if the sample spills into the motor.) These analyzers can normally test higher-modulus materials than torsional analyzers and can run TMA studies in addition to creep-recovery, stress-relaxation, and stress–strain experiments.

There is really considerable overlap between the type of samples run by axial and torsional instruments. With the proper choice of sample geometry and good fixtures, both types can handle similar samples, as shown by the extensive use of both types to study the curing of neat resins. Normally, axial analyzers cannot handle fluid samples below about 500 Pa·s.

APPLICATIONS

Thermoplastic Solids and Cured Thermosets

As mentioned above, the thermal transitions in polymers can be described in terms of either free-volume changes [35] or relaxation times. A simple approach to looking at free volume, which is popular in explaining DMA responses, is the crankshaft mechanism [27], where the molecule is imagined as a series of jointed segments. From this model, it is possible to describe simply the various transitions seen in a polymer. Other models exist that allow for more precision in describing behavior; the best seems to be the Doi-Edwards model. [36] Aklonis and Knight [37] give a good summary of the available models, as does Rohn [38].

The crankshaft model treats the molecule as a collection of mobile segments that have some degree of free movement. This is a very simplistic approach, yet it is very useful for explaining behavior (Fig. 13). As the free volume of the chain segment increases, its ability to move in various directions also increases. This increased mobility in either side chains or small groups of adjacent backbone atoms results in a greater compliance (lower modulus) of the molecule. These movements have been studied, and Heijboer [39] classified β and γ transitions by their type of motions. The specific temperature and frequency of this softening help drive the end use of the material.

Moving from very low temperature, at which the molecule is tightly compressed, to higher temperatures, the first changes are solid-state transitions. This process is shown in Fig. 14. As the material warms and expands, the free volume increases so that localized bond movements (bending and stretching) and side-chain movements can occur. This is the gamma transition, $T\gamma$, which may also involve association with water [27]. As the temperature and the free volume continue to increase, the whole side chains and localized groups of 4–8 backbone atoms begin to have enough space to move and the material starts to develop some toughness [40]. This transition, called the beta transition, T_β, is not as clearly defined as described here. Often it is the T_g of a secondary component in a blend or of a specific block

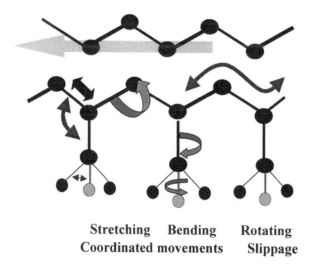

Stretching   Bending   Rotating
Coordinated movements   Slippage

Figure 13 The crankshaft mechanism is a simple way of considering the motions of a polymer chain permitted by increases in free volume. The molecule is visualized as a series of balls and rods, and these move as the free volume increases.

in a block copolymer. However, a correlation with toughness is seen empirically [41].

As heating continues, the T_g or glass transition appears when the chains in the amorphous regions begin to coordinate large-scale motions. One classical description of this region is that the amorphous regions have begun to melt. Since the T_g occurs only in amorphous material, in a 100% crystalline material there will not be a T_g. Continued heating drives the material through the T_α^* and T_{ll}. The former occurs in crystalline or semicrystalline polymers and is a slippage of the crystallites past each other. The latter is a movement of coordinated segments in the amorphous phase that relates to reduced viscosity. These two transitions are not universally accepted. Finally, the melt is reached, where large-scale chain slippage occurs and the material flows. This is the melting temperature, T_m. For a cured thermoset, nothing happens after the T_g until the sample begins to burn and degrade, because the cross-links prevent the chains from slipping past each other.

This quick overview provides an idea of how an idealized polymer responds. Now a more detailed description of these transitions can be provided, with some examples of their applications. The best general collection of this information is still McCrum's 1967 text [27].

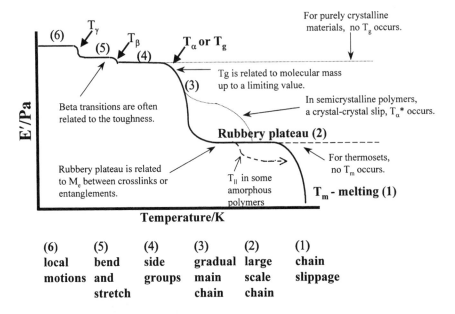

Figure 14 Idealized temperature scan of a polymer. Starting at low temperature, the modulus decreases as the molecules gain more free volume, resulting in more molecular motion. This shows the main curve divided into six regions which correspond to: local motions (6), bond bending and stretching (5), movements in the side chain or adjacent atoms in the main chain (4), the region of the T_g (3), coordinated movements in the amorphous portion of the chain (2), and the melting region (1). Transitions are marked as described in the text. (From Ref. 20a.)

Sub-T_g Transitions

The area of sub-T_g or higher-order transitions has been heavily studied [42], as these transitions have been associated with mechanical properties. These transitions can sometimes be seen by DSC and TMA, but they are normally too weak or too broad for determination by these methods. DMA, DEA, and similar techniques are usually required [43]. Some authors have also called these types of transitions [44] second-order transitions to differentiate them from the primary transitions of T_m and T_g, which involve large sections of the main chains. Boyer reviewed the T_β in 1968 [45] and pointed out that while a correlation often exists, the T_β is not always an indicator of toughness. Bershtein [46] reported that this transition can be considered the "activation barrier" for solid-phase reactions, deformation, flow or creep, acoustic damping, physical aging changes, and gas diffusion into polymers, as the activation energies for the transition and these processes are usually similar. The strength of these transitions is related to how strongly a polymer responses to those processes. These sub-T_g transitions are associated with the material properties in the glassy state. In paints, for example, peel strength (adhesion) can be estimated from the strength and frequency dependence of the subambient β transition [47]. For example, nylon 6,6 shows a decreasing toughness, measured as impact resistance, with declining area under the T_β peak in the tan δ curve. It has been shown, particularly in cured thermosets, that increased freedom of movement in side chains increases the strength of the transition. Cheng [48] reports in rigid-rod polyimides that the β transition is caused by the noncoordinated movement of the diamine groups, although the link to physical properties was not investigated. Johari has reported in both mechanical [49] and dielectric studies [50] that both the β and γ transitions in bisphenol-A-based thermosets depend on the side chains and unreacted ends, and that both are affected by physical aging and postcure. Nelson [51] has reported that these transitions can be related to vibration damping. This is also true for acoustical damping [52]. In both of this cases, the strength of the β transition is taken as a measurement of how effectively a polymer will absorb vibrations. There is a frequency dependence in this transitions, and this is discussed below.

Boyer [53] and Heijober [39] showed that this information needs to be considered with care, as not all β transitions correlate with toughness or other properties. This can be due to misidentification of the transition or because the transition does sufficiently disperse energy. A working rule of thumb [54] is that the β transition must be related to either localized movement in the main chain or very large side-chain movement to absorb enough energy. The relationship of large side chain movement and toughness has been extensively studied in polycarbonate by Yee [55] as well as in many other tough glassy polymers [56].

Less use is made of the T_γ transitions, and they are studied mainly to understand the movements occurring in polymers. Wendorff [57] reports that this transition in polyarylates is limited to inter- and intramolecular motions within the scale of a single repeat unit. Both McCrum [34] and Boyd [58] similarly limit the T_γ and T_δ to very small motions either within the molecule or with bound water. The use of what is called 2D-IR, which couples at Fourier transform infrared (FTIR) spectroscopy and a DMA to study these motions, is a topic of current interest [59].

The Glass Transition (T_g or T_α)

As the free volume continues to increase with increasing temperature, the glass transition, T_g, occurs where large segments of the chain start moving.

This transition is also called the alpha transition, T_α. The T_g is very dependent on the degree of polymerization up to a value known as the critical T_g or the critical molecular weight. Above this value, the T_g typically becomes independent of molecular weight [60]. The T_g represents a major transition for many polymers, as physical properties changes drastically as the material goes from a hard glassy to a rubbery state. It defines one end of the temperature range over which the polymer can be used, often called the operating range of the polymer. When strength and stiffness are needed, it is normally the upper limit for use. In rubbers and some semicrystalline materials such as polyethylene and polypropylene, it is the lower operating temperature. Changes in the temperature of the T_g are commonly used to monitor changes in the polymer such as plasticizing by environmental solvents and increased cross-linking from thermal or UV aging.

The T_g of cured materials or thin coatings is often difficult to measure by other methods, and more often than not the initial cost justification for a DMA is measuring a hard-to-find T_g. While estimates of the relative sensitivity of DMA to DSC or DTA vary, it appears that DMA is 10 to 100 times more sensitive to changes occurring at the T_g. The T_g in highly cross-linked materials can easily be seen long after the T_g has become too flat and broad to be seen in the DSC. This is also a problem with certain materials such as medical-grade urethanes and very highly crystalline polyethylenes.

The method of determining the T_g by DMA can be a manner for disagreement, as at least five ways are in current use (Fig. 15). Depending on the industry standards or background of the operator, the peak or onset of the tan δ curve, the onset of the E' drop, or the onset or peak of the E'' curve may be used. The values obtained from these methods can differ by up to 25°C from each other for the same run. In addition, a 10–20°C difference from the DSC is also seen in many materials. In practice, it is important to specify exactly how the T_g should be determined. For DMA, this means defining the heating rate, applied stresses (or strains), the frequency used, and the method of determining the T_g: for example, the sample will be run at 10°C/min under 0.05% strain at 1 Hz in nitrogen purge (20 cm^3/min) and the T_g determined from peak of the tan δ curve.

It is not unusual to see a peak or hump on the storage modulus directly preceding the drop that corresponds to the T_g. This is also seen in the DSC and DTA and corresponds to a rearrangement in the material to relieve stresses frozen in below the T_g by the processing method. These stresses are trapped in the material until enough mobility is obtained at the T_g to allow the chains to move to a lower energy state. Often a material will be annealed by heating it above the T_g and slowly cooling it to remove this affect. For similar reasons, some experimenters will run a material twice or use a heat–cool–heat cycle to eliminate processing effects.

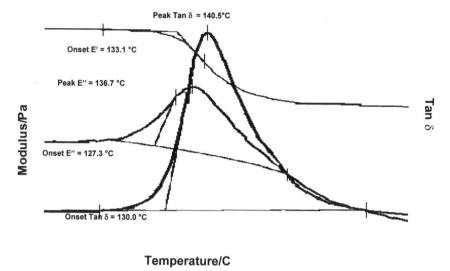

Peak Tan δ = 140.5°C

Onset E' = 133.1 °C

Peak E″ = 136.7 °C

Modulus/Pa

Tan δ

Onset E″ = 127.3 °C

Onset Tan δ = 130.0 °C

Temperature/C

Figure 15 Methods of determining the T_g are shown for the DMA. The temperature of the T_g varies by as much as 10°C in this example depending on the value chosen. Differences as great as 25°C have been reported. (From Ref. 20a.)

The Rubbery Plateau, T_α^* and T_{ll}

The area above the T_g and below the melt is known as the rubbery plateau, and its length as well as its viscosity are dependent on the molecular weight between entanglements (M_e) [61] or cross-links. The molecular weight between entanglements is normally calculated during a stress-relaxation experiment, but similar behavior is observed in the DMA. The modulus in the plateau region is proportional to either the number of cross links or the chain length between entanglements. This is often expressed in shear as

$$G' \simeq \frac{(\rho RT)}{M_e} \tag{21}$$

where G' is the shear storage modulus of the plateau region at a specific temperature, ρ is the polymer density, and M_e is the molecular weight between entanglements. In practice, the relative modulus of the plateau region shows the relative changes in M_e or the number of crosslinks compared to a standard material.

The rubbery plateau is also related to the degree of crystallinity in a material, although DSC is a better method for characterizing crystallinity than DMA. [62]. Also, as in the DSC, there is evidence of cold crystallization in the temperature range above the T_g (Fig, 16). That is one of several transitions that can be seen in the rubbery plateau region. This crystallization occurs when the polymer chains have been quenched (quickly cooled) into a highly disordered state. Upon heating above the T_g, these chains gain enough mobility to rearrange into crystallites, which causes a sometimes-dramatic increase in modulus. DSC or its temperature-modulated variant, StepScan^{TM} differential scanning calorimetry, can be used to confirm this [63]. The alpha star transition, T_α^*, the liquid–liquid transition, T_{ll}, the heat-set temperature, and the cold crystallization peak are all transitions that can appear on the rubbery plateau. In some crystalline and semicrystalline polymers, a transition is seen called the T_α^* [64]. This transition is associated with the slippage between crystallites and helps extend the operating range of a material above the T_g. This transition is very susceptible to processing induced changes and can be enlarged on decreased by the applied heat history, processing conditions, and physical aging [65]. Hence, the T_α^* has been used by fiber manufacturers to optimize properties in their materials.

In amorphous polymers, the T_{ll} transition is seen instead, this is a liquid–liquid transition associated with increased chain mobility and seg-ment-segment associations. [66]. This order is lost when the T_{ll} is exceeded and regained upon cooling from the melt. Boyer [67] reports that, like the

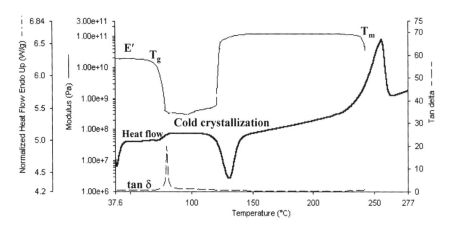

Figure 16 Cold crystallization in PET caused a large increase in the storage modulus, E', above the T_g. A DSC scan of the same material is included. (From Ref. 20a.)

T_g, the appearance of the T_{ll} is affected by the heat history. The T_{ll} is also dependent on the number-average molecular weight, M_n, but not on the weight-average molecular weight, M_w. Bershtein [68] suggests that this may be considered as quasi-melting upon heating or the formation of stable associates of segments upon cooling. While this transition is reversible, it is not always easy to see, and Boyer [69] spent many years trying to prove it was real. It is still not totally accepted. Following this transition, a material enters the terminal or melting region.

Depending on its strength, the heat-set temperature can also be seen in the DMA. While it is normally seen in a TMA experiment, it will sometimes appear as either a sharp drop in storage modulus (E') or an abrupt change in probe position. Heat set is the temperature at which some strain or distortion is induced into polymeric fibers to change its properties, such as to prevent a nylon rug from feeling like fishing line. Since heating above this temperature will erase the texture, and polyesters must be heated above the T_g to dye them, it is of critical importance to the fabric industry. Many final properties of polymeric products depend on changes induced in processing [70].

The Terminal Region

With continued heating, the melting point, T_m, is reached. The melting point is where the free volume has increased so the chains can slide past each other and the material flows. This is also called the terminal region. In the molten state, this ability to flow is dependent on the molecular weight of the polymer (Fig. 14). The melt of a polymer material will often show changes in temperature of melting, width of the melting peak, and enthalpy as the material changes [71], resulting from changes in the polymer molecular weight and crystallinity.

Degradation, polymer structure, and environmental effects all influence what changes occur. Polymers that degrade by cross-linking will look very different from those that exhibit chain scission. Very highly cross-linked polymers will not melt, as they are unable to flow.

The study of polymer melts and especially their elasticity was one of the areas that drove the development of commercial DMAs. Although a decrease in the melt viscosity is seen with temperature increases, the DMA is most commonly used to measure the frequency dependence of the molten polymer as well as its elasticity. The latter property, especially when expressed as the normal forces, is very important in polymer processing.

Frequency Dependencies in Transition Studies

The choice of a testing frequency or its effect on the resulting data must be addressed. A short discussion of how frequencies are chosen and how they affect the measurement of transitions is in order. Considering that higher frequencies induce more elastic-like behavior, there is some concern that a material will act stiffer than it really is if the test frequency is chosen to be too high. Frequencies for testing are normally chosen by one of three methods. The most common method is to use the frequency of the stress or strain to which the material is exposed in the real world. However, this is often outside the range of the available instrumentation. In some cases, the test method or an industry standard sets a certain frequency, and this frequency is used. Ideally, a standard method like this is chosen so that the data collected on various commercial instruments can be shown to be compatible. Some of the ASTM methods for TMA and DMA are listed in Table 1. Many industries have their own standards, so it is important to know whether the data is expected to match a Mil spec, an ASTM standard, or a specific industrial test. Finally, one can pick a frequency arbitrarily. This is done more often than not, so that 1 Hz and 10 rad/s are often used. As long as the data are run under the proper conditions, they can be compared to

Table 1   ASTM Tests for the DMA

D3386	CTE of Electrical Insulating Materials by TMA
D4065	Determining DMA Properties Terminology[a]
D4092	Terminology for DMA Tests
D4440	Measurement of Polymer Melts
D4473	Cure of Thermosetting Resins
D5023	DMA in Three Point Bending Tests
D5024	DMA in Compression
D5026	DMA in Tension
D5279	DMA of Plastics in Tension
D5418	DMA in Dual Cantilever
E228-95	CTE by TMA with Silica Dilatometer
E473-94	Terminology for Thermal Analysis
E831-93	CTE of Solids by TMA
E1363-97	Temperature Calibration for TMA
E1545-95(a)	T_g by TMA
E1640-94	T_g by DMA
E1824-96	T_g by TMA in Tension
E1867-97	Temperature Calibration for DMA

[a] This standard qualifies a DMA as acceptable for all ASTM DMA Standards.

highlight material differences. This requires that frequency, stresses, and the thermal program be the same for all samples in the data set.

Lowering the frequency shifts the temperature of a transition to a lower temperature (Fig. 17). At one time, it was suggested that multiple frequencies could be used and the T_g should then be determined by extrapolation to 0 Hz. This was never really accepted, as it represented a fairly large increase in testing time for a small improvement in accuracy. For most polymer systems, for very precise measurements, one uses a DSC. Different types of transitions also have different frequency dependencies, and McCrum et al. [27] has listed many of these. If one looks at the slope of the temperature dependence of transitions against frequency, one sees that in many cases the primary transitions such as T_m and T_g have a different dependence on

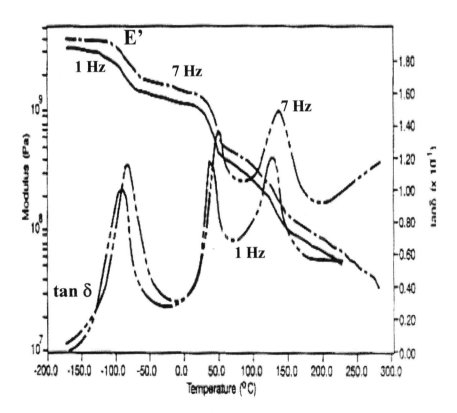

Figure 17 Effect of frequency on transitions: (a) the dependence of the T_g in polycarbonate on frequency. (Used with permission of Rheometric Scientific, Piscataway, NJ.)

frequency than the lower-temperature transitions. In fact, the activation energies are different for α, β, and γ transitions because of the different motions required, and the transitions can be sorted by this approach [20].

Polymer Melts and Solutions

A fluid or polymer melt responds to strain rate rather than to the amount of stress applied. The viscosity is one of the main reasons why people run frequency scans. As stress–strain curves and the creep-recovery runs show, viscoelastic materials exhibit some degree of flow or unrecoverable deformation. The effect is strongest in melts and liquids, for which frequency-versus-viscosity plots are the major application of DMA. Figure 18 shows a frequency scan on a viscoelastic material. In this example, the sample is a rubber above the T_g in three-point bending, but the trends and principles apply to both solids and melts. The storage modulus and complex viscosity are plotted on log scales against the log of frequency. In analyzing the frequency scans, trends in the data are more significant than specific peaks or transitions.

On the viscosity curve, η^*, a fairly flat region appears at low frequency, called the zero-shear plateau [72]. This is where the polymer exhibits Newtonian behavior and its viscosity is dependent on molecular weight,

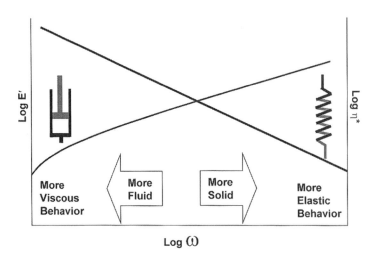

Figure 18 An example of a frequency scan showing the change in a materials behavior as frequency varies. Low frequencies allow the material time to relax and respond, hence, flow dominates. High frequencies do not, and elastic behavior dominates. (From Ref. 20a.)

not the strain rate. The viscosity of this plateau has been shown to be experimentally related to the molecular weight for a Newtonian fluid;

$$\eta \propto cM_v^1 \tag{22}$$

for cases where the molecular weight, M_v, is less than the entanglement molecular weight, M_e and for cases where M_v is greater than M_e;

$$\eta \propto cM_v^{3.4} \tag{23}$$

where η_o is the viscosity of the initial Newtonian plateau, c is a material constant, and M_v is the viscosity-average molecular weight. This relationship can be written in general terms, replacing the exponential term with the Mark-Houwink constant, a. Equation (23) can be used as a method of approximating the molecular weight of a polymer. The value obtained is closest to the viscosity-average molecular weight obtained by osmometry. [73]. In comparison with the weight-average data obtained by gel permeation chromatography (GPC), the viscosity-average molecular weight is between the number-average and weight-average molecular weights, but closer to the latter [74]. This technique was originally developed for steady shear viscosity, but it applies to complex viscosity as well. The relationship between steady shear and complex viscosity is fairly well established. Cox and Merz [75] found that an empirical relationship exists between complex viscosity and steady shear viscosity when the shear rates are the same. The Cox-Merz rule is stated as

$$\left|\eta(\omega)\right| = \eta(\dot{\gamma})\big|_{\dot{\gamma}=\omega} \tag{24}$$

where η is the constant shear viscosity, η^* is the complex viscosity, ω is the frequency of the dynamic test, and $d\gamma/dt$ is the shear rate of the constant-shear test. This rule of thumb seems to hold for most materials to within about ±10%. Another approach is the Gleissle [76] mirror relationship, which states the following:

$$\eta\dot{\gamma} = \eta^+(t)\big|_{t=1/\dot{\gamma}} \tag{25}$$

when $\eta^+(t)$ is the limiting value of the viscosity as the shear rate, $\dot{\gamma}$, approaches zero.

The low-frequency range is where viscous or liquid-like behavior predominates. If a material is stressed over long enough times, some flow occurs. As time is the inverse of frequency, this means that materials are expected to flow more at low frequency. As the frequency increases, the material will act in a more and more elastic fashion. Silly Putty^{TM}, the children's toy, shows this clearly. At low frequency Silly Putty flows like a liquid, while at high frequency it bounces like a rubber ball. This behavior is also similar to what happens with temperature changes. A polymer becomes softer and more fluid as it is heated, and it goes through transitions that increase the available space for molecular motions. Over long enough time periods, or small enough frequencies, similar changes occur. So one can move a polymer across a transition by changing the frequency. This relationship is also expressed as the idea of time–temperature equivalence [77]. Often stated as low temperature is equivalent to short times or high frequency, it is a fundamental rule of thumb in understanding polymer behavior.

As the frequency is increased in a frequency scan, the Newtonian region is exceeded and a new relationship develops between the rate of strain, or the frequency, and the viscosity of the material. This region is often called the power law zone and can be modeled by

$$\eta* \simeq \eta(\dot\gamma) = c\dot\gamma^{n-1} \tag{26}$$

where η^* is the complex viscosity, $\dot\gamma$ is the shear rate, and the exponent term n is determined by the fit of the data. This can also be written as

$$\sigma \simeq \eta(\dot\gamma) = c\dot\gamma^n \tag{27}$$

where σ is the stress and η is the viscosity. The exponential relationship is why the viscosity-versus-frequency plot is traditionally plotted on a log scale. With modern curve-fitting programs, the use of log-log plots has declined and is a bit anachronistic. The power law region of polymers shows shear-thickening or -thinning behavior. This is also the region in which the $E'-\eta^*$ or the $E'-E''$ crossover point is found. As frequency increases and shear thinning occurs, the viscosity (η^*) decreases. At the same time, increasing the frequency increases the elasticity (E'). This is shown in Fig. 18. The $E'-\eta^*$ crossover point is used an indicator of the molecular-weight and molecular weight distribution [78] Changes in its position are used as a quick method of detecting changes in the molecular weight and distribution of a material.

After the power law region, another plateau is seen, the infinite shear plateau. This second Newtonian region corresponds to where the shear rate is so high that the polymer no longer shows a response to increases in the shear rate. At the very high shear rates associated with this region, the polymer chains are no longer entangled. This region is seldom seen in DMA experiments and is usually avoided because of the damage done to the chains. It can be reached in commercial extruders and causes degradation of the polymer, which causes the poorer properties associated with regrind.

As the curve in Fig. 18 shows, the modulus also varies as a function of the frequency. A material exhibits more elastic-like behavior as the testing frequency increases and the storage modulus tends to slope upward toward higher frequency. The change in storage modulus with frequency depends on the transitions involved. Above the T_g, the storage modulus tends to be fairly flat, with a slight increase with increasing frequency, as it is on the rubbery plateau. The change in the region of a transition is greater. If one can generate a modulus scan over a wide enough frequency range, the plot of storage modulus versus frequency appears like the reverse of a temperature scan. The same time–temperature equivalence discussed above also applies to modulus, as well as compliance, tan δ, and other properties.

The frequency scan is used for several purposes that will be discussed in this section. One very important use, which is very straightforward, is to survey the material's response over various shear rates. This is important because many materials are used under different conditions. For example adhesives, whether tape, Band-Aids^{TM}, or hot melts, are normally applied under conditions of low frequency, and this property is referred to as tack. When they are removed, the removal often occurs under conditions of high frequency called peel. Different properties are required in these regimes, and to optimize one property may require chemical changes that harm the other. Similarly, changes in polymer structure can show these kinds of differences in the frequency scan. For example, branching affects different frequencies [38].

For example, in a tape adhesive, sufficient flow under pressure at low frequency is desired to fill the pores of the material to obtain a good mechanical bond. When the laminate is later subjected to peel, the material needs to be very elastic so it will not pull out of the pores [79]. The frequency scan allows measurement of these properties in one scan, thus ensuring that tuning one property does not degrade another. This type of testing is not limited to adhesives, as many materials see multiple frequencies in actual use. Viscosity-versus-frequency plots are used extensively to study how changes in polymer structure or formulations affect the behavior of the melt. Often changes in materials, especially in uncured thermosetting resins and molten materials, affect a limited frequency range and testing at a specific frequency can miss the problem.

It should be noted that since the material is scanned across a frequency range, there are some conditions in which the material–instrument system acts like a guitar string and begins to resonate when certain frequencies are reached. These frequencies are either the natural resonance frequency of the sample–instrument system or one of its harmonics. When the harmonics occur, the sample–instrument system is oscillating like a guitar string and the desired information about the sample is obscured. Since there is no way to change this resonance behavior (and in a free-resonance analyzer this effect is necessary to obtain data), it is required to redesign the experiment by changing sample dimensions or geometry to escape the problem. Using a sample with much different dimensions, which changes the mass, or changing from extension to three-point bending geometry, changes the natural oscillation frequency of the sample and may solve this problem.

Thermosets

The DMA's ability to give viscosity and modulus values for each point in a temperature scan allows estimation of kinetic behavior as a function of viscosity. This has the advantage of describing how viscous the material is at any given time, so as to determine the best time to apply pressure, what design of tooling to use, and when the material can be removed from the mold. Recent reviews have summarize this approach for epoxy systems [80].

Curing

The simplest way to analyze a resin system is to run a plain temperature ramp from ambient to some elevated temperature [81]. This *cure profile* allows collection of several vital pieces of information as shown in Fig. 19. Samples may be run "neat" or impregnated into fabrics in techniques that are referred to as "torsional braid." There are some problems with this technique, as temperature increases will cause an apparent curing of non-drying oils as thermal expansion increases friction. However, the "soaking of resin into a shoelace," as this technique has been called, allows one to handle difficult specimens under conditions where the pure resin is impos-

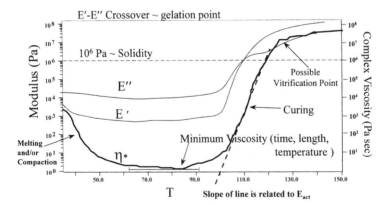

Figure 19 The DMA cure profile of a two-part epoxy showing the typical analysis for minimum viscosity, gel time, vitrification time, and estimation of the action energy. See discussion in text. (From Ref. 20a.)

sible to run in bulk (due to viscosity or evolved volatiles). Composite materials such as graphite-epoxy composites are sometimes studied in industrial situations as the composite rather than the "neat" or pure resin because of the concern that the kinetics may be significantly different. In terms of ease of handling and sample, the composite is a better sample. Another area of concern is paints and coatings [82], in which the material is used in a thin layer. This can be addressed experimentally by either a braid as above or by coating the material on a thin sheet of metal. The metal is often run first and its scan subtracted from the coated sheet's scan to leave only the scan of the coating. This is also done with thin films and adhesive coatings.

From the cure profile seen in Fig. 19, it is possible to determine the minimum viscosity (η^*_{min}), the time to η^*_{min}, and the length of time it stays there, the onset of cure, the point of gelation where the material changes from a viscous liquid to a viscoelastic solid, and the beginning of vitrification. The minimum viscosity is seen in the complex viscosity curve and is where the resin viscosity is the lowest. A given resin's minimum viscosity is determined by the resin's chemistry, the previous heat history of the resin, the rate at which the temperature is increased, and the amount for stress or stain applied. Increasing the rate of the temperature ramp is known to decrease the η^*_{min}, the time to η^*_{min}, and the gel time. The resin gets softer faster, but also cures faster. The degree of flow limits the type of mold design and when as well as how much pressure can be applied to the sample. The time spent at the minimum viscosity plateau is the result of a competitive relationship between the material's softening or melting as it heats and its rate of curing. At some point, the material begins curing faster than it softens, and that is where the viscosity starts to increase.

As the viscosity begins to climb, an inversion is seen of the E'' and E' values as the material becomes more solid-like. This crossover point also corresponds to where the tan δ equals 1 (since $E' = E''$ at the crossover). This is taken to be the gel point [83], where the cross-links have progressed to forming an "infinitely" long network across the specimen. At this point, the sample will no longer dissolve in solvent. While the gel point correlates fairly often with this crossover, it doesn't always. For example, for low initiator levels in chain-addition thermosets, the gel point precedes the modulus crossover [84]. A temperature dependence for the presence of the crossover has also been reported [81]. In some cases, when powder compacts and melts before curing, there may be several crossovers [85]. Then, the one following the η^*_{min} is usually the one of interest. Some researchers [86] believe the true gel point is best detected by measuring the frequency dependence of the crossover point. This is done by either by multiple runs at different frequencies or by multiplexing frequencies during the cure. At the gel point, the frequency dependence disappears [87]. This value is usually only a few degrees different from the one obtained in a normal scan and in most cases is not worth the additional time. During this rapid climb of viscosity in the cure, the slope for η^* increase can be used to calculate an estimated E_a (activation energy) [88]. This will be discussed below, but the fact that the slope of the curve here is a function of E_a is important. Above the gel temperature, some workers estimate the molecular weight, M_c, between cross-links as

$$G' = \frac{RT\rho}{M_c} \tag{28}$$

where R is the gas constant, T is the temperature in kelvin, and ρ is the density. At some point the curve begins to level off, and this is often taken as the vitrification point, T_{vf}.

The vitrification point is where the cure rate slows because the material has become so viscous that the bulk reaction has stopped. At this point, the rate of cure slows significantly. The apparent T_{vf}, however, is not always real: any analyzer has an upper force limit. When that force limit is reached, the "topping out" of the analyzer can pass as the T_{vf}. Use of a combined technique such as DMA-DEA [89] to see the higher viscosities or the removal of a sample from parallel plate and sectioning it into a flexure beam is often necessary to see the true vitrification point. Vitirfication may also be seen in the DSC if a modulated temperature technique such as StepScan is used [90]. A reaction can also cure completely without vitrifying and will level off the same way. One should be aware that reaching vitrification or complete cure too quickly can be as bad as reaching it too slowly. Often an overly aggressive cure cycle will cause a weaker material, as it does not allow for as much network development, but gives a series of hard (highly cross-linked) areas among softer (lightly cross-linked) areas. On the way to vitrification, an important value is 10^6 Pa-s. This is the viscosity of bitumen [91] and is often used as a rule of thumb for where a material is stiff enough to support its own weight. This is a rather arbitrary point, but it is chosen to allow the removal of materials from a mold and the cure is then continued as a postcure step. The cure profile is both a good predictor of performance as well as a sensitive probe of processing conditions. As discussed above under TMA applications, a volume change occurs during the cure [92]. This shrinkage of the resin is important and can be studied by monitoring the probe position of some DMAs as well as by TMA and dilatometry.

The above is based on using a simple temperature ramp to see how a material responds to heating. In actual use, many thermosets are actually cured using more complex cure cycles to optimize the trade-off between processing time and the final product's properties [93]. The use of two-stage cure cycles is known to develop stronger laminates in the aerospace industry. Exceptionally thick laminates often also require multiple-stage cycles in order to develop strength without porosity. As thermosets shrink on curing, careful development of a proper cure cycle to prevent or minimize internal voids is necessary. One reason for the use of multistage cures is to drive reactions to completion. Another is to extend the minimum viscosity range to allow greater control in forming or shaping of the material. The development of a cure cycle with multiple ramps and holds will be very expensive if done with full-sized parts in production facilities. The use of the DMA gives a faster and cheaper way of optimizing the cure cycle to generate the most efficient and tolerant processing conditions.

Because of the limits of industrial equipment and cost constraints, curing is done at a constant temperature for a period of time. This can be done both to cure the material initially and to "postcure" it. The kinetic models discussed in the next section also require data collected under isothermal conditions. It is how rubber samples are cross-linked, how initiated reactions are run, and how bulk polymerizations are performed. Industrially, continuous processes, as opposed to batch processes, often require an isothermal approach. UV light and other forms of nonthermal initiation also use isothermal studies for examining the cure at a constant temperature.

Photocuring

A photocure in the DMA is run by applying a UV light source to a sample (held to a specific temperature or subjected to a specific thermal cycle) [94] Photocuring is done for dental resin, contact adhesives, and contact lenses. UV exposure studies are also run on cured and thermoplastic samples by the same techniques as photocuring to study UV degradation. The cure profile of a photocure is very similar to that of a cake or epoxy cement. The same analysis is used and the same types of kinetics are developed as for thermal-curing studies.

The major practical difficulty in running photocures in the DMA is the current lack of a commercially available photocuring accessory, comparable to the photocalorimeters on the market. One normally has to adapt a commercial DMA to run these experiments. The PerkinElmer DMA-7e has been successfully adapted [95] to use quartz fixtures and a commercial UV source from EFOS, triggered from the DMA's software. This is a fairly easy process, and other instruments such as the RheoSci™ DMTA Mark 5 have also been adapted.

Curing Kinetics by DMA

Several approaches have been developed to studying the chemorheology of thermosetting systems. MacKay and Halley [96] reviewed chemorheology and the more common kinetic models. A fundamental method is the Williams-Landel-Ferry (WLF) model [97], which looks at the variation of T_g with degree of cure. This has been used and modified extensively [98]. A common empirical model for curing has been proposed by Roller [99]. In the latter approach, samples of the thermoset are run isothermally as described above and the viscosity-versus-time data collected. This is plotted as log η^* versus time in seconds, where a change in slope is apparent in the curve. This break in the data indicates that the sample is approaching the gel time. From these curves, the initial viscosity, η_o, and the apparent kinetic factor, k, can be determined. By plotting the log viscosity versus time for each isothermal run, the slope, k, and the viscosity at $t = 0$ are apparent. The initial viscosity and k can be expressed as

$$\eta_o = \eta_e^{\Delta E_\eta/RT} \tag{29}$$

$$k = k_\infty e^{\Delta E_k/RT} \tag{30}$$

Combining these allows setup of the equation for viscosity under isothermal conditions as

$$\ln \eta(t) = \ln \eta_\infty + \frac{\Delta E_\eta}{RT} + t k_\infty e^{\Delta E_k/RT} \tag{31}$$

By replacing the last term with an expression that treats temperature as a function of time, the equation becomes

$$\ln \eta(T, t) = \ln \eta_\infty + \frac{\Delta E_\eta}{RT} + \int_0^t k_\infty e^{\Delta E_k/RT} dt \tag{32}$$

This equation can be used to describe viscosity-time profiles for any run for which the temperature can be expressed as a function of time. The activation energies can now be calculated. The plots of the natural log of the initial viscosity (determined above) versus $1/T$ and the natural log of the apparent rate constant, k, versus $1/T$ are used to give us the activation energies, ΔE_η and ΔE_k. Comparison of these values to the k and ΔE calculated by DSC shows that this model gives larger values [82]. The DSC data are faster to obtain, but it does not include the needed viscosity information. Several corrections have been proposed, addressing different orders of reaction [100] (the above assumes first order) and modifications to the equations. [101]. Many of these adjustments are reported in Roller's 1986 review [102] of curing kinetics. It is noted that these equations do not work well above the gel temperature. This same equation has been used to predict the degradation of properties in thermoplastics successfully [103].

The Gillham-Enns Diagram

The most complete approach to studying the behavior of a thermoset was developed by Gillham [104] and is analogous to the phase diagrams used by metallurgists. The time–temperature–transformation diagram (TTT) or the Gillham-Enns diagram (after its creators) is used to track the effects of temperature and time on the physical state of a thermosetting material. Figure 20 shows an example. Running isothermal studies of a resin at various temperatures and recording the changes as a function of time can do this. One has to choose values for the various regions and Gillham has done an excellent job of detailing how one picks the T_g, the glass, the gel, the rubbery, and the charring regions [105]. These diagrams are also generated from DSC data [106], and several variants [107] such as continuous heating

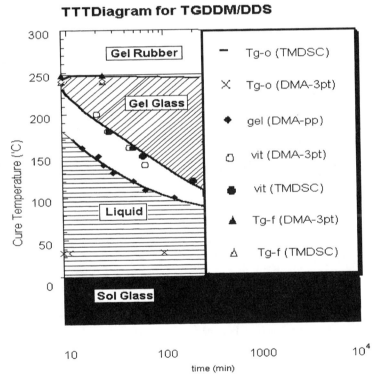

Figure 20 An example of the Gilham-Enns diagram generated by the author on a commercial epoxy resin system. The lines of gelation and vitrification are marked.

transformation and conversion–temperature–property diagrams, have been reported. Surprisingly easy to do, although a bit slow, they have not yet been accepted in industry despite their obvious utility. A recent review [108] may increase the use of this approach.

ACKNOWLEDGMENTS

I wish to acknowledge the PerkinElmer Instruments Thermal Business Unit and the Material Science Department of the University of North Texas (UNT) for their support and assistance. In addition, the help and advice of Prof. Witold Brostow of the Materials Science Department of UNT, Debra Kaufman of the PE Division Research Library, and especially my graduate students, John White and Bryan Bilyeu, are greatly appreciated.

SYMBOLS, ABBREVIATIONS, AND ACRONYMS

δ	phase angle
$\tan \delta$	tangent of the phase angle, also called the damping
Stress	
Shear strain	
Tensile strains	
$\dot{\gamma}$	shear strain rate
$\dot{\varepsilon}$	strain rate
Viscosity	
η^*	complex viscosity
η'	storage viscosity
η''	loss viscosity
E^*	complex modulus
E'	storage modulus
E''	loss modulus
J	compliance
k	deformation
Period	
ρ	density
G	shear modulus
M_e	entanglement molecular weight
M_c	molecular weight between cross-links
M_w	molecular weight
f	frequency
ω	frequency in hertz
k	rate constant
E_a	activation energy
DMA	dynamic mechanical analysis or analyser

DMTA	dynamic mechanical thermal analysis or analyzer
DEA	dielectric analysis or analyzer
DSC	differential scanning calorimeter
TBA	torsional braid analyzer
TGA	thermogravimetric analyzer
v^f	free volume
$T_{\alpha,\beta,\gamma}$	transition (subscript type)
Λ	logarithmic decrement
Γ	torque

NOTES AND REFERENCES

1. V. Kargin et al., *Dokl. Akad. Nauk SSSR 62*:239 (1948).
2. R. Bird, C. Curtis, R. Armstrong, and O. Hassenger, *Dynamics of Polymer Fluids*, 2nd ed., Wiley, New York (1987).
3. (a) J. Ferry, *Viscoelastic Properties of Polymers*, 3rd ed., Wiley, New York (1980). (b) J. J. Aklonis and W. J. McKnight, *Introduction to Polymer Viscoelasticity*, 2nd ed., Wiley, New York (1983).
4. (a) L. C. E. Struik, *Physical Aging in Amorphous Polymers and Other Materials*, Elsevier, New York (1978). (b) L. C. E. Struik, in *Failure of Plastics*, edited by W. Brostow, and R. D. Corneliussen, Hanser, New York (1986). (c) S. Matsuoka, in *Failure of Plastics*, edited by W. Brostow and R. D. Corneliussen, Hanser, New York (1986). (d) S. Matsuoka, *Relaxation Phenomena in Polymers*, Hanser, New York (1992).
5. J. D. Vrentas, J. L. Duda, and J. W. Huang, *Macromolecules 19*:1718 (1986).
6. W. Brostow and M. A. Macip, *Macromolecules 22*(6):2761 (1989).
7. Thermal expansivity is often referred to as the coefficient of thermal expansion or CTE by polymer scientists and in the older literature.
8. G. Curran, J. Rogers, H. O'Neal, S. Welch, and K. Menard, *J. Adv. Mater. 26*(3): 49 (1995).
9. W Brostow, A. Arkinay, H. Ertepinar, and B. Lopez, *POLYCHAR-3 Proc. 3*: 46 (1993).
10. R. Boundy, R. Boyer, and S. Stoesser, eds., *Styrene, Its Polymers, Copolymers, and Derivatives*, Reinhold, New York (1952).
11. A. Snow and J. Armistead, *J. Appl. Polymer Sci. 52*:401 (1994).
12. B. Bilyeu and K. Menard, *POLYCHAR-6 Proc.* (1998).
13. P. Zoller and Y. Fakhreddine, *Thermochim. Acta 238*:397 (1994).
14. P. Zoller and D. Walsh, *Standard Pressure-Volume-Temperature Data for Polymers*, Technomic, Lancaster, PA (1995).
15. J. Berry, W. Brostow, M. Hess, and E. Jacobs, *Polymers 39*:243, (1998).
16. (a) R. Cassel and R. Fyans, *Ind. Res*, August 1977, p. 44. (b) R. Cassel, *Thermal Appl. Study 20* (1977).
17. (a) P. Webber and J. Savage, *J. Mater. Sci. 23*:783 (1983). (b) H. Tong and G. Appleby-Hougham, *J. Appl. Polymer Sci. 31*:2509, (1986).
18. M. McLin and A. Angell, *Polymer 37:* 4703 (1996).

19. See, for example, the product literature for PerkinElmer's TMA 7 and PYRIS™ software. The optins exist to add both these function to the TMA. Other vendors have similar packages.
20. (a) K. Menard, *Dynamic Mechanical Analysis: A Practical Introduction*, CRC Press, Boca Raton, FL (1999). (b) M. Sepe, *Dynamic Mechanical Analysis for Plastic Engineering*, Plastic Design Library, New York (1998). (c) T. Murayama, *Dynamic Mechanical Analysis of Polymeric Materials*, Elsevier, New York (1977).
21. J. H. Poyntang, *Proc. Roy. Soc. Ser. A 82*:546 (1909).
22. A. Kimball and D. Lovell, *Trans. Am. Soc. Mech. Eng. 48*:479 (1926).
23. K. te Nijenhuis, in *Rheology, Volume 1, Principles*, edited by G. Astarita et al., Plenum Press, New York, p. 263 (1980).
24. M. L. Miller, *The Structure of Polymers*, Reinhold, New York, (1966).
25. J. Dealy, *Rheometers for Molten Plastics*, Van Nostrand Reinhold, New York, pp. 136–137 and 234–236 (1992).
26. J. Ferry, *Voscoelastic Properties of Polymers*, 3rd ed., Wiley, New York (1980).
27. N. McCrum, B. Williams, and G. Read, *Anelastic and Dielectric Effects in Polymer Solids*, Dover, New York (1991).
28. J. Gilham and J. Enns, *Trends Polymer Sci 2*:406 (1994).
29. C Macosko and J. Starita, *SPE J. 27*:38 (1971).
30. T. Murayama, *Dynamic Mechanical Analysis of Polymeric Materials*, Elsevier, New York (1977).
31. B. E. Read and G. D. Brown, *The Determination of the Dynamic Properties of Polymers and Composites*, Wiley, New York (1978).
32. (a) J. Gillham, in *Developments in Polymer Characterizations: Volume 3*, edited by J. Dworkins, Applied Science, Princeton, NI, pp. 159–227 (1982). (b) J. Gillham and J. Enns, *Trends Polymer Sci 2*(12):406 (1994).
33. U. Zolzer and H.-F. Eicke, *Rheol. Acta 32*:104, (1993).
34. N. McCrum et al., *Anelastic and Dielectric properties of Polymeric Solids*, Dover, New York, pp. 192–200 (1991).
35. P. Fory, *Principles of Polymer Chemistry*, Cornell University Press, Ithaca, NY (1953).
36. M. Doi and S. Edwards, *The Dynamics of Polymer Chains*, Oxford University Press, New York (1986).
37. Aklonis and Knight, *Introduction to Viscoelasticity*, Wiley, New York (1983).
38. C. L. Rohn, *Analaytical Polymer Rheology*, Hanser-Gardener, New York (1995).
39. J. Heijboer, *Int. J. Polymer Mater. 6*:11 (1977).
40. R. F. Boyer, Polymer Eng. Sci 8(3):161 (1968).
41. C. L. Rohn, *Analytical Polymer Rheology*, Hanser-Gardner, New York, pp. 279–283 (1995).
42. (a) J. Heijboer, *Int. J. Polymer Mater. 6*: 11 (1977). (b) m. Mangion and G. Johari, *J. Polymer Sci. B Polymer Phs 29*:437 (1991). (e) G. Johari, G. Mikoljaczak, and J. Cavaille, *Polymer 28*:2023 (1987). (d) S. Cheng et al.,

Polymer Sci. Eng. 33:21 (1993). (e) G. Johari, *Lecture Notes Phys 277*:90 (1987). (f) R. Daiz-calleja and E. Riande, *Rheol. Acta 34*:58 (1995). (g) R. Boyd, Polymer 26:323 (1985). (h) V. Bershtien, V. Egorov, L. Egorova, and V. Ryzhov, *thermochim. Acta 238*:41 (1994).

43. B. Twombly, *Proc. NATAS 20*:63 (1991).
44. C. L. Rohm, *Analytical Polymer Rheology*, Hanser-Gardner, New York (1995). (b) J. Heijboer, *Int. J. Polymer Mater. 6*:11 (1977).
45. R. Boyer, *Polymer Eng. Sci. 8*(3):161 (1968).
46. V. Bershtien and V. Egorov, *Differential Scanning Calorimetry in the Physical Chemistry of Polymers*, Ellis Horwood, Chichester, U.K. (1993).
47. B. Coxton, private communication.
48. S. Cheng et al., *Polymer Sci. Eng. 33*:21 (1993).
49. G. Johari, G. Mikoljaczak, and J. Cavaille, *Polymer 28*:2023 (1987).
50. M. Mangion and G. Johari, *J. Polymer Sci. B Polymer Phys. 29*:437 (1991).
51. F. C. Nelson, *Shock and Vibration Digest 26(2): 11 and 24 (1994)*.
52. W. Brostow, Private communication.
53. R. Boyer, *Polymer Eng. Sci. 8*(3):161 (1968).
54. (a) G. Johari, *Lecture Notes Phys. 277*:90 (1987). (b) J. Heijboer, *Intl. J. Polymer Mater 6*:11 (1977). (c) J. Heijboer et al., in *Physics of Non-crystalline Solids*, edited by J. Prins, Interscience, New York (1965). (d) J. Heijboer, *J. Polymer Sci. C16*:3755 (1968). (e) L. Nielsen et al., *J. Macromol. Sci Phys. 9*:239 (1974).
55. A. Yee and S. Smith, *Macromolecules 14*:54 (1981).
56. G. Gordon, *J. Polymer Sci. A2*(9):1693 (1984).
57. J. Wndorff and B. Schartel, *Polymer 36*(5):899 (1995).
58. R. H. Boyd, *Polymer 26*:323 (1985).
59. (a) I. Noda, *Appl. Spectrosc. 44*(4):550 (1990). (b) V. Kien, Proc. 6th Symposium on Radiation Chemistry 6(2):463 (1987).
60. L. H. Sperling, *Introduction to Physical Polymer science*, 2nd ed., Wiley, New York (1992).
61. C. Macosko, *Rheology*, VCH, New York (1994).
62. (a) F. Quinn et al., *Thermal Analysis*, Wiley, New York (1984). (b) B. Wunderlich, *Thermal Analysis*, Academic Press, New York (1990).
63. (a) J. Schawe, *Thermochim. Acta 261*:183 (1995). (b) J. Schawe, *Thermochim. Acta 260*:1 (1995). (c) J. Schawe, *Thermochim. Acta 271*:1 (1995). (d) B. Wunderlich et al., *J. Thermal Anal. 42*:949 (1994).
64. (a) R. H. Boyd, *Polymer 26*:323 (1985). (b) R. H. Boyd, *Polymer 26*:1123 (1985).
65. (a) S. Godber, private communication. (b) M. Ahmed, *Polypropylene Fiber Science and Technology*, Elsevier, New York (1982).
66. A. Lobanov et al., *Polymer Sci. USSR 22*:1150 (1980).
67. (a). R. Boyer, *J. Polymer Sci. B Polymer Phys. 30*:1177 (1992). (b) J. K. Gilham et al., *J. Appl. Polymer Sci. 20*;1245 (1976). (c) J. B. Enns and R. Boyer, in *Encyclopedia of Polymer Science*, Vol. 17, pp. 23–47 (1989).

68. V. Bershtien, V. Egorov, L. Egorova, and V. Ryzhov, *Thermochim. Acta* *238*:41 (1994).

69. C. M. Warner, Evaluation of the DSC for Observation of the Liquid-Liquid Transition, Master's thesis, Central Michigan State University (1988).

70. (a) J. Dealy et al., *Melt Rheology and its Role in Plastic Processing*, Van Nostrand Reinhold, Toronto (1990). (b) N. Chereminsinoff, *An Introduction to Polymer Rheology and Processing*, CRC Press, Boca Raton, FL (1993).

71. (a), E. (ed.), *Thermal Characterization of Polymeric Materials*, Academic Press, Boston (1981). (b) Turi, E., (ed.), *Thermal Analysis in Polymer Characterization*, Heydon, London (1981).

72. C. Rohn, *Analytical Polymer Rheology*, Hanser, New York (1995).

73. M. Miller, *The Structure of Polymers*, Reinhold, New York, pp. 611–612 (1966).

74. S. Rosen, *Fundamental Principles of Polymeric Materials*, Wiley Interscience, New York, pp. 53–77 and 258–259 (1993).

75. (a) W. Cox and E. Merz, *J. Polymer Sci.* 28:619 (1958). (b) P. Leblans et al., *J. Polymer Sci. 21*;1703 (1983).

76. W. Gleissele, in *Rheology*, Vol. 2, edited by G. Astarita et al., Plenum Press, New York, p. 457 (1980).

77. D. W. Van Krevelin, *properties of Polymers*, Elsevier, New York, p. 289 (1987).

78. C. Macosko, *Rheology*, VCH, New York, pp. 120–127 (1996).

79. This is a very simplified version of adhesion. For a detailed discussion, see L.-H. Lee (ed.), *Adhesive Bonding*, Plenum press, New York (1991) or L. H. Lee (ed.), *Fundamentals of Adhesion*, Plenum Press, New York (1991).

80. W. Brostow, B. Bilyeu, and K. Menard, *J. Mater. Educ.* in press (2001).

81. (a) G. Martin et al., *Polymer Characterization*, American Chemical Society, Washington, DC (1990). (b) M. Ryan et al., *ANTEC Proc 31*:187 (1973). (c) C. Gramelt, *Am. Lab.* January 1984, p. 26. (d) S. Etoh et al., *SAMPE J. 3*:6 (1985). (e) F. Hurwitz, *Polymer composites 4*(2):89 (1983).

82. (a) M. Roller, *Polymer Eng. Sci. 19*:692 (1979). (b) M. Roller et al., *J. Coating Technol. 50*:57 (1978).

83. (a). M. Heise et al., *Polymer Eng. Sci 30*:83 (1990). (b) K. O'Driscoll et al., *J. Polymer Sci. Polymer Chem 17*:1891 (1979). (c) O. Okay, Polymer 35:2613 (1994).

84. M. Hiese, G. Martin, and J. Gotro, *Polymer Eng. Sci. 30*(2):83 (1990).

85. K. Wissbrun et al., *J. Coating Technol 48*:42 (1976).

86. (a) F. Champon et al., *J. Rheol. 31*:683 (1987). (b) H. Winter, *Polymer Eng. Sci 27*:1698 (1987). (c) C. Michon et al., *Rheol. Acta 32*:94 (1993).

87. C. Michon et al., *Rheol Acta 32*:94 (1993).

88. I. Kalnin et al., *Epoxy Resins*, American Chemical Society, Washington, DC (1970).

89. DEA is dielectric analysis, in which an oscillating electrical signal is applied to a sample. From this signal, the non mobility can be calculated, which is then

converted to a viscosity. See McCrum for details. DEA will measure to significantly higher viscosities than DMA.

90. B. Bilyeu, W. Brostow, and K. Menard, in *Materials Characterization by Dynamic and Modulated Thermal Analytical Techniques*, ASTM STP 1402, edited by A. T. Riga and L. H. Judovits, American Society for Testing and Materials, West Conshohocken, PA (2001).

91. H. Barnes et al., *An Introduction to Rheology*, Elsevier, New York (1989).

92. A. W. Snow et al., *J. Appl. Polymer Sci. 52*:401 (1994).

93. (a) R. Geimer et al., *J. Appl. Polymer Sci 47*:1481 (1993). (b) R. Roberts, *SAMPE J. 5*: 28 (1987).

94. (a). T. Renault et al., *NATAS Notes 25*:44 (1994). (b) H. L. Xuan et al., *J. Polymer Sci. A 3*:769 (1993). (c) W. Shi et al., *J. Appl. Polymer Sci. 51*:1129 (1994).

95. J. Enns, private communication.

96. P. J. Halley and M. E. MacKay, *Polymer Eng. Sci 36*(5):593 (1996).

97. J. Ferry, *Viscoelastic Properties of Polymers*, 3rd ed., Wiley, New York (1980).

98. (a) J. Mijovic et al., *J. Comp. Met. 23*:163 (1989). (b) J. Mijovic et al., *SAMPE J. 23*:51 (1990).

99. (a) M. Roller, *Metal Finishing 78*:28 (1980). (b) M. Roller et al., *ANTEC Proc 24*:9 (1978). (c) J. Gilham, *ACS Symp. Ser. 78*:53 (1978). (d) M. Roller et al., *ANTEC Proc. 21*:212 (1975). (e) M. Roller, *Polymer Eng. Sci. 15*:406 (1975). (f) M. Roller, *Polymer Eng. Sci. 26*:432 (1986).

100. C. Rohn, *Problem Solving for Thermosetting Plastics*, Rheometrics, Austin TX (1989).

101. (a) J. Seferis et al., *Chemorheology of Thermosetting Polymers*, ACS, Washington DC, American Chemical Society, p. 301 (1983). (b) R. Patel et al., *J. Thermal Anal. 39*:229 (1993).

102. M. Roller, *Polyme Eng. Sci. 26*:432 (1986).

103. M. Roller, private communication (1998).

104. (a) J. Gillham et al., *Polymer composites 1*:97 (1980). (b) J. Enns et al., *J. Appl. Polymer Sci 28*: 2567 (1983). (c) L. C. Chan et al., *J. Appl. Polymer Sci. 29*:3307 (1984). (d) J. Gillham, *Polymer Eng. Sci. 26*:1429n (1986). (e) S. Simon et al., *J. Appl. Polymer Sci 51*:1741 (1994). (f) G. Palmese et al., *J. Appl. Polymer Sci. 34*:1925 (1987). (g) J. Enns et al., in *Polymer Characterization*, edited by C. Craver, American Chemical Society, Washington, DC (1983).

105. (a) J. Gillham et al., *J. Appl. Polymer Sci. 53*:709 (1994). (b) J. Enns and J. Gillham, *Trends Polymer Sci 2*(12):406 (1994).

106. A. Otero et al., *Thermochim. Acat 203*:379 (1992).

107. (a) J. Gillham et al., *J. Appl. Polymer Sci 42*:2453 (1991). (b) B. Osinski, *Polymers 34*:752 (1993).

108. J. Enns and j. Gillham, *Trends Polymer Sci 2*(12):406 (1994).

7

Infrared and Raman Analysis of Polymers

Koichi Nishikida

Thermo Electron Corporation, Madison, Wisconsin, USA

John Coates

Coates Consulting, Newtown, Connecticut, USA

1. INTRODUCTION

Infrared (IR) and Raman are both well established as methods of vibrational spectroscopy. Both have been used for decades as tools for the identification and characterization of polymeric materials; in fact, the requirement for a method of analysis synthetic polymers was the basis for the original development of analytical infrared instrumentation during World War II. It is assumed that the reader has a general understanding of analytical chemistry, and a basic understanding of the principles of spectroscopy. A general overview of vibrational spectroscopy is provided in Sec. 5 for those unfamiliar with the infrared and Raman techniques.

Sample preparation and measurement procedures are very important, especially for infrared methods of analysis. A brief discussion of instrumentation and sample handling accessories, along with a summary of the most common sample handling methods, is provided in Sec. 5. Raman spectroscopy is quite different from infrared spectroscopy, insofar as there is generally very little sample preparation required, and in fact most samples are studied in their native form. Sample preparation methods are very diverse for infrared spectroscopy, and the selection of method is dependent on the nature and form of the sample, the type of information required, and the sample handling accessories available. It is important to appreciate that the appearance of the spectrum is dependent on both the spectral acquisition parameters and the sample handling method, with the latter often having the largest impact. These issues are covered in Sec. 5.

The first step in a polymer analysis is to identify the specific type of polymer in a given sample. This may be complicated in a formulated sample by the presence of additives. Infrared spectroscopy will usually provide information on both the base polymer(s) and the additive(s) present. The second step, if possible, is to determine details of the chemical and physical characteristics, which define the quality and properties of the polymer. The chemical properties that can be determined are stereo specificity, any irregularities in the addition of monomer (such as 1,2- versus 1,4-addition and head-to-head versus head-to-tail addition), chain branching, any residual unsaturation, and the relative concentration of monomers in the case of copolymers. Other important characteristics include specific additives in a formulated product, and the physical properties, which include molecular weight, molecular-weight dispersion, crystallinity, and chain orientation. Some properties such as molecular weight and molecular-weight dispersion are not determined directly by infrared and Raman spectroscopy, except in some special cases.

In this chapter, we describe how to classify polymers into 21 groups from their infrared spectra as outlined in Sec. 2. This is followed by a more detailed identification and characterization of selected polymers by IR and Raman methods in Sec. 3. Both sections are intended to provide an introduction to polymer analysis.

The polymers described in this chapter are industrial-grade materials, and consequently some of the examples may contain additives and/or may be chemically modified. Polymers in various morphological forms may be analyzed, and these include films, fibers, solid pelletized and powdered products, and dissolved/dispersed materials in liquids such as paints and latex products. Also, the same base polymer, such as a styrene–butadiene copolymer, for example, may exist in a rubber, a resin, or a plastic. In general, reference will not be made to the original source of the polymer samples. Because infrared spectroscopy is more widely used than the Raman method, the authors will focus more on the applications of this technique. However, the Raman method, which is complementary to the IR method, does have important and unique applications in the polymer analysis, especially with regard to the determination of the fundamental polymer structure and its morphology. Examples of the use of Raman are used as appropriate to reinforce the benefits of this spectral method for polymer characterization.

A set of Reference Spectra is also provided in the Appendix, featuring matched pairs of IR and Raman spectra of nearly 60 common polymers. The IR spectra were recorded on a commercial FT-IR at spectral resolution of 4 cm^{-1}. For most of the spectra, a transmission method was used, featuring KBr pellets, capillary films (liquids), and self-supporting films. The Raman spectra were obtained with near-IR Fourier transform (FT)-Raman instrument with 1064-nm excitation (Nd:YAG laser). It should be noted that the cutoff frequency of the InGaAs detector used for the measurements corresponds to an estimated Raman shift of 3500 cm^{-1}. This results in reduced performance (detectivity) in this higher Raman shift region of the spectrum, and as a consequence, the intensities of any NH and OH stretching bands are usually affected. Raman spectra by nature are single-beam spectra; in an attempt to compensate for the instrument function an energy throughput correction is performed, using a tungsten blackbody source and an assumed theoretical emission curve. For reference purposes, the source used for the data presented was not NIST traceable and so, the intensity correction may only be considered approximate.

2. PRINCIPLES OF POLYMER IDENTIFICATION FROM INFRARED AND RAMAN SPECTRA

In this section, we demonstrate how to identify and characterize a polymer from its infrared spectrum. This section is intended for people who are not particularly experienced in spectral interpretation. There are several well-established reference texts that may be used for both general compound identification [1–8] and for polymers [9–12]. However, in order to interpret infrared spectra by the assignment of absorption peaks in terms of specific functional groups or microstructures, it is necessary to gain practical experience, and it is not a task well suited for beginners. Consequently, a simple method to identify polymers based on the position of the most intense peak in the spectrum is proposed [13]. This method is intended to provide a broad-based characterization of the sample. Another method, which is also described in this section, classifies natural and synthetic polymers into 21 subgroups based on the presence and absence of specific absorption bands [14]. Because each subgroup may contain quite different polymer types, such as poly(styrene) and poly(phenylene oxide), the final identification must be performed by a comparison of the sample spectrum with the reference spectra of the polymers within the group.

In order to initiate the characterization process, the reader must follow the flow of the diagram shown in Fig. 1. Starting from the top left box of the

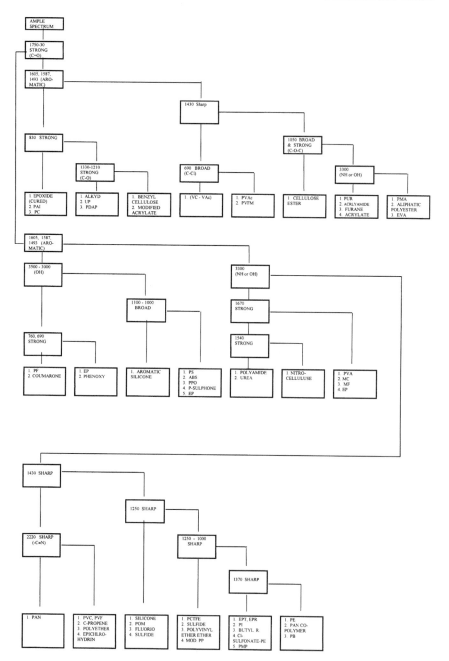

figure, one proceeds, step by step, to one of the 21 groups by assessing the presence or absence of the absorption band(s) described in each box.

The first important characteristic feature is the presence of a strong carbonyl group absorption. If the absorption is weak compared with those in the fingerprint region of the spectrum, the carbonyl group must be regarded as a minor component such as an additive or even a degradation product. In this situation, the peak is disregarded and is not used in the decision process. If the absorption is present, the next step is to differentiate the polymer on the basis of being an ester or a ketone, or some similar carbonyl functionality.

A next differentiating feature is the presence of absorptions associated with an aromatic ring, at 1605, 1587, or 1493 cm^{-1}. The bands at 1605 and 1587 cm^{-1} often exist as an absorption-band doublet with an uneven intensity ratio, the 1587 cm^{-1} absorption typically being weaker than the 1605 cm^{-1} absorption. This step helps to differentiate ester polymers into aliphatic versus aromatic esters.

The existence of an absorption at around 1430–1420 cm^{-1} indicates the presence of an α-methylene (sometimes known as an active methylene), adjacent to the ester or ketone carbonyl. This, in conjunction with the presence of a normal aliphatic methylene sequence (1450–1470 cm^{-1}) differentiates the vinyl polymers, such as polyvinyl alcohol (PVA) and polyvinyl chloride (PVC), from compounds such as PMA and PMMA. Since this step may be confusing to the reader, it is suggested that attention is paid to the assignments for vinyl acetate-type and acrylate-type polymers.

Figure 1 Polymer interpretation chart. PAI, polyamideimide; PC, polycarbonate; UP, unsaturated polyester; PDAP, diarylate phtalate resin; VC-VAc, vinyl chloride-vinyl acetate copolymer; PVAc, polyvinyl acetate; PVFM, polyvinyl formal; PUR, polyurethane; PA, polyamide; PMA, methacrylate ester polymer; EVA, ethylene–vinyl acetate copolymer; PF, phenol resin; EP, epoxide resin; PS, polystyrene; ABS, acrylonitrile–butadiene–styrene copolymer; PPO, polyphenylene oxide; P-SULFONE, polysulfone; PA, polyamide; UF, urea resin; CN, nitrocellulose; PVA, polyvinyl acetate; MC, methyl cellulose; MF, melamine resin; PAN, polyacrylonitrile; PVC, polyvinyl chloride; PVF, polyvinyl fluoride; CR, polychloroprene; CHR, polyepichlorohydrin; SI, polymethylsiloxane; POM, polyoxymethylene; PTFE, polytetrafluoroethylene; MOD-PP, modified PP; EPT, ethylene–propylene terpolymer; EPR, ethylene–propylene rubber; PI, polyisoprene; BR, butyl rubber; PMP, poly(4-methyl pentene-1); PE, poly(ethylene); PB, poly(butene-1). (Adapted from Ref. 22, p. 50.)

All of the other inputs to this chart are based on the presence and absence of the other important functional groups. Such groups are the hydroxy (OH) group, which gives rise to a broad band in the region 3000–3500 cm^{-1} (assigned to the O−H stretching frequency), amino (NH) around 3300 cm^{-1}, siloxy (Si−O) between 1200 and 1000 cm^{-1} (intense, usually doublet), chloro (C−Cl) at 690 cm^{-1} (intense, often multiple bands), nitrile or cyano (−C≡N) located at 2220 cm^{-1} (sharp), fluoro (C−F) at 1250 cm^{-1} (intense), amido (CONH) group, with absorptions at 1670 cm^{-1} and 1550 cm^{-1}, and the 1,4-disubstituted (*para*-)benzene ring at 830 cm^{-1}, versus the mono-substituted benzene ring located at 760 and 690 cm^{-1}.

It is important to show an example in order to explain how to utilize this chart. For example, if the answers are all "yes" to the 1730 cm^{-1} carbonyl band, the 1605, 1587, and 1493 cm^{-1} aromatic bands, and the 830 cm^{-1} 1,4-disubstituted benzene ring, the diagram will lead to the selection containing "cured epoxide," "polyimide," or "polycarbonate." These three categories of polymers are completely different types of polymers, yet they share the same basic functional components, that is, a 1,4-substituted benzene ring and a carbonyl group, as indicated in Fig. 2. The most frequently encountered of these polymer types is poly(carbonate), which is based on the bisphenol A moiety (a) with carbonate functionality for the carbonyl groups (b). The polyimide is similar with imide functionality (c) and either bisphenol A-type or phenyl ether (d)-type moieties. The most common epoxide resin also features a bisphenol A-type epoxide, and so all three polymers feature a 1,4-disubstituted benzene structure with a carbonyl functional group. Although uncured epoxide resins do not necessarily have a carbonyl

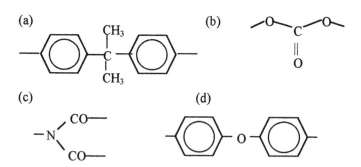

Figure 2   Example answers that fit the "1730 cm^{-1}" and the "1,4 aromatic disubstitution" question (see Fig. 1). (a) Bis-phenol A; (b) carbonate; (c) imide; (d) phenyl ether moieties.

group, the cured resins, especially those cured with acids, include an ester-based carbonyl group.

As indicated in the next section, there are several different types of epoxy resins. One type, classified as "aliphatic epoxides," by definition do not feature an aromatic ring. The term "cured epoxide" as used in Fig. 2 refers to a resin formed from a *bisphenol A-type epoxide cured with a carboxylic acid*. Polycarbonate polymers also have different chemical compositions. Similar to the epoxide example, for example, there are also aliphatic-based polycarbonate resins, with one of the most famous being a polymer used for optical lenses. The polycarbonate resin example provided in Fig. 2 is an *aromatic polycarbonate*, which includes the one based on the bis-phenol A moiety, commonly known as Lexan.

This method of classification is useful for "pure" polymer samples without significant modification or in the absence of additives. In the presence of a polymer modification or blended additives, a misinterpretation may result because of interference from other components. Usually the amounts of additives used in a formulated product are relatively low, and their presence is seldom a major interference. An exception is experienced with certain plasticizers, in which the concentration is often high. A common example is plasticized poly(vinyl chloride), which is a mixture of poly(vinyl chloride), a stabilizer, and a plasticizer such as dioctyl phthalate (often diisooctyl isomer). In this example, features associated with the plasticizer dominate the infrared spectrum. Certain additives, such as fillers (calcium carbonate, for example) may also be misleading, and can confuse the spectral interpretation. For example, products fabricated from poly(vinyl chloride) are used for construction and piping, and these are typically formed from a blend of poly(vinyl chloride) and calcium carbonate. The two examples provided are the common cases where the additives dominate the infrared spectra, and these are sufficiently popular combinations that the spectra are easily recognized.

Once the criteria for one of the 21 categories is met, the final assignment must be confirmed by comparing the sample spectrum with reference spectra, either those provided within this chapter or spectra published in reference books or commercial digital spectral libraries offered by most of the instrument vendors, and specialist companies, such as Sadtler Laboratories, Division of BioRad. The characteristic spectral peaks that should be used for identification of a polymer sample are described earlier in this section, and later in Sec. 4. We wish to stress that it is unwise to place blind trust in answers provided solely from a commercial computer search program without challenging the result, and without comparing the sample spectrum with reference spectra in published spectral databases.

3. IDENTIFICATION AND CHARACTERIZATION OF POLYMERS USING INFRARED AND RAMAN SPECTROSCOPY

In this section, we discuss the identification process and the chemical structures of polymers as obtained from their infrared and Raman spectra. Many polymers have common features, and it is convenient to segregate polymers into groups, such that the characterization of the polymers in a group can be discussed together. A popular method for classifying polymers is by their modes of application. For instance, some polymers, such as polyvinyl acetate, polystyrene, and nylons, are classified as *thermoplastics*, while urea, melamine, and epoxide resins are classified as *thermosets* or thermosetting resins. In this chapter we will use a different approach to classify the polymers, based on their similarity of chemical structure. This enables us to utilize the correlation between the functional groups of polymers and their characteristic infrared and Raman frequencies.

The purpose of this section is not only to confirm the identification, but also to characterize certain polymers and polymer types in detail. Although methods to determine microstructures and impurities, such as chemical inversions, modifications, and multiple bond formations, are different from polymer to polymer and are discussed separately, the methods used for the determination of density and crystallinity, as well as polymer orientation, are common to most polymers. Thus, the determination of crystallinity and density will be covered in this section, in Sec. 3.1, and likewise, the orientation of the polymer chain will be described in Sec. 3.2. The use of absorption coefficients to calculate properties, such as crystallinity, double-bond content, chain branching, and monomer ratios, is described in reference texts [14,15]. Today most work is performed by Fourier transform infrared (FTIR), and so an attempt has been made to feature coefficients from the latest reference sources, which include data acquired by FTIR.

The differences in the IR and Raman spectra of random copolymers, block copolymers, and polymer mixtures, A_n and B_n, will be covered in a moment. It should be appreciated that it is difficult to distinguish between polymer mixtures of the form $A_n + B_m$ and block copolymers defined as A_n-B_m from either IR or Raman spectra, because the chemical bonding between species A and B is only one of many bonds within a polymer chain. Column chromatographic separation followed by IR or Raman spectral identification or a GPC-IR method [16] is needed to determine whether a sample is a block copolymer or a mixture. In the case of a random copolymer in which component B is very small, B is mixed into the A–A chain sequence in the form $-A_n-B-A_m-$. While the IR and Raman spectra of the A_n sequences may stay essentially the same as the A_n homopolymer, the spectra of B in an

A–B–A sequence would be different than that of the B_n homopolymer. Consequently, the IR and Raman spectra of the random copolymer from monomers A and B may differ from those of a mixture or a block copolymer [17]. Individual examples will be discussed later in each section.

3.1. Determination of Crystallinity and Density in Polymer Systems

In a spectrum of a polymer of mixed morphology, some absorption bands are assigned to the crystalline parts of the polymer; others to the amorphous parts, and the remainder may be common to both. The intensities of the morphologically specific absorption bands correspond to the relative contributions of crystalline and amorphous components. If the intensities are correlated with either X-ray crystallographic or thermal analysis data of the polymer, the crystalline and the amorphous contributions can be determined. An alternative approach for certain polymer systems takes advantage of the difference in density between the crystalline and amorphous phases of a polymer. Density may be expressed as

$$\frac{100}{r} = \frac{X}{r_c} + \frac{100 - X}{r_a} \tag{1}$$

where X is the degree of crystallinity (in %), and r_c and r_a are the densities of the crystalline and amorphous phases, respectively. The degree of crystallinity, X, may be expressed as

$$X(\%) = \frac{(1/r_a - 1/r)}{(1/r_a - 1/r_c)} \cdot 100 = \frac{r - r_a}{r_c - r_a} \cdot \frac{r_c}{r} \cdot 100 \tag{2}$$

In several cases, the relationship between the experimentally determined sample densities and the intensities of the crystalline sensitive peaks can be defined by the following relationship:

$$r = C \cdot \frac{A(c)}{t} \tag{3}$$

where C is the experimentally deduced constant, $A(c)$ is the absorbance of the selected crystalline peak, and t is the sample thickness. If an absorption band is selected from the backbone of the polymer, which is independent of the morphology, it is possible to restate the relationship without the sample thickness term:

$$r = C' \cdot \frac{A(c)}{A(r)} \tag{4}$$

where $A(r)$ is the absorbance of the internal reference band.

Replacing r in Eq. (2) using Eq. (3) or (4), the crystallinity can be obtained from the intensity of a peak of the crystalline phase (refer to the section dealing with polyethylene). It is important to remember that this method assumes that there are only amorphous and crystalline phases in the semicrystalline polymer. As discussed in detail by D'Esposito and Koenig [18] in their studies of polyethylene terephthalate (PET) crystallinity, some absorption bands which are apparently sensitive to thermal treatment, such as annealing, are not necessarily attributed to the three-dimensional crystalline lattice, but rather to local regularity. For example, if a certain chain conformation, which is predominant in the crystalline phase, is isolated in a random chain and thermal annealing increases the content of that conformation, some peaks due to this conformation may appear to correlate with crystallinity. However, these peaks will remain associated with the amorphous environment. In such cases, these bands only indicate changes in the microstructure, and these are known as "conformational bands" and are not a measure of crystallinity.

Locally ordered chains having an extended conformation, with the same conformation as the crystalline phase, will give rise to "regularity bands." Vibrational coupling will occur between these locally ordered chains. Such systems may produce a series of bands present within a single band envelope (refer to the section on polyvinyl chloride). An absorption band originating from the three-dimensional lattice of the polymer is defined as a true crystalline band. Truly "crystalline bands" are observed when absorptions are assigned to interchain interactions, which produce characteristic splitting of bands, such as the 730 cm^{-1} and 720 cm^{-1} splitting of $-CH_2-$ rocking vibration of polyethylene. It is not easy to observe thermal treatments of polymers, such as the direct IR/Raman measurements of melts, and during film preparation after high-temperature annealing. In practice, we tend to choose bands that change in their intensity by annealing. However, it is unwise to try to correlate crystallinity and band intensity when measurements in melts are not performed.

3.2. Orientation of Polymer Chains

Some polymer sheets and fibers are prepared in such a way that polymer chains are spatially oriented. The degree of orientation of the polymer chain is related to physical properties. For instance, certain polypropylene and polyethylene films are partially oriented in one direction because of the way the films are drawn during manufacture. Such oriented films may be easily torn to one direction but not the other, or they may preferentially yield to stretch in one direction. In the case of fibers, the drawn fiber is more resistant to yielding from stretching.

The degree of polymer chain orientation may be calculated from measurements with polarized light. The infrared absorption intensity of a band is strongest when the direction of the transition moment of the vibration and the electric vector of the light are parallel to each other, and it becomes a minimum when their directions are perpendicular. For polymer fibers and films that are drawn in one direction, the intensities of the parallel and perpendicular absorptions are given by $A_{\parallel} \propto m(\mu)^2 \cos^2 \theta$ and $A_{\perp} \propto m(\mu)^2 \sin^2 \theta$. In this case, the parallel and perpendicular absorptions correspond to the electric vector of the IR radiation being parallel and perpendicular to the direction of the draw axis, respectively. The terms m, μ, and θ are the mass of the oscillator, the transition moment, and the angle between the dipole moment and the draw axis, respectively. The dichroic ratio, R_0, for a perfectly oriented sample is given by

$$R_0 = \frac{A_{\parallel}}{A_{\perp}} = 2 \cot^2 \theta \tag{5}$$

As in the case of crystallinity, the sample is assumed to be composed of the oriented and nonoriented fractions, defined by f and $1-f$, respectively. Using the observed dichroic ratio of a partially oriented polymer film or fiber, the fraction of oriented material is given by [19]

$$f = \left(\frac{R-1}{R+2}\right) \bigg/ \left(\frac{3\cos^2 \theta - 1}{2}\right) = \left(\frac{(R-1)(R_0+2)}{(R+2)(R_0-1)}\right) \tag{6}$$

3.3. Correlation of IR and Raman Frequencies of Functional Groups with Microstructures

In this section, the correlation between IR/Raman frequencies of functional groups relevant to polymer analysis and polymer microstructures is described. The functional groups considered are the C=C double bond and substituted benzenes. The microstructure featuring the C=C double bond is important because the physical properties of materials such as polybutadiene and related polymers depend on the double-bond content of the overall structure. Also, a small percentage of residual C=C double bonds are formed in polyethylene chains as a result of side reactions. Substituted benzene compounds are used as starting materials such as bis-phenol A, phthalates, and benzoates.

3.3.1. Carbon–Carbon Double Bonds

It is well known that the out-of-plane C−H bending vibration frequencies of olefins, and in particular ethylene derivatives, are indicative of the double-bond configuration of the olefin species [1–8]. These vibrational bands are useful because the frequencies are generally insensitive to the presence of different substituents, but they are sensitive to the number of substituents and their positions. As indicated in Fig. 3, it is possible to correlate the microstructures in common polymers to these out-of-plane CH and CH_2 vibrations. The top structure represented in the Fig. 3 is a terminal vinyl group of polyethylene or can equally represent the side chain in the 1,2-addition polymer of butadiene. This structure gives rise to two characteristic peaks at 990 and 910 cm^{-1}, the latter peak being up to four times stronger than the other. The second structure is the *trans*-RHC=CR′H structure, which is observed in polybutadiene formed from 1,4-*trans*-addition polymerization, or the *trans*-vinylene bond in a polyethylene chain. The third structure corresponds to *cis*-RHC=CR′H unsaturated compounds, which give rise to a broad diffuse absorption band at 740 cm^{-1}. This structure is featured in polybutadiene obtained via 1,4-*cis*-addition polymerization.

Figure 4 presents the IR spectra in the out-of-plane bending vibration region of two different forms of polybutadiene. These are in agreement with the nature of sample (a), which has between 96% and 98% of the 1,4-*cis* component, as indicated by the major broad *cis*-vinylene absorption band at 740 cm^{-1}. The weak peaks at 910 cm^{-1} and 994 cm^{-1} are both assigned to the vinyl group, resulting from 1,2-addition. The *trans*-vinylene absorption band, which is observed at 965 cm^{-1}, is also weak. The out-of-plane CH and CH_2 bending vibrations of sample (b), whose major component is the 1,2-addition component, are also consistent with the assigned microstructure. It

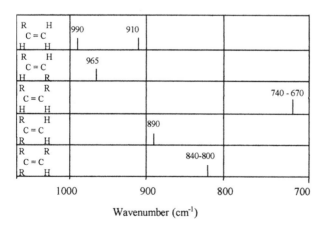

Figure 3   Characteristic IR frequency of C—H out-of-plane vibration of ethylene derivatives.

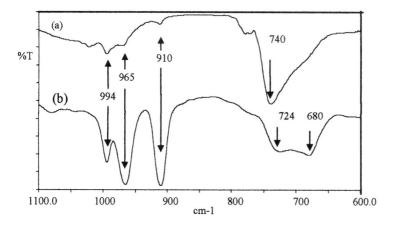

Figure 4 Comparison of IR spectra of (a) 1,4-*cis*- only (96–98%) and (b) low *cis*-poly(butadiene) samples.

is evident that the vinyl group bands are strong and so is the *trans*-vinylene band. The *cis*-vinylene band is weak and appears to be split into two bands at 724 cm^{-1} and 680 cm^{-1}. For future reference within this text, the band at 680 cm^{-1} is utilized for the calculation of the 1,4-*cis* component concentration in styrene–butadiene copolymers, as reported by Binder [20]. It is known that the frequency of the 1,4-*cis* double-bond out-of-plane CH bending vibration shifts toward lower frequencies with a decrease in its concentration. The observed trend agrees with this expectation.

In the case of Raman analysis of the microstructures of polybutadiene and related structures, the C=C stretching vibration frequency is observed around 1600 cm^{-1} – 1700 cm^{-1}. This is a direct measure of the double bond, and is used instead of CH out-of-plane bending mode as used in the IR at around 1000–700 cm^{-1}. Figure 5 is equivalent to Fig. 3 and provides Raman frequencies of microstructures featuring the > C=C < double bond. The frequencies of the > C=C < stretching vibration in the Raman spectra of these two copolymers are located between 1660 and 1630 cm^{-1}, and their assignments [21] agree with the chemical structures provided in Fig. 5. For example, the C=C double bond stretching Raman bands of three example compounds are shown in Fig. 6. Sample (b), with 96%–98% of the 1,4-*cis* component, shows a single C=C band at 1653 cm^{-1} (cf. 1654 cm^{-1} for IR), while sample (c), with the major component being the 1,2-addition component, shows the major component C=C band at 1642 cm^{-1} (cf. 1640 cm^{-1} for IR). Acrylonitrile–butadiene–styrene copolymer (ABS) sample (a),

General characteristic Raman wavenumbers from carbon-carbon double bond stretching		
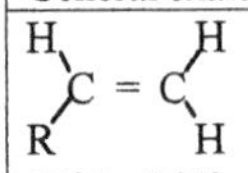 H C = C H / R \ H 1635 - 1650	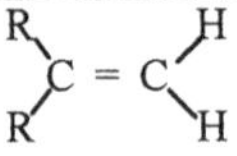 R C = C H / R \ H 1640 - 1660	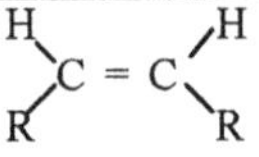 H C = C H / R \ R 1635 - 1660
Vinyl group (1,2-addition)	Vinylidene group	*cis*-vinylene (cis 1,4-addition)
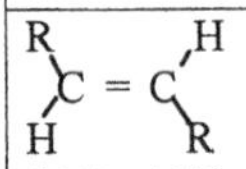 R C = C H / H \ R 1665 - 1680	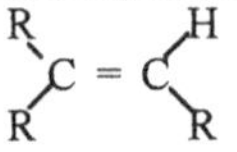 R C = C H / R \ R 1665 - 1695	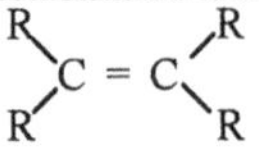 R C = C R / R \ R 1665 - 1685
trans-vinylene (trans 1,4-addition)		

Figure 5 Characteristic frequencies in Raman spectra from carbon–carbon double-bond stretching (11). (Adapted from Ref. 22.)

which is known to have a 1,4-*trans*-addition as its major component, gave the most intense peak at 1668 cm^{-1} (cf. 1668 cm^{-1} for IR).

3.3.2. Substituted Benzenes

The correlation between the vibrational frequencies of benzene derivatives with the number and positions of substituents is well documented in most texts dealing with infrared group frequencies [1,2,5–7]. The main diagnostic frequency ranges used to correlate the structures and the corresponding vibrations are the combination band region (2000–1700 cm^{-1}) and the

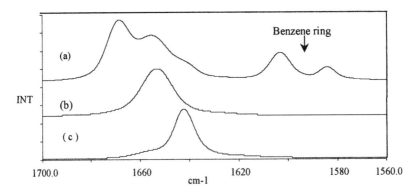

Figure 6 Raman C═C stretching vibration of (a) ABS (trans 1,4-addition predominant in butadiene moiety); (b) 97% 1,4-*cis*-polybutadiene, and (c) 1,2-polybutadiene.

C−H out-of-plane vibrations of the substituted benzene ring located between 900 and 700 cm^{-1}. A summary of the relevant vibrational frequencies is provided in Table 1. Judicious use of these frequencies can indicate the number and position of substituents on the aromatic ring.

The frequencies shown in Table 1 will not be always applicable, and certain substituents, the carboxylate ester, COOR group, for example, will deviate from the normal values. Phthalates, isophthalates, terephthalates, and benzoates are good examples. Table 2 and Fig. 7 indicate the characteristic IR frequencies of these specific groups. Raman spectroscopy is also useful for the determination of the number and position of substituents on a benzene ring. A method to analyze the number and position of the benzene substituents based on existence and absence of Raman band in the 1100– 600 cm^{-1} region is summarized in Fig. 8 [22].

Table 1 IR Bands of Substituted Benzenes

	cm^{-1}	
Benzene	671	–
Monosubstituted benzene	770–730	710–690
1,2-disubstituted	770–735	–
1,3-disubstituted	810–750	710–690
1,4-disubstituted	833–810	–
1,2,3-trisubstituted	780–760	745–705
1,2,4-trisubstituted	825–805	885–870
1,3,5-trisubstituted	865–810	730–675
1,2,3,4-tetrasubstituted	810–800	–
1,2,3,5-tetrasubstituted	850–840	–
1,2,4,5-tetrasubstituted	870–855	–
Pentasubstituted	870	–
Five adjacent free hydrogen atoms	770–730	710–690
Four adjacent free hydrogen atoms	770–735	–
Three adjacent free hydrogen atoms	810–750	–
Two adjacent free hydrogen atoms	860–800	–
One free hydrogen atom	900–860	–

Source: Ref. 23.

Table 2 IR Characteristic Bands of Benzoates, Phthalates, Isophthalates, and Terephthalates

Benzoates	Phthalates	Isophthalates	Terephthalates
1725 (C=O)	1725 (C=O)	1725 (C=O)	1725 (C=O)
1600 (ring)	1600 (ring)	1610 (ring)	1610 (weak or none)*
1580 (ring)	1580 (ring)	1590 (ring)	1590 (weak)*
1280 (C−O)	1280 (C−O)	1250 (C−O)	1410 (ring)
1110 (OCH$_2$)	1120 (OCH$_2$)	1100 (OCH$_2$)	1280 (C−O)
1070 (ring)	1080	—	1110 (OCH$_2$)
1030 (ring)	1040 (ring)	—	1020 (ring)
710 (ring)	745 (ring)	730 (ring)	730 (ring)

Source: *Adapted from Ref. 24.

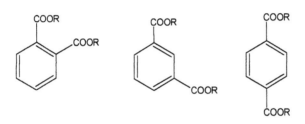

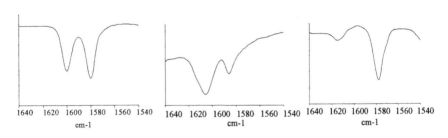

Figure 7 Chemical structures of (left to right) phthalates, isophthalates, and terephthalates. IR spectral features of *ortho*-, *iso*-, and *tere*-phthalates in 1600 cm^{-1} region. Spectral features of the phthalate benzene ring at 1600 cm^{-1} are sometimes useful. Note that these rules apply only for the phthalates, not for general disubstituted benzene compounds.

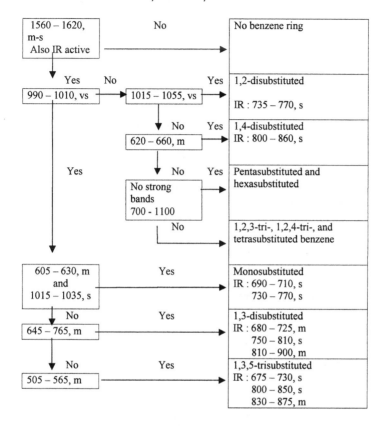

Figure 8 Scheme for the determination of substitution pattern of benzene derivatives from Raman spectra: m, medium; s, strong; vs, very strong; IR, characteristic IR band.

4. CHARACTERIZATION OF POLYMERS

4.1. Polyolefins

In this section, we are concerned with a polyolefin class of polymers, which includes polyethylene, polypropylene, polybutene, polybutadiene resins, and their various copolymers.

4.1.1. Polyethylene (PE)

The IR and Raman spectra of polyethylene (PE) are shown in the Reference Spectra in the Appendix. If we consider the total macromolecule as a single linear chain, we are theoretically dealing with a long chain of methylene groups, $-(CH_2)_n-$, with two terminating groups (usually a methyl group). Because of this simple chemical structure, with only carbon and hydrogen atoms, predominantly as methylene, the IR [25–28] and Raman [29] spectra are consequently very simple. The main stretching vibrations appear in IR spectrum at 2927 cm^{-1} (asymmetric CH$_2$ stretch) and 2852 cm^{-1} (symmetric CH$_2$ stretch). The CH stretching vibrations in the Raman spectrum of the reference sample appear at 2920 and 2882 cm^{-1} (asymmetric), and at 2848 cm^{-1} (symmetric).

The main bending modes of the CH$_2$ groups are located in the IR spectrum at 1475 and 1463 cm^{-1} (the CH$_2$ scissors vibration, exhibiting a 12-cm^{-1} splitting due to the crystalline nature of this specific polymer sample) and at 730 and 720 cm^{-1} (CH$_2$ rocking, again with splitting associated with the crystallinity of the sample). The Raman active bands of the bending vibrations are the CH$_2$ scissors vibration at 1460 and 1440 cm^{-1}, the CH$_2$ wagging at 1417 cm^{-1}, and the CH$_2$ twist at 1295 cm^{-1}. A weak band at 1169 cm^{-1} is a backbone vibration assigned to CH$_2$ rocking. In addition, $C-C$ skeletal vibrations are also observed in the Raman spectrum in the form of the $C-C$ bend/stretch at 1130 cm^{-1} and the $C-C$ stretch at 1061 cm^{-1}. Weak fingerprint region IR absorption bands appear at 1370 cm^{-1} (methyl $C-H$ bend), 1353 cm^{-1}, and 1303 cm^{-1}.

The occurrence of "impurities" and structural deviations, such as branching and double-bond formation, depend on the method of polymerization. The degree of branching is related to the crystallinity and consequently influences the density of the polyethylene. The more branching the PE has, the more amorphous (i.e., the lower the density; *vide infra*) and the more transparent the PE becomes. Three types of polyethylene are differentiated: high-pressure, low-density polyethylene (LDPE), linear low-density polyethylene (LLDPE), and high-density polyethylene (HDPE). All three of these PE types are manufactured commercially [30].

LDPE is polymerized in radical reactions at elevated pressure and temperature. Owing to the nature of radical reaction, it has a significant amount of both short- and long-chain branching. As a result, LDPE is the most amorphous among the classes of polyethylene. HDPE is polymerized via an ionic reaction mechanism featuring Ziegler-Natta-type catalysts or solid acidic catalysts. HDPE has a very small degree of branching, and as a result it is a highly crystalline polymer. Consequently, HDPE has high density and sometimes the sample is not transparent, but appears milky white or translucent.

LLDPE is polymerized via an ionic reaction, similar to HDPE, and involves small amounts of an α-olefin such as butene-1, hexene-1,4-methylpentene, or octene-1, which are added as a comonomer. When an α-olefin such as butene-1 (CH$_3$CH$_2$CH=CH$_2$) is inserted into the polymer chain, the

CH_3CH_2- group becomes a short side chain of the PE molecule. Because the side chain reduces the crystallinity of polyethylene, the α-olefin content is controlled to match to the intended crystallinity or density. Needless to say, LLDPE has more short-chain branching than HDPE.

4.1.1.1. General Characterization

The first step in the characterization of polyethylene is to determine the class or morphology of the material: HDPE, LLDPE, or LDPE (also high-pressure, low density PE, HPLDPE). When the IR spectra of polyethylene are thoroughly examined, subtle but definite differences are observed among with these different forms of polyethylene [31–33]. Conversely, the Raman spectra of these materials do not show any significant variation, except for differences in line width [34]. For this reason, the infrared spectra of the polyethylenes will be the main focus of this section, and only a brief review of the Raman spectra will be provided later.

In order to classify the sample polyethylene relative to the three types, the infrared spectral regions at 1400–1350 cm^{-1} and 1000–850 cm^{-1} are examined. Figures 9a and 9b show the spectra of LDPE and HDPE, respectively. The main difference is seen at ca. 1380 cm^{-1}. In the case of LDPE, the band in the region 1400–1330 cm^{-1} consists of three peaks, while that of the HDPE consists of two peaks. A peak at 1377 cm^{-1} is assigned to the CH_3- groups terminating the short- and long-chain branching, and the main polyethylene chain. If no individual peak is observed at 1377 cm^{-1}, then the material is a HDPE. LDPE and LLDPE both produce this peak, although the intensity is generally higher for LDPE than for LLDPE. One must be

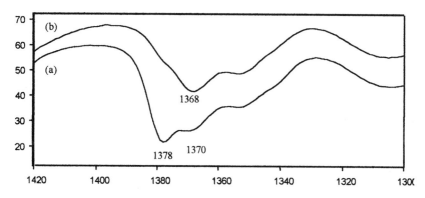

Figure 9 IR spectra of (a) LDPE and (b) HDPE [ordinate of (a) offset −10%T].

careful with the interpretation as the intensity of the 1377 cm^{-1} increases, because this peak will obscure the 1370 cm^{-1} peak, leaving what appears to be a two-peak arrangement. Therefore, one must evaluate the peak position and confirm the presence or absence of a peak at 1377 cm^{-1}. In this way, HDPE is differentiated from LDPE and LLDPE.

The next step is to differentiate LDPE from LLDPE. In order to do this, it is necessary to examine the peaks in the region from 990 to 890 cm^{-1} to determine the presence and nature of residual unsaturation. A pair of the absorbances at 990 and 910 cm^{-1} (the 910 cm^{-1} peak is approximately four times more intense than the 990 cm^{-1} peak) are assigned to a terminal vinyl group; $CH_3-(CH_2)_n-\underline{CH=CH_2}$ (see Sec. 3.3.1). Likewise, an absorption at 890 cm^{-1} is assigned to a vinylidene group $\underline{RR'C=CH_2}$. An absorption at 965 cm^{-1} represents trans unsaturation within the polyethylene main chain (*trans*-$\underline{RCH=CHR'}$). In the case of LLDPE, the intensities of the vinylidene and the terminal vinyl group (910 cm^{-1}) are almost identical, and both are weak as indicated in Fig. 10a. This relationship is common to the Ziegler-Natta-type ($AlEt_3$–$TiCl_4$–$MgCl_2$) polymerization [35]. HDPE produced by polymerization with a Ziegler-Natta catalyst will also show this feature, although a peak at 1377 cm^{-1} will not be observed. As noted in Fig. 10b, the vinylidene group is the most prominent group in the case of LDPE [35]. In this way, all three forms of polyethylene are differentiated.

4.1.1.2. Different Forms of HDPE

If the spectrum of an HDPE sample shows noticeably strong peaks associated with a terminal vinyl group (990 and 910 cm^{-1}) compared with other HDPE spectra, and without the trans unsaturation (965 cm^{-1}), as shown in

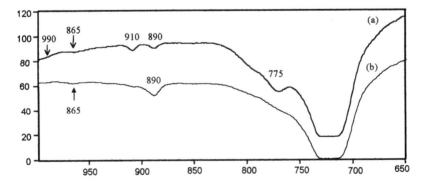

Figure 10 IR spectra of LLDPE with (a) butene-1 comonomer and (b) LDPE. Spectrum (b) ordinate offset 18%.

Fig. 11a, the HDPE is probably polymerized via a catalyst developed by Phillips Petroleum Company, which is a $SiO_2–Al_2O_3–CrO_3$-type material. A listing of different catalyst types and the corresponding presence of double-bond species, and short-chain branchings associated with those catalysts, is provided in Table 3. Comparing the spectra in Figs. 11a and 11b with Table 3, the catalysts used in the formation of the HDPE samples are assigned to Phillips and Ziegler-Natta or Wacker types, respectively. If the HDPE under investigation shows a single absorption assigned to trans unsaturation (965 cm^{-1}), the HDPE is most likely formed from a $\gamma–Al_2O_3–MoO_2$-type catalyst (originally developed by Standard Oil Company). Note that this evaluation procedure is approximate, and the relative intensities of the assigned double bonds can change depending on other polymerization factors such as temperature, pressure, and solvent.

4.1.1.3.   Different Forms of LLDPE

Once it is confirmed that the sample is a LLDPE, it is necessary to identify the α-olefin used as the comonomer. The following comonomers are commonly utilized in the polyethylene industry: butene-1, hexene-1, 4-methylpentene-1, and octene-1. The cost of LLDPE increases with the order of the comonomers as listed, and LLDPE formed from butene-1 as comonomer is the most common form. Butene-1 will produce ethyl branching, $-CH_2-CH_2-CH_2-CH(\underline{CH_2CH_3})CH_2-CH_2-$, which produces an

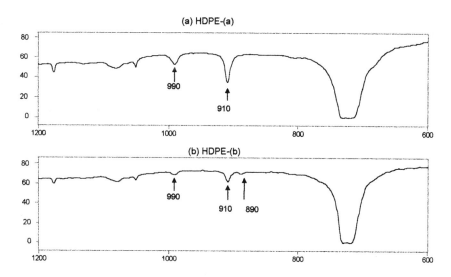

Figure 11   IR spectra of two different kinds of HDPE.

Table 3 Characteristics of Various Polyethylenes

Branching	Total methyl	Methyl	Ethyl	Terminal	Catalyst
LDPE	10–30	~5%	~35%	~25%	Free radical
HDPE					
Ziegler	3–5	0	~20%	~80%	$Et_3Al–TiCl_4$
Wacker	<5	—	—	—	$SiO_2–Al_2O_3–CrO_3$
Phillips	<1.5	—	—	—	$\gamma–Al_2O_3–MoO_2$
Standard	<5	—	—	—	—
Double bond	Total (C=C)	Vinylidene	Vinyl	t-Vinylene	
LDPE	0.6–0.8	~70–90%	~4–14%	~6–16%	
HDPE					
Ziegler	0.4–0.7	~30–35%	~40–50%	~20–25%	
Wacker	~0.7	~20%	~80%	Trace	
Phillips	~1.5	~1–2%	~90–95%	~4–8%	
Standard	~0.8	Trace	Trace	~90%	

Total methyl groups: numbers of methyl groups per 1000 carbon atoms.
Total double bonds: numbers of double bonds per 1000 carbon atoms.
Source: Ref. 33.

absorption at 775 cm^{-1}, as seen in Fig. 10a. The LLDPE in this example is produced from butene-1 as the comonomer, and is polymerized with a Ziegler-Natta-type catalyst. The butene-1 content can be determined from the intensity of 1380 cm^{-1} absorption (Table 4).

Propylene may be used as a comonomer (cf. Sec. 4.1.3, ethylene–pro-pylene copolymer). In this case, it should be pointed out that propylene comonomer produces the methyl branching, $-CH_2-CH_2-CH(CH_3)-CH_2-CH_2-$. In order to confirm that propylene is used as a comonomer, one must confirm the existence of methyl branching by absorptions at 1150.7 and 936.7 cm^{-1}. Note that the increased intensity of the 1380 cm^{-1} band is diagnostic, but other chain branching will also cause an increase in the intensity. If 4-methyl pentene-1 is used as a comonomer, the side chain is an isobutyl group, $-CH_2CH(CH_3)_2$, which has a geminal dimethyl config-uration. This produces a characteristic doublet for the CH$_3$ bending, with one peak at 1383 cm^{-1} and the other at approximately 1370 cm^{-1}, over-lapping the 1368 cm^{-1} peak. The higher-frequency peak at 1383 cm^{-1} is evident in the spectrum in Fig. 12a, which may be compared to the spectrum of a LLDPE, produced from a linear comonomer that exhibits the peak at 1377 cm^{-1}, as illustrated in Fig. 12b.

Table 4 provides the assignments for the short-chain branching, and indicates how to determine the comonomer content. The adjacent peak may change the position and intensity of the CH$_3$ $-$ peak, and a compensa-tion method based on absorbance subtraction of a polymethylene spectrum as reference is recommended. Using this procedure, the peaks at 1350 and

Table 4 Positions of CH$_3$ Groups and Particular Chain Branchings

Branching	δ-CH$_3$ cm^{-1}	Coefficient	Peak position	Coefficient
Methyl	1377.3	0.39	1150.7	5.9
			936.7	19
Ethyl	1379.4	0.59	772.2	4.4
n-Butyl	1378.1	0.70	894.3	8.9
i-Butyl	1383.6	0.42	1169.4	1.4
	1365.6		919.5	6.4
n-Hexyl	1377.9	0.76	889.8	9.3
n-Decyl	1378.3	0.67	893.7	9.3

Coefficient: degree of branching per 1000 carbons as expressed.
SCB/1000C = [coefficient] $\cdot$ $A/(t \cdot r)$; (A = absorbance; t = thickness (cm), r = density (g/cm^3)].
Source; Refs. 36, 37.

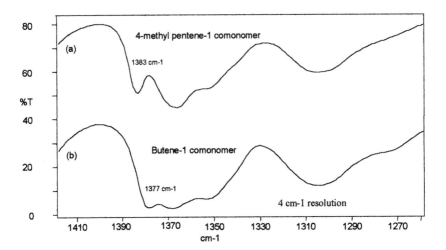

Figure 12 Methyl band at 1380 cm^{-1} region of (a) 4-methylpentene-1 comonomer and (b) butene-1 comonomer.

1303 cm^{-1} should be completely removed, leaving a residual CH$_3-$ peak. It is recommended practice in infrared spectroscopy that one should use more than the single peak around 1378 cm^{-1} for the characterization.

4.1.1.4. Determination of Crystallinity and Density

A traditional method used in the polyethylene industry utilizes the peak intensities of absorptions at 1901 and 1303 cm^{-1} as described below. The 1901 and 1303 cm^{-1} absorptions are assigned to crystalline and amorphous material, respectively. Teranishi and Sugahara [38] derived the following empirical relationships between absorption intensities and the degree of crystallinity ($X\%$) and density (r):

$$X = 18.9\frac{A_{1901}}{t} \tag{7}$$

$$X = 100 - \frac{5.61 A_{1303}}{t} \tag{8}$$

$$X = \frac{337}{(A_{1303}/A_{1901}) + 3.37} \tag{9}$$

$$\frac{100}{r} = \frac{X}{r_c} + \frac{100 - X}{r_a} \tag{10}$$

where $r_c = 1.0070$ and $r_a = 0.8599\,\text{g/cm}^3$. Baselines for the 1901 cm^{-1} band and the 1303 cm^{-1} band are constructed between 1910 and 1890 cm^{-1}, and 1389 and 1290 cm^{-1}, respectively.

4.1.1.5. Determinations of Long-Chain Branching (LCB) and Terminal Methyl Content

The evaluation of short-chain branching (SCB) was mentioned above. The measurement of LCB has traditionally been considered impractical. The term "long" in this case means $-(CH_2)_n-$, where $n \geq 6$. Hosoda [39] found that the orientation of the methyl group of the branch is chain length-dependent in the case of LLDPE, and that a qualitative estimation of the chain length is possible.

4.1.1.6. Screening the High-Environment Stress-Crack Resistance (ESCR) of Polyethylene

It is known that polyethylene has higher mechanical strength and higher ESCR when the degree of short-chain branching is constant over the molecular weight distribution, or in other words is independent of the molecular weight. In order to achieve this SCB behavior, a modification of the method of catalysis is required. GPC-FTIR is an excellent tool for establishing whether a polyethylene sample has constant SCB over the entire molecular-weight distribution. Since polyethylene is insoluble in solvents at ambient temperature, a high-temperature (130°C) GPC instrument, equipped with a high-temperature flow cell, is required for the analysis [40]. The use of a combined GPC-FTIR system provides a means of complete sample analysis: the GPC separates the polyethylene as a function of molecular weight, and the FTIR provides an on-line method for the characterization of the GPC effluent. The degree of SCB is determined as the number of methyl groups per 1000 carbon atoms, minus one terminal methyl group, as a function of molecular weight. A couple of examples are provided here to illustrate analysis. The first, a commercially available LLDPE, exhibits a high degree of SCB in the low-molecular-weight components, and low degree of SCB in the high-molecular-weight components, as illustrated in Fig. 13a. In the second example, Fig. 13b, a different LLPDE exhibits a SCB almost independent of molecular-weight distribution. The latter LLDPE would typically have better physical properties than the previous example. It is known that a traditional Ziegler-Natta-type catalyst will produce the former type of LLDPE, with a SCB that exhibits molecular-weight dependency [41]. The most successful HDPE for thin-film use showed very small SCB, and the degree of SCB was constant over the wide range of molecular-weight components, as shown in Fig. 13c.

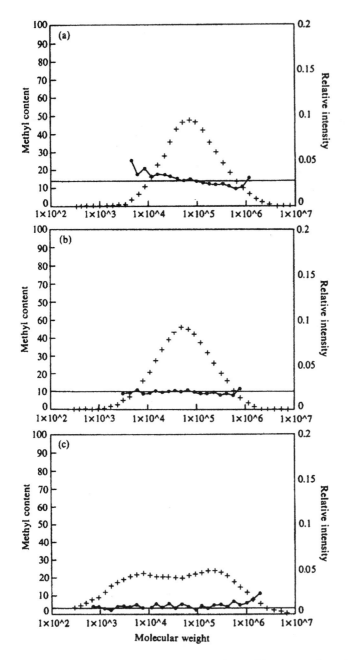

Figure 13 Degrees of short-chain branching as a function of molecular weight determined by GPC-IR: (a) LLDPE 1; (b) LLDPE 2; (c) HDPE.

4.1.1.7. Characterization of Polyethylene from Raman Spectra

In the molten form, the three bands at 1463, 1441, and 1418 cm^{-1}, observed in the highly crystallized forms of polyethylene, change to a broad asymmetrical band with its maximum at 1440 cm^{-1}. In a similar manner, the narrow band at 1296 cm^{-1} of crystalline polyethylene is broadened and shifted to the band at 1303 cm^{-1} in the melt. Finally, the bands at 1130 and 1064 cm^{-1} disappear and coalesce to form a single weak, broad band centered at 1080 cm^{-1} [34]. The band decomposition of the Raman spectrum from a partially crystallized polyethylene produces a fractional sum of the crystalline and amorphous forms of polyethylene, as illustrated in Fig. 14.

Many of the Raman bands are sensitive to the crystallinity of the polyethylene, and the examples provided are useful in the determination of the degree of crystallinity. The intensity of the broad weak band at around 890 cm^{-1} is an approximate inverse indicator of the molecular weight [42].

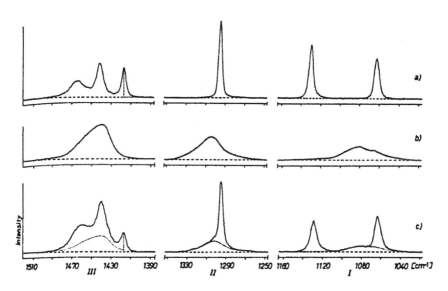

Figure 14 Raman spectra, measured in the spectral ranges for the C−C stretching vibration (I), the −CH$_2$− twisting vibration (II), and the −CH$_2$− bending vibration (III) for a sample of an extended-chain polyethylene at 25°C (a); the melt at 150°C (b); and a branched, partially crystalline polyethylene at 25°C (c). The dotted line represents the amorphous component.

Nishikida and Porro [43] correlated the Raman spectra from controlled
LLDPE standards to the crystallinity and degree of short-chain branching
of the material, by the use of multivariate analysis. Six different levels of
SCB and density of the three different copolymers (butene-1, octene-1, and
4-methyl pentene-1 as comonomers) were used as the training set. The estimated values of SCB and density are compared to the certified values (as
determined by the gradient column method and ^{13}C NMR for density and
SCB, respectively) in Figs. 15a and 15b. Although it is not possible to
characterize the samples via the Raman spectra by visual comparison, the
multivariate analysis was able to determine both the crystallinity and
degree of short-chain branching, as well as providing an identification of
the comonomer used in the LLDPE, as shown in Fig. 16.

4.1.1.8. Ethylene Copolymers: Ionomers

Ionomer is a general name for copolymers made of α-olefin and an unsaturated carboxylic acid, where the acid hydrogens of the copolymer are
partially replaced with metal ions, such as sodium or zinc, as illustrated
below. Example industrial ionomers include copolymers of ethylene and
acrylic acid or methacrylic acid.

$$CH_2=CH_2 + CH_2=C(CH_3)COOH \longrightarrow -(CH2)_n-CH_2-\underset{\underset{COOH}{|}}{C}(CH_3)-(CH_2)_{n'}-$$

$$\underset{\underset{\substack{COONa \\ or\ COO(Zn)_{1/2}}}{|}}{\overset{NaOH}{\underset{or\ ZnO}{\longrightarrow}}}\ -(CH_2)_n-C(CH_3)-(CH_2)_{n'}-\underset{\underset{COOH}{|}}{C}(CH_3)-(CH_2)_{n''}-$$

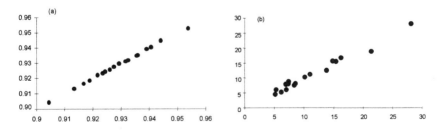

Figure 15   Estimated density (a) and SCB (b) plotted versus certified values
for three different comonomers and six different SCBs for each comonomer.

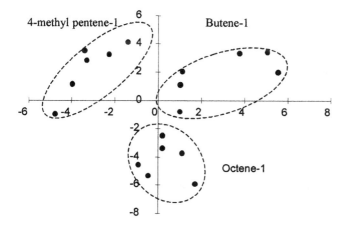

Figure 16 Discrimination of LLDPE with three different comonomers by plotting loadings of factor 4 (ordinate) and factor 3 (abscissa).

The IR and Raman spectra of this example ionomer are provided in Reference Spectrum 5. In addition to the spectral features of polyethylene, characteristic bands of COOH (C=O stretch at 1698 cm^{-1}), COONa (COO$^-$ stretch at 1558 cm^{-1}) or COO(Zn)$_{1/2}$ (COO$^-$ stretch at 1590 cm^{-1}), and C−O stretch at 1262 cm^{-1} are observed in the infrared spectrum. However, given the low acid content, and the weak Raman scattering cross section of the C=O group, the Raman spectrum does not provide characteristic C=O-related spectral features. As a result, the Raman spectrum is almost identical to that of polyethylene, except that the line widths are slightly broader.

As noted from the frequencies quoted above, the counterion, sodium or zinc, is correlated to the position of COO$^-$ stretching frequency. The degree of neutralization of the carboxylic acid is determined from the peak intensity at 1698 cm^{-1} obtained from the sample, compared to the intensity of the same peak following treatment of the polymer with hydrochloric acid (to convert the salt form to the acid form), from the following equation:

$$\text{Salt}(\%) = 100 - 100\frac{A_{1698}(\text{ionomer})/t}{A_{1698}(\text{acidified})/t'} \tag{11}$$

where t and t' are the thicknesses of the samples. When the thickness of the sample is unknown, a band related to the polyethylene backbone may be used as an internal standard for correlation with thickness [44].

4.1.2. Polypropylene (PP)

Commercially available polypropylene, in the form of pellets, films, and fibers, exists as isotactic polypropylene, and this is produced by well-controlled stereoregular head-to-tail addition polymerization reaction with Ziegler-Natta-type catalysts. Formed in this manner, isotactic polypropylene is a crystalline polymer. Commercial samples typically contain small amount of atactic and/or syndiotactic polypropylene. Furthermore, blocks with different stereoregularities are also observed. The reference spectra (Reference Spectrum 2) in the Appendix provide the IR and Raman spectra of isotactic polypropylene.

The IR and Raman spectra of isotactic polypropylene in the fingerprint region are shown in Figs. 17a and 17b, respectively. Readers will observe a close correspondence between the IR and Raman spectra, with obvious exceptions in band intensity. This results from the symmetry properties of the polypropylene crystal lattice. The IR bands, highlighted by arrows, at 1167, 998, 899, and 842 cm^{-1} of isotactic polypropylene are sensitive to crystallinity. A shoulder located on the lower frequency side of the 1168 cm^{-1} absorption (or 1153 cm^{-1} band in Raman) is assigned to the methyl groups of the regular head-to-tail sequence, as illustrated in Table 5. Sequences produced by the head-to-head and/or tail-to-tail additions, as

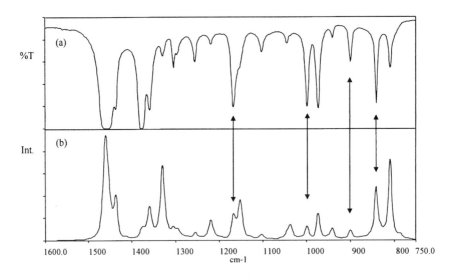

Figure 17 IR and Raman spectra of isotactic polypropylene.

Table 5 Ethylene–Propylene Sequence and Peak Position

Monomer sequence	P–P sequence	Position (cm^{-1})	Chemical structures
P-P	HT	815	$-CH_2-CH(CH_3)-CH_2-CH(CH_3)-CH_2-$
P-P	TT	750	$-CH_2-CH(CH_3)-CH(CH_3)-CH_2-$
P-P	HH	752	$-CH_2CH(CH_3)-CH_2-CH_2-CH(CH_3)-$
P-E-P	TT	752	$-CH_2CH(CH_3)-CH_2-CH_2-CH(CH_3)-$
P-E-P	HT	736	$-CH(CH_3)-CH_2-CH_2-CH_2-CH(CH_3)-$
P-E-P	HH	726	$-CH_2-CH(CH_3)-(CH_2)_4-CH(CH_3)-CH_2-$
P-E-E-P	HT	722	$-CH(CH_3)-(CH_2)_5-CH(CH_3)-$

P, propylene; E, ethylene; HT, head-to-tail; TT, tail-to-tail; HH, head-to-head.

shown in Figs 18b and 18c, caused by side reactions, give rise to the peaks at 752 and 1133 cm^{-1}, respectively (refer to Sec. 4.1.3).

The crystallinity of isotactic polypropylene may be calculated [45] from

$$X(\%) = 109 \cdot \frac{A_{998} - A_{920}}{A_{974} - A_{920}} - 31.4 \tag{12}$$

The infrared band at 875 cm^{-1} of the syndiotactic polypropylene is used to estimate crystallinity. These three crystallinity sensitive bands (1167, 998, and 875 cm^{-1}) do not appear in the IR spectrum of atactic polypropylene. If a sample of polypropylene is drawn to one direction—that is, it is uniaxially oriented—some of the absorption band intensities become dependent on the angle between the drawn axis and the angle of polarization of the infrared radiation.

The IR spectra of (a) isotactic, (b) syndiotactic, and (c) atactic PP are compared in Fig. 19. The poor stereoregularity of atactic polypropylene (c) is reflected in the broadening of the absorption bands compared to the corresponding bands in isotactic and syndiotactic polypropylene.

4.1.3. Ethylene-Propylene Copolymers

In terms of IR and Raman spectroscopy, an ethylene–propylene copolymer may be viewed as an extreme case because both ethylene and propylene may exist in crystalline forms. The IR and Raman spectra of a random copolymer with one major component will reflect the crystalline nature of that

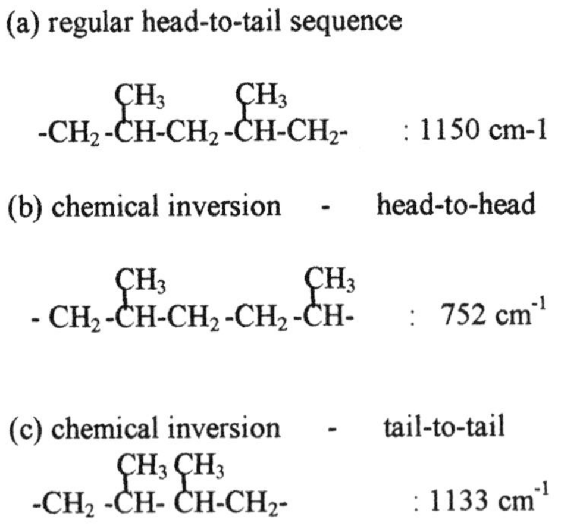

(a) regular head-to-tail sequence

$$\underset{\text{-CH}_2\text{-CH-CH}_2\text{-CH-CH}_2\text{-}}{\overset{\text{CH}_3 \qquad\quad \text{CH}_3}{}} \qquad : 1150 \text{ cm-1}$$

(b) chemical inversion - head-to-head

$$\underset{\text{- CH}_2\text{-CH-CH}_2\text{-CH}_2\text{-CH-}}{\overset{\text{CH}_3 \qquad\qquad \text{CH}_3}{}} \qquad : \ 752 \text{ cm}^{-1}$$

(c) chemical inversion - tail-to-tail

$$\underset{\text{-CH}_2\text{ -CH- CH-CH}_2\text{-}}{\overset{\text{CH}_3\ \text{CH}_3}{}} \qquad : 1133 \text{ cm}^{-1}$$

Figure 18 Regular addition sequence and chemical inversions of polyolefins.

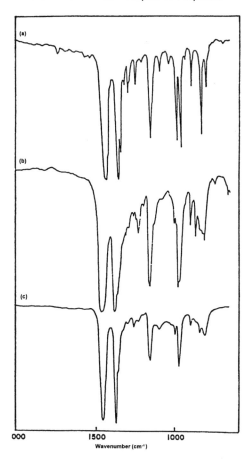

Figure 19 The IR fingerprint region of (a) isotactic, (b) syndiotactic, and (c) atactic polypropylene.

polymer of the major component and will produce a relatively weak spectrum of the isolated minor component. As the fraction of the minor component increases, the features of isolated minor component increase and the features of the major component decrease. A transition occurs where the original minor component dominates as a crystalline entity, and the other component is reduced in significance to become the isolated entity. In this way, the crystallinity of both components pass through a minimum value as the composition changes. The IR spectra of ethylene–propylene copolymers with low (below 10%) and high levels of ethylene are provided in Figs. 20b and 20c, respectively. The copolymer with low ethylene content shows the

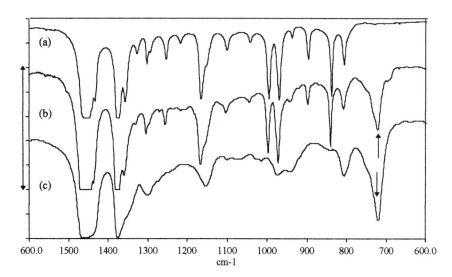

Figure 20 IR spectra of (a) polypropylene, (b) propylene–ethylene (low-ethylene) copolymer, and (c) propylene–ethylene (32:68) copolymer.

dominant features of the polypropylene homopolymer (Fig. 20a), and the copolymer with high ethylene content loses the crystalline bands completely (Fig. 20c), as expected from the previous discussion. An extra peak at around 726 cm^{-1} (denoted by an arrow in each spectrum) is used to determine whether the material is a random copolymer or block copolymer.

The band locations within the frequency range of 815–720 cm^{-1}, and the corresponding chemical sequences, are summarized in Table 5. A random copolymer is evaluated in terms of methylene $-(CH_2)_n-$ rocking band [46–48]. When $n = 1$, the sequence corresponds to the regular head-to-tail addition of the propylene homopolymer, and the peak position is located at 815 cm^{-1}. A $-(CH_2)_2-$ group, produced as a result of head-to-head sequence of the propylene homopolymer or the insertion of an ethylene moiety into a tail-to-tail propylene sequence, is identified from the peak at 752 cm^{-1}. A polypropylene tail-to-tail addition sequence is characterized from a peak at 750 cm^{-1}. The $-(CH_2)_3-$ group, which results from the insertion of one ethylene molecule into the regular HT addition polymerization sequence of propylene, is identified by a band at 736 cm^{-1}. The isolated ethylene monomer units due to insertion of a single ethylene unit in a chain are indicative of a random copolymerization. The $-(CH_2)_4-$ group is reported to show the peak at 726 cm^{-1}. Methylene sequences $-(CH_2)_n-$ where $n \geq 6$, give rise to the normal long-chain methylene rocking band at

720 cm^{-1}. An insertion of two or more ethylene molecules in the regular head-to-tail and head-to-head polymerization result in the $-(CH_2)_m-$ ($m = 5, 7, 9, \ldots$) and $-(CH_2)_n-$ ($n = 6, 8, 10, \ldots$) sequences, respectively. These long-chain sequences are indicative of block copolymerization. If the polyethylene block is crystalline, the band is observed as a doublet at 730 and 720 cm^{-1}.

It is useful to use resolution enhancement or deconvolution techniques to investigate the underlying structure of this band. The ~ 720 cm^{-1} band of an ethylene–propylene copolymer and its deconvolved pattern from underlying absorptions are provided in Fig. 21. Three peaks are clearly separated from the original broad line. The most intense peak at 722 cm^{-1} is assigned to the P–E–E–P sequence with five consecutive CH$_2$ groups, resulting from the insertion of two ethylene units into the normal head-to-tail sequence. The 733 and 751 cm^{-1} bands are assigned to P–E–P with three and two consecutive CH$_2$ groups, respectively. These are rationalized in terms of the insertion of a single ethylene unit into a head-to-tail sequence and a tail-to-tail sequence (or a head-to-head sequence without any ethylene insertion), respectively. This is defined as an ethylene–propylene random copolymer, where the ethylene-to-propylene ratio may be determined from the intensity ratio of peaks at 722, 736, and 751 cm^{-1}.

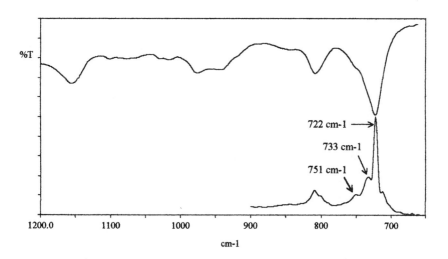

Figure 21 Deconvolved peak at 726 cm^{-1} of ethylene–propylene copolymer.

4.1.4. Polybutene-1 and Poly 4-Methyl Pentene-1 (TPX)

Polyethylene, polypropylene, poly(butene-1), and poly(4-methyl pentene-1) may be represented as $-(CH_2-CHX)_n-$ with $X = -H$, $-CH_3$, $-CH_2CH_3$, and $-CH_2-CH(CH_3)_2$, respectively. Similar to the previous polyethylene and polypropylene polymer systems, butene-1 and 4-methyl pentene-1 are normally polymerized with an ionic polymerization mechanism using a Ziegler-Natta-type catalyst. Peaks assigned to a methyl group at 1381 cm^{-1} and an ethyl group at 772 cm^{-1} are expected to be observed in poly (butene-1); absorption bands with these assignments were discussed in terms of the short-chain branching of LLDPE. This polymer is both isotactic and crystalline, and three different crystal structures are known to exist, each of which may be differentiated from the IR spectrum [49].

Poly(4-methyl pentene-1) or TPX is also an isotactic and crystalline polymer; the IR and Raman spectra of material are provided in Reference Spectrum 6 in the Appendix. As expected from the discussion of LLDPE produced from 4-methyl pentene-1 as comonomer, the geminal dimethyl groups of the isopropyl moiety give rise to a doublet line at 1384 and 1366 cm^{-1}, plus peaks at 1169 and 918 cm^{-1}. It is reported that peaks at 948, 844, and 810 cm^{-1} are indicative of isotacticity, and that peaks at 844 and 810 cm^{-1} are related to crystallinity [50]. The isopropyl group (or geminal dimethyl groups) of TPX show a weak doublet near 1380 cm^{-1} in the Raman spectrum, and notably, the methyl peak at 1384 cm^{-1} is almost indiscernible. However, two other peaks are attributed to the isopropyl group attached to an alkyl carbon: a sharp peak at 1169 cm^{-1} and a peak at 918 cm^{-1}. Both peaks are also observed as absorptions in the IR spectrum. A weak Raman band at 1301 cm^{-1} is assigned to the CH$_2$ twisting vibration. The bands assigned above are consistent with the known $-CH_2CH(CH_2CH(CH_3)_2)-$ structure of this polymer.

4.2. Vinyl Polymers with Ester Groups

4.2.1. Polyvinyl Acetate (PVAc) and the Vinyl Acetate-Ethylene Copolymer (EVA)

The IR and Raman spectra of polyvinyl acetate are provided in Reference Spectrum 10. Key features of the IR spectrum of PVAc are the C=O stretching vibration at 1740 cm^{-1}, the C$-$O stretching vibration of the ester group at 1230 and 1020 cm^{-1}, an absorption due to methyl CH bending at 1375 cm^{-1}, and CH$_2$ bending (scissors) vibration of the methylene groups at 1440 cm^{-1}. An important copolymer of vinyl acetate is ethylene–vinyl acetate copolymer, EVA. For this copolymer, an increase in the number of $-CH_2-$ groups, due to the insertion of ethylene molecules into the polymer chain, results in a corresponding increase in the intensity of CH$_2$-related bands in both the infrared and Raman spectra. The ethylene content of the copolymer may be estimated by a comparison of the intensity of the "ethylene" C—H stretch at 2916 cm^{-1} with the intensity of either the C=O stretch at 1735 cm^{-1} or the C—O stretch at 1237 cm^{-1}. This method is particularly convenient when the ATR method of sample handling is used, because the CH stretching vibration is measurable, even with a high ethylene content. The Raman spectra of typical EVA samples covering a range of compositions of EVA for both the CH stretching and fingerprint regions are provided in Figs. 22 and 23, respectively. Although it is evident from Fig. 22 that the frequency of the peak around 2936 cm^{-1} changes with monomer composition, the intensity ratio, I_{2936}/I_{2883}, correlates well with composition.

From Fig. 23, it is obvious that the similar band ratio relationships can be established, for the CH$_2$ bending mode (1440 cm^{-1}) or other polyethylene-related bands (1295, 1129, and 1062 cm^{-1}) when compared with PVAc-related carbonyl stretch (1740 cm^{-1}) or the acetate peak at 630 cm^{-1}. In Fig. 24, a relationship between certified ethylene content of EVA and the Raman intensity ratio, I_{2936}/I_{2883}, is demonstrated. A well-defined curve indicates that the determination of the ethylene content is possible by the Raman method.

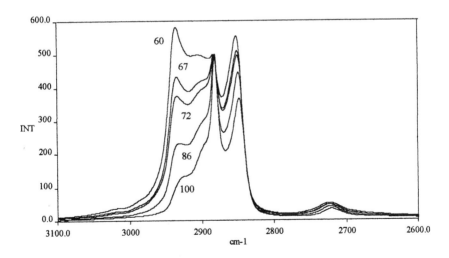

Figure 22 Raman CH stretching region of EVA. (Numbers show the percent ethylene composition, intensities normalized to the peak at 2882 cm^{-1}.)

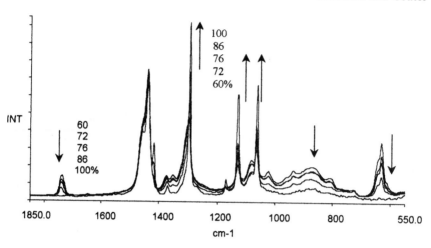

Figure 23 Raman spectra of various ethylene–vinyl acetate copolymers. Raman intensities corrected referring blackbody radiation intensity and normalized at the 1430 cm^{-1} peak. Arrows indicate the intensity of the peak changes to the arrow direction as ethylene contents change.

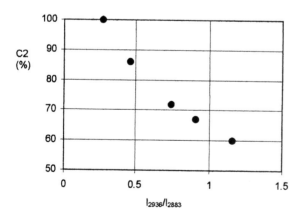

Figure 24 Relationship between ethylene content of EVA and Raman band intensity ratio, I_{2936}/I_{2883}.

Another important copolymer of vinyl acetate, the vinyl acetate–vinyl chloride copolymer [P(VAc-VC)], is discussed in Sec. 4.3.2. However, for reference purposes, the combination of bands for the characterization and identification of P(VAc-VC), in addition to PVAc, and EVA, are included in Table 6.

4.2.2. Polyvinyl Alcohol (PVA) and Related Polymers

Polyvinyl alcohol (PVA) is made from polyvinyl acetate by hydrolysis of the ester group with potassium hydroxide, a process also known as *saponification*. The IR spectrum of PVA features bands due to the hydroxyl group, such as the broad, strong OH stretching vibration band at 3400 cm^{-1}, the OH bending at 1330 cm^{-1}, and the CO stretching (coupled to $C-C$) at 1100 cm^{-1}, which is assigned to the secondary alcohol. The sharp band at 1143 cm^{-1} is assigned to crystalline material, and this may be used to estimate the degree of crystallinity of the polymer. The Raman spectrum of PVA reveals a distinctive band associated with crystallinity at 1148 cm^{-1}, and the secondary alcohol band at 1095 cm^{-1}. Comparing the IR bands at 1144 and 1100 cm^{-1} with those of Raman bands, as shown in Fig. 25, the Raman method appears to have an advantage for the determination of crystallinity. The IR band at 920 cm^{-1} and a shoulder at 822 cm^{-1}, and the Raman band at 923 cm^{-1} and a shoulder at 820 cm^{-1}, are reported to indicate a syndiotactic sequence in the PVA.

Any unhydrolyzed vinyl acetate units will be detected in both the IR and Raman from the presence of the carbonyl stretch at 1738 and 1729 cm^{-1}, respectively. For comparison, the IR and Raman spectra of 100% and 88% hydrolyzed material are shown in Reference Spectra 50 and 51, respectively.

Polyvinyl butyral (PVB) and polyvinyl formal (PVFM) are produced from reactions between polyvinyl alcohol and butylaldehyde and formaldehyde, respectively, as indicated below. An alternative, and favored, method of production of these polymers involves the addition of the aldehyde to the

Table 6 Characteristic IR and Raman Frequencies for Polyvinyl Acetate, Polyvinylacetate–Polyvinylchloride Copolymer, and Ethylenevinyl Acetate

	IR	Raman
VAc	1738, 1373, 1126, 948, 795	1733, 1377 & 1357, 1131, 804, 632
EVA	1736, 1373, 1025, 959, 800	1738, 1373 & 1351, 1131, 804, 632
VAc-VC	1740, 1373, 1330, 1028, 696–689	1741, 1334 & 1311, 1104, 814, 695

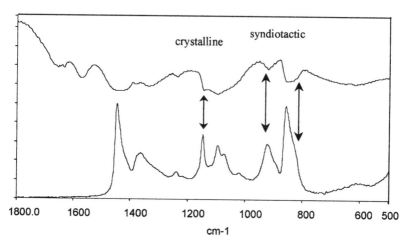

Figure 25 Fingerprint region of IR and Raman spectra of 100% hydro-lyzed PVA.

saponification reaction of polyvinyl acetate. Consequently, these polymers often contain residues of unreacted polyvinyl acetate, which manifests itself as ester carbonyl bands in the IR and Raman spectra. Example IR and Raman spectra of PVFM are provided in Reference Spectrum 58. The concentration of unreacted starting material, such as PVA, may be determined from the IR spectra.

$$
\left[\begin{matrix} -CH_2-CH-CH_2-CH- \\ | \quad\quad | \\ OH \quad\quad OH \end{matrix}\right]_n + nRCHO \rightarrow \left[\begin{matrix} -CH_2-HC-CH_2-CH- \\ | \quad\quad | \\ O-CH-O \\ | \\ R \end{matrix}\right]_n
$$

4.2.3. Methacrylate and Acrylate Polymers

Methacrylate and acrylate polymers used for industrial applications are obtained from the free-radical polymerization of methacrylate ester and acrylate ester monomers shown below, respectively. Both exist as atactic polymers.

(a) Methacrylate ester   (b) Acrylate ester

$$CH_2=C\begin{smallmatrix} CH_3 \\ \\ COOR \end{smallmatrix}$$

$$CH_2=C\begin{smallmatrix} H \\ \\ COOR \end{smallmatrix}$$

Several types of methacrylate and acrylate polymers are based on variations of the alcohol (alkyl substituent) moiety. With the structural similarity of these polymers, it is important to be able to distinguish between methacrylate and acrylate polymers, and it is also important to identify the alcohol moiety. Example IR spectra in the fingerprint region for typical methacrylate and acrylate polymers are illustrated in Fig. 26 [51]. Methacrylate and acrylate polymers may be differentiated from the band profiles in the 1300–1150 cm^{-1} region. The ester bands in this region are essentially a doublet of a doublet for the methacrylate, and a doublet of a singlet for the acrylate. Polymethyl acrylate and polyethyl acrylate are exceptions, in which one of the ester bands splits into two. The alcohol moiety of the ester can also be identified from the features in the IR fingerprint region (Fig. 26). The most common methacrylate and acrylate polymers are polymethyl methacrylate (PMMA) and polymethyl acrylate (PMA). Reference Spectrum 12 provides IR and Raman spectra of atactic polymethyl methacrylate. It is reported that the IR spectra of atactic and syndiotactic PMMA are qualitatively the same, and in the case of isotactic PMMA, the absorption at 1063 cm^{-1} does not occur [52].

In the Raman spectrum of PMMA the carbonyl-related features of the ester grouping are weak, and this applies to the carbonyl stretch band at 1729 cm^{-1} and the band profile in the 1300–1150 cm^{-1} region. The strongest and most characteristic band is the band at 813 cm^{-1}, and this is related to the methyl group on the carbon adjacent to the ester functionality.

Important copolymers of PMMA and PMA are those with styrene. These polymers are described further in the section dealing with polystyrene. It is understood that the "doublet of doublet" features in the IR fingerprint region are associated with the long backbone sequence of the methacrylate chain. Correspondingly, methacrylate copolymers will also exhibit the "doublet of doublet" feature in the IR spectrum only when the methacrylate sequence within the copolymer is long.

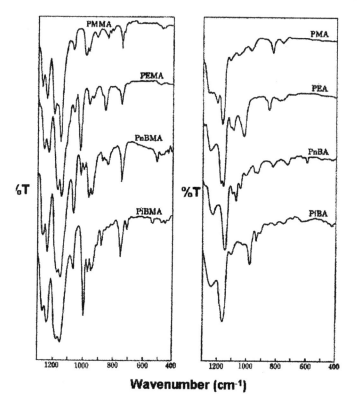

Figure 26 Fingerprint region of methacrylate and acrylate polymers. (From Ref. 51, p. 225.)

4.3. Polymers of Vinyl Chloride and Related Compounds, and Poly "Halogenated" Ethylenes

4.3.1. Polyvinyl Chloride (PVC)

The IR spectrum of PVC shows a distorted doublet-like convoluted band profile in the region between 700 and 600 cm^{-1} assigned to the C$-$Cl stretching vibration. In addition to this characteristic band structure, the major features of the PVC IR spectrum are a partially resolved doublet band due to CH$_2$ scissors [δ(CH$_2$)] observed at 1435 and 1427 cm^{-1}, the CH bending [δ(CH)] of the $-$CHCl$-$ group at 1331 and 1254 cm^{-1}, the backbone $-$C$-$C$-$ stretching [ν(CC)] at 1098 cm^{-1}, and the CH$_2$ rocking [r(CH$_2$)] at 961 cm^{-1}. The Raman spectrum of PVC features a narrow, partially resolved doublet structure (1437 and 1429 cm^{-1}) assigned to the

CH$_2$ scissors vibration, and broad doublet-like band envelope in the region 600–700 cm^{-1}, attributed to C–Cl stretching vibrations.

Most commercially available PVC is polymerized with radical initiators by a liquid-phase suspension process. In other production methods, the material is formed with a free-radical initiator via mass polymerization, emulsion polymerization, or solution polymerization processes. PVC is mostly a pure head-to-tail polymer with predominantly a syndiotactic structure, but exhibiting low crystallinity.

Pohl and Hummel [53] have investigated the C–Cl stretching vibration in the region 700–600 cm^{-1} for PVC samples prepared under different reaction conditions, with infrared measurements acquired at different sample temperatures. Stereoregular sequences were determined by means of a curve-fitting technique (a band decomposition method). They demonstrated that the convoluted envelope with the distorted doublet-like structure is actually composed of eight overlapping bands, with centers at 603, 613, 624, 633, 639, 647, 677, and 695 cm^{-1}). The peaks at 603 and 639 cm^{-1} are assigned to the long syndiotactic chain sequences. This enables the syndiotactic stereoregularity of PVC to be determined from the sum of the intensities of these two peaks, as a function of the total band intensity. The Raman C–Cl band is very similar in appearance to the IR spectrum, and may also be utilized for the determination of the stereoregularity of PVC [54]. The low-frequency component of the CH$_2$ scissors vibration band in the IR and Raman spectra is also attributed to the crystalline component of PVC [55].

The IR spectrum of pure PVC is shown in Fig. 27a together with the spectrum of a plasticizer treated PVC (Fig. 27b). Although the IR spectrum of the plasticized PVC looks similar to that of a polyester or an alkyd resin because of the dominant spectral features of the plasticizer, such as dioctyl phthalate (DOP), one can confirm the presence of PVC from the characteristic pattern of the C–Cl stretching vibration band at around 700–600 cm^{-1} (marked by arrows). Strong features attributed to the carbonate ion often appear in the IR spectrum of many formulated PVC samples that feature materials such as calcium carbonate, which are used as fillers or extenders.

4.3.2. Vinyl Chloride–Vinyl Acetate Copolymers [P(VC-VAc)]

The IR and Raman spectra of a vinyl chloride–vinyl acetate copolymer are presented in Reference Spectrum 21. The IR spectrum of the copolymer is essentially equivalent to the sum of the individual component, PVC and PVAc, spectra, as noted in Fig. 28. The doublet-type structure of the IR absorption at 1434 and 1426 cm^{-1} (CH$_2$ bending), and the absorption at 1331 cm^{-1} (CH bending) are attributed to PVC, while the peaks at 1372 and 1025 cm^{-1} are assigned to PVAc. Obviously, the carbonyl band at 1739

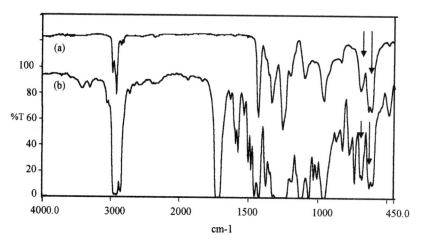

Figure 27 IR spectra of pure (a) PVC and (b) PVC with plasticizer. The C−Cl stretching vibration is marked by arrows.

cm^{-1} originates from the PVAc component, and the characteristic distorted doublet peaks at 693 and 615 cm^{-1} belong to PVC, with some overlap from the weak acetate absorption of PVAc. The strong peak at 1238 cm^{-1} has contributions from both PVC and PVAc.

The composition of the vinyl chloride–vinyl acetate copolymer may be determined from the peak areas of C=O at 1730 cm^{-1} and C−Cl with

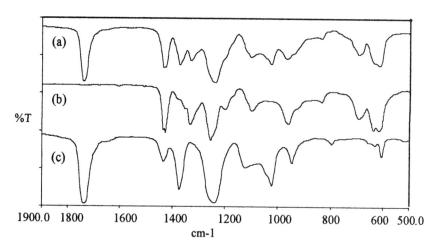

Figure 28 IR spectra of (a) polyvinyl acetate–vinyl chloride copolymer, (b) polyvinyl chloride, and (c) polyvinyl acetate.

the 600–700 cm^{-1} band profile. The integrated absorbance ratio, $A(C=O)/A(C-Cl)$ is dependent on the percent of the vinyl acetate moiety, as indicated in Fig. 29. The vinyl acetate content, for concentrations below 20%, follow, the following empirical relationship:

$$y(\%) = 34.9\left(\frac{x}{x+1}\right) - 1.432 \tag{13}$$

where x is the ratio of the peak areas, $x = A(C=O)/A(C-Cl)$.

Because the line width of the C=O stretching vibration increases significantly with the concentration of vinyl acetate moiety, the traditional peak-height method of measurement should not be used.

4.3.3. Vinylidene Chloride Copolymers

Polyvinylidene chloride (PVDC), a polymer formed from $CH_2=CCl_2$, is difficult to formulate and is not used for any type of product by itself. It is usually manufactured as a copolymer with various monomers, including acrylonitrile, vinyl chloride, acrylic esters, or methacrylic esters. A unique property of PVDC copolymers is to prevent water and gases from diffusing through the film. In this way, PVDC copolymer films provide a vapor barrier, and are commonly used to wrap foods for storage. The commercial name Saran refers to both the PVDC homopolymer and its copolymers. Reference Spectrum 22 provides the IR and Raman spectra of a vinylidene

Composition of VAc (%)

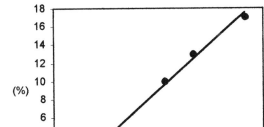

Figure 29 Relationship between vinyl acetate contents and ratio of peak area relevant to vinyl acetate and vinyl chloride.

chloride–methyl acrylate copolymer. Although the carbonyl stretching frequency is observed at 1743 cm^{-1}, attributed to the acrylate comonomer of this film, the most dominant features of the IR spectrum of this polymer are the doublet-like bands at 1071 and 1044 cm^{-1}, and the band at 1404 cm^{-1}. The doublet band (1071 and 1044 cm^{-1}) is indicative of a crystalline phase from a polyvinylidene chloride sequence.

The 1404 cm^{-1} peak is assigned to a CH$_2$ bending, the frequency of which is lowered from the usual frequency at approximately 1460 cm^{-1}, because it exists as an α-methylene, adjacent to the CCl$_2$ groups. Consequently, the band at 1404 cm^{-1} is assigned to −CCl$_2$−CH$_2$−CCl$_2$− sequence, which is a normal head-to-tail sequence for the vinylidene chloride units. It is worth noting that the −CCl$_2$−CH$_2$−CHCl− sequence of the vinyl chloride–vinylidene chloride copolymer appears at 1420 cm^{-1}. If an increased frequency for this CH$_2$ bending is observed for a polyvinyl–vinylidene chloride copolymer, it is interpreted as originating from a random or block copolymer [56]. The presence of the 1420 cm^{-1} band suggests the random copolymerization, and a peak at 1404 cm^{-1} suggests block copolymerization for the polyvinyl–vinylidene chloride system.

The CCl$_2$ stretching vibrations are observed in the IR and Raman spectra of these polymers, as illustrated in Fig. 30. Peaks at 531, 600, 653, and 688 cm^{-1} are assigned to the CCl$_2$ stretching vibrations [57].

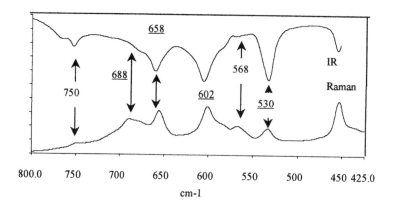

Figure 30 IR and Raman bands of vinylidene chloride copolymer. CCl$_2$ stretching frequencies are underlined.

4.3.4. Polytetrafluoroethylene (PTFE) and Related Polymers

Polytetrafluoroethylene, $-(CF_2-CF_2)_n-$, is also known commercially as Teflon—see Reference Spectrum 23 for the IR and Raman spectra. The CF_2 stretching vibration occurs as the most intense IR absorption band, near 1200 cm^{-1}. This band is a multiplet, and consists of three peaks at 1240, 1215, and 1150 cm^{-1}. Other major bands are located at 641, 554, and 515 cm^{-1} and are assigned to the CF bending modes. The IR absorption band at 2366 cm^{-1} is the overtone of the CF_2 stretching vibration.

The Raman spectrum shows three weak CF_2 stretching bands between 1400 and 1200 cm^{-1}, and two bands at 386 and 292 cm^{-1}, both of which are assigned to CF_2 rocking vibration. The most intense band of Teflon is at 734 cm^{-1}, and this is assigned to the skeletal $C-C$ stretching band. This band is observed as a very weak and complex band in the IR spectrum.

The crystallinity of Teflon may be determined from the intensity ratio of the IR bands at 785 and 2366 cm^{-1} [58] from the following relationship:

$$\text{Crystallinity}\,(\%) = 100 - \left(\frac{A_{782}}{A_{2366}}\right) \times 25 \qquad (14)$$

Note that the overtone of the $C-F$ band is used in preference to the fundamental, because the $C-F$ band is too intense to measure accurately.

A phase transition occurs in Teflon at 19°C, and this involves a change in the degree of helical structure. It may be observed [59] as a spectral change in a weak doublet-type Raman band at 598 and 577 cm^{-1}. A few copolymers of tetrafluoroethylene exist, including tetrafluoroethylene-hexafluoropropylene (FEP), tetrafluoroethylene-perfluoroalkyl-vinylether (PFA), and tetrafluoroethylene-hexafluoropropylene-perfluoroalkylvinylether (EPE). An example of the IR and Raman spectra of EPE is provided in Reference Spectrum 24. Although these three copolymers have similar IR spectra, the bands around 990 cm^{-1} are used to distinguish these copolymers [60]. The hexafluoropropylene moiety of FEP produces a band at 982 cm^{-1}, while the perfluoroalkyl vinylether moiety of PFA has a band at 993 cm^{-1}. As expected, EPE exhibits both the 993 and the 982 cm^{-1} bands. The IR spectrum in Reference Spectrum 24 shows the 993 cm^{-1} band as a shoulder on the 982 cm^{-1} band.

Other fluorine-containing polymers, such as polychlorotrifluoroethylene, polyvinylidenefluoride, polyvinylfluoride, ethylene-tetrafluoroethylene coplymer, and ethylene-chlorotrifluoroethylene copolymer, will not be covered in this chapter.

4.4. Polybutadiene (PBD or BR) and Polymers of Substituted Butadiene

4.4.1. Polybutadiene and Polyisoprene

The polymerization of butadiene monomer proceeds with chain propagation via 1,2-, 1,4-*trans*- or 1,4-*cis*-additions. If the polymerization is controlled to form mostly the 1,2-addition product, the polymer has a $-CH_2-$ chain with a terminal vinyl, $-CH=CH_2$, substituent, at alternating carbon atoms. However, if 1,4-addition dominates the polymerization proceeds to form a polymer chain with a molecular structure of $-(CH_2-CH=CH-CH_2)_n-$, normally with a *trans* configuration at the double bond. 2-Chloro-1,3-butadiene ($CH_2=CCl-CH=CH_2$ chloroprene) and 2-methyl-1,3-butadiene (isoprene) are polymerized in a similar manner. With these compounds, the asymmetry of the carbon atoms at positions 1 and 4 produces a variety of addition products with 1,2-, 1,4-*cis*, 1,4-*trans*, and 3,4-configurations. In the case of polyisoprene, which in nature occurs as natural rubber, the 1,4-cis configuration is the dominant structure. A summary of the polymerization products of butadiene, isoprene, and chloroprene is provided in Fig. 31.

The IR and Raman spectra of polybutadiene and polyisoprene are provided in Reference Spectra 7, 8, and 9, respectively.

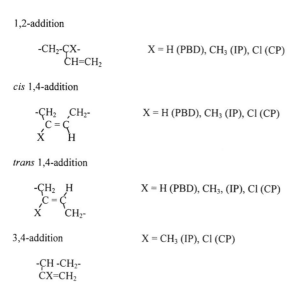

1,2-addition

-CH₂-CX-
  CH=CH₂   X = H (PBD), CH₃ (IP), Cl (CP)

cis 1,4-addition

-CH₂   CH₂-
   C = C   X = H (PBD), CH₃ (IP), Cl (CP)
  X   H

trans 1,4-addition

-CH₂   H
   C = C   X = H (PBD), CH₃, (IP), Cl (CP)
  X   CH₂-

3,4-addition   X = CH₃ (IP), Cl (CP)

-CH -CH₂-
 CX=CH₂

Figure 31 Various double-bond sequences of substituted polybutadiene.

Morero [61] provided the following empirical equations for the IR determination of the microstructure of these polymers:

$$C = (1.7455A_{741} - 0.0151A_{910})$$

$$T = (0.4292A_{967} - 0.0129A_{910} - 0.0454A_{741}) \qquad (15)$$

$$V = (0.3746A_{910} - 0.0070A_{741})$$

cis-1,4-addition component	$(\%) = 100 \cdot C/(C + T + V)$
trans-1,4-addition component	$(\%) = 100 \cdot T/(C + T + V) \qquad (16)$
1,2-addition component	$(\%) = 100 \cdot V/(C + T + V)$

It is evident that the 1,2-addition of butadiene, isoprene, and chloroprene all produce the vinyl group substituent on the polymer chain, which is assigned to the bands at 910 and 990 cm^{-1}. A vinylidene substituent results from 3,4-addition in the case of isoprene and chloroprene, and is assigned to the band at 890 cm^{-1} (this form of pendant methylene was previously noted in the section dealing with polyethylene).

Important copolymers of butadiene include styrene–butadiene copolymers (SBR), acrylonitrile–butadiene-styrene copolymers (ABS), and methyl methacrylate–butadiene–styrene copolymers.

4.4.2. Butadiene Resins

Butadiene resin is manufactured by means of a 1,2-addition polymerization of butadiene, being differentiated from the well-known 1,4-addition product of butadiene used to make butadiene rubber. The material exists as a syndiotactic polymer, and is partially crystalline. It contains both *cis* and *trans* 1,4-addition polymers as minor components. Features of the IR and Raman spectra were addressed in the preceding sections.

4.5. Polystyrene (PS)

Industrially manufactured polystyrene, polymerized with radical initiators, exists mainly as the atactic polymer. The IR spectrum of this form of polystyrene features bands consistent of a mono-substituted aromatic compound. In addition to the normal aromatic absorptions, such as the C−H stretching between 3110 and 3000 cm^{-1}, and the characteristic ring vibrations at approximately 1600 and 1500 cm^{-1} (1601 and 1493 cm^{-1} for polystyrene), the spectrum also contains the combination bands characteristic of monosubstitution (at 1942, 1868, 1802, and 1741 cm^{-1}), and the out-of-plane C−H ring deformations for monosubstitution (757, 699, and 541 cm^{-1}). Note that the C−H stretching reflects the alternating methylene components in the backbone with the bands at 2923 and 2850 cm^{-1}. The asymmetry on the low-frequency side of the 2923 cm^{-1} band is attributed to the methyne (C$-$H) on the backbone, which normally produces an unresolved band situated close to 2900 cm^{-1}.

The Raman spectrum of PS shows a couple of distinctive doublets at 1603 and 1584 cm^{-1} and at 1033 and 1002 cm^{-1}, both ring-mode vibrations, and both characteristic of a monosubstituted aromatic compound. The Raman band at 622 cm^{-1} is also indicative of the substituted benzene ring. In addition, there are a few IR absorption peaks due to functional groups such as terminal vinyl group at 907 and 980 cm^{-1} (refer to 910 and 990 cm^{-1} peaks of PE) and the *trans* C=C double bond at 967 cm^{-1} resulting from termination reactions. Note also that these bands can also be assigned to trace amounts of butadiene that are sometimes added to polystyrene when it is used for making thin films. This addition adds some flexibility and reduces the brittle nature of the polystyrene film. Atactic polystyrene has bands related to its atactic nature at 1370, 1328, 1306, 1070, and 943 cm^{-1}, and the isotactic form exhibits bands at [62] 1364, 1314, 1297 cm^{-1} and a doublet at 1075 and 1056 cm^{-1} (see Fig. 32). As isotactic polystyrene crystallizes, the doublet is seen to shift to 1080 and 1048 cm^{-1}, and a new band appears at 985 cm^{-1}. The crystallinity may be determined from the intensity of the 985 cm^{-1} band, and the ratio A_{566}/A_{543} is used to evaluate the content of isotactic sequence. For refer-

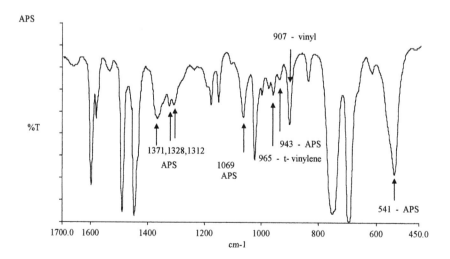

Figure 32   Spectral features of atactic polystyrene.

ence, the IR and Raman spectra of atactic PS are provided in Reference Spectrum 14.

Atactic polystyrene is sometimes known as general-purpose polystyrene (GPPS). There are other commercial grades of polystyrene, such as medium-impact polystyrene (MIPS) and high-impact polystyrene (HIPS). These impact-resistant forms of polystyrene are specially formulated by grafting to a rubber, such as polybutadiene or a styrene–butadiene copolymer, and dispersed in the base polystyrene. The inclusion and detection of small amounts of butadiene-based materials into the base polystyrene has been discussed.

Important copolymers of styrene are SBR (styrene–butadiene rubbers), ABS, a copolymer with butadiene and acrylonitrile, and various copolymers with methacylic and acrylic esters, such as styrene–methyl methacrylate and styrene–methyl acrylate copolymers. Additionally, copolymers formed with one or more of the following monomers also exist: ethylene, α-methyl styrene, vinyl acetate, maleic anhydride, and acrylonitrile.

4.5.1. Styrene–Butadiene Copolymers (SBR)

There are two types of styrene–butadiene copolymer: a random copolymer and a block copolymer. Spectra for the block and random styrene–butadiene copolymers are provided in Reference Spectra 15 and 16, respectively. For a full characterization of SBR it is necessary to determine the composition in terms of the monomer ratios, including the three configurations of the butadiene addition reaction components, and to be able to differentiate random and block copolymerization.

One infrared spectral method used for the determination the monomer units of SBR is known as Hampton's method [63]. This method utilizes the characteristic monosubstituted aromatic band intensity at 699 cm^{-1} (A_{699}) for the determination of styrene unit composition, and the scheme defined as Morero's method (see section on polybutadiene and polyisoprene) is used to determine the butadiene moiety. As noted in the section on polybutadiene, the 1,4-cis, 1,4-$trans$, and 1,2-addition components appear in the IR spectrum at 724, 965, and 910 cm^{-1}.

Hampton's method is defined by the following set of equations:

$$
\begin{aligned}
S &= 0.374A_{699} - 0.260A_{724} - 0.014A_{910} \\
C &= -0.025A_{699} + 1.834A_{724} - 0.027A_{910} - 0.004A_{967} \\
V &= -0.007A_{699} - 0.015A_{724} + 0.314A_{910} - 0.007A_{967} \\
T &= -0.007A_{699} - 0.036A_{724} - 0.011A_{910} + 0.394A_{967}
\end{aligned}
\tag{17}
$$

$P = S + C + V + T$ (where S = styrene, $C = cis$-1,4, V = vinyl, 1,2 and $T = trans$-1,4)

Percent styrene $= 100\,S/P$, cis-1,4 (%) $= 100\,C/P$,

$trans$-1,4 (%) $= 100\,T/P$, and 1,2(%) $= 100\,V/P$

As is evident from the IR spectrum of SBR (Reference Spectra 15 and 16), the 1,4-cis component at 740 cm^{-1} overlaps with the styrene bands at 757 and 699 cm^{-1}. In order to avoid an error from the styrene interference, Binder [64] used the band at 680 cm^{-1} for the determination of the 1,4-cis component instead of the 740 cm^{-1} band. Whether the SBR is a random copolymer or a block copolymer is evaluated using the IR polystyrene band at around 540 cm^{-1}. As indicated above, atactic polystyrene shows this band at 541 cm^{-1}. Although block SBR copolymers retain the peak frequency at 541 cm^{-1}, this band shifts to 558 cm^{-1} for the random SBR copolymers, as observed in Fig. 33. The Raman C=C stretching vibration also provides information on the microstructure of butadiene moiety, as discussed earlier.

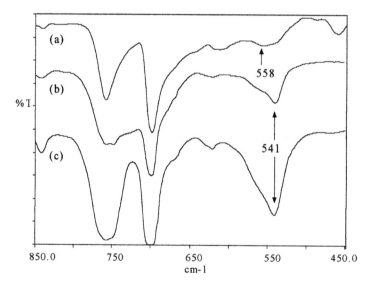

Figure 33 IR spectra of (a) styrene–butadiene random copolymer, (b) styrene–butadiene block copolymer, and (c) polystyrene (ordinate scale offset).

4.5.2. Acrylonitrile–Butadiene–Styrene Copolymers (ABS)

The IR and Raman spectra of ABS are shown in Reference Spectrum 18. Both the IR and Raman spectra show characteristic peaks for the three components: the nitrile band of acrylonitrile component at 2237 cm^{-1} (IR) and 2238 cm^{-1} (Raman), the aromatic ring of the styrene component at 1602, 1494, 761, and 699 cm^{-1} (IR) and 1603, 1584, and 1003 cm^{-1} (Raman), and the double bond of the butadiene component 967 cm^{-1} *trans* and 911 cm^{-1} vinyl (IR) and 1667 cm^{-1} *trans*, 1653 cm^{-1} *cis* and 1641 cm^{-1} vinyl (Raman).

Iwamoto's method for the determination of the monomer compositions is described here [65]. Three baselines are constructed, A, B or B', and S for each of the components, as indicated in Fig. 34. It is necessary to establish a linear relationship for the composition/absorbance ratio from the IR spectra of ABS standards of known composition. This is required because the sample to be analyzed may have a different distribution of *cis*-1,4-, *trans*-1,4-, and 1,2-addition ratios than that of the standard materials. The aromatic band at 1603 cm^{-1} is free from interference, and is utilized as an internal standard. Peak ratios of were evaluated for three components, where the band at 2237 cm^{-1} was taken for acrylonitrile (a) and bands at 967 and 911 cm^{-1} were taken for *trans*-1,4- and 1,2-addition components of butadiene

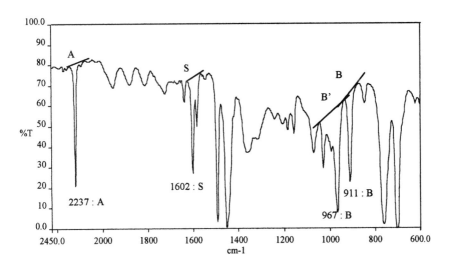

Figure 34 Selection of bands for the ABS composition determination. A, acrylonitrile; B, butadiene; S, styrene.

(b), respectively. The composition in terms of C_A (% acrylonitrile), C_B (% butadiene), and C_S (% styrene) are given by

$$C_A = 100 \times \frac{a}{a+b+1}(\%)$$

$$C_B = 100 \times \frac{b}{a+b+1}(\%) \tag{18}$$

$$C_S = 100 \times \frac{1}{a+b+1}(\%)$$

where the relationships $a = C_A/C_S$ and $b = C_B/C_S$ are obtained from the absorbance ratios of the standards with known composition.

If the sample gives the same C_B/C_S from the following intensity ratios,

$$D^{967}_{1602} = \frac{A(967)}{A(1602)} \quad \text{and} \quad D^{911}_{1602} = \frac{A(911)}{A(1602)} \tag{19}$$

the sample has the same configurations for the butadiene moieties as that of the standards, and this is the "b value" used in the equations. If different C_B/C_S values are obtained from these intensity ratios, then the sample has different configurations for the butadiene moieties, and an approximate value for b value should be estimated as described below.

In Fig. 35, the values obtained for the sample are: D^{967}_{1602} at point I, giving $D = C_B/C_S$, D^{911}_{1602} at point G giving $A = C_B/C_S$. Also, the values D^{911}_S and D^{967}_S, which are the spectral intensity contributions at zero butadiene content, are obtained from the extrapolation of the calibration measurements to zero butadiene concentration. The approximate C_B/C_S

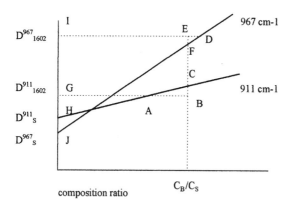

Figure 35 IR procedures to determine composition ratio, when sample has different microstructure from those of standards.

value is obtained when a vertical line is drawn to give the following relationship:

$$AB : DE = K : 1 \qquad \text{where } K = (BC/EF)(AG/DI)(IJ/GH) \qquad (20)$$

where BC, EF, AG, DI, IJ, and GH are all observed and measured values.

It is known that the content of 1,4-*cis* component is very low in the case of ABS and its concentration is assumed to be zero. If the concentration of the 1,4-*cis* component is small but nonzero, the determination of the 1,4-*cis* component is difficult because the weak band of the 1,4-*cis* component at 724 cm^{-1} is buried beneath the strong absorptions for the monosubstituted aromatic ring. Consequently, the need to neglect the contribution from the 1,4-*cis* component is inevitable.

A Raman method can be utilized for the same analysis. The acrylonitrile and styrene components are determined using the same bands as the IR method. Butadiene, however, is determined using C=C stretching vibration at around 1650 cm^{-1}, an easier analysis than the IR method. It should be noted that the C=C Raman band is composed of three bands at 1668 cm^{-1} (*trans*-1,4 addition), 1654 cm^{-1} (*cis*-1,4-addition), and 1441 cm^{-1} (1,2-addition) as described in Sec. 3.3.1. A peak area method of measurement is recommended for the nitrile, C=C, and aromatic ring Raman scattering bands, as shown in Fig. 36. In a similar manner to the IR method proposed by Iwamoto, the relationship between Raman intensity and composition must be established for B/A and S/A as shown in Fig. 37a and 37b, respectively, using known standards. The B/A and S/A values are determined for the sample from its Raman spectrum, and each component is determined from the previous equations (18).

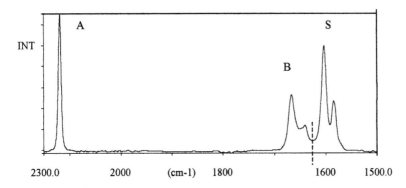

Figure 36   Raman bands relevant to composition analysis. A, acrylonitrile; B, butadiene; S, styrene.

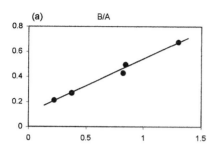 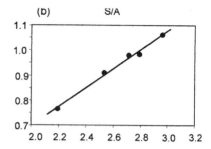

Figure 37 Observed Raman peak intensity ratio plotted versus compositions. (a) butadiene–acrylonitrile and (b) styrene–acrylonitrile.

4.5.3. Styrene–Methacrylate and Styrene–Acrylate Copolymers

The monomer ratio of styrene–methacrylate/acrylate copolymers is determined from the intensity ratio of the carbonyl band at 1730 cm^{-1} for the methacrylate and aromatic ring band at 699 cm^{-1} for the styrene component. In the measurement, a baseline is drawn between 1990 and 630 cm^{-1}, and the intensity ratio of the 1730 and 699 cm^{-1} bands is determined. Styrene content is determined from the empirical equation, percent styrene $= 71.4 \times (A_{699}/A_{1730})$ [66,67]. If the methacrylate and acrylate are both present in the same copolymer, the methacrylate content may be determined from the intensity ratio of the ester–ether bands at 1032 cm^{-1} (acrylate) and 1021 cm^{-1} (methacrylate). Figure 38 illustrates the procedure [68] for the measurement of styrene and styrene–methacrylate in methyl styrene–acrylate copolymers and methyl methacrylate-modified styrene–acrylate copolymers, respectively.

4.6. Polyamides

The most well known polyamide resin is Nylon, which is based the secondary amide functionality, $-$CONH$-$. As a result, the IR spectra of polyamides share common features with materials such as proteins. The main features of the infrared spectra of polyamides are the N$-$H stretching band at around 3300 cm^{-1}, accompanied by a weak peak at around 3050 cm^{-1}, the amide-I band at 1630 cm^{-1}, and the amide-II band at 1550 cm^{-1}, as shown in Reference Spectrum 25, the IR and Raman spectra of Nylon-6. Although the broad band at 3300 cm^{-1} accompanied by the weak one at 3050 cm^{-1} is assigned to Fermi split pair of NH stretch, there is a suggestion that these bands may suggest the *cis-trans* isomers of the Nylon amido group as follows:

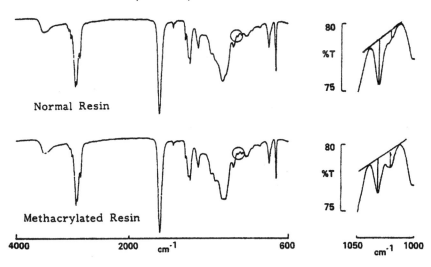

Figure 38 Absorption bands used to calculate styrene in styrene–methyl acrylate, and styrene methyl methacrylate in methyl methacrylate modified styrene–methyl acrylate. Bands at 1729 cm^{-1} and 700 cm^{-1} represent methyl acrylate and styrene, respectively; methyl acrylate and methyl methacrylate are measured at 1032 cm^{-1} (acrylate) and 1021 cm^{-1} (methacrylate).

$-(NHRCO)_n-$ (structure I) I   or

$-(NHR_1NHCOR_2CO)_n-$ (structure II)

where R, R$_1$, and R$_2$ = (CH$_2$)$_{x,y,\text{or } z}$.

$$\begin{array}{cc} O \diagdown \quad R' & O \diagdown \quad H \\ \quad C-N & \quad C-N \\ R \diagup \quad H & R \diagup \quad R' \end{array}$$

Trans Amido   Cis Amido

The naming convention for polyamides is based on the numbers of carbon atoms in the acid and amino components used to form the amide. Nylons may be formed using two approaches: one with a single reactant, a lactam, and the other from two reactants, a dicarboxylic acid and a diamine. The lactam yields structure I, a material designated Nylon-$(x + 1)$, where x is the number of carbon atoms assigned to the methylene groups and $+1$ plus one for the CO group. For example, Nylon-6, Nylon-7, Nylon-11, and Nylon-12 belong to this group. The combination of a diamine and a dicarboxylic acid yields structure II, with examples being Nylon-66, Nylon-610, and Nylon-612. The numbers for the carbon atoms in the NHR_1NH and the COR_2CO moieties are concatenated to one number. For example, Nylon-66, which is made from hexamethylenediamine and adipic acid, has the chemical form $R_1 = (CH_2)_6 = 6$ and $COR_2CO = (CH_2)_4 + 2$ (for the carbonyl CO) = 6. Nylon-610 has the substructure of $(-NH(CH_2)_6NHCO(CH_2)_8CO)_n-$.

The application of IR to the identification of Nylon samples relative to a specific polyamide structure is important yet can be quite difficult. The difference between these polyamides is associated with the relative number of methylene groups. The difficulty can be compounded because certain polyamides exhibit polymorphism. Ogawa published the following table for the identification of polyamides from IR spectra [69].

For the purpose of identification of polyamides, Raman spectra of a few polyamides are shown in Figs. 39 and 40. Maddams [70] noticed that the amide-I (~ 1650 cm^{-1}) intensity relative to the intensity of the $CH_2 -$ bending mode (~ 1440 cm^{-1}), $I_{Amide-I}/I_{CH_2\text{-bend}}$, produced a linear relationship, for the diamine/diacid-based Nylons, versus the number of CH_2 groups flanked by two C=O groups, as shown in Fig. 41. For example, in the case of Nylon-610, the number of CH_2 group flanked by two C=O groups is equal to 8, and similarly, it is 6 for Nylon-66. Based on Fig. 41, it is possible to determine the Nylon (note: the slope shown in Fig. 41 may not hold for all Raman instruments). Typically, the slope must be calibrated for each instrument, although the deviation from the value presented in Fig. 41 will not be large because the frequencies of the amide-I and the CH_2 bending bands are relatively close to each other).

In the general formula of polyamides, as given by $-(NHRCO)_n-$ or $-(NHR_1NHCOR_2CO)_n-$, the substituent R, R_1 or R_2 can also be an aromatic ring. Such aromatic polyamide (aromatic amide = aramid) are commercially available, a well-known example being Kevlar. The basic form of the structure of Kevlar is provided below. It should be noted that the aromatic rings are substituted in the para position. Other aromatic

Table 7 Characteristic IR Frequencies for Common Nylon-Based Polymers

Nylon	Characteristic band (IR) (cm^{-1})
Nylon-6	1465, 1265, 960, 925
Nylon-66	1480, 1280, 935
Nylon-610	1480, 1245, 940
Nylon-11	1475, 940, 720

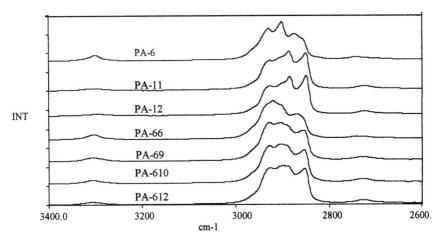

Figure 39 NH and CH stretching region of various polyamide Raman spectra.

polyamides have been reported with meta substitution on the aromatic ring(s).

The IR spectrum of Kevlar features a strong NH stretch at 3319 cm^{-1}, with a shoulder at higher frequency, the amide-I and -II bands at 1648 and 1544 cm^{-1}, respectively, and bands at 1609 and 1514 cm^{-1} assigned to the aromatic ring. Absorption bands for the aromatic ring also appear at 1017 cm^{-1} (in-place C$-$H bending) and at 825 cm^{-1} for the 1,4-disubstituted ring (out-of-plane C$-$H bending). The Raman spectrum exhibits the amide-I band clearly at 1650 cm^{-1} accompanied by a strong ring vibration at 1611 cm^{-1}. The amide-II band is too weak to be detected in the 1550 cm^{-1} region, but a peak at 1517 cm^{-1} assigned to the aromatic ring is observed. A lack of a strong Raman band in the 1055–990 cm^{-1} region, and the presence of a medium-intensity band at 631 cm^{-1} indicate 1,4-substitution on the aromatic ring. The sharp band at 1183 cm^{-1} and the band at 788 cm^{-1}

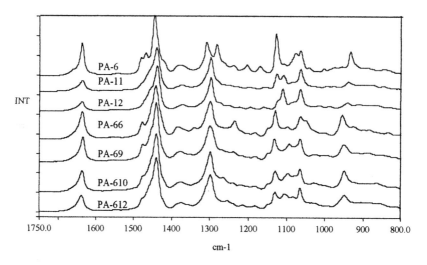

Figure 40   Fingerprint region of Raman spectra of various polyamides.

$I_{\text{Amide-I}}/I_{\text{CH2-bend}}$

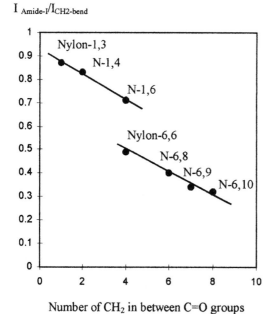

Number of CH_2 in between C=O groups

Figure 41   Relationship between Raman intensity ratio, $I_{\text{Amide-I}}/I_{\text{CH2-bend}}$ and the number of CH_2 group flanked by two C=O groups. (Graph adapted from Ref. 70; data from K. Nishikida.)

are also both assigned to 1,4-disubstitution. The $N-H$ stretching vibration, seen at 3319 cm^{-1} in the IR spectrum, is also present in the FT-Raman spectrum as a weak feature, in part due to the use of a 1064-nm Nd:YAG laser source/InGaAs detector combination.

4.7. Polycarbonates (PC)

The general structure of a polycarbonate is written $-[O-R-O-(C=O)-]_n-$, as explained in Sec. 2 (Fig. 2). R may represent either an aromatic or an aliphatic base structure, and in practice both homopolymers and copolymers can exist, in which mixed aromatic/aliphatic functionality occurs. The most common polycarbonate is the "bis-phenol A type" poly-carbonate, known commercially as Lexan, which has the following chemical structure:

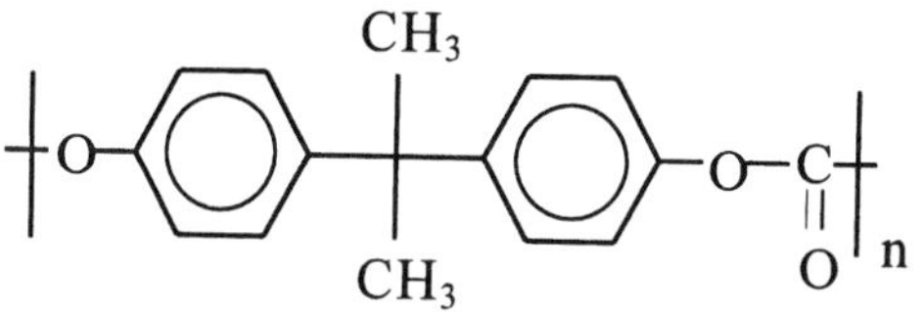

The bis-phenol A type of polycarbonate features a higher carbonyl C=O stretching frequency than simple esters (1774 cm^{-1} IR and 1775 cm^{-1} Raman) and an aryl ester band at 1230 cm^{-1} (IR) and 1236 cm^{-1} (Raman). The other bands in the spectra of this material are assigned to the aromatic ring and the dimethyl branched substituent. These are as follows: the aromatic ring bands, a doublet at 1602 and 1594 cm^{-1}, and a band at 1506 cm^{-1} for the IR; a doublet at 1605 and 1594 cm^{-1} and a band at 1510 cm^{-1} for the Raman; 1,4-disubstituted aromatic ring bands at 1081, 1015, and 831 cm^{-1} for the IR and at 1181, 1114, 709, and 638 cm^{-1} for the Raman; and a weak doublet at 1386 and 1366 cm^{-1} for the IR and at 1388 and 1366 cm^{-1} for the Raman. In the latter case, the intensity of the lower-frequency methyl band is slightly greater than that of the high-frequency one, due to $R-C$ $(CH_3)_2R$ structure [similar to $-CH(CH_3)_2$ structure, as stated in the section on polyethylene]. High-frequency carbonyl absorptions are also observed in acid anhydrides, such as maleic anhydride–ethylene copolymers and poly-amide-imides. The polycarbonate can be differentiated from these polymers by comparison of the IR spectra: Reference Spectrum 28 provides the IR and Raman spectra of Lexan.

Two methods are in common use for the synthesis of Lexan: trans-esterification and a phosgene-based method. Polycarbonates produced via transesterification typically contain trace amounts of free phenol, which gives rise to a weak absorption at 687 cm^{-1}, as shown in Fig. 42.

Commercial aliphatic polycarbonates exist, diethyleneglycol *bis*-allylcarbonate (see structure below) being an example of a material used for the manufacture of optical lenses. The IR and Raman spectra of this compound are provided in Reference Spectrum 29. In this case, the carbonyl C =O of the aliphatic polycarbonate at 1745 cm^{-1} is only slightly higher than that of most other esters. Consequently, it is not possible to uniquely assign the carbonyl group of the sample spectrum to a polycarbonate. The most intense IR band is located at 1263 cm^{-1}, and this is assigned to the $-$C$-$O$-$C$-$ group, asymmetric stretching vibration, and the weaker band at 1028 cm^{-1} is assigned to the symmetric $-$C$-$O$-$C$-$ stretching vibration. A pair of bands at 1456 and 1403 cm^{-1} indicates the presence of CH$_2$ groups, with the lower frequency assigned to the $-$O$-$CH$_2-$ group (1403 cm^{-1}). Unreacted double bonds are detected by the $-$C=C$-$ vibration at 1649 cm^{-1} readily in the Raman spectrum, but are barely discernable in the IR spectrum.

$$CH_2{=}CH\text{-}CH_2\text{-}O\text{-}\underset{\underset{O}{\|}}{C}\text{-}O\text{-}CH_2\text{-}\underset{\underset{\underset{\underset{CH_2{=}CH\text{-}CH_2\text{-}O\text{-}\underset{\underset{O}{\|}}{C}\text{-}O\text{-}CH_2\text{-}CH_2}{|}}{O}}{|}}{CH_2}$$

4.8. Aliphatic and Aromatic Polyethers

4.8.1. Polyoxymethylene, Polyethylene Oxide, and Polypropylene Oxide

There is a series of polymers having a chemical structure $-[(CHR)_n-O-]_m$, which are derived as polyacetal resins, and are known as polyalkyene oxides or polyalkylene glycols. In the above structure, the polymer with R=H and $n = 1$ is polyoxymethylene, which is known as Delrin. This material is a high polymer of formaldehyde, which is terminated by an ether or ester function added to stabilize the final product. Other manufactured products include copolymers with ethylene oxide or propylene oxide. The IR and Raman spectra of polyoxymethylene are shown in Reference Spectrum 55. A strong peak at 1098 cm^{-1} and a doublet at 936 and 900 cm^{-1} in the IR spectrum are assigned to the C$-$O$-$C stretching vibration. It is not possible to determine if the sample is a homopolymer or copolymer from this spectrum.

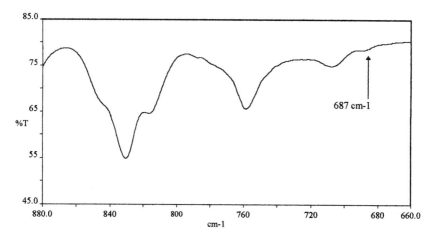

Figure 42 Trace amount of free phenol in polycarbonate, indicating that the production method is transesterification of bis-phenol A and phenyl carbonate.

Crystallinity may be estimated [71] from the intensity ratio of the 632 cm^{-1} band relative to the 1434 cm^{-1} band.

Polyethylene oxide (or polyethylene glycol) and polypropylene oxide (or polypropylene glycol) have the following repeating units: R = $-CH_2-CH_2-$ and $-CH(CH_3)-CH_2-$, respectively. These polymers are naturally terminated with OH groups, and unlike the polyoxymethylene, it is unnecessary to modify this terminal group to stabilize the polymer. However, modified and unmodified versions of these polymers do exist. Reference Spectra 56 and 57 provide IR and Raman spectra of polyethylene glycol. The strong $C-O-C$ ether band at around 1100 cm^{-1} in the IR spectrum is a broad single feature when amorphous (Reference Spectrum 56), and splits into three peaks when the material becomes crystalline. Reference Spectrum 57 clearly indicates that the sample is crystalline (three peaks at 1149, 1101 cm^{-1}, and 1061 cm^{-1}) polyethylene glycol. It is known that the low-molecular-weight polyethylene glycol is amorphous and it becomes crystalline as the molecular weight increases. Infrared may be used to determine the OH group concentration of polyalkylene glycols, modeling a wet-chemical method known as hydroxyl value. Moisture is an interferant, but this may be compensated for by taking into account the $H-O-H$ bending vibration in the sample spectrum. Alternatively, a method based on the measurement of an absorption at 4760 cm^{-1} (near-infrared region) has also been suggested to minimize the water interference [72].

4.8.2. Polyphenylene Oxide (PPO)

Polyphenylene oxide, or polyphenylene ether, is an amorphous polymer for which the IR and Raman spectra are presented in Reference Spectrum 45. As expected from the chemical structure, bands relevant to the aromatic ring system are observed at 1601, 1492, and 858 cm^{-1} (IR), and 1603 and 835 cm^{-1} (Raman)—the low-frequency band in the IR being associated with the substitution of the aromatic ring. The other important feature of the IR spectrum of polyphenylene oxide is the intense absorption band at 1188 cm^{-1}, associated with the ether bonding.

Polyphenylene oxide may be "alloyed" with polystyrene at any ratio. Significant overlap of the aromatic spectral features will be experienced, as noted in Fig. 43. The 760 and 699 cm^{-1} bands of polystyrene are readily observed, even at low concentrations. Estimation of the polyphenylene oxide content of the alloys with polystyrene may be obtained from the intensity ratio of the 1021 cm^{-1} (PPO) and 699 cm^{-1} (PS) bands as follows [73]:

$$\text{PPO (\% w/w)} = 60.0 \times \frac{A_{1021}}{A_{699}} \tag{21}$$

or

$$\text{PPO (\% w/w)} = \frac{100R}{R+4.3} \quad \text{where } R = \frac{A_{960}}{A_{906}} \tag{22}$$

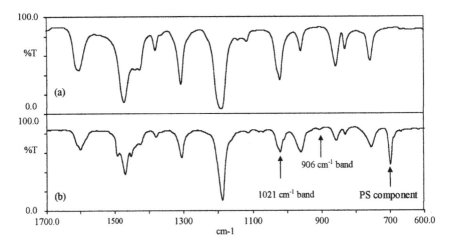

Figure 43 Fingerprint region of (a) pure polyphenylene oxide and (b) an alloy with polystyrene.

The latter equation is used when a thin film of the polymer alloy is not available.

4.9. Polysulfones (PSF), Polyethersulfones (PES), and Polyarylsulfones (PAS)

All three sulfone polymers share common structures featuring the $Ar-SO_2-$ and $Ar-O-$ functionalities—see example structures. Example IR and Raman spectra of polysulfone (PSF) and polyethersulfone (PES) are provided in Reference Spectra 37 and 36, respectively. The aromatic ether band, the asymmetric $-C_6H_4-O-C_6H_4-$ vibration appears at 1245 cm^{-1}, and the SO_2 group exhibits asymmetric O=S=O and symmetric O=S=O stretching vibrations at 1325 and 1152 cm^{-1}, respectively. Note that the asymmetric SO_2 vibrations of these polymers are split into three bands: 1324, 1299, and 1289 cm^{-1} (shoulder) for the polyethersulfone, and 1324, 1307, and 1294 cm^{-1} for the polysulfone. A band at 1585 cm^{-1}, with a weak band at around 1605 cm^{-1}, a peak at 1488 cm^{-1}, and 1169 cm^{-1} are indicative of the aromatic ring. Weak absorptions in the combination band/overtone region of these polymers, with bands at 1905 cm^{-1} and a doublet at 1775 and 1739 cm^{-1}, are consistent with a 1,4-disubstituted aromatic ring features. Bands in the 900–600 cm^{-1} region, normally used to assess the nature of aromatic substitutions, are relatively complex for these compounds, and are not readily interpreted.

Although polyethersulfone as detailed does not have any aliphatic hydrocarbon moiety, the example polysulfone does feature two methyl groups of the bis-phenol A moiety. Consistent with the presence of these methyl groups, the polysulfone exhibits a sharp band at 2968 cm^{-1} together with two weak absorption methyl bands at 2928 and 2871 cm^{-1}, as well as the more characteristic geminal dimethyl group absorptions at 1387 and 1365 cm^{-1}.

The Raman spectra of these polymers, as expected, look alike. A doublet or a distorted band at around 1600 cm^{-1}, and bands around 1175, 1117, and 1013 cm^{-1}, as well as a band at 791 cm^{-1}, are assigned to an aromatic ring indicating the presence of 1,4-substitution. A pair of bands, one weak (1300 cm^{-1}) and the other a sharp strong band (1150 cm^{-1}), are characteristic of the Ar$-$SO$_2$$-$Ar structure. Although the IR intensity of the aromatic ether absorption is very strong, it is very weak in the Raman and is observed around 1245 cm^{-1}. The polysulfone, featuring the bis-phenol A moiety, gives the aliphatic C$-$H stretching vibration around 2850–3000 cm^{-1}.

4.10. Polyphenylene Sulfide (PPS)

Polyphenylene sulfide is a partially crystalline polymer featuring an aromatic ring bridged by sulfur atoms in the 1,4-positions, as shown below. The form presented has the trade name of Ryton. Sample preparation of this material can be difficult, and the traditional KBr pellet method produces a poor-quality spectrum with Christiansen-type distortion (see Fig. 44b). A CsI pellet may be used instead to obtain an improved IR spectrum, suitable for quantitative analysis (see Fig. 44a). This polymer is usually used in conjunction with structural reinforcement additives, such as glass fibers or fillers such as PTFE (Teflon), and caution is necessary when attempting

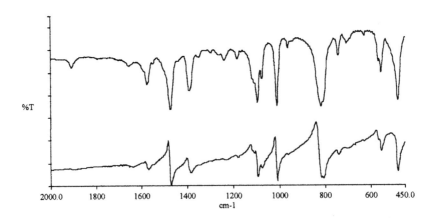

Figure 44 Comparison of CsI (upper) and KBr (lower) pellets of polyphenylene sulfide.

to identify the polymer. The IR (CsI pellet) and Raman spectra of a "pure" polyphenylene sulfide are shown in Reference Spectrum 44. Characteristic bands of the polyphenylene sulfide are observed at 1473, 1093, 1010, and 820 cm^{-1} (IR). In the Raman spectrum, the strongest band occurs at 1077 cm^{-1}, with additional bands at 1572, 1183, 743, and 708 cm^{-1}. The band at 743 cm^{-1} may be assigned to the 1,4-disubstituted aromatic ring. The Raman frequency of the Ar−S (Ar = aryl, or aromatic) stretching vibration is not clearly defined because Ar−S stretching and the ring vibration undergo vibrational coupling.

4.11. Polyimide (PI) and Polyamidimide (PAI)

Polyimide is a general name of series of polymers, each possessing the imide functional group,

As a repeating imide unit, a polyimide has the general structure

where Ar and Ar_1 are aromatic rings. Although there are several different types of polyimide, the most well known is the aromatic polypyromelliti-mide. The IR spectrum of this polymer features a pair of carbonyl groups in the imide ring, which produce a doublet structure at 1777 and 1722 cm^{-1}, as clearly shown in Reference Spectrum 42 (differentiated from the single band at 1780 cm^{-1} for the aryl polycarbonate). Sharp, but weak, peaks at 1595 and 1500 cm^{-1} are assigned to the aromatic rings, and the 1236 cm^{-1} band is consistent with an aromatic ether structure, $-C_6H_4-O-C_6H_4-$. The five bands indicated may be used to confirm the following type of polyimide structure:

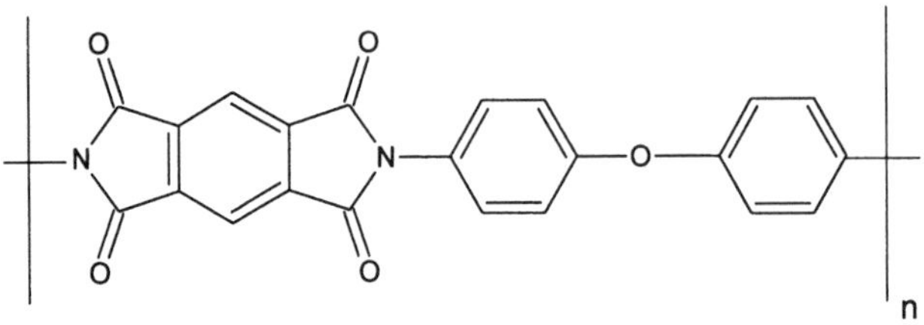

The Raman spectrum of the polypyromellitimide also produces the two peaks assigned to the carbonyl groups at 1789 and 1722 cm^{-1}, with an intensity ratio opposite to that observed in the IR spectrum. Also, one is cautioned that the 1722 cm^{-1} band is very weak, and may be overlooked (see Reference Spectrum 42). Bands at 1613 and 1600 cm^{-1}, and at 1514 cm^{-1} for the aromatic rings, are better defined than in the IR spectrum.

A polyamide-imide (PAI) features alternating imide and amide substructure sequences, an example being a pyromellitimide and amide, with a general structure as indicated below, where Al is an alkyl group. The IR spectrum of a polyamide-imide is similar to that of a polyimide, except that amide grouping provides an additional carbonyl stretching vibration at around 1670 cm^{-1}.

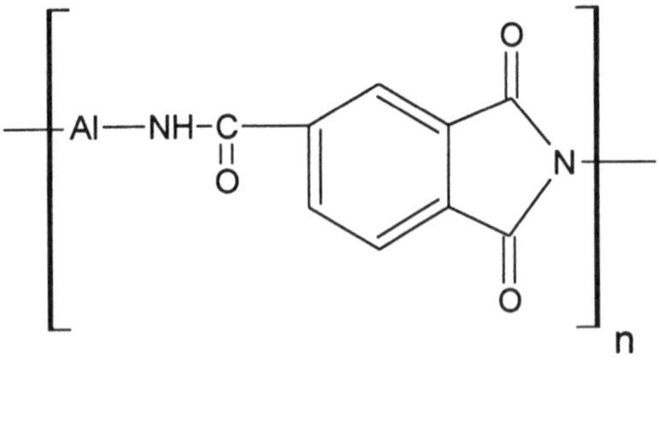

4.12. Polyurethanes

Polyurethane is a general name for polymers featuring the urethane bond as indicated below:

The standard method for the synthesis of polyurethane polymers features the reaction between an diisocyanate and an dihydroxy compound (usually an glycol or bifunctional phenol). The reaction is summarized as follows:

$$n\text{OCN}-\text{R}-\text{NCO}+n\text{HO}-\text{R}'-\text{OH}$$

$$\rightarrow -(\text{OCNH}-\text{R}-\text{NHCOO}-\text{R}'-\text{O})_n-$$

These materials are often used in a partially polymerized form, known as prepolymers. These are reacted with glycols to form high-molecular-weight polymers, or with mixtures containing controlled amounts of water to form rigid foams (note that water reacts with isocyanates to form CO_2, which is generated in situ to form the foam). R and R$'$ may be aromatic or aliphatic, to form aromatic, aliphatic, or mixed polyurethanes. Also, different polyurethane prepolymers are produced from diisocyanates and diamines, giving rise initially to a urea bond ($\text{NH}-\text{CO}-\text{NH}-$) instead of urethane:

$$(n+1)\ \text{OCN}-\text{R}-\text{NCO} + n\text{H}_2\text{N}-\text{R}'-\text{NH2}$$

$$\rightarrow \text{OCN}-\text{R}-(\text{NHCONH}-\text{R}'-\text{NHCONH})_n-\text{R}-\text{NCO}$$

This material reacts further to become a polyurethane. As a result, a wide range of polyurethanes may be formed from different combinations of diioscyanates and compounds with OH or NH_2 groups. In this chapter, we deal only with the final products, which have high molecular weights. The characteristic features of polyurethanes in the IR spectra are the 3330 cm^{-1} assigned to NH stretch, the 1730 cm^{-1} band of the urethane C=O group, the 1530 cm^{-1} band of the N$-$H bend, and the 1250 or 1220 cm^{-1} band of the C$-$O stretch. Most common polyurethanes are the aromatic, and the consecutive peaks at 1730, 1600, and 1535 cm^{-1}, and the NH band between 3340 and 3320 cm^{-1}, are usually sufficient to characterize an aromatic polyurethane. The Raman spectrum of a polyurethane similarly features the NH stretch at 3340–3320 cm^{-1}, the C=O at around 1730 cm^{-1}, the aromatic ring at 1615 cm^{-1}, and the N−H bend at 1540 cm^{-1}, although the band intensities are opposite to those in the IR spectrum. Reference Spectra 53 and 54 illustrate two types of aromatic polyurethane.

Reference Spectrum 53 provides the IR and Raman spectra of a polyester urethane, in which the terminal hydroxy groups of a low-molecular-weight polyester are reacted with isocyanate groups of the polyurethane prepolymer. Two types of carbonyl groups exist in this material, polyester and urethane carbonyl groups, and these are well differentiated (two bands observed) in the IR and Raman spectra. The other example of a polyurethane features a polyether urethane, which is formed between a di-isocyanate and a diol. Terminal isocyanate groups are reacted with water to produce NH$_2$ groups, which further react with di-isocyanate to form a urea bond. The resultant polymer contains both urea and urethane bonds. If the diol is a polyether, such as a polyethylene glycol or a polypropylene glycol, multiple ether bonds exist. Reference Spectrum 54 are the IR and Raman spectra of a polyether urethane, which has its strongest absorption band at 1109 cm^{-1}, assigned to the aliphatic ether group. Bands in the region 900–600 cm^{-1} depend on the aromatic moiety of diisocyanate starting material. In the example illustrated, the medium intensity bands at 818 and 772 cm^{-1} compare well with those of a phenyl Novolac resin.

Polymer formation may be incomplete, in particular if the labile isocyanate is prematurely decomposed at the time of polymerizaton. Also, a slight excess of isocyanate may be used in the manufacture of the polyurethane, and this is detected as a weak IR absorption band at 2280 cm^{-1} due to the isocyanate group. The isocyanate group is a strong absorber, and even low levels of isocyanate can be detected in polyurethane. Monitoring of the isocyanate level is often used to assess the degree of polymerization and to measure the quality of the final product [74].

4.13. Polyesters

Polyesters are high-molecular-weight compounds formed from polyfunctional alcohols and organic acids, and they include a wide range of compounds. This section discusses three types of polymers in this class: polyethylene terephthalate and its analogs, aklyd resins, and unsaturated polyesters.

4.13.1. Polyethylene Terephthalate (PET) and Polybutylene Terephthalate (PBT)

The chemical structures of PET and PBT are shown later in this section, and corresponding IR and Raman spectra of PET are provided in Reference Spectrum 38. The IR spectrum (see Fig. 45) shows the strong features asso-

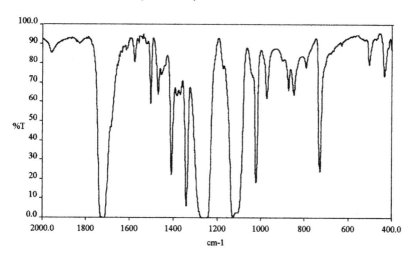

Figure 45 Fingerprint region of thin polyethylene terephthalate film IR spectrum.

ciated with ester functionality (1718 cm^{-1}, 1252 cm^{-1}, and a doublet at 1126 and 1099 cm^{-1}) and the aromatic ring (3054, 1615, 1578, 1505, 1021, and 728 cm^{-1}). Bands assigned to the "ethylene" CH$_2$ group of $-O-(CH_2CH_2)-O-$ moiety are also observed at 1134 and 848 cm^{-1}. The Raman spectrum of PET shows aromatic $C-H$ stretching vibration at 3084 cm^{-1}, a sharp strong band due to benzene ring at 1615 cm^{-1}, and features related to the para-disubstituted benzene ring at 1178 (weak), 705 and 633 cm^{-1}. The carbonyl band at 1726 cm^{-1} combined with the $C-O-C$ ester bands at 1289 and 1119 cm^{-1} represent the features of the terephthalate ester. Bands attributed to the CH$_2$ groups in the $O-CH_2CH_2O-$ sequence are observed at around 1450, 1030 (very weak), and 885 cm^{-1}.

Band intensities and frequencies of these CH$_2$-related bands are frequently discussed in terms of the trans- and gauche-rotational isomers of the $-O-CH_2CH_2-O-$ groups in relation to the orientation of polymer chain and the crystallinity of PET. Ueda and Nishiumi [75] reported the following relationship between the crystallinity and the absorbance of a band assigned to the crystalline form at 848 cm^{-1}, using the absorbance of a peak at 794 cm^{-1} as an internal standard:

$$\alpha = 12.13 - \left\{ 147.25 \middle/ \left[\left(\frac{D_{848}}{D_{794}}\right) + 11.27 \right] \right\} \tag{23}$$

A similar relationship was also given using the 972 cm^{-1} band as a crystalline index:

$$\alpha = 12.13 - \left\{288.78 \middle/ \left[\left(\frac{D_{972}}{D_{794}}\right) + 21.26\right]\right\} \tag{24}$$

The absorbance values determined from the assigned crystalline bands are correlated with the density of different PET samples, with the assumption that the polymers are composed of only amorphous and crystalline forms.

A Raman method for the determination of PET density has also been proposed [76]. In this case, the line width of the carbonyl band at 1726 cm^{-1}, $\Delta\bar{v}_{1/2}$, was correlated with density ρ:

$$\Delta\bar{v}_{1/2} = 305 - 209\rho \tag{25}$$

Bands associated with the crystallinity of PET are identified to 988 (chain folding in crystalline part), 1020, 1340, 1280, and 1260 cm^{-1}.

Reference Spectra 39–41 show the IR and Raman spectra of poly-1,4-butylene terephthalate (PBT), a polyester elastomer and a liquid crystal-type polyester, respectively. The butylenes glycol-based compound features a longer methylene sequence than the PET as seen in the $-O-CH_2CH_2CH_2CH_2-O-$ moiety. The elastomeric material is a block copolymer of PBT and poly(tetrabutyl glycol terephthalate)—structure indicated below.

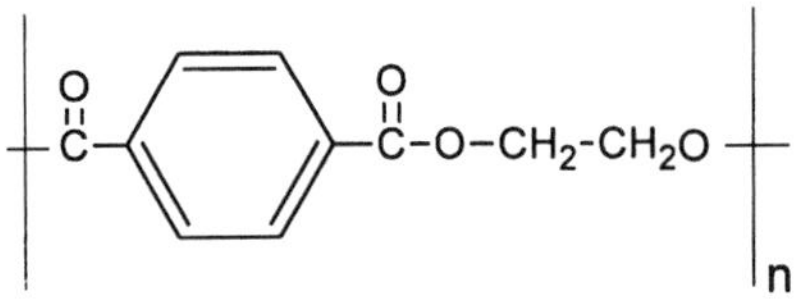

Polyethylene terephthalate

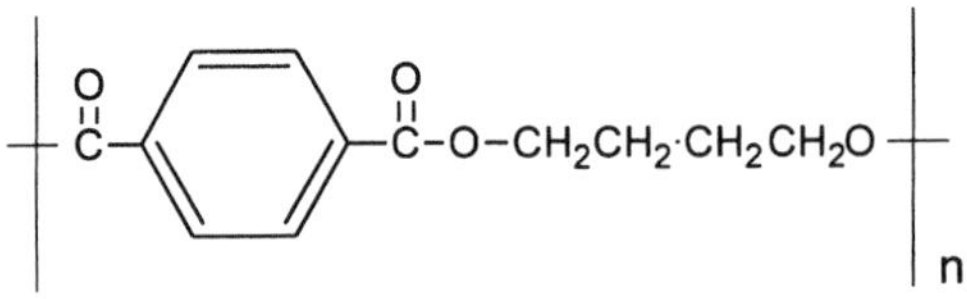

Polybutylene terephthalate

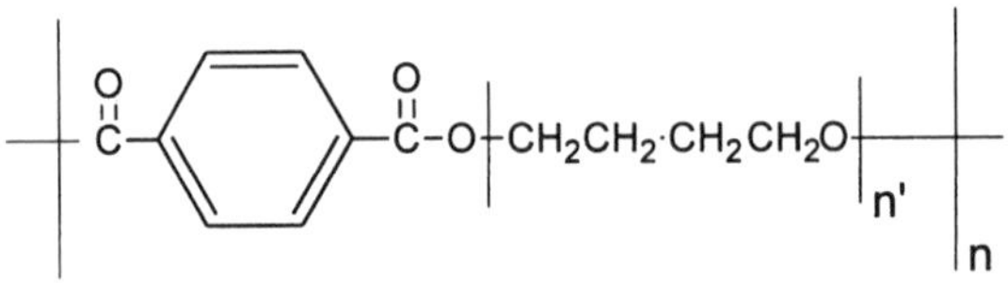

Polybutyl glycol terephthalate—block copolymer

The C−H stretching vibrations and the fingerprint regions of the IR spectra of PET, PBT, and the polyester elastomer are provided in Figs. 46 and 47, respectively.

4.13.2. Alkyd Resins

Alkyd resins are a group of esters produced from of polyols and carboxylic acids. They feature a range of different starting materials, including monobasic acids such as benzoic acid, p-butyl benzoic acid, and fatty acids and bibasic acids such as phthalic acids as the acid component, and propylene glycol (diol), glycerol (triol), and pentaerythritol (tetrahydric polyol) as typical polyols. Alkyd resins are formed from these polyesters with modified fatty acids and esters, including naturally occurring materials from seed and vegetable oils (soy, cotton-seed, tall, and linseed are example oils used in this context). A good spectral reference source for these materials is the infrared atlas published by the Chicago Coating Society [77], which includes IR spectra taken from 50 different alkyd resins. Peacock and Pross [78] estimated the oil length* from absorbance ratio of the methylene CH_2 group at 2950 cm^{-1} and the ester carbonyl C=O group at 1725 cm^{-1}. The example IR spectrum in Fig. 48 shows a characteristic doublet band at around 1600 cm^{-1}, in addition to the CH stretching vibration at 3000 cm^{-1}, with addi-

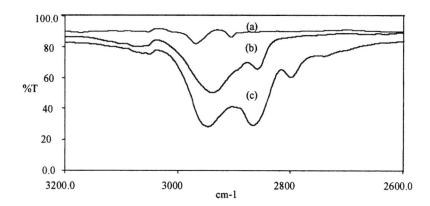

Figure 46   C−H stretching vibration of (a) PET, (b) PBT, and (c) polyester elastomer. Intensities of CH stretching region are matched to the intensities of the fingerprint region shown in Fig. 48.

---

*Defined as the ratio of oil to resin, expressed as percent of oil in resin by weight.

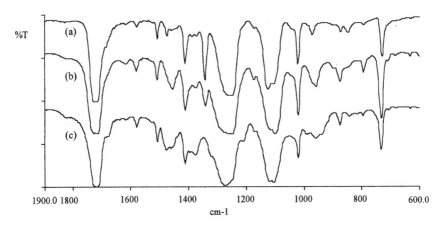

Figure 47 Fingerprint region IR spectra of (a) PET, (b) PBT, and (c) polyester elastomer.

tional absorption bands in a region from 650 to 850 cm^{-1} and in the region of 1100 cm^{-1}. These bands, which originate from an aromatic component, are highly characteristic of a phthalate, confirming the use of phthalic acid as the primary acid. A method for differentiating phthalates, isophthalates, and terephthalates was described earlier, in Sec. 3.3.2.

In the example discussed above, the phthalate used in the sample alkyd resin is identified as the normal *ortho*-phthalate. This is confirmed from the absorptions in the 850–650 cm^{-1} region: the two peaks at 740 and 700 cm^{-1}

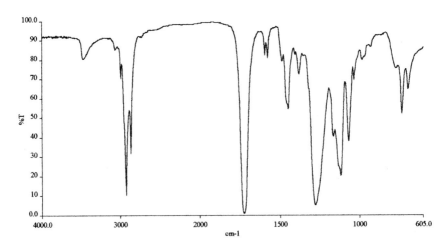

Figure 48 Typical IR spectrum of a long oil alkyd resin.

confirm the ortho substitution. If benzoic acid is added to modify the resin, the weight ratio of benzoic acid to the o-phthalic anhydride is determined [79] from

$$x(\%) = 21.8\sqrt{D_{710} - 0.513D_{748}} \qquad (26)$$

In this equation, for a resin length of 40% (a short oil alkyd), D_{710} and D_{748} are the absorbance values for the benzoic acid and o-phthalate bands, respectively.

There is the potential for three absorption bands around the 1470 cm^{-1} band: at 1490, 1470, and 1450 cm^{-1}. The weak but distinct absorption at 1490 cm^{-1} and a major absorption band at 1450 cm^{-1}, combined with a weak overlapped absorptions at 1100 cm^{-1}, indicates the presence of secondary alcohol function, originating from a material such as propylene glycol and glycerol. Alternatively, if the material is based on pentaerythritol, trimethylolethane, or trimethylolpropane as the polyol function of the alkyd resin, where the OH groups are from a primary alcohol, the strong peak at 1470 cm^{-1} dominates and overlaps the bands at 1490 and 1450 cm^{-1}, reducing them to shoulders on the main absorption [79].

The intensity of the methyl CH$_3$ deformation band at $\sim$1380 cm^{-1} depends on the nature of the alcohol/polyol. For example, propylene glycol and trimethylolethane, CH$_3$CH(OH)CH$_2$OH and CH$_3$C(CH$_2$OH)$_3$, respectively, have methyl groups, while glycerol does not have methyl group and this is reflected in the intensity of the absorption. Furthermore, propylene glycol-based alkyd resins exhibit two additional bands at 1352 and 1390 cm^{-1}, in addition to the 1378 cm^{-1} band [79].

The spectral region between 1000 and 850 cm^{-1} is used to determine the nature and configuration of the $-$C$=$C$-$double bonds of unsaturated fatty acids. The relative intensity ratio of $-$C$=$C$-$ absorptions in the 1000–900 cm^{-1} region (990, 970, and 930 cm^{-1}) are dependent on the fatty acids and oils that are used, such as oleic acid, ricinoleic acid, linoleic acid, and stearic acid (an example of a saturated naturally occurring acid). A direct comparison of the sample spectrum with reference spectra is recommended to determine the nature of the fatty acids [79].

4.13.3. Unsaturated Polyesters (UP)

Unsaturated polyesters are produced from a mixture of esters made from unsaturated fatty acids, cross-linked vinyl monomer, and various polyols. The mixture is polymerized to form a cross-linked resin, with styrene monomer being used as a common vinyl cross-linking agent. Methacrylate esters, diallyl phthalate, α-methyl styrene, and vinyl toluene have also been used for this purpose. Certain dibasic unsaturated acids (or derivatives), such as maleic anhydride and fumaric acid, have also been used. The degree of unsaturation is adjusted by the amount of phthalic anhydride, isophthalic acid, terephthalic acid, or tetrahydrophthalic acid that is added to the mixture. The polyols used are typically propylene glycol, ethylene glycol, diethylene glycol, dipropylene glycol, or neopentyl glycol. From the ingredients used, it is appreciated that the appearance of the spectra from the polyester component of the resin is very similar to that of an alkyd resin, the main exceptions being some spectral contribution from the vinyl cross-linking agent.

4.14. Bisphenol A Polyester Resins

The polyesters formed from bisphenol A and isophthalic/terephthalic acids have the general chemical structure provided below. In these compounds, the carboxylic acid COOH groups are in located in para and meta positions in terephthalates and isophthalate, respectively. Example IR and Raman spectra are presented in Reference Spectrum 27.

Similarity between these IR and Raman spectra and those of other polyesters, such as PET, PBT, etc., are expected. Also, the presence of the bisphenol A and paradisubstituted aromatic ring components have a close similarity to those of Lexan (a polycarbonate). Consequently, the IR and Raman spectra are expected to be close in appearance to those of Lexan, although the frequency of the carbonyl group is expected to be different. The similarities can be appreciated by comparing the spectra from these materials Fig. 49 with those of Lexan (Reference Spectrum 28).

The IR absorption band at 1740 cm^{-1} and the Raman band at 1744 cm^{-1} of this bis-phenol A-based material indicate that an ester type carbonyl group is present, differentiating it from the higher-frequency carbonate carbonyl. The intensities of aromatic ring-related bands are greater than for PET, as indicated by the absorptions at 1600 and 1505 cm^{-1}. The bands assigned to the methyl groups of the bisphenol A component appear at 1387 and 1364 cm^{-1}, similar to the case of Lexan as noted in Figs. 49a and 49b.

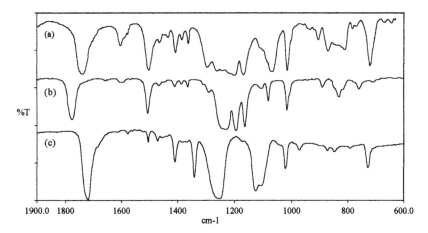

Figure 49 Fingerprint region of (a) bis-phenol A polyester, (b) Lexan, and (c) PET.

4.15. Phenolic Resins

Phenolic resins, which are formed as condensation products of formaldehyde and different phenols, form the basis of a unique class of rigid, cross-linked polymers. From a structural point of view, these resins are categorized into two types, one-step (resole) resins and two-step (Novolac) resins. One-step (resole) phenolic resins are formed from phenol and an excess of formaldehyde, in an aqueous catalytic medium. The reaction is terminated before the completion to produce a resin material that may be cross-linked in its final application. Both liquid and solid forms of these resins are available, and are formed by removing water or retaining water from the reaction. Phenolic resole resins are a mixture of the following compounds:

Table 8   Examples of Different Types of Phenolic Resins

Group	Type	Starting materials
Phenol resin	Novolac	Phenol, formaldehyde
	Resole	Phenol, formaldehyde
Cresol resin	Novolac	o-, m-, p-Cresol, formaldehyde
	Resole	o-, m-, p-Cresol, formaldehyde
Xylenol resin	Resole	Xylenol, formaldehyde
p-t-butyl phenol resin	Novolac	p-t-Butyl-phenol, formaldehyde
	Resole	p-t-Butyl-phenol, formaldehyde
p-phenyl-phenol resin	Novolac	p-Phenyl-phenol, formaldehyde
Resorcinol resin	Novolac	Resorcinol, formaldehyde

Note that generally $n1 = 0–1$ and $n2 = 1–2$. The positions of the methylene CH_2 and the methylol CH_2OH groups are ortho and para to the phenolic OH group.

The two-step (Novolac) phenolic resins are formed with phenol, but with less formaldehyde in an aqueous catalyst, so that the solid product has capacity for further cross-linking. With the addition of a curing agent containing formaldehyde, the product becomes fully cross-linked high polymer.

The IR and Raman spectra of a phenol-based Novolac resin are provided in Reference Spectrum 46. In the IR spectrum, the broad strong band at around 3340 cm^{-1} is attributed to the phenolic OH stretching. Bands at 1610, 1596, and 1510 cm^{-1} are the aromatic ring bands, and the broad intense band at 1230 cm^{-1} is the phenoxy C−O. The two bands at 819 and 757 cm^{-1} are linked to the aromatic substitution and are assigned to 2,4-substituted and p-substituted phenols, and 2,6-substituted and 2-substituted phenols, respectively [80]. The ortho-ortho phenol absorption may be assigned to a shoulder at around 780 cm^{-1}. The existence of the shoulder may be substantiated from the second derivative spectrum (Fig. 50) in which a resolved feature at 780 cm^{-1} corresponds to the shoulder and indicates the presence of a band in the original spectrum. The Bender method for the quantitative analysis of the ortho components utilizes the intensity of the 757 cm^{-1} instead of the 780 or 740 cm^{-1} (another band attributed to the ortho substituents) because of spectral overlap and interference. The assignment for the band at 820 cm^{-1} partially to the 1,2,4-trisubstituted aromatic (ortho-para component) is supported by the presence of a band at 888 cm^{-1}. The remaining fraction of the intensity of 819 cm^{-1} peak is assigned to a 1,4-disubstituted, aromatic para component. The ratio of *ortho*-methylene content to *para*-methylene content is calculated from the intensity ratio of the

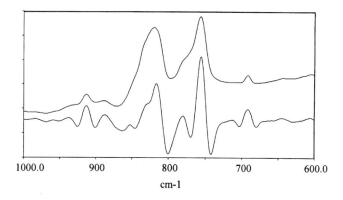

1000.0   900   800   700   600.0
cm-1

Figure 50 Observed and deconvolved spectra of a Novolak-type phenol resin.

757 cm^{-1} band and the 819 cm^{-1} band, measured from a tetrahydrofuran (THF):carbon disulfide (1:2) mixed solution from the following [81]:

$$\text{Ortho/para} = \frac{A_{757}}{A_{819} \times 1.4} \qquad (27)$$

A weak IR band at 690 cm^{-1} is assigned to free phenol. Another band associated with free phenol is located around 750 cm^{-1} and overlaps the 757 cm^{-1} band. The substituted aromatic frequencies of representative phenolic resins are summarized in Fig. 51 [82].

4.16. Amino Polymers

Amino polymers are derived from various amines and formaldehyde. Example amines used for these polymers are urea, benzoguanamine, melamine, and aniline, with the most common forms being urea and melamine resins. The formation of amino resins is complex, because the addition and condensation products of the amine and formaldehyde take multiple pathways involving by-products and side reactions. There is no general formula for a urea-based resins. The early reaction products are similar to those formed in phenolic resins or polyurethanes, but these early reaction products react further to form the final high polymer. Only the final reaction products will be considered in this section.

4.16.1. Urea–Formaldehyde Resins

The initial addition reaction of urea and formaldehyde takes the following form:

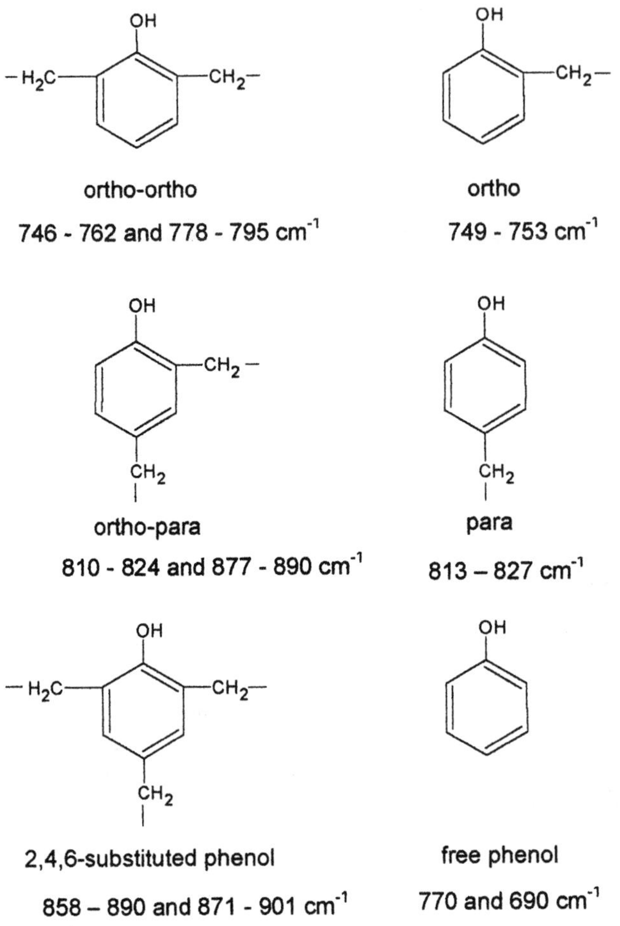

Figure 51   Assignment of phenolic resin benzene absorption frequencies [82].

$$NH_2CONH_2 + HCHO \rightarrow NH_2CONHCH_2OH$$

The monomethylol formed from this reaction reacts further with formaldehyde to form a dimethylol component (see below), which continues to react with the formaldehyde.

$$NH_2CONHCH_2OH + HCHO \rightarrow HOCH_2NHCONHCH_2OH$$

As a result, several different functional groups are expected in the final urea-based resin, such as $-CONH-$ (amides), $-NHCONH-$ (urea), and OH groups. Thus, features of amide and OH groups appear in the IR and

Raman spectra. Hydroxyl OH group and the amido NH bands overlap in the 3300 cm^{-1} region, producing a single broad and convoluted band. The normal amide I, II, and III bands appear in the IR spectrum at 1640, 1530, and between 1230 and 1270 cm^{-1}, respectively. These resins are differentiated from the polyamides on the basis of the spectral line shapes.

4.16.2. Melamine-Based Resins

Melamine

The triazine ring of melamine is featured in the IR and Raman spectra of the melamine-based resins, in addition to the normal OH and amide features described above; IR and Raman reference spectra are provided in Reference Spectrum 35. The most characteristic feature is the sharp IR band at 812 cm^{-1}, which is assigned to the triazine ring. The triazine ring structure also produces sharp IR absorption bands at 1570 and 1450 cm^{-1}, overlapping the amide-I and -II bands. The IR spectrum of a melamine resin is often dominated by the two bands at 1570 and 1490 cm^{-1}, thereby obscuring normal amide bands at 1640 and 1530 cm^{-1}. In the example shown in Reference Spectrum 35, the 1450 cm^{-1} band is observed as a shoulder. Melamine itself is often used as a curing reagent for alkyd and epoxide resin, and its presence as a curing agent is indicated by the sharp band at 810 cm^{-1} (vis-à-vis melamine-cured resins).

The Raman spectrum of the melamine resin has a weak band at 3400 cm^{-1} assigned to both NH and OH groups (note in the Raman spectrum that the NH will often be more intense). Bands at 819 cm^{-1} and at 977 and 675 cm^{-1} are attributed to the melamine moiety [83]. The amide-I and triazine ring bands are observed at 1630 and 1560 cm^{-1}, respectively.

4.17. Silicone-Based Polymers, Polysiloxanes

Silicone polymers feature the siloxane backbone ($-Si-O-$) with organic groups attached to silicon atoms, either directly or via an oxy linkage. Silicone polymers exist in a wide variety of forms and are used as oils, rubbers, sealants, and resins. The molecular structure of the basic organo-polysiloxane is represented as [84]

$$\left(R_m SiO_{(4-m)/2}\right)_n$$

where the R are normally hydrogen or alkyl and aryl groups such as $-CH_3$, $-C_6H_5$, $-C_3H_7$, and $-C_3H_4F_3$. However, certain reactive groups are introduced into the backbone, such as $-CH=CH_2$, $-OH$, $-OCH_3$ and $-OC_2H_5$, to provide sites for cross-linking. Substituting $m = 3$ in the above structure gives $R_3SiO_{1/2}$, representing a terminating group in the molecule, and $m = 2$ gives the (R_2SiO) unit, which is the repeating unit of the polymer chain. Note that $m = 1$ forms $(RSiO_{3/2})$, which represents chain branching.

$$
\begin{array}{ccc}
\text{R} & \text{R} & \text{R} \\
| & | & | \\
\text{R--Si--O} & \text{O--Si--O} & \text{O--Si--O} \\
| & | & | \\
\text{R} & \text{R} & \text{O} \\
\\
m = 3 & m = 2 & m = 1
\end{array}
$$

All polysiloxanes feature a strong IR absorption associated with the asymmetric $Si-O-Si$ stretching vibration in the region 1130–1000 cm^{-1}, where the observed frequencies are dependent on the mass and properties of substituent groups [85]. The band position and shape of this band are also influenced by the molecular weight of the polymer. When the molecular weight is low, the band is viewed as a singlet; but as the molecular weight increases, the band separates into a fairly symmetric doublet pattern, with absorption maxima around 1100 and 1020 cm^{-1}. The Raman bands of the asymmetric $Si-O-Si$ stretching vibration are usually very weak and almost unobservable—this is not surprising because it is common that Raman spectra are recorded from samples placed in glass or quartz containers, both materials possessing the same basic form as the $Si-O$ backbone.

The most common silicone material is polydimethyl siloxane, $-(Si(CH_3)_2-O)_n-$. This compound has a very simple IR spectrum, featuring four distinctive lines at 1262 cm^{-1} (symmetric CH$_3$ deformation of the $CH_3-Si-CH_3$ group), 1098 and 1023 cm^{-1} ($Si-O-Si$ and $Si-O-C$), and 802 ($Si-CH_3$) cm^{-1} as shown in Fig. 52 (see also Reference Spectrum 32).

The asymmetric CH$_3$ deformation of $CH_3-Si-CH_3$ and the symmetric $Si-O-Si$ stretch appear as weak bands at 1414 cm^{-1} and around 480 cm^{-1}, respectively. The Raman spectrum of polydimethyl siloxane features a strong band at 491 cm^{-1}, attributed to the symmetric stretching vibration of $Si-O-Si$. As noted above, asymmetric vibrations of $Si-O-Si$, expected

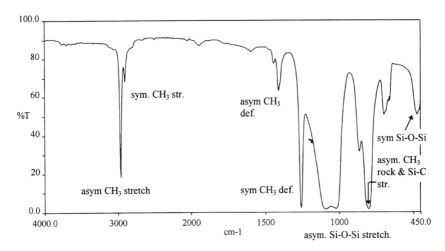

Figure 52 Assignment of polymethylsiloxane IR bands.

to be located in the region between 1100 and 1020 cm^{-1}, are very weak and are difficult to locate. The symmetric and asymmetric CH$_3$ deformations of the CH$_3$—Si—CH$_3$ group appear at 1412 and 1264 cm^{-1}, respectively, with intensities inverse to the corresponding bands in the IR spectrum.

Other important polysiloxanes are polymethyl hydrogen siloxane, $-(HSi(CH_3)-O)_n-$, polydiphenylsiloxane, $-(Si(C_6H_5)_2-O)_n-$, and polymethyl phenyl siloxane, $-(CH_3Si(C_6H_5)-O)_n-$. The IR and Raman spectra of the first and last of these polysiloxanes are presented in Reference Spectra 33 and 34, respectively. In the case of polymethyl hydrogen siloxane, intense bands located at 2177 cm^{-1} (IR) and 2180 cm^{-1} (Raman) are characteristic of the Si—H absorption. The combination of this band in the Raman with the 529 cm^{-1} band of the symmetric Si—O—Si vibration provide positive identification of a polymethyl hydrogen siloxane. The aromatic polysiloxane has characteristic IR bands associated with the phenyl group at 1600, 1490, and 1435, 1120, 741, and 699 cm^{-1}. The region between 1100 and 1000 cm^{-1} of the polymethyl phenyl siloxane IR spectrum has three bands, two assigned to the Si—O—Si and one assigned to the phenyl group, as indicated in Reference Spectrum 34. In the Raman spectrum of polymethyl phenylsiloxane distinctive bands assigned to the phenyl group are observed at 3055, 1594, and 1570, 1032, 1001, and 700 cm^{-1}. The intense doublet around 1000 cm^{-1} is characteristic of the phenyl group and is not associated with the Si—O—Si asymmetric vibration.

Polymethyl phenylsiloxane exhibits bands associated with both Si—CH$_3$ and Si—C$_6$H$_5$, and this is comparable to the spectral data obtained from a copolymer or a mixture of dimethyl siloxane and diphenyl siloxane. Three bands are observed around 700 cm^{-1} for the copolymer [86], at 741, 720, and 699 cm^{-1}. However, the homopolymer of methylphenyl siloxane has only the 741 and 699 cm^{-1} bands, as shown in Reference Spectrum 34.

The frequencies of the terminal and the branching groups of polydimethyl siloxane have been studied. The $-Si(CH_3)_3$ group terminates the polymer chain and displays two bands assigned to the $Si(CH_3)_3$ rocking vibrations at 845 and 760 cm^{-1}. The repeating unit in the chain, $-(CH_3)_2Si-$, also produces a pair of bands at 885 and 805 cm^{-1}. Finally, at the chain branch, the $(CH_3)Si\equiv$ produces only one absorption, linked to the single $SiCH_3$, at 775 cm^{-1} [85].

The bending vibration of the terminal $-Si(CH_3)_3$ group appears in the IR spectrum at a slightly different frequency than that of the repeating unit. This fact was used by Lipp [87] to correlate the peak ratio of the $(CH_3)_3Si-$ (band at 1270 cm^{-1}) to $-(CH_3)_2Si-$ (band at 1262 cm^{-1}), with the molecular weight of polydimethyl siloxane. Molecular weights for polydimethyl siloxane of up to 15,000 can be determined by this method.

4.18. Epoxides

Epoxide resins are formed from compounds containing the following functional group:

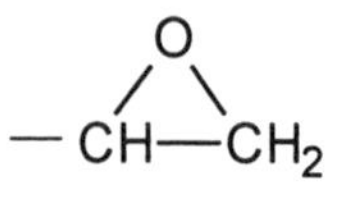

These compounds are formed into a prepolymer, which in turn is converted into a series of high-molecular-weight compounds when reacted with a suitable curing agent. Polymers featuring the epoxide group as a part the chemical structures will react with other chemical species to form a three-dimensional cross-linked network. One of the most popular forms of epoxide resin is derived from bisphenol A, and the associated epoxide resin has the following chemical structure:

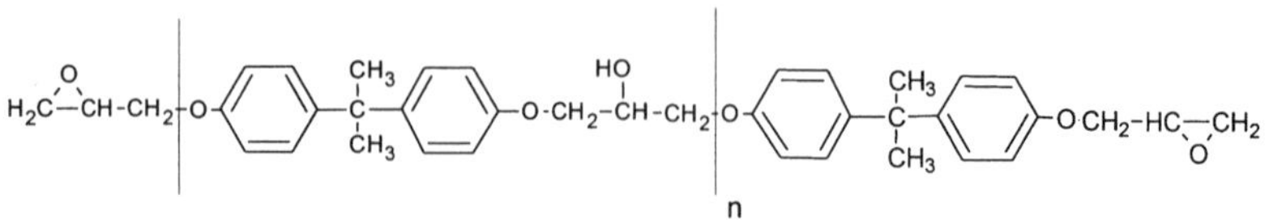

When the degree of polymerization, defined by n in the above structure, is large, the bisphenol A epoxide is solid and is used as a polymer in its own right, without additional curing to form a three-dimensional cross-linked resin. The spectral features of bisphenol A-derived compounds were discussed earlier under polyesters and polycarbonates, and like the previous examples, the IR and Raman spectra have features common to bisphenol A; see Reference Spectrum 30. The IR spectrum features a doublet band due to the aromatic ring at 1608 and 1582 cm^{-1}, an aromatic ether stretching vibration at 1246 and 1039 cm^{-1}, and absorptions assigned to the para-disubstituted aromatic ring at 830 and 566 cm^{-1}, the latter being split into two bands, at 574 and 559 cm^{-1}. The geminal dimethyl groups of bis-phenol A, $R-C(CH_3)_2R-$, appear as a doublet at 1385 and 1363 cm^{-1}, and the presence of CH_2 groups is indicated by a peak at 1460 cm^{-1}. In this form, the epoxide groups remain unreacted, and are observed as the characteristic band at 916 cm^{-1}.

The Raman spectrum features a doublet band assigned to the aromatic ring at 1610 and 1583 cm^{-1}, and the aromatic ether band at 1232 cm^{-1}. The absence of strong bands in the region 1055–990 cm^{-1} and the presence of a 641 cm^{-1} medium-intensity band, combined with bands assigned to aromatics at 1186 and 1114 cm^{-1}, confirm the presence of a 1,4-disubstituted aromatic ring. The band at 1253 cm^{-1}, which decreases in intensity as a function of curing, is assigned to the epoxide group, but it is uncertain whether the band at 917 cm^{-1} is also assigned to this group. A broad, medium-intensity band at 1460 cm^{-1} is assigned to the CH_2 groups.

Although the band close to 910 cm^{-1} is well documented as being characteristic for an epoxide group, a total of three absorption bands have been proposed for the group. These are sometimes called "8μ band," "11μ band," and "12μ band," appearing at 1253, 911, and 839 cm^{-1}, respectively [88,89]. All three bands are evident in the case of the polyalcohol epoxide prepolymer illustrated in Fig. 53. For bisphenol A-type epoxides, the 8μ band is expected to occur at around 1233 cm^{-1}. This overlaps the aromatic ether absorption band at 1246 cm^{-1}, leaving the 910 cm^{-1} absorption as the main band for characterization, and for monitoring the kinetics of curing and the determination of unreacted epoxide groups.

There are many types of epoxide resin; the *Atlas for the Coating Industry*, produced by the Federation of Societies for Coating Technology, cites 47 IR spectra of different resins [77]. Some typical example structures are provided below for Novolac-type phenol (and Novolac-type *o*-cresol) epoxides, aromatic glycidyl ester-type epoxides, alicyclic epoxides (18-4) , and polyalcohol glycidyl epoxides (polyglycol and polyol based). The IR spectra of some epoxides are show in Fig. 54.

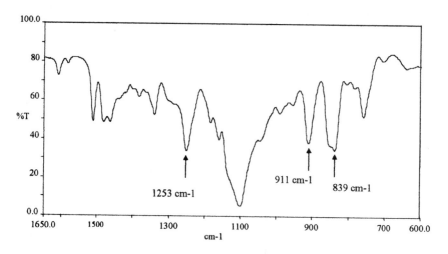

Figure 53   Three characteristic absorption bands of the epoxide group.

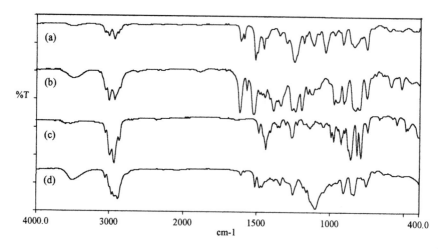

Figure 54   IR spectra of (a) phenol Novolac-type epoxide, (b) Ciba-Geigy's tetrafunctional epoxy resin, (c) alicyclic epoxide, and (d) polyglycol epoxide.

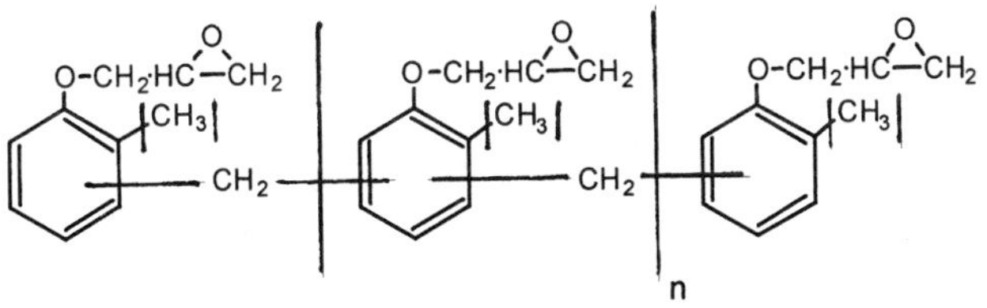

phenol Novolac type (and o-cresol Novolac-type) epoxide

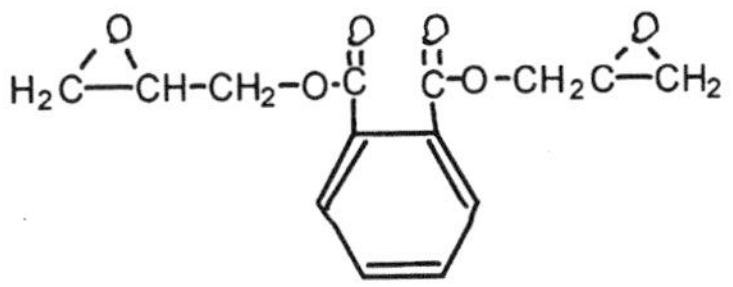

aromatic glycidyl ester-type epoxide 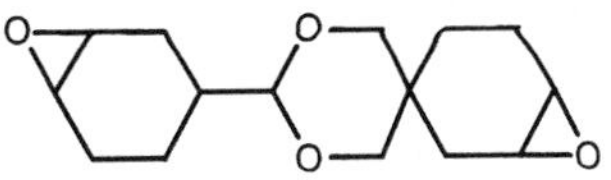   alicyclic epoxide

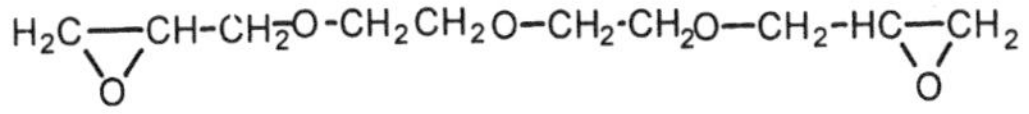

polyalcohol glycidyl epoxide

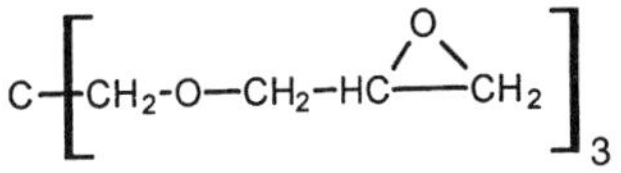

polyalcohol glycidyl epoxide, pentaerythritol-based

4.19. Cellulose Derivatives

Cellulose itself, with the chemical structure shown below, is a natural polymer. Most of the important commercial polymers are derivatives of cellulose, such as nitrocellulose or cellulose nitrate $(R-ONO_2)$, cellulose acetate $(R-OCOCH_3)$, cellulose acetate propionate $(C_2H_5OCO-R-OCOCH_3)$, cellulose acetate lactate $(CH_3OCO-R-OCOC_3H_7)$, and ethyl cellulose $(R-OCH_2CH_3)$. Most of these derivatives are cellulose esters, formed by the reaction of acids with some or all of the three available hydroxyl groups of the repeating sugar units; the exception is ethyl cellulose, which is an ether.

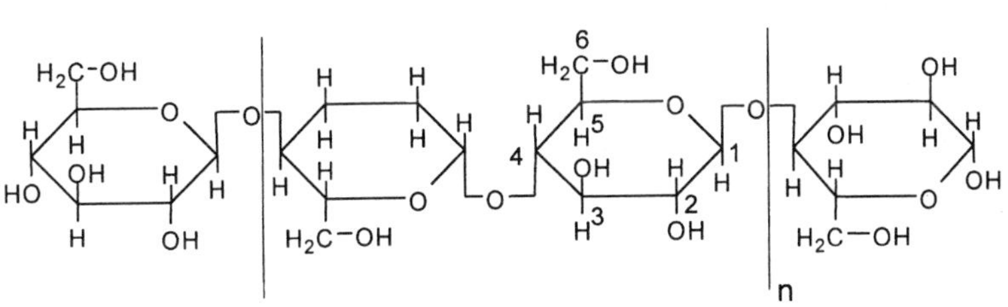

4.19.1. Cellulose

The IR and Raman spectra of cellulose are provided in Reference Spectrum 59. As expected, the IR spectrum is dominated by absorptions associated with the hydroxy and ether groups. It features a strong broad band due to OH stretching at 3450 cm^{-1}, and corresponding bending vibrations of the C$-$OH groups around 1300–1400 cm^{-1}, which are coupled to other vibrations and appear relatively weak in intensity. Intense bands associated with C$-$O stretching of the C$-$OH groups appear in the 1100–1000 cm^{-1} region, alongside absorptions assigned to the C$-$O$-$C ether linkages inside each glucopyranose ring, and the ether linkage connecting neighboring rings at 1061 cm^{-1}. The overlap of these ether bands and "alcohol" C$-$O bands generates a unique shape to the overall band envelope in the 1100–1000 cm^{-1} region. A weak band at 895 cm^{-1}, assigned to the C$_1$$-$H bending vibration, indicates that the hydrogen atom attached to the C$_1$ atom of the glucopyranose ring is axial (thereby indicating that the OH is equatorial). Consequently, the neighboring pyranose rings are β-1,4-linked, confirming that the chemical structure of cellulose is a polymer of β-glucose linked at 1 and 4′ positions as indicated above. Differences between α- and β-glucose forms are shown below.

α-D-glucopyranose   β-D-glucopyranose

The Raman spectrum of cellulose has a pair of bands at 1122 and 1097 cm^{-1}, and a few other peaks below 500 cm^{-1}. As is normal with Raman spectra, frequencies associated with the hydroxy functionality are extremely weak in intensity. The most informative band is located at 899 cm^{-1}, which confirms the β-1,4-linkage of pyranose rings are, as discussed above. In the case of α-pyranose compounds, such as D-glucose and sucrose, the C$_1$$-$H vibration appears at around 825 cm^{-1} (glucose) and 850 cm^{-1} (sucrose).

4.19.2. Cellulose Acetate

The IR and Raman spectra of cellulose acetate are presented in Reference Spectrum 60. A decrease (or even disappearance) in band intensity of the cellulose OH stretch at 3450 cm^{-1} is observed as the ester bonds are formed. Likewise, a change in the C$-$O stretch region around 1100–1000 cm^{-1} is observed as the C$-$OH content is reduced by esterification, and C$-$O$-$C bonds of the ester are formed. Both sets of intensity changes (OH and C$-$O) may be monitored to ascertain the degree of esterification of the cellulose OH groups. Both intra- and interring ether bondings are unaffected by esterification, and so the pyranose ether band at around 1050 cm^{-1} remains unchanged.

The IR spectrum of cellulose acetate shows a significantly weakened OH stretch at 3450 cm^{-1}, the presence of an ester carbonyl, with a C=O stretching vibration at 1744 cm^{-1}, and a strong C$-$O$-$C stretching vibration at 1237 cm^{-1}. A well-defined band at 1370 cm^{-1} suggests a notable contribution from methyl groups, and this is assigned to the methyl group on the acetate group. An IR spectrum of cellulose acetate is provided in Fig. 55, which bears a strong similarity to the IR spectrum of polyvinyl acetate. One measurable difference is noted in the position of the band at around 1050 cm^{-1}; as indicated by an arrow, cellulose acetate has the pyranose ring ether band at 1050 cm^{-1}, while polyvinyl acetate has a well-defined band at 1025 cm^{-1}. The 1050 cm^{-1} band is considered to be characteristic of the pyranose ring, and this allows positive identification of cellulose derivatives. A weak doublet at 1601 and 1580 cm^{-1} and a peak at 748 cm^{-1} are indicative of an o-phthalate ester, used as a plasticizer additive.

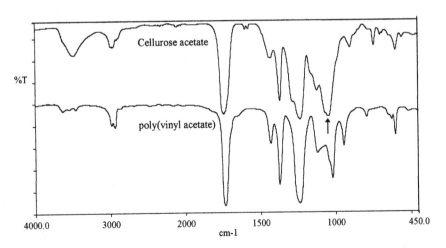

Figure 55 Comparison of ethyl cellulose and polyvinyl acetate. Interring ether band is marked by an arrow.

5. EXPERIMENTAL PROCEDURES

Infrared spectroscopy is a recognized standard method of analysis for characterization and identification, and for the measurement of composition of a wide range of materials. The infrared spectrum is a unique physical property of a molecular species, and the spectral information obtained can be correlated directly to the chemical structure of the material, and the underlying composition in the case of admixtures. Infrared spectroscopy is also one of the most widely utilized analytical methods for the identification and characterization of plastics and polymers. The first sections of this chapter have focused on the specific attributes of the technique as they relate to the analysis of polymeric materials. In this final section, the focus is on the more general and fundamental aspects of infrared measurements, and the methods used for the handling of plastics and polymers.

5.1. Infrared and Raman Spectroscopy: General Principles

The principle of optical spectroscopy involves the measurement of the amount of light (radiation) that is absorbed by the sample when the radiation interacts with the sample. The most basic method involves the determination of the fraction of the radiation that is actually transmitted through a sample. The aspects of the measurement, and their relationship to the actual absorption of radiation are illustrated in Fig. 56. In this example, I_0 is the power of the incident radiation from the infrared light source, and I is the actual amount of radiation transmitted through the sample. The fundamental relationships are provided with Fig. 56, and these form the basis of a fundamental expression that is used to correlate the analytical spectrum with the amount(s) of material(s) present in a sample. This fundamental expression is a simple rendering of the Beer-Lambert-Bouguer law, which is used in one form or another in the quantitative determination of material composition.

This chapter has discussed the use of infrared and Raman spectroscopies for the characterization of polymer systems. Both of these techniques form the basis of vibrational spectroscopy. Before proceeding with the practical aspects of instrumentation and the methods used for sample handling and preparation, it is important to review the origins of the vibrational spectrum. The first step is to define the spectral region covered, and to place this in perspective relative to other forms of optical spectroscopy, such as visible and near-infrared (NIR) spectroscopy. Figure 57 presents the electromagnetic spectrum from the visible to the mid-infrared. As noted from the scale presented, the wavelength range covered for the fundamental infrared is from 2,500 to 25,000 nm (2.5 to 25 μm). In practice, it is more common to use the term frequency, expressed in wavenumber (4000 to

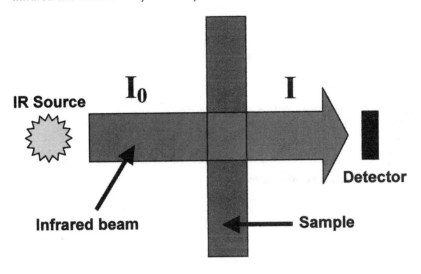

I_0 = Original intensity of source
I = Intensity after passing through sample

%Transmittance (%T) = 100 x I/I_0
Absorbance = $\log_{10}(1/T)$ = $\log_{10}(I_0/I)$
Absorbance = α.b.c
(where α is a constant, b is the sample thickness
or pathlength, and c is the analyte concentration)

Figure 56 Light transmission through a material plus the fundamental relationships in terms of the absorption of radiation by the material.

$400\ \text{cm}^{-1}$), because this is related directly to the vibrational energy involved in the energy transitions of the molecule. The upper limit is more or less arbitrary, and was originally based on the performance of early instruments. Also, it is adequate to cover all fundamental absorptions, with the exception of the H−F stretching vibration, which is observed between 4100 and 4200 cm^{-1}. The lower limit, in many cases, is defined by a specific optical component, such as, the beam splitter that is used in a modern FTIR instrument, which is made from a potassium bromide (KBr) substrate with a transmission cutoff just below $400\ \text{cm}^{-1}$. The range may be extended, but special optical materials are required for this, and this facility is generally reserved for high-end instruments.

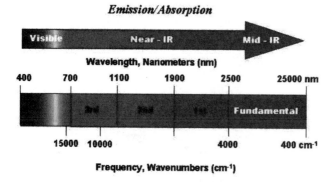

Figure 57 Wavelength and frequency ranges of electromagnetic radiation, from the visible to the mid-infrared.

Raman spectroscopy has somewhat different frequency or wavelength limits imposed on it. The concept of Raman spectroscopy, which is based on a scattering phenomenon, involves the use of a monochromatic light source, typically a laser, as the incident or "exciting" radiation. Most of the radiation obtained from a Raman measurement remains unchanged with respect to the energy of the original incident radiation, and is known as Rayleigh scattering. The Raman spectral information is contained within a small fraction of the radiation emanating from the sample, which is scattered at longer (Stokes) or shorter (anti-Stokes) wavelengths than the original incident radiation (see Figs. 58 and 59). Theoretically, one can extend Raman down to a region close to zero wavenumbers and to an upper limit which may be as high as 4000 cm^{-1} but is typically less—both are dependent on the optical system used. The latter is often limited by the performance of the detection system in use. As implied, the Raman effect is extremely weak, with observed intensities several orders of magnitude less than the intensity of the illuminating monochromatic light source. The intensity of the Stokes lines is more intense than that of the anti-Stokes lines, and is utilized for most analytical applications.

At the molecular level, vibrational spectroscopy involves energy transitions measured from the ground state to an excited, higher-energy state. The various transitions occurring in infrared and Raman spectra are summarized in Fig. 60. In the fundamental region of the mid-infrared the measurement involves transitions from the ground state ($v = 0$) to the first excited stated ($v = 1$). Transitions to higher energy levels do occur, and these give rise to what are known as overtones in the spectrum (occurring nominally as higher multiples of the fundamental frequency). These absorptions are weak

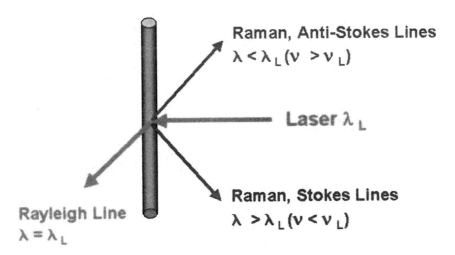

Figure 58 The Raman effect: Rayleigh scattering and the Stokes and anti-Stokes scattering.

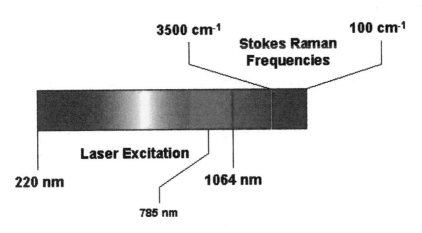

Figure 59 The Raman scattering measurement range in relation to the excitation wavelengths.

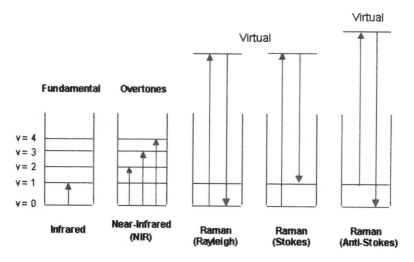

Figure 60 Vibrational transitions for the mid-infrared, near-infrared, and Raman spectroscopy.

compared to that of the fundamental. They may be observed in the main spectral region of the mid-infrared, but they are also observed in higher frequency ranges, in a region of the spectrum known as the near-infrared. The Raman effect is the result of the molecule undergoing vibrational transitions, usually from the ground state ($v = 0$), to the first vibrational energy level ($v = 1$), during the scattering process. This transition gives rise to the Stokes lines, observed as spectral lines occurring at lower frequency (longer wavelength) than the incident beam. Because the effect involves a vibrational transition, with a net gain in energy, it is comparable to the absorption of photon energy experienced in the generation of the infrared vibrational spectrum. Consequently, both techniques, infrared and Raman spectroscopy, are derived from effectively the same energy transitions, and the information content of the spectra should be similar. However, the spectra are not identical because not all energy transitions are allowed by quantum theory, and different selection rules govern which transitions are allowed. The following are fundamental rules that define what is observed.

1. An infrared vibration is active, or absorbs infrared energy, when the molecular vibration induces a net change in dipole moment during the vibration.
2. A Raman vibration is active when the molecular vibration induces a net change in the bond polarizability during the vibration.

The result of these rules is that the two sets of spectral data are complementary. In practice, for most molecular compounds, reality is somewhere in between these two extremes and as a consequence, most compounds produce unique infrared and Raman spectra that feature a mixture of strong, medium, and weak spectral bands. Typically, a strong feature in one spectrum will either not show, or may be weaker in the other spectrum, and vice versa.

Unlike traditional infrared spectroscopy, Raman is an emission phenomenon, and does not follow the laws of absorption defined earlier (Fig. 56). The intensity of a recorded spectral feature is a linear function of the concentration of the species present, and the intensity of the incident radiation. The measured intensity is not constant across the spectrum, and is limited by the response of the detector at the absolute wavelength of the point of measurement. As a result, with some detectors it is not possible to measure much beyond 3000 or 3500 cm^{-1}. The InGaAs detector used in combination with the Nd:YAG laser in a FT-Raman experiment is a case in point (the actual limit being influenced by the detector used, and whether it is cooled). Also, Raman scattering is not a linear effect, and varies as a function of λ^4, where λ is the absolute wavelength of the scattered radiation. While Raman intensity may be used analytically to measure the concentration of an analyte, it is necessary to standardize the output in terms of the laser intensity, the detector response, and the Raman scattering term.

It is assumed that the reader has some basic understanding of spectroscopy and organic chemistry. For a further understanding of underlying theories of infrared and Raman spectroscopy, the reader is directed to standard reference texts on the structure/spectra relationships of molecular compounds [90–94].

5.2. Instrumentation

Infrared spectroscopy was developed as an analytical technique back in the 1940s as a method for the quality control of petroleum products, and in particular polymers. One of the first practical applications involved the determination of the different monomer species used the production of synthetic rubbers. The very first instruments used the principle of light dispersion, originally from a prism, to generate the spectrum. This has in recent years led to the name "dispersive instrumentation." From the 1940s to the early 1980s these dispersive instruments evolved, and the use of the prism optics gave way to the use of the diffraction grating, the latter providing a higher-resolution spectrum. In the 1960s, another infrared measurement technology was developed, based on the optical measurement known as interferometry. This technique generates a unique signature, in which each frequency in the spectrum is uniquely modulated. This signature, known as an interferogram, is converted into the normal analytical infrared spectrum by a mathematical function called a Fourier transform, hence the technique, Fourier transform infrared or FTIR [95].

Today, FTIR forms the mainstay of analytical infrared instrumentation [96]. All of the spectra presented in this chapter were produced on FTIR instrumentation. However, the older traditional dispersive instruments are still adequate for most polymer applications. FTIR offers some unique advantages in terms of sample handling, and as such is more versatile for polymer analysis. Applications that take full advantage of the properties of FTIR, which extend the capabilities of infrared spectroscopy for polymer characterization, include infrared microscopy, GC-IR (in the form of pyrolysis GC-IR), GPC-IR (gel permeation chromatography-IR combination), TGA-IR (thermal gravimetric analysis-IR combination), and step scan, for dynamic-mechanical property measurements.

Figure 61 provides a general schematic diagram indicating the main features of a FTIR instrument. In this system, the key components are the source, the beam splitter, and the detector. These may be modified to customize the instrument to meet the needs of specific applications. Different sources, beam splitters, and detectors may be used to extend the

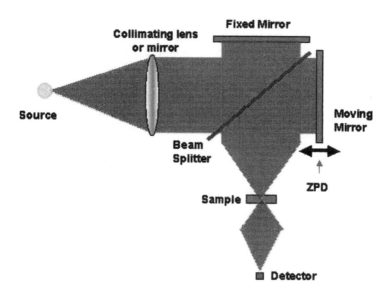

Figure 61 Typical optical layout for a Michelson interferometer as used in a traditional FTIR instrument.

range of the instrument—either to shorter wavelengths (the near-infrared and visible) or to longer wavelengths (far-infrared). The most practical combination in the mid-infrared is a thermal source, such as an electrically heated metal filament, or a ceramic or silicon carbide rod. Typical source temperatures range from 1100 to 1500 K. The standard beam splitter in the mid-infrared is germanium, deposited as a thin film on a potassium bromide (KBr) substrate. Several detector choices exist, the most popular being the DTGS (deuterated triglycine sulfate) pyroelectric detector. Other options include the lower-cost, and slightly lower-sensitivity, lithium tantallate detector, and the higher-performance (speed and sensitivity) photon-detecting MCT (mercury cadmium telluride) detector. The latter requires cooling, normally cryogenically with liquid nitrogen.

Raman spectroscopy may be performed on either a dispersive instrument or a FT instrument. All of the spectra provided in this chapter were obtained from a FT-Raman instrument (see Fig. 62), featuring a Nd:YAG solid-state laser, and a InGaAs (indium–gallium–arsenide) detector, combined with a silicon on quartz beam splitter. Note that in the FT-Raman experiment, the sample effectively becomes the source to the FT spectrometer. Dispersive Raman instruments are also popular, and these usually feature a silicon-based array detector (CCD array) in combination with either a visible laser (doubled YAG or HeNe) or a short-wavelength solid-state NIR laser.

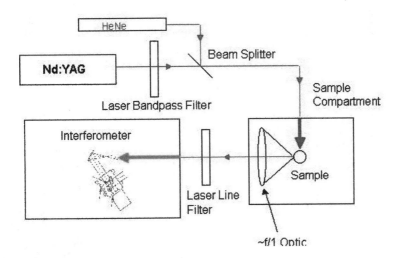

Figure 62   General schematic for a FT-Raman spectrometer.

The instrument parameters used in the measurement of a sample are influenced by the hardware used in the instrument and the sampling method adopted. For a given instrument, the sampling accessory often defines the spectrum quality, and in turn tends to define the instrument parameters that are used. For most analytical methods used for polymers, the normal parameters are 2 or 4 cm^{-1} spectral resolution (for some applications, such as microscopy, 8 cm^{-1} resolution will often suffice), and 32 to 128 averaged scans (the latter sometimes being defined by the desired timeframe for the analysis).

5.3. Sample Handling Methods

The quality of an infrared spectrum is dependent on the method of sample preparation and the nature of the optical interface between the sample and the infrared instrument—usually defined by a sampling accessory. The physical state and how the sample is treated influences the appearance of the spectrum. Also, different sampling accessories, which use different methods of optical measurement, generate spectra that may differ in appearance for a given sample. Care and attention at the moment of sample preparation, and an understanding of the impact of the measurement techniques, can save time and effort. Raman is very different, and in general, other than options for the method of sample illumination (and scattered light collection), the sampling procedure is simple, with few variations. With Raman, the sample is usually studied directly, usually as a solid or liquid, in its natural state, and unmodified. Solids are usually the easiest, and are measured directly, either as an extended sample (sheet, rod, foam, etc.) or as a powder. Liquids (and powders) are normally measured in a container, such as a glass or quartz vial, or a glass capillary tube. Having indicated the relative simplicity of sample handling with Raman spectroscopy, the remainder of this section will be dedicated to the more complex and involved subject of sample handling for infrared spectral measurements.

Most infrared interactions with the sample are recorded as absorptions of energy, either as a direct or an indirect measurement. Emission measurements may be performed, but these are less frequent for analytical chemistry applications. Measurements may be made via light transmission through the sample, or by reflection directly from the sample surface or from an interface, and these are highlighted in Fig. 63. Absorption in the mid-infrared can be extremely high, and this often dictates which sampling method is most practical.

Polymers normally exist as viscous liquids or solids, and so most of the focus, from the point of view of sampling, is on methods for such samples. The only exceptions are when monomers are being studied, and these can

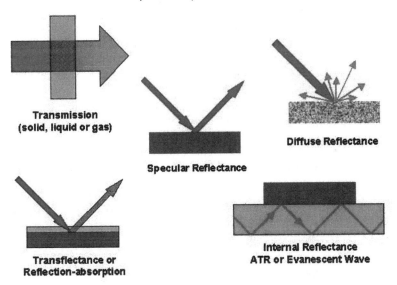

**Transmission
(solid, liquid or gas)**

Specular Reflectance

Diffuse Reflectance

**Transflectance or
Reflection-absorption**

**Internal Reflectance
ATR or Evanescent Wave**

Figure 63   The common methods for infrared sample measurement.

exist as vapors or low-viscosity liquids. Also, if a procedure such as pyrolysis is used, the materials formed are also vapors and/or low-viscosity condensate.

5.3.1.   Gases and Vapors

The sample handling for gases and vapors is usually simple, and the most important issue is the selection of the optical path length required for the measurement. For a concentrated sample, it is normal to use a transmission cell with a short path length, typically in the range of 1–10 cm. Two specialist techniques for vapor analysis involve the interfacing of the IR instrument to a gas chromatograph (GC) or a thermogravimetric analysis (TGA) instrument. In both cases, a relatively high concentration of gas or vapor is generated within a small volume. For such applications, it is important to minimize dead volume between the two instruments (GC-FTIR or TGA-FTIR), and to have good optical coupling with the sample while maintaining a small sample volume.

5.3.2.   Liquids

Liquids are usually considered relatively easy for sample handling, with the exception of low-molecular-weight compounds. These are often volatile and exhibit extremely high absorptivity, which typically requires a short optical path length (10 μm or less). The traditional approach for liquids is to use a transmission cell, either in the form of a sealed or semipermanent cell, or as a thin sandwich film between windows. ATR-based (internal reflectance) methods [97,98] have gained in popularity, and have become the preferred method for many applications because of ease of use, ease of cleaning, and a good match to the required sample path length. For volatile liquids, there is always the option to handle the sample in vapor form, within a heated gas cell.

One important issue relative to knowledge of the sample is to know when the sample contains water. A good example is polymer latex emulsions. Among the most common forms of windows used for IR measurements are those made from salt optics, which are usually made from KBr, NaCl, or CsI. These are very soluble in water, and must not be used for aqueous media. Table 9 provides a list of common optical materials that

Table 9   Optical Properties of Common IR Window Materials

Material	Useful range, cm^{-1} (transmission)	Refractive index at 1000 cm^{-1}
Sodium chloride, NaCl	40,000–590	1.49
Potassium bromide, KBr	40,000–340	1.52
Cesium iodide, CsI	40,000–200	1.74
Calcium fluoride, CaF$_2$	50,000–1140	1.39
Barium fluoride, BaF$_2$	50,000–840	1.42
Silver bromide, AgBr	20,000–300	2.2
Zinc sulfide, ZnS	17,000–833	2.2
Zinc selenide, ZnSe	20,000–460	2.4
Cadmium telluride, CdTe	20,000–320	2.67
AMTIR[a]	11,000–625	2.5
KRS-5[b]	20,000–250	2.37
Germanium, Ge	5,500–600	4.0
Silicon, Si	8,300–660	3.4
Cubic zirconia, ZrO$_2$	25,000–~1600	2.15
Diamond, C	45,000–2,500, 1650–< 200	2.4
Sapphire	55,000–~1800	1.74

[a] AMTIR: infrared glass made from germanium, arsenic, and selenium.
[b] Eurectic mixture of thalium iodide/bromide.
The spectral ranges cited represent the optical transmission through a thin film of the material. In practice, the optical range may be less than listed because of the actual thickness of the material. For example, zinc selenide, which is commonly used as an IRE for ATR measurements, has a more practical lower limit of around 650 cm^{-1} with the extended path length of material actually encountered with an ATR accessory.

transmit infrared radiation. Not all of these materials are available as infrared window materials. Also, one must be careful with materials of high refractive index when used as a window for transmission measurements. The high reflectivity of the window can cause annoying fringes (an interference pattern superimposed on the spectrum) to occur. Also, back reflections from the front window surface into the interferometer of the FTIR can cause double modulation, which can result in serious photometric errors. Figure 64 illustrates the back-reflection problem, as well as other sampling-related problems that can lead to photometric or other measurement errors.

Latex emulsions, containing water, are best handled with either barium or calcium fluoride, or silver halide windows, or with an ATR accessory (see later). It is important to ensure that separation does not occur at the window or the IRE interface. Also, care must be exercised with materials that contain ammonia—at high concentration this will etch many of the window materials.

Many polymer products exist as viscous liquids. A simple and convenient method for such nonvolatile materials is to produce a capillary film between a pair of transmission windows. If the material is very viscous, another approach is to form a thin-smear film on a single window. If the sample is extremely viscous, such as a tar or a resinous material, it is possible to dissolve the sample in a suitable volatile solvent, and evaporate a film of the material from solution onto the surface of a single infrared window. This approach may also be used for solid polymers that are easily dissolved in a

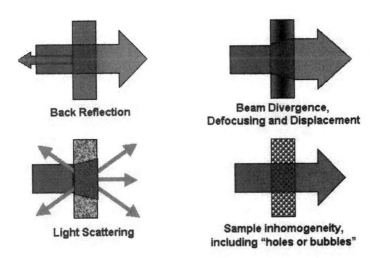

Figure 64   Sampling problems that can result in photometric errors.

volatile solvent, such as hexane, toluene, or acetone. If this method is used, it is important to ensure that all traces of solvent are removed before analyzing. There is a tendency for solvent residues to be trapped by viscous materials. Heating under an infrared lamp, or in a vacuum oven, will assist removal of the solvent.

5.3.3. Solids

The standard approaches for solids that are easily ground is the preparation of a compressed potassium halide (KBr or KCl) pellet or a mineral oil mull. In both cases it is necessary to reduce the particles to a submicrometer particle size, and to disperse the finely ground material in a matrix with a closely matched refractive index. This approach is intended to reduce the impact of light scattering from the sample particles (see Fig. 64 for error sources). If the sample particle size is the same as or greater than the wavelength of the radiation, a large amount of the radiation is lost (not transmitted or absorbed) because of light scattering. Reducing the particle size below this level will reduce the light losses; this is further helped by dispersing the material in a matrix of similar refractive index (the alkali halide or the mulling agent).

These techniques are not widely used these days, and the utility for polymers may be limited. If the polymer is brittle and/or friable, it may be abraded with sandpaper or a file, and the powder produced may be analyzed by one of these techniques. Outside of this, the application may be limited to a few polymer powders and/or certain solid additives used in polymer formulations.

Coarse or hard powders are not well served by either the compressed pellet or mull technique, mainly because of difficulties associated with grinding. In such situations, the best approaches require the use of an accessory, such as a diffuse reflectance or photoacoustic detector. Both diffuse reflectance and photoacoustic methods [99,100] may be applied to most forms of powdered solids. As a rule, photoacoustic measurements, which are the only form of true absorption measurement, are not significantly influenced by sample morphology. An alternative procedure for powders is ATR, especially a horizontal accessory, preferably equipped with a pressure applicator. Note that the use of pressure is recommended to ensure intimate contact between the sample and the IRE (internal reflectance element) surface. Normally, the sample must conform to the surface of the IRE, and because the strength of the IRE is typically limited, the procedure is recommended only for soft powders. However, with the introduction of diamond-based ATR accessories [101–103], it is possible to handle most types of powdered material.

Moldable materials may be prepared as self-supporting films for transmission or ATR measurements. If one is measuring a polymer film, it is important to be aware that interference fringes in the spectrum may result from internal reflections of the front and back surfaces of the film. Materials not preexisting as films, such as polymer pellets, may be molded or hot-pressed; accessories are available for producing such films. Another method is to cast a film of the polymer from solution on a transmission window or on the surface of a IRE of an ATR accessory. As noted earlier, care must be taken to ensure that solvent residues are removed. Also, rapid evaporation of the solvent must be avoided, and if this is carried out too fast, small bubbles will form in the film. If this occurs, photometric distortions can occur in the recorded spectrum—see Fig. 64 for error sources.

There are several benefits to the use of ATR as a sampling procedure, including the fact that it is capable of handling samples in various physical forms. The fundamental relationships for light (IR energy) absorption are indicated in Fig. 56. It is noted that the absorbance (amount of light absorbed) is proportional to the pathlength (b) through the sample, which in a transmission measurement is defined by the thickness of the sample. In an ATR measurement, as indicated in Fig. 65, thickness is defined by the term depth of penetration (D_p), which is influenced by several factors, including the wavelength of light. The consequence of this is that the measured absorption across the spectrum changes as a function of wavelength (λ), and the observed result is that the sample spectrum intensity increases as a result of an effective path length increase with increase in wavelength. This leads to differences in appearance between normal transmission spectra and ATR spectra.

Polymer pellets can also be handled by ATR. Although it is not always the case, new accessories featuring a diamond-sampling surface make this

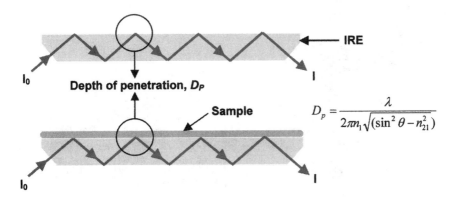

$$D_p = \frac{\lambda}{2\pi n_1 \sqrt{(\sin^2 \theta - n_{21}^2)}}$$

Figure 65   The internal reflectance or ATR method of infrared sample handling.

measurement practical (see Fig. 66 for an example system). Certain materials already exist as a sheet or film, and these include plastics or plastic-coated materials (such as metals and paper). If the material is transparent (to IR radiation), it may be analyzed directly as a self-supported film by transmission. Alternatively, one of the reflectance methods may be used. For polymers and polymer coatings, the ATR approach is usually preferred. When dealing with a coating on a metal surface, either ATR or external reflectance/specular reflectance may be used. The latter measurement is sometimes called a reflection-absorption measurement. This method works well if the film has thin to medium thicknesses. Very thick films cause problems because of anomalous optical effects at the surface, resulting in spectral distortions. If distorted spectra occur, an alternative method of sampling, such as ATR, is recommended. However, one alternative is to apply a mathematical function known as the Kramers-Kronig transformation [104] to generate a traditional absorbance-like spectrum.

If the polymer is hard, such as a thermosetting resin, one method to consider is gentle abrasion of the surface with silicon carbide paper. This action will transfer a small quantity of the polymer to the abrasive paper,

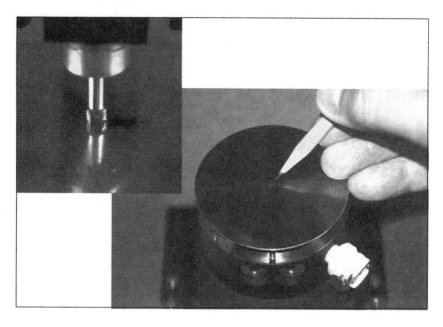

Figure 66 A diamond ATR sampling system, single-reflection configuration. Inset, top left: sample press for polymer pellets.

and this in turn may be examined with the aid of a diffuse reflectance accessory.

Elastomeric or rubber-like materials may be difficult to handle by a transmission-based technique. However, depending on the ingredients, they are usually ideal for ATR-based measurements. Most elastomers conform well to the IRE surface, providing good intimate contact. Elastomers with a high filler content, and in particular those containing dispersed carbon black, may cause problems because of the absorption characteristics of the carbon. In such cases, there are two possibilities: either ATR with a high refractive index material, such as germanium, or a photoacoustic measurement. In the event that neither is available, a method of last resort might be a destructive method, such as pyrolysis.

Fibers may be studied by several methods, including the standard compressed halide pellet methods. In this case, the fiber structure is often destroyed during sample preparation. Alternative methods include diffuse reflectance, photoacoustic, or infrared microscopy [105,106]. Infrared microscopy is a preferred technique for fibers because the properties of the fiber are retained. In some cases it is possible to study the orientation characteristics of the fiber, and to correlate this information to mechanical and/or physical properties.

This has been a brief review of sampling techniques as applied to polymers. For more information, the reader is directed to more complete reference texts [107–111].

ACKNOWLEDGMENTS

We wish to express gratitude to the individuals who supplied many of the polymer specimens. Specific mention is made to the following: Dr. E. D. Lipp of Dow Research Center for the polymethyl hydrogen siloxane and polymethyl phenyl siloxane samples; Mr. Tatsunami of Nihon Zeon Co. Ltd for the polyisoprene, polybutadiene with 98% *cis*-1,4-addition, and the SBR random copolymer; Mr. T. Yoshioka of Sumitono Bakelite Company for the Novolac based on phenol, PVC, EPDM, ethylene–propylene copolymers, polyimide samples; Dr. K. Kubodera of Toray Research Laboratory for certain epoxide samples.

REFERENCES

1. L. J. Bellamy, Infrared Spectra of Complex Molecules, Vol. 1. Chapman & Hall, New York (1975).
2. L. J. Bellamy, Advances in Infrared Group Frequencies, Infrared Spectra of Complex Molecules, Vol. 2. Chapman & Hall, New York (1980).

3. D. Steele, Interpretation of Vibrational Spectra. Chapman & Hall, London (1971).
4. K. Nakanishi and P. H. Solomon, Infrared Absorption Spectroscopy, 2nd ed. Holden-Day, San Francisco (1977).
5. N. B. Colthrup, L. H. Daly, and S. E. Wiberley, Introduction to Infrared and Raman Spectroscopy. Academic Press, San Diego, CA (1990).
6. D. Lin-Vien, N. B. Colthup, W. G. Fateley, and J. G. Grasselli, Infrared and Raman Characteristic Frequencies of Organic Molecules. Academic Press, San Diego, CA (1991).
7. G. Socrates, Infrared Characteristic Group Frequencies. Wiley, New York (1994).
8. B. Smith, Infrared Spectral Interpretation, A Systematic Approach. CRC Press, Boca Raton, FL (1999).
9. J. Haslam and H. A. Willis. Identification and Analysis of Plastics. Iliffe, London (1965).
10. D. O. Hummel and F. K. Scholl, Atlas of Polymer and Plastics Analysis, Vols 1–3. Verlag Chemie (1980/1).
11. P. C. Painter, M. M. Coleman, and J. L. Koenig, The Theory of Vibrational Spectroscopy and Its Application to Polymer Materials. Wiley, New York (1982).
12. J. L. Koenig, Spectroscopy of Polymers. American Chemical Society, Washington, DC (1992).
13. T. Shimanouchi, Analysis Method of IR Spectra. Nankodo, Tokyo (1980).
14. S. Fujiwara (ed.), Handbook of Polymer Analysis. Japan Society for Analytical Chemistry, Tokyo (1985).
15. J. Haslam, H. A. Willis, and D. C. Squirrel, Identification and Analysis of Plastics. Wiley, New York (1979). Original publisher: Iliffe, London (1972).
16. K. Nishikida, E. Nishio, and R. W. Hannah, Selected Applications of Modern FT-IR Techniques, pp. 225–234. Gordon & Breach, Luxembourg (1995).
17. D. O. Hummel, Infrared Analysis of Polymers, Resins and Additives. An Atlas, Vol. 1, Part 1, Text. Wiley-Interscience, New York (1971).
18. L. D'Esposito and J. L. Koenig, J. Polymer Sci., 14, 1731 (1976).
19. R. D. B. Fraser, J. Chem. Phys., 21, 1511 (1953); J. Chem. Phys., 29, 1428 (1958).
20. J. L. Binder, Anal. Chem., 26, 1877 (1954).
21. K. D. O. Jackson, M. J. R. Loadman, C. H. Jones, and G. Ellis, Spectrochim. Acta., 46A, 217 (1990).
22. J. G. Grasselli, M. K. Snavely, and B. J. Bulkin, Chemical Applications of Raman Spectroscopy, p. 40, Wiley-Interscience, New York (1981).
23. R. N. Jones, NRC Bulletin, No. 6, Table IV, National Research Council, Ottawa, Canada (1959).
24. N. B. Colthrup, L. H. Daly, and S. E. Wiberley, Introduction to Infrared and Raman Spectroscopy, p. 305, Academic Press, San Diego, CA (1990).
25. R. G. Snyder, J. Mol. Spectrosc., 4, 411 (1960).
26. R. G. Snyder and J. H. Schachtschneider, Spectrochim. Acta, 19, 85 (1963).

27. S. Krimm, C. Y. Liang, and G. B. B. M. Sutherland, J. Chem. Phys., 25, 549 (1956).
28. M. Tasumi, T. Shimanouchi, and T. Miyazawa, J. Mol. Spectrosc., 9, 261 (1962).
29. M. J. Gall, P. J. Hendra, C. J. Peacock, M. E. A. Cudby, and H. A. Willis, Spectrochim. Acta, 28A, 1485 (1972).
30. Modern Plastics Encyclopedia '91, Vol. 67 (11), pp. 55–70 (1990).
31. A. H. Willbourn, J. Polymer Sci., 34, 569 (1959).
32. S. Fujiwara (ed.), Handbook of Polymer Analysis, pp. 235–247. Japan Society for Analytical Chemistry, Tokyo (1985).
33. K. Nishikida and R. Iwamoto, Material Analysis by Infrared Method (in Japanese), pp. 208–213. Kodansha, Tokyo (1986).
34. G. R. Strobl and W. Hagedorn, J. Polymer Sci., Polymer Phys. Ed., 16, 1181 (1978).
35. N. G. Gaylord et al., Polymer Rev., Vol II, 69/70 (1959).
36. T. Usami and T. Goto, in S. Fujiwara (ed.), Handbook of Polymer Analysis, p. 238. Japan Society for Analytical Chemistry, Tokyo (1985).
37. Tests for Absorbance of Polethylene due to Methyl Groups at 1378 cm^{-1}, ASTM D 2238-68, Section 8: Plastics, Vol. 08.01 Plastics (I). American Society for Testing and Materials, West Conshohocken, PA (2000).
38. K. Teranishi and K. Sugahara, Kobunshi Kagaku, 23, 512 (1966).
39. S. Hosoda and M. Furuta, Macromol Chem. Rapid Commun., 2, 577 (1981).
40. K. Nishikida, T. Housaki, M. Morimoto, and T. Kinoshita, J. Chromatog., 512, 209 (1990).
41. S. Hosoda, Polymer J., 20, 383 (1988).
42. J. G. Grasselli, M. K. Snavely, and B. J. Bulkin, Chemical Applications of Raman Spectroscopy, p. 69. Wiley-Interscience, New York (1981).
43. K. Nishikida and T. Porro, preprint of FACSS (1996).
44. K. Takiyama, in S. Fujiwara (ed.), Handbook of Polymer Analysis, p. 248. Japan Society for Analytical Chemistry, Tokyo (1985).
45. R. G. Quynn, J. L. Riley, D. A. Young, and H. D. Hoether, J. Appl. Polymer Sci., 2, 166 (1959).
46. A. Zambelli, C. Tosi, and C. Sacchi, Macromolecules, 5, 649 (1972).
47. G. Bucci and T. Simonazzi, J. Polymer Sci., C7, 203 (1964).
48. Y. Takegami, T. Suzuki, T. Kondo, and K. Mitani, Kobunshi Kagaku, 29, 199 (1972).
49. J. P. Luongo and R. Salovey, J. Polymer Sci., A-2, 4, 997 (1966).
50. S. M. Gabbey et al., Polymer, 17, 121 (1976).
51. K. Nishikida and R. Iwamoto, Material Analysis by Infrared Method (in Japanese), p. 225. Kodansha, Tokyo (1986).
52. D. O. Hummel, Infrared Analysis of Polymers, Resins and Additives. An Atlas, Vol. 1, Part 1, p. 168, Wiley-Interscience, New York (1971).
53. H. U. Pohl and D. O. Hummel, Makromol. Chem., 113, 190 (1968); Makromol. Chem., 113, 203 (1968).

54. M. E. R. Robinson, D. I. Bower, and W. F. Maddams, Polymer, 19, 773 (1978).
55. M. Tasumi and T. Shimanouchi, Spectrochim. Acta, 17, 731 (1961).
56. H. Garner, Makromol. Chem., 84, 36 (1965).
57. M. S. Wu, P. C. Painter, and M. M. Coleman, J. Polymer Sci. Polymer Phys. Ed., 18, 111 (1980).
58. M. Noshiro and Y. Akatsuka, in S. Fujiwara (ed.), Handbook of Polymer Analysis. Japan Society for Analytical Chemistry, Tokyo (1985).
59. J. K. Agbenyega, G. Ellis, P. J. Hendra, W. F. Maddams, C. Passingham, H. A. Willis, and J. Chalmers, Spectrochim. Acta, 46A, 197 (1990).
60. Y. Tabata, K. Ishigure, and H. Sobue, J. Polymer Sci., A2, 2235 (1964).
61. D. Morero, A. Santambrogio, L. Porri, and F. Ciampelli, Chem. Ind., 41, 758 (1959).
62. H. Tadokoro, Y. Nishiyama, S. Nozakura, and S. Murahashi, Bull. Chem. Soc. Jpn., 34, 381 (1961).
63. R. R. Hampton, Anal. Chem., 20, 923 (1949).
64. J. L. Binder, Anal. Chem., 26, 1877 (1954).
65. R. Iwamoto, Annual Report on Testing of Polymer Materials, Vol. 10, 2-1, MITI Osaka Laboratories, Osaka, Japan (1982).
66. J. Haslam, H. A. Willis, and D. C. M. Squirrell, Identification and Analysis of Plastics, p. 263. Iliffe Books, London (1972).
67. T. Hayashi, in S. Fujiwara (ed.), Handbook of Polymer Analysis. Japan Society for Analytical Chemistry, Tokyo (1985).
68. R. A. Spragg and A. D. Williams, J. Mol. Struct., 80, 497 (1982).
69. T. Ogawa, in S. Fujiwara (ed.), Handbook of Polymer Analysis, p. 321. Japan Society for Analytical Chemistry, Tokyo (1985).
70. W. F. Maddams and I. A. Royaud, Spectrochim. Acta, 47A, 1327 (1991).
71. J. Mejar and O. Hainova, Kolloid-Z. Z. Polymer, 201, 23 (1965).
72. J. Haslam, H. A. Willis, and D. C. M. Squirrell, Identification and Analysis of Plastics, p. 523, Iliffe Books, London (1972).
73. A. R. Mukherji, M. A. Butler, and D. L. Evans, J. Appl. Polymer Sci., 25, 1145 (1980).
74. K. Nishikida, E. Nishio, and R. W. Hannah. Selected Applications of Modern FT-IR Techniques, p. 104, Gordon & Breach, Luxembourg (1995).
75. N. Ueda and S. Nishiuni, Kobunshi Kagaku, 21, 166 (1966).
76. A. J. Melveger, J. Polymer Sci., A-2, 10, 317 (1972).
77. D. G. Anderson, J. K. Duffer, J. M. Julian, R. W. Scott, T. M. Sutliff, M. J. Vaikus, and J. T. Vandeberg (eds.), An Infrared Spectroscopy Atlas for the Coating Industry. Infrared Spectroscopy Committee, Chicago Society for the Coating Industry, Chicago (1980).
78. N. M. Peacock and A. W. Pross, Offic. Dig., 27, 702 (1961).
79. M. Nagakura, Y. Ogawa, and K. Yoshitomi, Shikizai, 41, 542 (1968).
80. H. L. Bender, Mod. Plastics, 30, 136 (1958).
81. M. Kaminaka, J. Naokawa, T. Watabe, H. Kamata, and S. Tanaka, Koubunshi Kagaku, 13, 93 (1956).

82. D. O. Hummel, Infrared Analysis of Polymers, Resins and Additives. An Atlas, Vol. 1, Part 1, p. 117. Wiley-Interscience, New York (1971).
83. Bernhard Schrader, Raman/Infrared Atlas of Organic Compounds, 2nd ed., p. I 13-01. VCH, Weinheim (1989).
84. J. Haslam, H. A. Willis, and D. C. M. Squirrell, Identification and Analysis of Plastics, p. 542. Iliffe, London (1972).
85. D. R. Anderson, Infrared, Raman and Ultraviolet Spectroscopy in A. L. Smith (ed.), Analysis of silicones, chap. 10. Wiley, New York (1974).
86. D. O. Hummel, Infrared Analysis of Polymers, Resins and Additives. An Atlas, Vol. 1, Part 1, p. 192. Wiley-Interscience, New York (1971).
87. E. D. Lipp, Appl. Spectrosc., 40, 1009 (1986).
88. K. Nakanishi and P. H. Solomon, Infrared Absorption Spectroscopy, 2nd ed., p.192, Holden-Day, San Francisco (1977).
89. D. O. Hummel, Infrared Analysis of Polymers, Resins and Additives. An Atlas, Vol. 1, Part 1, p. 174. Wiley-Interscience, New York (1971).
90. J. M. Hollas, Modern Spectroscopy. Wiley, Chichester, UK (1996).
91. A. L. Smith, Infrared Spectroscopy, in J. W. Robinson (ed.), Practical Handbook of Spectroscopy, pp. 481–535, CRC Press, Boca Raton, FL (1991).
92. N. B. Colthup, L. H. Daly, and S. E. Wiberley, Introduction to Infrared and Raman Spectroscopy, 3rd ed., chap. 1. Academic Press, New York (1990).
93. K. Feinstein, Guide to Spectroscopic Identification of Organic Compounds. CRC Press, Boca Raton, FL (1995).
94. R. M. Silverstein, G. C. Bassler, and T. C. Morrill, Spectrometric Identification of Organic Compounds, 5th ed. Wiley, New York, 1991.
95. P. R. Griffiths and J. A. de Haseth, Fourier Transform Infrared Spectrometry (Volume 83 of Chemical Analysis). Wiley-Interscience, New York (1986).
96. J. P. Coates, Vibrational Spectroscopy: Instrumentation for Infrared and Raman Spectroscopy, in Galen Wood Ewing (ed.), Analytical Instrumentation Handbook, pp. 393–555. Marcel Dekker, New York (1997).
97. N. J. Harrick, Internal Reflection Spectroscopy, Harrick Scientific Corporation, Ossining, NY (1987). (Original publisher: Wiley, New York, 1967.)
98. F. M. Mirabella, Jr. (ed.) Internal Reflection Spectroscopy: Theory and Applications. Marcel Dekker, New York (1993).
99. J. F. McClelland, Anal. Chem., 55, 1, 89A–105A (1983)
100. J. F. McClelland, R. W. Jones, S. Luo, and L. M. Seaverson, in P. B. Coleman (ed.), Practical Sampling Techniques Techniques for Infrared Analysis, pp. 107–144. CRC Press, Boca Raton, FL (1993).
101. J. P. Coates, Spectrosc. Showcase, March 1997.
102. J. P. Coates, Am. Lab., April 1997, pp. 22C–22J.
103. J. P. Coates and J. Reffner, Spectroscopy, 14, 4, 34–45 (1999).
104. K. Ohta and H. Ishida, Appl. Spectrosc., 42, 952–957 (1988).
105. M. A. Harthcock and R. G. Messerschmidt (eds), Infrared Microspectroscopy Theory and Applications. Marcel Dekker, New York (1988).

106. H. J. Humecki (ed.), Practical Guide to Infrared Microspectroscopy, Practical Spectroscopy Series, Vol. 19. Marcel Dekker, New York (1995).

107. D. O. Hummel and F. Scholl, Atlas of Polymer and Plastics Analysis. Verlag Chemie, Weinheim (1981).

108. T. J. Porro and S. C. Pattacini, Spectroscopy, 8, 7, 40–47 (1993).

109. T. J. Porro and S. C. Pattacini, Spectroscopy, 8, 39–44 (1993).

110. P. B. Coleman (ed.), Practical Sampling Techniques Techniques for Infrared Analysis. CRC Press, Boca Raton, FL (1993).

111. J. P. Coates, in J. Workman and A. W. Springsteen (eds.), Applied Spectroscopy: A Compact Reference for Practitioners. Academic Press, San Diego, CA (1997).

APPENDIX: REFERENCE SPECTRA FOR THE INFRARED AND RAMAN ANALYSIS OF POLYMERS

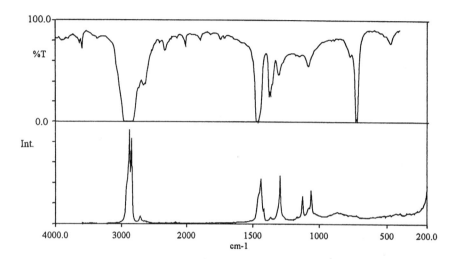

RS 1 Linear low-density polyethylene (LLDPE) with butene^{-1} as comonomer.

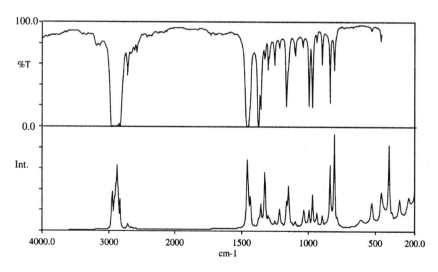

RS 2 Isotactic polypropylene (PP).

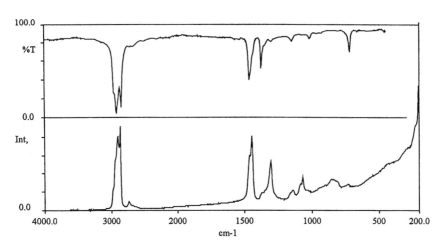

RS 3   Ethylene–propylene copolymer (ethylene = 68%, propylene = 32%).

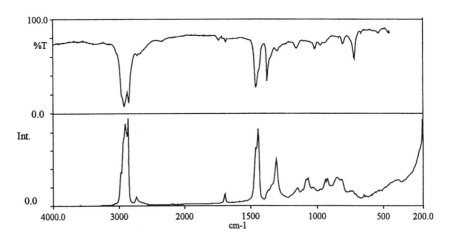

RS 4   EPDM rubber (ethylene = 56.1%, propylene = 34.4%, norbornene = 9.5%).

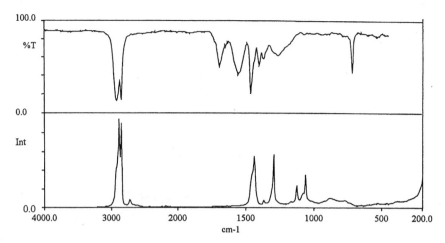

RS 5 Ionomer: ethylene–methacrylic acid copolymer, partially neutralized by alkali.

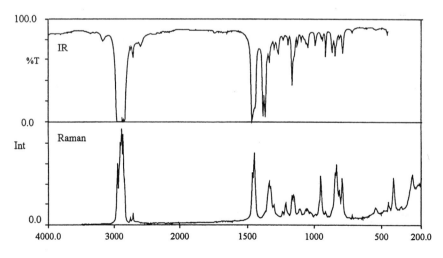

RS 6 Poly-4-methyl pentene-1 TPX.

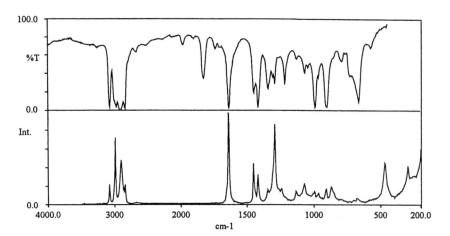

RS 7   Poly-1,2-butadiene: 29% crystalline.

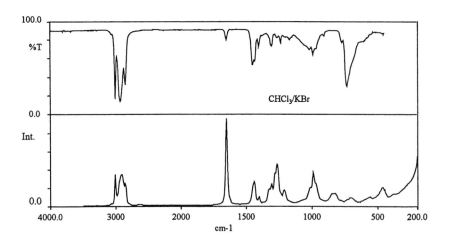

RS 8   Polybutadiene (1,4): 96–98% 1,4-cis addition (BDR).

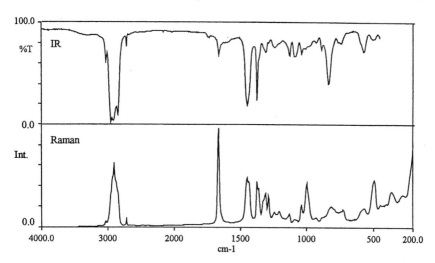

RS 9   Isoprene rubber.

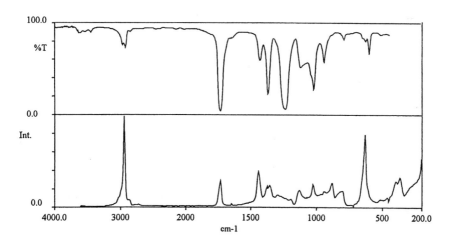

RS 10   Polyvinyl acetate (PVAc).

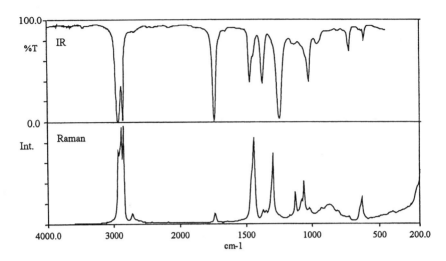

RS 11   Ethylene–vinyl acetate (EVA): 70% ethylene.

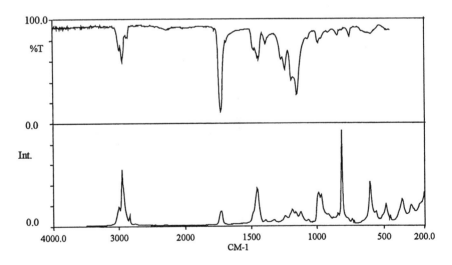

RS 12   Polymethyl methacrylate (PMMA).

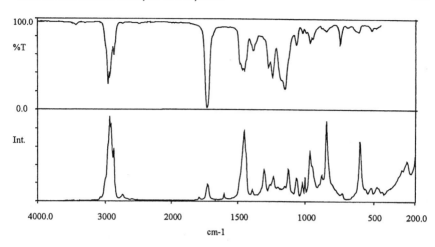

RS 13   Poly-*n*-butyl methacrylate (PBMA).

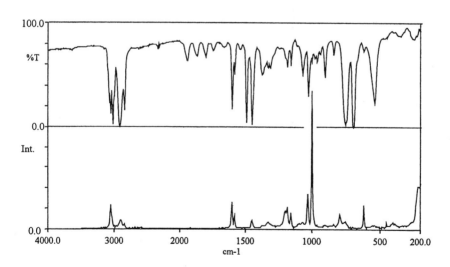

RS 14   Atactic polystyrene (PS).

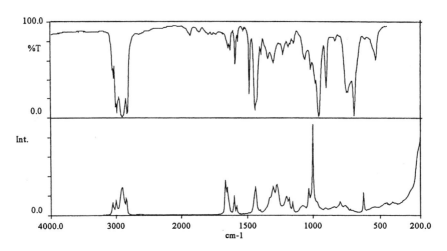

RS 15 Styrene–butadiene rubber (SBR): block copolymer, 30% styrene.

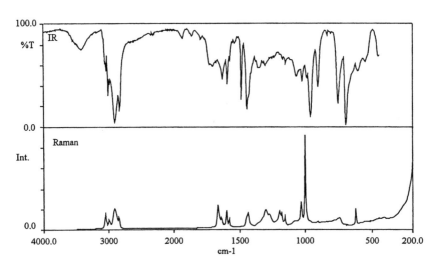

RS 16 Styrene–butadiene rubber (SBR): random copolymer, 48% styrene.

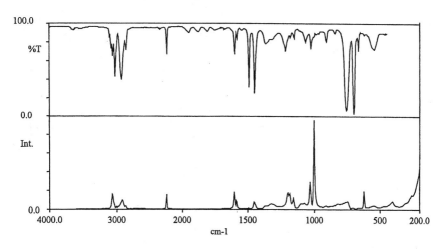

RS 17 Styrene acrylonitrile copolymer (SAN).

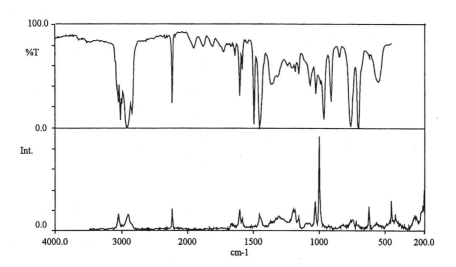

RS 18 Acrylonitrile–butadiene–styrene copolymer (ABS): 22% acrylonitrile, 18.9% butadiene, 58% styrene.

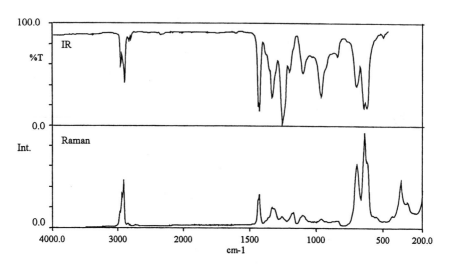

RS 19 Polyvinyl chloride (PVC).

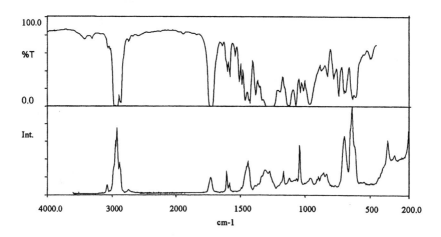

RS 20 Plasticized polyvinyl chloride: phthalate plasticized PVC.

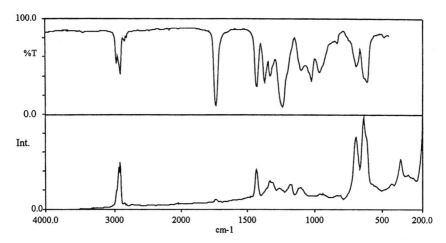

RS 21 Vinyl chloride–vinyl acetate copolymer (PVC-VAc): 13% vinyl acetate.

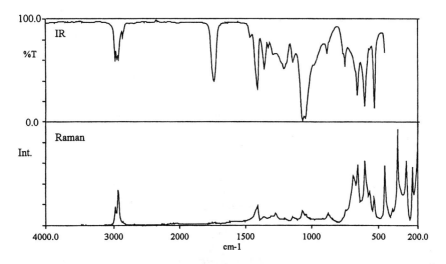

RS 22 Vinylidene chloride–methyl acrylate copolymer.

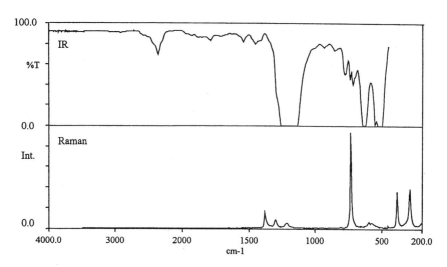

RS 23   Polytetrafluoroethylene (PTFE): Teflon.

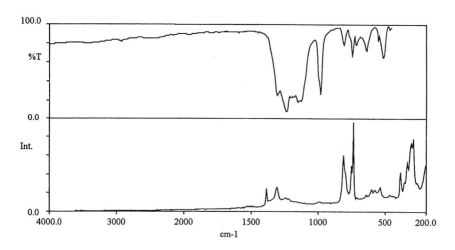

RS 24   Perfluoroalkyl vinyl ether–tetrafluorethylene–perfluoropropylene copolymer (EPE).

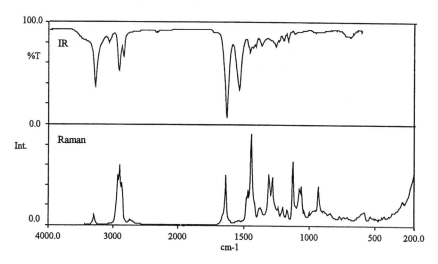

RS 25 Polyamide: Nylon-6.

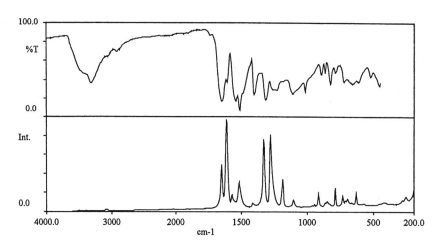

RS 26 Aromatic polyamide: Aramid, Kevlar.

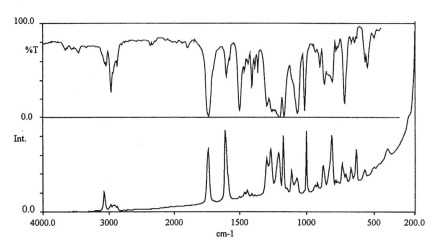

RS 27   Bis-phenol A polyester.

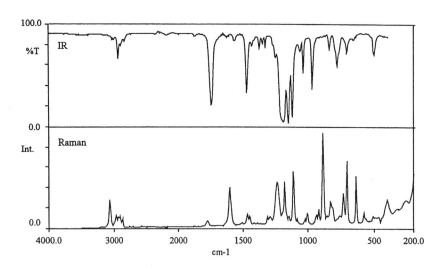

RS 28   Polycarbonate, bis-phenol A type: Lexan.

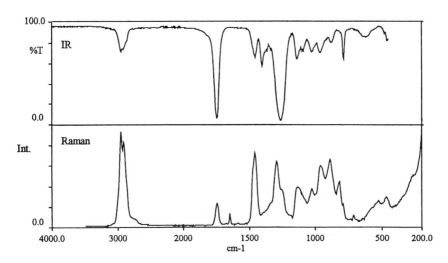

RS 29   Aliphatic polycarbonate.

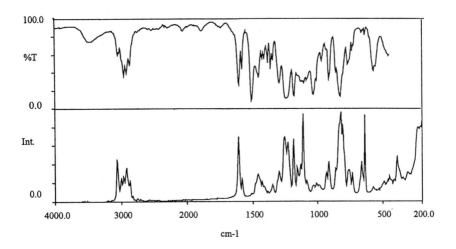

RS 30   Epoxide, bis-phenol A/epichlorohydrin type: before curing.

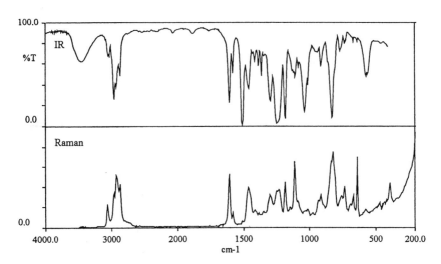

RS 31   Epoxide, bis-phenol A type: after amine curing.

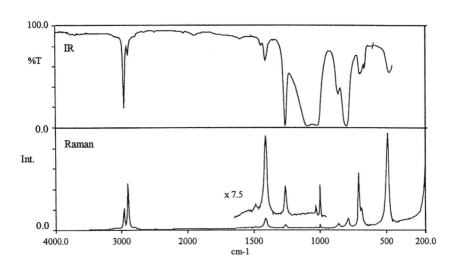

RS 32   Polymethyl siloxane.

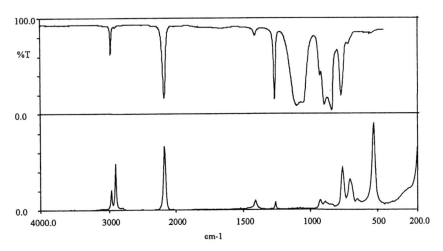

RS 33 Polyhydrogen methyl siloxane.

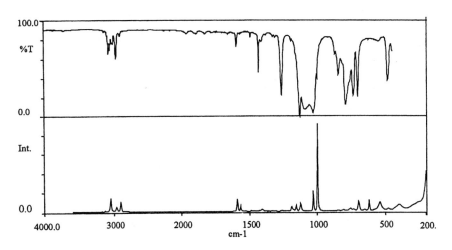

RS 34 Polymethylphenyl siloxane.

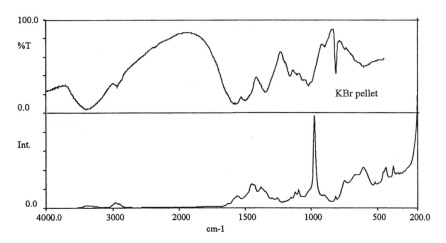

RS 35   Melamine resin.

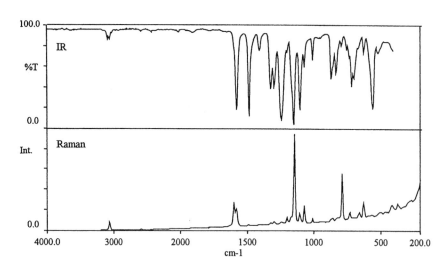

RS 36   Polyethersulfone (PES).

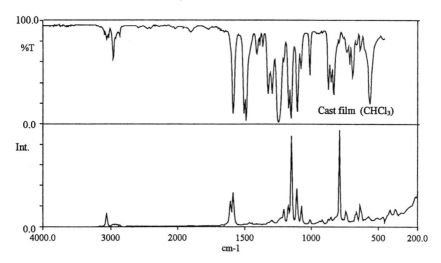

RS 37   Polysulfone (PSF).

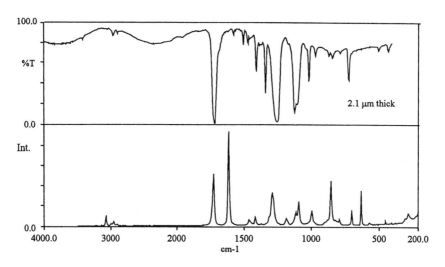

RS 38   Polyethylene terephthalate (PET).

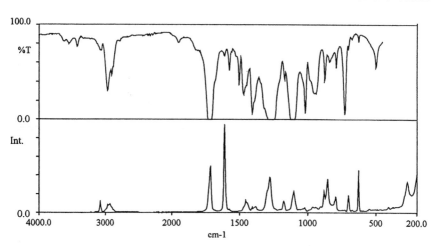

RS 39   Polybutylene terephthalate (PBT).

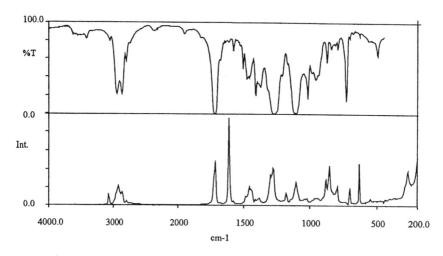

RS 40   Polyester elastomer.

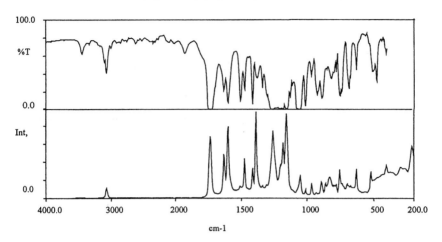

RS 41   Liquid crystal type polyester.

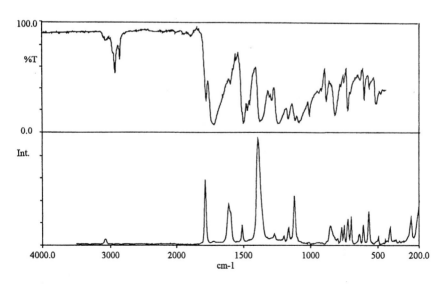

RS 42   Polyimide.

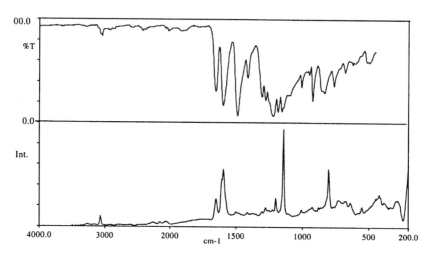

RS 43   Polyether–ether–ketone (PEEK).

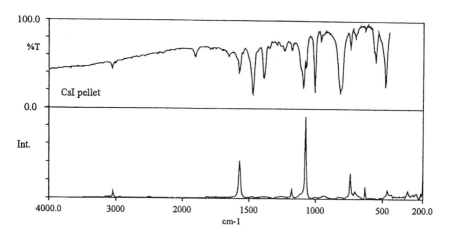

RS 44   Polyphenylsulfide.

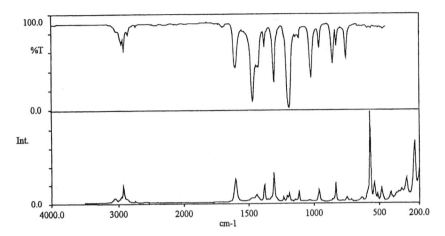

RS 45 Poly-2,6-dimethyl *p*-phenylene oxide (PPO).

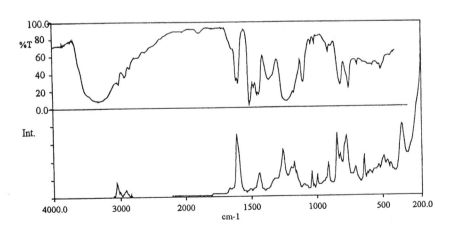

RS 46 Phenolic resin: Novolac.

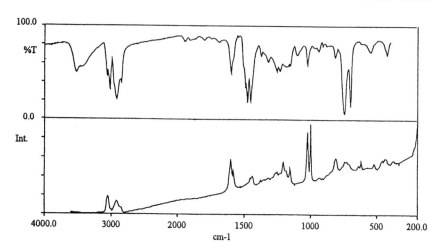

RS 47   Coumaron resin: KBr pellet spectrum.

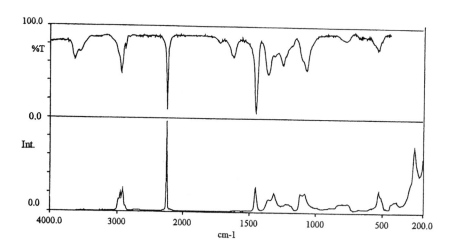

RS 48   Polyacrylonitrile (PAN).

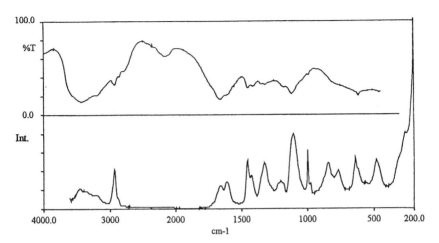

RS 49 Polyacrylamide.

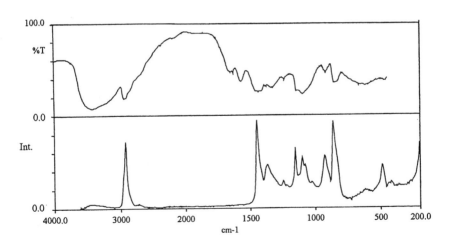

RS 50 Polyvinyl alcohol: 100% hydrolyzed.

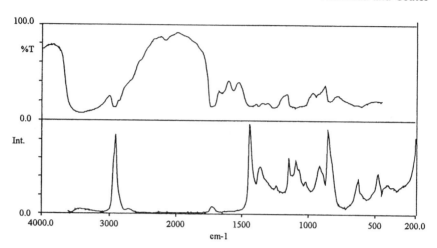

RS 51   Polyvinyl alcohol: 88% hydrolyzed.

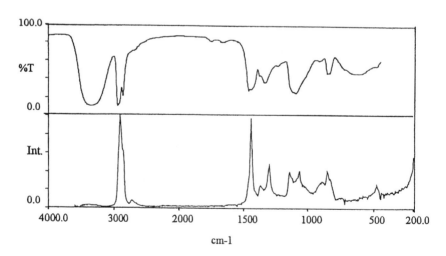

RS 52   Ethylene–vinyl alcohol copolymer (EVAL).

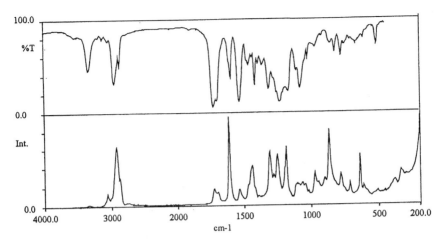

RS 53   Polyurethane (polyester type).

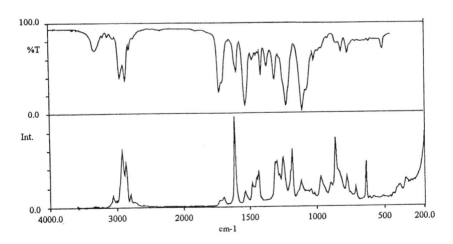

RS 54   Polyurethane (polyether type).

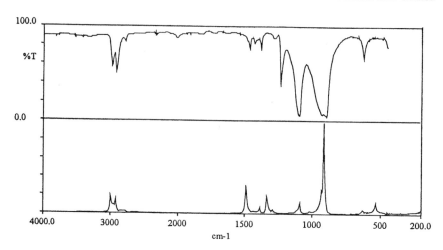

RS 55   Polyoxymethylene (POM).

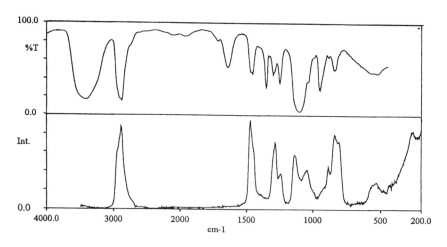

RS 56   Polyethylene glycol (PEG): amorphous, MW = 1000.

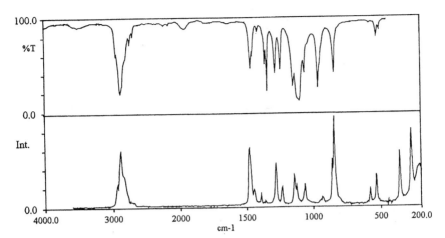

RS 57 Polyethylene glycol (PEG): MW = 14,000.

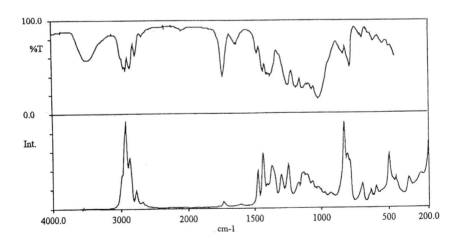

RS 58 Polyvinyl formal (PVFM).

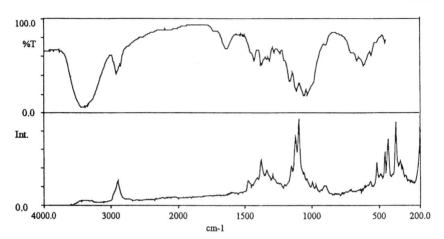

RS 59   Cellulose.

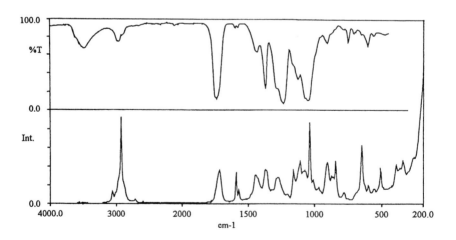

RS 60   Cellulose acetate.

8

Plastics Analysis by Gas Chromatography

Jose V. Bonilla

ISP Chemicals, Calvert City, Kentucky, USA

INTRODUCTION

The main purpose of this chapter is to present the power of gas chromatography (GC) for the analysis of plastic materials—its simplicity, speed, and broad range of applicability. Generally speaking, plastic materials are mixtures of macromolecules (polymers) and smaller chemical entities (normally in minor concentrations), such as residual monomers, residual solvents, and additives. In the materials which we call plastics there is actually a wide variety of molecules, ranging from low-molecular-weight (relatively high-volatility) monomers and solvents to medium-molecular weight (medium-volatility) additives, to the very-high-molecular-weight (nonvolatile) polymers themselves. Each of these components plays a special role in the plastic material. Accurate information about the chemical composition plays a crucial role in predicting and controlling the physical as well as chemical properties of plastic materials. These properties in turn determine the processing behavior, mechanical properties, and end-use performance of the plastics materials.

Gas chromatography is a highly versatile analytical technique used widely in the plastics industry for the qualitative and quantitative determination of a broad variety of compounds ranging from small gaseous molecules to very-high-molecular-weight polymers. The many options available in gas chromatography offer great flexibility and capability in terms of the types of compounds that can be analyzed by this technique. GC offers a wide variety of options in terms of sample introduction, types of columns available, a great selection of detectors, and a host of other options to solve many types of plastics analysis requirements. The remainder of this chapter will present the different options available to the analyst to explain different ways to optimize analysis by this technique.

Numerous additives are used in plastic formulations to control end-use properties, processability, and stability. UV stabilizers for example, provide weatherability, antioxidants provide protection during processing, and other additives provide other desired end-use properties. Because knowledge of the concentrations and types of additives is essential to proper control of the compounding process, there is a need for reliable methods to analyze for these chemicals in resin formulations. Traditionally, gas chromatography has been the analytical technique of choice for the analysis of volatile and soluble additives. High-performance liquid chromatography (HPLC) has been used for a wider variety of additives but is less attractive than GC, mainly due to generation of large volumes of liquid waste, lower chromatographic resolution, and fewer available specific detectors for identification of different components.

Recent developments in gas chromatography have considerably enhanced the power and attractiveness of this technique to solve analytical problems. A special section at the end of this chapter highlights the nature of these developments and the advantages they bring to the analyst. New technical developments deserving special attention include: High-temperature gas chromatography, high-speed GC (fast GC), retention-time locking, and high-volume injectors.

This chapter is intended primarily to serve as a practical tool to the everyday user of gas chromatography. Specific applications, detailed explanations of theoretical principles, and mathematical equations can be found in the books and articles listed in the reference section.

THE GAS CHROMATOGRAPH

Since its introduction in the early 1950s, gas chromatography and later capillary gas chromatography has grown at a fast rate. Basically, any substance, organic or inorganic, which exhibits a vapor pressure low enough to elute from a GC column at the operating temperature can be analyzed by GC. The major limitation of GC is that sample mixtures, or their derivatives, must be volatile at the column operating temperature.

The GC instrument is a rather simple, yet very powerful. It is one of the most common analytical tools used in plastics analysis. When used properly, it can provide both qualitative (identification) and quantitative (amount) information about the individual components in sample mixtures. For a mixture to be suitable for gas chromatographic analysis it should be relatively volatile at temperatures below 350°C (450°C for high-temperature GC). In other words, the components of interest must become a gaseous form by rapid heating without any degradation or destruction of their chemical structure. This does not mean that other components are not amenable to GC analysis. In theory, most components can be analyzed by GC if a proper sample pretreatment or proper sample introduction technique is used (e.g., pyrolysis GC, sample derivatization) [1–5].

In general, the gas chromatographic system is comprised of six major components: gas supply and flow controllers, injector, detector, oven, column, and a recording device (Fig. 1). A high-purity gas flows into the injector, through the column, and through the detector. A liquid (or gas mixture) containing the dissolved sample components is introduced into the

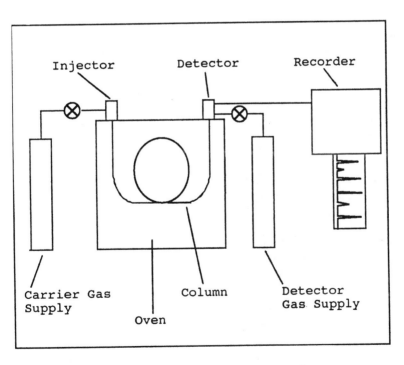

Figure 1 General diagram of a gas chromatograph.

heated injector, normally with a syringe. The sample components vaporize, and the gas flowing through the injector sweeps them into the column which is located inside the oven. The temperature of the oven is precisely and reproducibly controlled. Each compound in the sample interacts with the column to a different degree and so travels through the column at a different rate. The rate of travel through the column for each individual compound depends on the type of column, the gas flow rate, and the temperature of the oven. As each compound exits the column it is detected and an electrical signal is sent to a recording device (integrator or computer data system, etc.). The recorded signal will appear as a series of peak plotted versus time. This is called a *chromatogram*.

Normally, each peak represents an individual compound in the sample. In reality, however, it is not unusual for a peak to represent more than one compound when two peaks overlap. Using the proper column and operating conditions, this problem can be resolved or at least minimized. The time at which a compound exits from the column after introduction of the sample upon injection is called retention time. Most compounds have a unique *retention time* under particular conditions. This unique property is used for compound identification by GC.

The size of the peak provides quantitative information because the amount of the compound in the sample is proportional to the size of the corresponding peak in the chromatogram. By injecting a standard containing a known amount of the compound of interest, its retention time can be determined for the particular analysis conditions; the size of the peak (height or area) can then be compared to the corresponding peak of the known sample and a simple ratio used for quantitation. Like any other analytical technique, GC has some limitations and shortcomings, and even under various analysis conditions and columns some compounds will have identical retention times. This makes positive identification and quantitation of these compounds nearly impossible by gas chromatography. As the technique of gas chromatography expands and more technical innovations are introduced, however, the range of possible separations will definitely expand.

Numerous factors influence the successful identification and quantitation of compounds by GC. Understanding how all of these parameters can affect chromatographic analyses is critical if one is to obtain successful results.

One of the most important variables in GC is the GC column. The diameter, length, film thickness, and type of stationary phase depend on the column being used. With the exception of length, none of these factors can be changed without changing the column itself. On the other hand, oven temperature, carrier gas, and flow rate are factors which are readily controlled by the GC operator. The proper setup and operation of the GC injector and detector is critical. It is important to remember that the quality of chromatographic separation is limited by the weakest component of the chromatographic system, whether it be the column, injector type, operational parameters, detector type, or the instrument itself.

Understanding the function of each component of the GC system is key to successful use of this technique in identification and quantitation of components in plastic and related materials. Furthermore, understanding the function of each component in the system will allow the analyst to optimize the system to best solve specific problems.

CAPILLARY COLUMNS: CHARACTERISTICS AND PROPER SELECTION

The GC column is considered the heart of the chromatographic system. Although there are many different types of GC columns (packed and capillary) available in the market today, the most widely used type in modern chromatography are the capillary columns. Highly efficient capillary columns do not tolerate the imprecision allowed with packed columns. For example, small deviations in gas flow from the optimal carrier gas flow rate in a packed column are acceptable, whereas small changes in flow rates for a capillary column will lead to potential errors in results. Capillary (open tubular or fused silica) columns are by definition very narrow in diameter. The diameter of the fused silica tubing used to make these columns is precisely controlled during the manufacturing process. Accurate control of the tubing's diameter is critical to the production of columns with consistent performance. Capillary columns can range in length from 5 to over 100 m. The most common lengths are from 10 to 60 m.

The outer surface of the finished column is coated with polyimide. This polymer coating serves a dual purpose. It fills flaws in the tubing which prevents breakage of the glass tubing, and it acts as a waterproof barrier to prevent corrosion coating. The polyimide coating has nearly indefinite stability at temperatures of 350°C or below. Fused silica tubing is very flexible, but it is inherently straight. This straightness and the protection provided by the polyimide coating allows the tubing to be wound on cages for easy handling and storage. A limitation of the polyimide coating is its upper temperature of about 360°C. Recently, aluminum-coated or -clad tubing has been used to manufacture capillary columns. This tubing has a much higher upper temperature limit. One drawback to aluminum-clad tubing, however, is its slight brittleness when exposed to continuous changes in temperature. It is satisfactory for isothermal temperature work since the tubing becomes weak upon cycling of the oven temperature during temperature programming situations [1].

The Stationary Phase

The distinguishing feature of a column is its stationary phase. Capillary stationary phases are polymers that are deposited on the inner walls of the tubing in a thin, uniform film. The thickness and chemical nature of the stationary phase are critical to the overall performance of the column. High-efficiency capillary columns require relatively few types of stationary phase to achieve the separations necessary for complex samples. For this reason, only about three or four different stationary phases are actually necessary to accomplish most analyses of plastic materials. The first capillary columns had the stationary phase coated on the tubing walls, without any type of physical attachment. The stationary phase was easy to disrupt with solvents or contaminants. The advent of bonded and cross-linked phases substantially increased the stability and lifetime of capillary columns. The stationary phase is chemically bonded to fused silica capillary columns. Cross-linking is the joining of individual "strands" of polymer so that a larger and more stable stationary phase is formed. Practically every non-bonded liquid stationary phase has a bonded counterpart. There is virtually no difference in separation behavior between bonded and nonbonded equivalent phases. It is recommended, however, to use bonded and cross-linked phases if they are available. These stationary phases include polysiloxanes, polyethylene glycols, and solid adsorbents.

Polysiloxanes

The most common stationary phases are the polysiloxanes. These phases are considered to be the most resistant to abuse and have superior lifetimes. The type and amount of substitution groups on the polysiloxane backbone distinguishes each stationary phase and its properties. The most widely used phase is dimethyl polysiloxane. The interactions between solutes and this phase are limited to dispersive forces, thus solute elution occurs in the order of increasing solute boiling points. Compounds that cannot be differentiated on the basis of their boiling points (i.e., they have very similar or equal boiling points) require a different stationary phase for separation. Different stationary phases are made by changing the functionalities on the polysiloxane backbone. To obtain the differentiation of solutes by forces other than dispersion, a more selective stationary phase is required. Stationary-phase selectivity is controlled by the substitution of phenyl, cyanopropyl, and/or trifluoropropyl groups in place of some of the methyl groups on the polysiloxane backbone. This substitution enables the solutes to use dipole, acidic, and/or basic interactions in addition to dispersive interactions. The relative magnitude of each of the four interactions determines the selectivity of a stationary phase. Injecting the same sample into two columns with different selectivities will result in two chromatograms with the corresponding peaks differing in retention times and often in elution order as well. The degree of substitution of the siloxane polymer affects the selectivity of the stationary phase as well as the thermal stability and "bleed level" of the column. In general, as the amount of polar substitutions on the polysiloxane backbone increases, there is a corresponding decrease in the upper temperature the stationary phase will tolerate. Less substituted stationary phases (i.e., more methyl substitution) usually offer longer lifetimes and lower bleed levels. Columns coated with bonded and cross-linked stationary phases will usually exhibit longer lifetimes, especially for splitless and on-column injection applications.

Polyethylene Glycols

Polyethylene glycol is the second most common stationary phase after polysiloxane. Carbowax 20M is the most widely used for gas chromatography. The major disadvantage of the Carbowax phase is its extreme sensitivity toward oxygen, especially at high temperatures. Phase solubility in water and low-molecular-weight alcohols and a lower temperature limit are other drawbacks to columns coated with Carbowax 20M. Upon analysis of acidic or alkaline compounds, the column mimics their behavior. For example, if a large number of amine samples are injected into the column, it will become slightly alkaline. Subsequent injections of an acidic compound will give a poor peak or no peak at all. This behavior is reversible by solvent-rinsing the column with water, acetone, and hexane. Bonded and cross-linked Carbowax phases eliminate the phase and water/alcohol solubility problems, but the high sensitivity to damage by oxygen is still a problem. Nevertheless, the superior selectivity for certain solutes renders the Carbowax 20M stationary phases very useful in some applications such as hindered amines and phenolic additives.

Porous-Layer Phase

Several gas–solid adsorption capillary columns are available. They are commonly called porous-layer open tubular or PLOT columns. These columns contain a layer of adsorbent particles coated on the inner wall of the fused silica tubing. Phases of aluminum oxide (alumina), molecular sieves, and porous polymers (Poraplot-like) are commercially available. Gas–solid adsorption rather than a gas–liquid partition is the separation mechanism involved. PLOT columns are well suited for the analysis of light hydrocarbons, sulfur gases, permanent gases, or other very volatile solutes at or above ambient temperature conditions. Some of the disadvantages of PLOT columns are lower efficiencies, some loss of inertness, and problems with stability and reproducibility over time.

General Guidelines for Column Selection

The following guidelines are helpful for the selection of the appropriate stationary phase, depending on the type of analytes present in the mixture of interest.

1. For general purposes (screening purposes) use a nonpolar phase.
2. Use a stationary phase with polarity closely matching that of the solutes (i.e., nonpolar phase for nonpolar mixtures).
3. Use the least polar phase that will provide satisfactory separation; nonpolar phases have higher lifetimes than polar phases.
4. For solutes with dipoles or hydrogen bonding capabilities (i.e., amines and hydroxy-type additives), use a cyanopropyl or Carbowax stationary phase. Always consider polarity and temperature performance.
5. For light hydrocarbons or inert gases, use packed columns or PLOT columns (alumina, or molecular sieve).
6. Whenever possible, avoid using phases containing the specific element which interferes with specific detectors [e.g., do not use cyanopropyl phases with a nitrogen-phosphorous detector (NPDs) or trifluoropropyl phases with an electron-capture detector (ECD)].
7. In general, the widest range of compounds can be analyzed by using any of three or four columns: nonpolar, medium polarity, and polar (wax) columns. More than 99% of all analyses can be adequately performed with these three columns.

Column Diameter Considerations

The most popular column diameters available for fused silica capillaries are 0.18, 0.25 narrow bore), 0.32 and 0.52 mm (magabore or wide bore). Other diameters are also available from various manufacturers. The following guidelines apply to selection of a column diameter.

1. Use a 0.25-mm-I.D. column for split and splitless injections when sample overloading is not a problem. High column efficiencies (higher resolution) are achieved with small-diameter columns.

2. Use 0.32-mm-I.D. columns for splitless and on-column injections, especially when injecting large amounts of sample.
3. Use 0.53-mm-I.D. (megabore) columns as replacements for packed columns or for high sample loading applications. Use of these columns also permits lower split ratios, resulting in carrier gas savings.
4. A 0.18-mm-I.D. column can be very useful for GC/MS systems with low pumping capacities or when very high resolution is needed.

Film Thickness Selection

Column capacity is highly affected by the film thickness and column diameter. The capacity of a column is defined as the maximum amount of sample that can be injected into a column before significant peak distortion occurs. Capacity is related to film thickness, column diameter, and the solubility or polarity match between the solute and the stationary phase. Capacity increases as the column's film thickness or diameter is increased. The more soluble a salute is in the stationary phase, the greater is the column capacity for the solute. For example, a polar solute (e.g., an alcohol) will have greater solubility in a polar stationary phase (e.g., Carbowax) than in a nonpolar phase (e.g., dimethylsilicone). Exceeding column capacity or *overloading* is indicated by peak broadening or asymmetry.

Fused silica columns are available in different film thicknesses. Depending on the column diameter, film thicknesses are available from 0.10 to 5.0 μm. In most cases, a "standard" film thickness column is satisfactory. These columns have phase ratios of between 100 and 400. The phase ratio is defined as the column radious divided by 2 times the film thickness. In general, a standard film column is 0.25 μm for 0.25-mm- and 0.32-mm-I.D. columns, and 1.0–1.5 μm for 0.53 mm diameter columns.

The following guidelines can be used to select the optimal film thickness:

1. Use of a standard film thickness column is desirable for most applications.
2. Thin film columns are useful for high-boiling compounds, trace analysis, and for instances when column bleeding must be minimized at high temperatures.
3. Thick film columns should be used when higher capacity and longer column lifetime is desirable. Normally, thick films are used with wide-bore columns.

Column Length Selection

Column length has a direct effect on retention times when using isothermal analysis conditions. A nearly linear increase in retention time with increase in column length is seen. However, only about 40% increase in resolution is achieved. For temperature program analysis conditions, the increases in retention and resolution with column length are much less than those for isothermal conditions. Only a small increase in resolution is realized with longer columns. This is especially true for volatile compounds. Using thick-film columns is the best method to obtain better separation of volatile or early-eluting compounds. In general, the following guidelines can be used to select an optimum column length.

1. A 30-m column is suitable for most applications.
2. Fifteen-meter columns have been used conventionally for simple mixtures (less than 10 components) or for sample screening purposes. However, the advent of "fast chromatography" technology is opening new possibilities for the use of short columns for more complex mixtures. This issue is discussed in more detail later in this chapter.
3. Use of a column 60 m or longer is recommended for very complex samples or for situations requiring the highest possible number of theoretical plates. The use of long columns could provide resolution of components not possible with shorter columns. This normally results in the cost of a much longer run time.

Overall Considerations for Column Selection

If all other chromatographic conditions remain the same, it is the structure of the stationary phase that determines the relative retention time and elution order of compounds. The stationary phase determines the relative amount of time required for two compounds to travel through the column. It retains the compounds as they move through the column. If any two compounds take the same amount of time to migrate through the column, these two compounds are not separated; they co-elute. If two compounds take different times, the two compounds will be separated. The stationary phase is often selected on the basis of its polarity. Polarity of the stationary phase is determined by the chemical structure of the resin (polymer). Polarity affects several column characteristics. Some of the most important are column lifetime, temperature limits, bleed level, and sample capacity.

The selectivity of the stationary phase directly influences column separation properties. Stationary phase selectivity is not well understood, but selectivity can be thought of as the ability of the stationary phase to differ-entiate between two compounds by a difference of their chemical and/or physical properties. If there is a difference in the properties of two com-pounds, the amount of interaction between the compounds and the phase will be different. If there is a significant difference in the interactions, one compound will be retained more than the other and separation occurs. If there are no differences, co-elution occurs. The compounds may have dif-ferent structures or properties, but if a particular stationary phase cannot distinguish between the compound differences, co-elution takes place. Most analyte properties, such as hydrogen-bonding dipole strength, are not easily determined. The column stationary phase increases resolution by increasing the relative time the compound spends in the stationary phase. *Resolution*, the degree of separation of two adjacent peaks, is defined as the distance between the peaks centers divided by the average bandwidth. Strictly speak-ing, this is valid only when both peaks have the same height.

SELECTING THE PROPER INJECTOR

There are two main purposes when introducing the sample into the GC instrument. One purpose is to introduce the sample into the column in a short band. The smaller the sample band at injection, the sharper and more narrow the peaks will be on the chromatogram. The end result is more sensitivity and better resolution. The second goal is to have the composition of the sample introduced into the column be as representative as possible of the sample injected. There should be no sample degradation or adsorptive losses occurring during injection. With the exception of on-column injec-tion, all injectors utilize vaporization to introduce the sample into the capil-lary column. The injected sample is rapidly vaporized in the heated injector, and the carrier gas flowing through the injector carries the sample into the column. A significant problem with vaporization injection techniques is backflash. Upon sample vaporization, the gaseous sample expands to fill the injector liner volume. Backflash occurs when the vaporized sample expands beyond the capacity of the liner volume and into the injector body. Since the sample now occupies a larger volume, it takes longer for the sample to be carried into the column. A large (and tailing) solvent peak is obtained. If the vaporized sample comes in contact with cold spots such as the septum and gas inlets of the injector, small amounts of the sample may condense. This condensation may result in carryover problems on subse-quent injections. Backflash problems can be minimized by:

1. Using a septum purge with split/splitless injectors
2. Using small injection volumes
3. Using large-volume injector liners
4. Using the optimal injector temperature

The following discussion should serve as a general guide to the selection of injector configuration, proper injector setup and proper injection technique.

Injector Temperatures

The injector temperature should be just hot enough to ensure quick vaporization of the entire sample without thermally degrading the sample components. If the injector temperature is too low, carryover problems, incomplete sample vaporization, or broad peaks (especially the solvent front) will result. If the injector temperature is too high, excessive backflash or sample thermal degradation may occur. Using an injector temperature above the upper temperature limit of the column does not damage the column. For most samples, 200–250°C is a good injector temperature. Some screening may be needed to obtain the smallest amount of backflash and the maximum amount of sample throughput.

Inlet Discrimination

Upon injection, the less volatile sample components will not vaporize as rapidly as the more volatile sample components. Immediately following injection, the vaporized sample has a greater proportion of the more volatile compounds than the less volatile compounds. Therefore, more of the volatile compounds are introduced into the column. This effect is called inlet discrimination. The peaks for the less volatile compounds will be smaller than the more volatile compounds. The longer the sample spends in the heated injector, the less severe this type of discrimination is experienced by the sample. If inlet discrimination needs to be eliminated, an on-column injector should be used.

Septum Purge

Most split/splitless capillary injectors have a septum purge function. The septum purge minimizes the amount of septum bleed materials that may contaminate the GC system. The septum purge gas sweeps the bottom face of the septum and carries the contaminants out through the septum purge vent. The septum purge flow is usually between 0.5 and 5 ml/min. Higher than optimum septum purge flows may result in the loss of some of the more volatile sample components. The septum purge function is not essential in order to obtain good chromatographic results; however, any septum bleed and inlet contamination problems will be minimized.

INJECTION TECHNIQUES

There are four major capillary injection techniques: split, splitless, on-column, and megabore or direct injection. Nearly every standard capillary injector is capable of split and splitless injections. On-column injections require a dedicated capillary on-column, injector and is a required injector for high-temperature gas chromatography and to minimize inlet discrimination [1,2].

Split Injection

Split injection is very simple and the most common of the capillary injection techniques. The highest resolution and system efficiencies are obtained with split injections (Fig. 2). Split injections are used for highly concentrated samples with typical per-component concentrations of 0.1–10 µg/µl. Injection volumes of 0.5–2 µl are normally used, but volumes up to 5 µl can be used without significant problems. Split injection is a vaporization

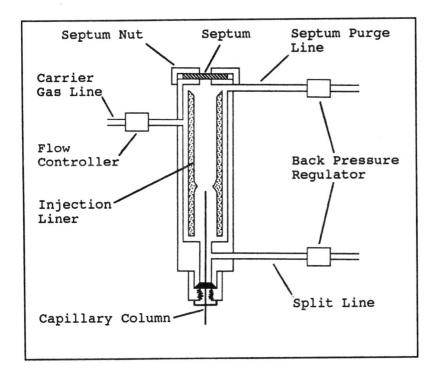

Figure 2 Split injector.

technique (Fig. 2). The sample is vaporized upon injection and rapidly mixed with carrier gas. A small amount of the carrier gas enters the column, and a much larger amount leaves the injector through the split vent. Since the vaporized sample is mixed with the carrier gas, only a small amount of the injected sample actually enters the column. The total gas flow through the injector at the moment of injection is quite high (the sum of the column and split vent flows). The sample is rapidly swept into the column, which accounts for the high efficiency of split injections. This also accounts for the severe discrimination obtained with split injections. The less volatile compounds do not have sufficient time to vaporize fully before they are discarded via the split vent.

Split Ratio

The amount of sample entering the column depends on the carrier gas flows into the column and out of the split vent. By measuring the column flow and the split vent flow, the amount of sample going into the column relative to the amount of sample being split can be calculated. This value is called the *split ratio*. The split ratio is normally reported with the column flow rate normalized to 1. A split ratio of 1:50 indicates that one part of the sample enters into the column and 50 parts are discarded out of the split vent. Therefore, 1/51 of the total sample theoretically makes it into the column. Typical split ratios range from 1:10 to 1:100, depending on the column diameter being used and the column loading desired for the analysis. Applications involving highly concentrated samples or very-small-diameter columns may require the use of higher split ratios. Split ratio is measured using a flow meter. The split ratio is equal to vent flow/column flow. For example: Column flow = 5 ml/min split vent flow = 100 ml/min split ratio = 100/5 = 20. Therefore, the split ratio is 1:20.

Splitless Injection

Splitless injections are used for trace analyses or when the component concentration in the mixture of interest is about 200 ng (Fig. 3). The injected sample is vaporized and carried into the column by the carrier gas. At the moment of injection, the flow through the injector is the same (1–2 ml/min) as the column flow. About 15–60 s after injection, additional carrier gas flow is introduced into the injector. This extra gas purges the injector of any remaining sample that has not entered the column. The time at which the extra gas flow is introduced is called the *purge activation time* (or *purge on*).

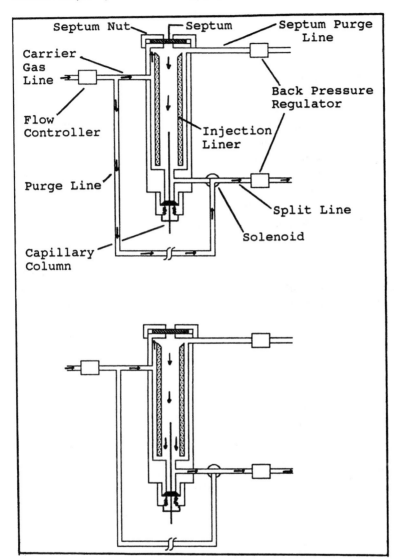

Figure 3 Splitless injector.

Solvent Effect and Cold Trapping

With splitless injections, the sample is introduced into the column at a much slower rate than for split injections. To prevent peak broadening, the sample needs to be refocused before starting the chromatographic process. One requirement of most splitless injections is that the initial temperature of the column oven be at least 10°C below the boiling point of the sample solvent. When the vaporized solvent leaves the injector and enters the cooler column, the solvent rapidly condenses at the front of the column and will trap and refocus the sample. This is called the *solvent effect*. Starting at too high a column temperature can result in broad peaks. If refocusing does not take place, the earlier-eluting peaks will suffer greater peak shape degradation than the later-eluting peaks. If this occurs, either a lower initial column temperature or a higher-boiling solvent should be used. If the sample components boil at 150°C or above the initial column temperature, the solvent effect does not have to occur for good peak shapes. The high-boiling compounds cold-trap in the column and refocus into a short band without help from the solvent effect. Injection sizes are usually limited to 2 μl or less for splitless injectors. Large injection volumes will normally result in broader peaks. Another limitation of splitless injections is that peaks that elute before the solvent front do not show good peak shapes. Changing to a lower-boiling solvent may help this situation.

On-Column Injection

On-column injection offers great advantages. This technique provides the optimum in capillary column performance by eliminating discrimination and degradation effects that can result from using a vaporization technique and is highly recommended for high-temperature GC applications (Fig. 4). With this technique, the sample is deposited directly into the column with a syringe. On-column injections are particularly useful for high-boiling compounds such as high-molecular-weight additives (e.g., waxy mold release agents) and thermally labile compounds. On-column injection requires the solvent effect or cold trapping to obtain acceptable peak shapes. Some on-column injectors use a secondary cooling function to eliminate the need to cool the entire column down to the appropriate temperature for the solvent effect. Only a small portion of the front of the column is cooled at the moment of injection. Megabore columns can tolerate large injection volumes (1–2 μl) and high sample concentrations (1–10 μg). Megabore columns are used for converted packed column instruments or for situations where large sample capacities are needed. Megabore injections are well suited for trace-level analyses. Highly concentrated samples may have to be diluted to avoid overloading the column. Small-diameter capillary

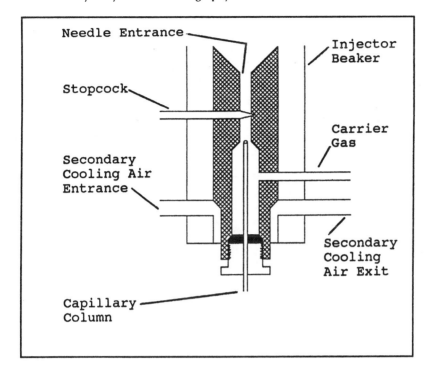

Figure 4 On-column injector.

columns (0.32 mm I.D. or less) are not compatible with megabore injectors. Some on-column injectors are temperature-programmable, permitting vaporization of solvent while the analytes still remain focused in a narrow band at the head of the column.

Direct Injection

Direct injection relies on vaporization processes to introduce sample into the column. However, it lacks a purge activation function, a septum purge flow, or secondary cooling (Fig. 5). The direct injection mode is used with packed columns or wide-bore (0.53-mm-I.D.) capillary columns. Wide-bore columns are used as higher-efficiency replacements for packed columns without the need to extensively modify the packed-column injector.

The process to convert a packed-column injector for use with wide-bore columns is relatively simple. The packed column is removed from the GC along with any injector (or detector) fittings attached to the base injector (or detector) body. A glass, direct-injection liner is placed into the packed

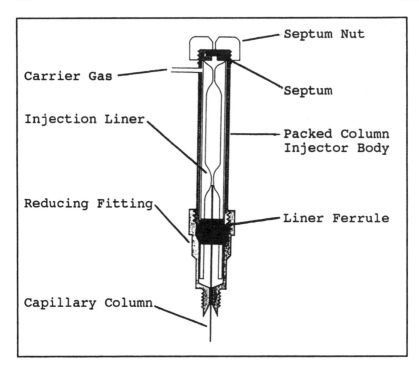

Figure 5 Direct injector.

injector. This liner serves two purposes. First, it decreases the volume of the injector so that excessive dead volumes and the resulting peak broadening effects are significantly reduced. Second, it serves as the vaporization site and means to transfer the vaporized sample directly into the wide-bore column. A ferrule and metal fitting is used to seal the liner and hold it in place. The metal fitting reduces the larger packed injector fitting down to the size normally used for capillary columns. Another fitting is used to reduce the packed detector fitting and to add makeup gas, if desired. A variety of different and specially designed conversion kits can be purchased for nearly every packed column

Injector Port Liners

Injector liners provide an inert environment in which the sample can vaporize and be properly introduced into the column. The liner design will vary depending on the type of injection technique being used. Liners that are dirty, poorly installed, or incorrectly selected will contribute to poor chromatographic results.

Split Injector Liners

The split injector liner must have a large enough volume to accommodate the expansion of the vaporized sample. However, the volume must be small enough for the gas flow to quickly sweep the vaporized sample into the column. Liners with too small a volume will be subject to severe backflash problems. Larger liner volumes become less important at high split ratios, since the high carrier gas flows rapidly sweep the injector. Flow disruption within the liner ensures thorough mixing of the sample, thus minimizing discrimination problems. Various types of flow disruption liners are available. The greatest discrimination is obtained with straight-bore liners, while the inverted-cup liners discriminate the least. Hourglass-shaped liners provide good sample mixing, and they are easier to clean than inverted-cup liners. Packing a split injection liner with silanized glass wool is another way to ensure flow disruption. The packed liner has a higher thermal mass, which aids in the rapid volatilization and mixing of the less volatile components. Additionally, the glass wool acts as a filter to help trap some of the nonvolatile materials in the injected sample. The glass wool should be lightly packed so that unnecessary peak broadening will not occur.

Splitless Injector Liners

Splitless injector liners are normally straight tubes without any flow-disruption devices. Any flow disruption in the injector usually causes peak broadening. Therefore, it is not recommended to pack a splitless liner with glass wool. Sometimes there is a restriction at the bottom of the liner. This is to keep the column centered in the liner and away from the liner walls. Use of small-volume liners will result in greater inlet efficiency, since the sample will be transferred into the column over a shorter period of time. However, smaller-volume liners are more subject to backflash problems. For small injection volumes (< 0.5 µl), the 2-mm-I.D. splitless liner is recommended; for larger injection volumes, the 4-mm liner is recommended to minimize backflash problems.

Megabore Injector Liners

The type of direct injection liners used greatly affects the quality of megabore chromatography. There are three general types of direct injection liners: the straight tube, the direct flash vaporization liner, and the hot on-column.

Straight-Tube Liners

Using a straight-tube type of megabore liner is not recommended. It typically gives a very broad and tailing solvent front, which may interfere with some of the early-eluting peaks of interest. Due to the lack of any type of restrictions, the vaporized sample can readily backflash out of the liner. The severity of solvent front tailing will be more pronounced for large injection volumes, volatile sample solvents and solutes, low carrier gas flows, and excessively hot injectors.

Direct Flash Vaporization Liners

The direct flash vaporization liner has a restriction at the top of the liner and another restriction several centimeters below the upper restriction. The sample is injected into the chamber formed by the two restrictions. When the syringe needle is inserted into the liner for injection, the needle blocks most of the upper restriction. This prevents the vaporized sample from escaping from the top of the liner. The lower restriction is tapered so that the megabore column becomes lightly wedged in the taper. Carrier gas (or sample) will not escape around the column. Backflash is greatly reduced, so the solvent front is narrow and without significant tailing. Direct flash vaporization liners can be packed with silylated glass wool, provided there is no top restriction. The glass wool should be lightly packed so that unnecessary peak broadening will not occur. The column end should be cleanly cut and checked with a magnifying lens. The column is inserted into the liner until it fits snugly in the restriction. Too much force will crush the column end, and too little force will not seal the column in the restriction. If the column is not sealed properly, the solvent peak will show tailing. Also, any debris in the injection liner taper will prevent a proper seal.

Hot On-Column Liners

The hot on-column injection-port liner has a single tapered restriction at the top of the liner. The megabore column seals in this region in the same manner as for the direct flash vaporization liner. The needle of a standard GC syringe can be inserted directly into the megabore column. The syringe needle should be straight, and the end should not have burrs or hooks. Injection volumes are limited to 0.5 μl or less. Hot on-column megabore injections are suitable for high-boiling samples that are difficult to vaporize by standard vaporization techniques. Since the injector temperatures are lower than those used for vaporization techniques, thermally labile samples can also be chromatographed with the on-column megabore liner. A direct flash vaporization liner system is more suitable and better in greater than

95% of all megabore applications. Hot on-column megabore liners should be used only when direct flash vaporization liners are not suitable.

THE CARRIER GAS

Selecting/Setting the Carrier Gas

The carrier gas is chosen for its inertness. Its only purpose is to transport the analyte vapors through the chromatographic system without interaction with the sample components. The gas is obtained from a high-pressure gas cylinder and should be free from oxygen and moisture. High-purity grades of carrier gas are usually less expensive in the long run. Helium is the most popular GC carrier gas in the United States.

The effects of the type of carrier gas and its flow rate upon a capillary column performance can be significant. The flow rate has to be accurately set, measured, and reproduced to fully experience the high efficiencies available with capillary columns. Unlike packed columns, small variations in the carrier gas flow rate can have significant effects on the separations obtained with capillary columns. Even the type of carrier gas can affect separation quality [2].

Instead of volumetric flow rate, a better way of assessing carrier flows is the average linear velocity. It can be thought of as the rate of travel for a nonretained compound through the column. It is the speed of the carrier gas through the column. The average linear velocity is a measure of carrier gas flow that is independent of the diameter of the column. Linear velocities are ideal for comparing carrier gas flows when the columns differ in diameter. There are small differences between the optimal linear velocities for columns of different diameters, film thickness, and length. In the laboratory environment, usually only one linear velocity per carrier gas type must be known, regardless of the column dimensions. van Deemter curves are usually given in terms of the linear velocity; in many cases, therefore, linear velocities are of the greatest practical use to the analyst (Fig. 6).

The volumetric flow rate of the carrier gas is still important for a number of applications and GC configurations. Typically a flow rate that is greater than the value corresponding to the minimum of the van Deemter curve is recommended. This is especially important for analyses using a temperature program, because at a constant inlet pressure, as the temperature of the oven increases, the viscosity of the carrier gas increases with a subsequent decrease in the carrier gas flow rate. This means that the carrier gas velocity and flow rate do not remain constant throughout the course of a temperature-programmed run. If the flow is set at the exact minimum in the van Deemter curve, a decrease in efficiency will result as the carrier gas flow decreases with the increasing oven temperature. If the flow rate is set slightly

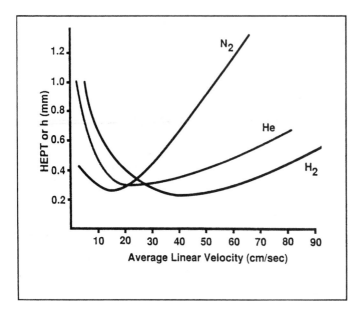

Figure 6   van Deemter curves for three main carrier gases: N_2, He, and H2.

above the minimum, an increase in separation efficiency with increasing temperature will result, since a shift toward the minimum in the van Deemter curve occurs. For helium and hydrogen as carrier gases, the minimum in the efficiency curves occurs over a much broader range and at higher linear velocities than for nitrogen. Using nitrogen provides the greatest column efficiency, but the minimum efficiency value occurs over a very narrow range at low linear velocities. Efficiency drops sharply with increasing linear velocities. Substantial analysis speed must be sacrificed for optimal resolution. For helium and especially hydrogen, high flow rates can be used for fast analysis times without sacrificing a large amount of separation efficiency. Large changes in oven temperatures will not grossly affect the column's efficiency due to flow-rate changes with temperature. In general, helium and hydrogen will provide nearly equivalent separations to nitrogen, but within a much shorter period of time. Also, helium and hydrogen provide the best separations when the analytes of interest elute over a wide temperature range. Although nitrogen is not normally recommended as a carrier gas for capillary columns, it can perform as well as helium, depending on the conditions chosen to run the analysis. There may be reluctance to use hydrogen as a carrier due to perceived explosion hazard. However, the ratio of air to hydrogen must be within a very narrow range (4–10%) before the possibility of an explosion exists. Hydrogen is very diffusive in air, so it is very unlikely that such a buildup of hydrogen might occur.

Before analysis and after proper conditioning of the column, the carrier gas flow rate has to be accurately set. The carrier gas flow rate depends on the column temperature. It is important, therefore, to set the carrier gas at the same column temperature for a given analysis. Significant changes in the resolution can occur with small changes in the carrier gas flow rate. The carrier gas is normally set at the initial temperature of the temperature program. For capillary columns, carrier gas flow is normally expressed as an average linear velocity (cm/s) instead of a volumetric flow rate (ml/min), as discussed previously. The average linear velocity can be thought of as the average rate at which a nonretained compound travels through the column, or the "speed" of the carrier gas. The linear velocity is determined by injecting a highly volatile compound that is not retained by the column. From the retention time of the nonretained peak, the average linear velocity can be calculated. The following compounds are recommended for the determination of average linear velocity for various detectors.

For FID and TCD, methane or butane
For ECD, methylene chloride or other halogenated methanes
For PID, ethylene or acetylene
For NPD, acetonitrile
For MS, butane, air, or halogenated methanes

It is not recommended to inject the neat liquid sample directly into the column for estimation of carrier gas velocity, as overloading may take place. Instead, make a headspace injection by placing a small volume of the appropriate compound in a septum-capped vial. Shake the vial, then insert the syringe needle into the headspace above the liquid. Pull up about 1 µl of the headspace vapors and inject. The nonretained peak should be very sharp and symmetrical. Leaks in the injector, a poorly installed column, or a lack of sufficient makeup gas flow will cause tailing or broadening of nonretained peaks. The effect of the carrier gas linear velocity on column efficiency is described using the van Deemter curves (Fig. 6). The curves show that there is an optimal linear velocity that provides the highest efficiency. This point is where the curve reaching the smallest value is used (the point of greatest column efficiency). The best resolution is obtained when a linear velocity that generates the highest efficiency for a column is used. Figure 6 also shows that using a linear velocity that is too low or too high will result in a rapid loss of column efficiency. Usually a linear velocity that is greater than the value corresponding to the minimum in the van Deemter curve is used. Setting the linear velocity at a higher value also compensates for the decrease in linear velocity with increasing column temperature, as encountered when using a temperature program. With helium and hydrogen as carrier gases, the minimum in the van Deemter curves occurs over a much broader range and at higher linear velocities than with nitrogen. Using nitrogen provides the greatest column efficiency, but the minimum in the van Deemter curves occurs over a very narrow range and at a low linear velocity. Substantial analysis speed may be sacrificed for optimal resolution when using nitrogen. For helium and hydrogen, high linear velocities can be used to reduce analysis times without sacrificing a lot of efficiency. The faster flow rates also sweep the injector faster, which improves the sample introduction process. Helium, and especially hydrogen, provide the best resolution when the analytes elute over a wide temperature range. Use of the highest-purity carrier gas will maximize column life. The use of impurity traps (water and oxygen) on the gas lines is highly recommended to extend column lifetime and to improve detector sensitivities.

Gas Purity and Gas Purification

Gas traps are strongly recommended, even if high-purity carrier gases are used. Trace amounts of contaminants can cause baseline problems, detector noise, and possible column damage. Contaminated detector gases can lead to artificially high baseline readings and poor performance. Individual traps are designed to remove moisture, oxygen, hydrocarbons, and other contaminants from the gas supply. The trap system recommended is a moisture trap, a high-capacity oxygen trap, and an indicating oxygen trap. The moisture trap should be first in line, because moisture will quickly deactivate the oxygen traps. The high-capacity oxygen trap should be installed second, followed by an indicating oxygen trap. An expired trap should be replaced immediately.

Moisture Traps

A moisture trap should be installed in the carrier and detector gas line(s). It will remove trace levels of water, some nonpermanent gases, and light hydrocarbons.

Oxygen Traps

Trace levels of oxygen in the carrier gas can dramatically shorten the life of a gas chromatographic column. Generally, the polyethylene glycol-based stationary phases (i.e., Carbowax) are readily oxidized, especially at elevated temperatures. Polysiloxane phases can also be irreversibly damaged by oxidation at high temperatures, but much more slowly than polyethylene glycol-based phases. Oxidized phases exhibit poor chromatographic performance and higher than normal bleed. Oxygen has a negative effect on many

GC detectors: it degrades the performance of electrolytic conductivity detectors (ELCDs), reduces filament lifetime in thermal conductivity detectors (TCDs), and reduces the linearity and sensitivity of electron capture detectors (ECDs). Both the column and detector benefit when oxygen is removed from the carrier gas supply. Oxygen contamination can usually be attributed to small leaks in the gas lines, septum leaks, or low-purity carrier gas. The most efficient oxygen traps are 99% efficient at reducing oxygen in carrier gases such as helium, nitrogen, hydrogen, argon, argon–methane, or CO_2 at 99.999% purity. A 20- to 40-fold reduction in oxygen in the gas is often obtained. These traps will effectively remove oxygen from argon–methane mixtures (used with ECDs) without disturbing the ratios of these gases. Metal-bodied traps avoid the signal noise associated with any plastic-bodied trap. The use of an indicating oxygen trap is recommended to determine when the high-capacity oxygen trap needs replacing. A glass indicating oxygen trap should be placed downstream of the high-capacity oxygen trap as a means of indicating when to replace the high-capacity nonindicating trap. This will prevent premature disposal of the high-capacity oxygen trap. The indicating material undergoes color change as it is depleted. The high-capacity oxygen trap must be replaced immediately upon the first indication of color change in the indicating trap. Fully expired oxygen traps will contaminate the gas with previously trapped materials.

Use of Gas Generators

Traditionally, compressed-gas tanks have been used as the source of laboratory gases. However, laboratory-size generators capable of producing ultrapure and ultradry (99.999%) hydrogen gas, nitrogen gas, and high-purity air are now available and are highly convenient and much less expensive in the long run compared to the traditional gas tanks. Hydrogen is conveniently produced by the electrolysis of deionized or distilled water. Separation of hydrogen from other electrolysis products is done by permeation through a palladium membrane. Nitrogen and high-purity air are produced from an air source. Because of the high-purity gas obtained from gas generators, further purification for use in gas chromatography is not normally required.

SELECTING THE RIGHT DETECTOR

In addition to high resolution and speed, gas chromatography offers the great advantage of interfacing with a large variety of detection methods. The detector is the critical link between the separation of the sample components by the column and the generation of a chromatogram. The signal generated by the detector is relayed to a recording device, such as an integrator or a computer which produces a corresponding chromatogram.

Capillary columns are compatible with most of the commonly available detectors used in gas chromatographic systems. However, each type of detector has a different set of requirements for optimal performance. In theory, the ideal situation is where the compounds are detected as they exit the column, thus preserving the separation achieved by the column.

Makeup Gas

Most detectors require a high volume of gas flow for optimal sensitivity and peak shape. For capillary columns, these gas volumes are much greater than those delivered by carrier gas alone. Even megabore (0.53-mm-I.D.) columns used at their highest recommended flow rates (10–20 mi/min) do not deliver the necessary flows (30–40 ml/min) required by most detectors for optimal performance. The difference in gas flows is even much greater for smaller-diameter columns. For optimum detector performance this difference is made up by the addition of makeup gas. The makeup gas is added at the detector side independently of the carrier gas. The carrier gas flow is not affected by the flow rate or type of makeup gas. In many cases the best makeup gas may be different than the carrier gas.

Detector Sensitivity, Selectivity, and Linear Range

Sensitivity

The detector response should is proportional to the mass of solute passing through the detector per unit time. In theory, these detectors are not affected by changes in the carrier gas flow rate. However, in practice, large changes in carrier-gas flows have a significant effect on detector behavior. For mass flow-rate detectors, sensitivity is defined as the peak area divided by the sample weight. The peak area can be obtained directly from the integrator. Concentration-dependent detectors are affected by the amount of gas flowing through the detector. The greater the amount of carrier and makeup gas is mixed with the sample, the less is the response from the detector. A concentration-dependent detector responds to changes in the concentration of the solute in the gas within the detector rather than to the presence of the substance itself; no response is obtained unless the composition of the flowing gas mixture changes. Sensitivity is defined as the product of the peak area and flow rate divided by the sample weight.

Selectivity

Some detectors respond to almost every compound in the column effluent. These detectors are called general or universal detectors. Flame ionization detectors (FIDs) and thermal conductivity detectors (TCDs) are examples of these types of nonspecific detectors. Other detectors respond to the presence of organic compounds containing certain functional groups or chemical structure. These detectors are called specific or selective detectors and are able to discriminate between compounds containing and not containing the particular functionality. Some specific detectors may respond to high concentrations of compounds that lack the particular functionality. Examples of specific detectors include the electron capture detectors (ECDs), nitrogen–phosphorus detectors (NPDs) and flame photometric detectors (FPDs).

Linear Range

Any change in the analyte concentration in a sample should result a corresponding change in the compound's peak size. If the amount of compound introduced into the detector is doubled, the size of the resulting peak should double. This normally occurs only over a certain range of compound concentrations—this is called the *linear range* of the detector. If the concentration of compound is outside this range, the response of the detector does not reflect the amount of analyte passing through the detector. To find the linear range of a detector, a simple test is performed. Increasing concentrations of the compounds of interest are injected and the response plotted versus the compound concentration. The linear range is the region where a straight line can be drawn through the resulting points.

Ideally, a GC detector should have the following attributes:

1. High sensitivity
2. Low noise level (background level)
3. Linear response over a wide dynamic range
4. Good response to all components being analyzed
5. Insensitivity to flow variations and temperature changes
6. Stability and ruggedness
7. Simplicity of operation
8. Positive compound identification

The Thermal Conductivity Detector

The thermal conductivity detector (TCD) is the most common universal detector used in GC. It is rugged, versatile, and relatively linear over a wide range and is highly useful in the analysis of components which do not give a signal by FID (water, CO_2, etc.). In operation it measures the difference in the thermal conductivity between the pure carrier gas and the carrier gas plus components in the gas stream from the separation column. The detector uses a heated filament (often rhenium–tungsten) placed in the emerging gas stream. The amount of heat lost from the filament by conduction to the detector walls depends on the thermal conductivity of the gas. When substances are mixed with the carrier gas, its thermal conductivity goes down (except for hydrogen in helium); thus, the filament retains more heat, its temperature rises, and its electrical resistance goes up. Monitoring the resistance of the filament provides a means of detecting the presence of the sample components. Of all the detectors, only the thermal conductivity detector responds to anything mixed with the carrier gas. Since it is non-destructive, the eluent may be passed through a thermal conductivity detector and then into a second detector or a fraction collector (i.e., preparative GC). The linearity of the detector is good at the lower concentration range but not in the high-percent range. In the high-percent range, a multipoint calibration is a way to ensure accurate measurements. Even gold- and nickel-coated tungsten–rhenium hot wires are susceptible to oxidation, which may unbalance the detector to the point where it cannot be zeroed properly. Oxide formation on the hot-wire surface will minimize the detector's ability to sense changes in thermal conductivity and thus decrease its sensitivity. Thus, it especially important to remove oxygen from the carrier gas when using a TCD.

The Flame Ionization Detector

The flame ionization detector (FID) is the most popular detector because of its high sensitivity and wide linear dynamic range [1]. The FID is an ionization detector that exhibits a nearly universal response to all organic compounds. The sensitivity, stability, excellent linear range, ease of operation and maintenance, along with wide applicability and low cost has made this detector the most popular gas chromatographic detector in use today [2].

A voltage is applied to the flame jet and the collector. Carrier gas exiting from the column is mixed with hydrogen and burned at the tip of the jet. Excess oxygen, usually as an air mixture, is supplied to the combustion chamber to ensure efficient ionization of the column effluent. Movement of the generated ions from the flame to the positively charged collector produces a small current. A background or baseline signal is always present and is a result of trace levels of gas impurities, system contamination, and normal column bleed. The introduction of an organic compound into the flame results in the compound's ionization and an increase in the detector current above the background level.

Most commercially available FIDs require 30–40 ml/min of total gas flow for optimal performance. Carrier gas flows for capillary columns are usually 0.5–10 mi/min; therefore, makeup gas must be added to supplement the column carrier flow. Nitrogen is a better makeup gas than helium.

Detector sensitivity will be increased by about 20% using nitrogen rather than helium as the makeup gas.

Sensitivity also depends on the flow of makeup gas. The sum of carrier and makeup gas flows should be 30–40 ml/min. The flows of the hydrogen and air combustion gases are critical for optimal sensitivity. Usually the air flow is about 10 times the flow of hydrogen (30–40 ml/min hydrogen and 300–400 ml/min for air).

The Nitrogen–Phosphorus Detector

A very widely used selective detector for the analysis of compounds containing nitrogen or phosphorus is the nitrogen–phosphorus detector (NPD). This detector is used in a number of novel applications including trace analysis of nitrogen-containing or phosphorous-containing additives, monomers, or residual solvents. Special applications requiring specificity for components present in complex mixtures take advantage of this detector, since it responds only to compounds that contain nitrogen or phosphorus. Its fabrication is similar to that of a flame ionization detector and, consequently, NPD equipment is usually designed to mount on an existing FID-type detector base. The thermionic source has the shape of a bead or cylinder centered above the flame tip. This bead is composed of an alkali metal compound impregnated in a glass or ceramic matrix. The body of the source is molded over an electrical heating wire. A typical operating temperature is between 600 and 800°C. A fuel-poor hydrogen flame is used to suppress the normal flame ionization response of compounds that do not contain nitrogen or phosphorus. With a very small hydrogen flow, the detector responds to both nitrogen and phosphorus compounds. Enlarging the flame size and changing the polarity between the jet and collector limits the response to phosphorus compounds only. Located in proximity to the ionization source is an ion collector. The thermionic source is also polarized at a voltage that causes ions formed at the source to move toward the ion collector. Compared with the flame ionization detector, the thermionic emission detector is about 50 times more sensitive for nitrogen and about 500 times more sensitive for phosphorus.

The Electron-Capture Detector

The electron-capture detector (ECD) is highly sensitive and selective for halogenated and other electronegative compounds, and as such remains one of the most widely used GC detectors. Applications in the plastics industry are relatively wide, especially in trace analysis or analyses of residual halogenated solvents or monomers and halogenated additives such as flame retardants. As with the NPD detector, one of the main advantages of the ECD in the analysis of complex mixtures is its specificity. The ECD responds only to electrophilic species such as oxygenated and halogenated compounds. The ECD is a selective electrode that is capable of providing extremely sensitive responses to specific substances that might be present in a sample containing a large excess of little or not responsive substances. It consists of two electrodes. On the surface of one electrode is a radioisotope (usually nickel-63 or tritium) that emits high-energy electrons as it decays. Argon mixed with 5–10 % methane is added to the column effluent. The high-energy electrons bombard the carrier gas (which must be nitrogen when this detector is used) to produce a plasma of positive ions, radicals, and thermo electrons. A potential difference applied between the two electrodes allows the collection of the thermo electrons. The resulting current when only carrier gas is flowing through the detector is the baseline signal. When an electron-absorbing compound is swept through the detector, there will be a decrease in the detector current, a negative excursion of the current relative to the baseline as the effluent peak is traced. The potential is applied as a sequence of narrow pulses with a duration and amplitude sufficient to collect the very mobile electrons but not the heavier, slower negative ions. The ^{63}Ni sources can be safely heated up to 400°C with no loss of activity. The carrier gas and makeup gases must be free from residual oxygen and water [6–8].

After the TCD and FID, the ECD has the highest usefulness in the GC field. Unlike the FID, the ECD has neither ease of operation nor dynamic range. What it does have is detectability on the order of picograms and good specificity. Additives containing amines or acid functionalities can be converted to perfluoro derivatives, which give a signal with this detector.

The Flame Photometric Detector

The flame photometric detector (FPD) is a highly useful detector in the analysis of components containing sulfur and phosphorous functionalities. In this detector, the column effluent passes into a hydrogen-enriched, low-temperature flame contained within a shield. Air and hydrogen are supplied as makeup gases to the carrier gas. Two flames are used to separate the region of sample decomposition from the region of emission. Flame blowout is no problem because the lower flame quickly reignites the upper flame. Phosphorus compounds emit green band emissions at 510 and 526 nm that are due to HPO species. Sulfur compounds emit a series of bands from excited diatomic sulfur; the most intense is centered around 394 nm. Phosphorus and sulfur can be detected simultaneously by attaching a photomultiplier tube and an interference filter for sulfur on one side of the flame, and a photomultiplier tube with an interference filter for phosphorus on the opposite side of the flame. The detector response to phosphorus is linear, and the response to sulfur depends on the square of its concentration. Carbon dioxide and organic impurities in the makeup and carrier gases must be limited to below 10 ppm. The quenching effect of carbon dioxide is very significant.

The Photoionization Detector

The photoionization detector (PID) is a highly selective and sensitive detector for chemical compounds containing aromatic components. The PID is a concentration-sensitive detector with a response that varies inversely with the flow rate of the carrier gas. A typical PID has two functional parts: an excitation source and an ionization chamber. The excitation source may be a discharge lamp excited by direct current, radio frequency, microwave, or a laser. The discharge lamp passes ultraviolet radiation through the column effluent from one of several lamps with energies ranging from 8.3 to 11.7 eV. Photons in this energy range ionize most organic species, but not the permanent gases. A potential of 100–200 V is applied to the accelerating electrode to push the ions formed by UV ionization to the collection electrode at which the current is measured. The most popular PID lamp is the 10.2-eV, which has the highest photon flux and therefore the greatest sensitivity. There are certain applications in which the 9.5-eV lamp is preferable to the 10.2-eV lamp; these include aromatics in an aliphatic matrix, mercaptans in the presence of H_2S, and amines in the presence of ammonia.

The discharge ionization detector (DID) uses far-ultraviolet photons to ionize and detect sample components. Helium gas is passed through a chamber in which high-voltage electrodes generate a glow discharge and cause it to emit a high-energy emission line at 58.84 nm This energy passes through an aperture to a second chamber in which it ionizes all gas or vapor species present in the sample stream that have an ionization potential less than 21.2 eV (which embraces practically all compounds including hydrogen, argon. oxygen, nitrogen, methane, carbon monoxide, nitrous oxide, ammonia, water, and carbon dioxide). A polarizing electrode directs the resulting electrons to a collector in which they are quantitated with a standard electrometer.

The Tandem PID/FID Combination Detector

The tandem PID/FID combination detector system is available commercially and incorporates the photoionization and flame ionization detectors. With both detectors in tandem, dual detector traces for benzene, toluene, ethyl benzene, and xylenes are possible, eliminating the need for two separate analyses. Both detectors can be used separately if so desired, and excellent results can be obtained with either packed or capillary columns. The first commercially available tandem PID/FID detector with no transfer lines eliminates the transfer line and improves peak shape and performance and enables the analyst to obtain screening and confirmatory information on sample in only one injection. The sample stream elutes from the column through the detector's reaction chamber, where it is continuously irradiated with high-energy ultraviolet light. When compounds that have a lower ionization potential than that of the irradiation energy (10.2 eV with a standard lamp) are present, they are ionized. The ions formed are collected in an electric field, producing an ion current that is amplified and output by the gas chromatograph's electrometer. The sample stream flows from the PID into the FID, which uses a flame produced by the combustion of hydrogen and air. As the analytes pass through this flame, they are ionized and attracted to the collector electrode due to an applied electric field in the ionization chamber. The collected ions produce a current proportional to the amount of sample in the flame. The PID shows the aromatic compounds, while the FID will detect all organic compounds present in the sample.

The Electrolytic Conductivity Detector

Because of its high versatility, the electrolytic conductivity detector (EICD) detector is used in analysis of plastic additives and solvents requiring high sensitivity and selectivity. When compared to other selective detectors, such as the nitrogen–phosphorus detector or the electron-capture detector, electrolytic conductivity detector chromatograms typically are much cleaner. In the electrolytic conductivity detector, also called the Hall detector, organic compounds in the effluent are first converted to carbon dioxide by passing the column eluent through a high-temperature reactor in which the hetero atoms of interest (halogen-, sulfur- and nitrogen-containing compounds) are converted to small inorganic molecules. The reaction-product stream is then directed into a flow-through electrolytic conductivity cell. Changes in electrolytic conductivity are measured. Ionic material is removed from the system by water that is continuously circulated through an ion-exchange column. The combustion products may be mixed with hydrogen gas and hydrogenated over a nickel catalyst in a quartz-tube furnace. Ammonia is formed from organic nitrogen, HCl from organic chlorides, and H_2S from sulfur compounds. If one wants to detect halogen compounds, a nickel reaction tube, hydrogen reaction gas, a reactor temperature of 850–1000°C, and 1-propanol are used. Under these conditions, compounds containing chlorine will be converted to HCl, methane, and water. The HCl will dissolve in 1-propanol and change its electrolytic conductivity, whereas the nonhalogen products and will not dissolve in the alcohol and not change its conductivity to any significant degree. In the detection of sulfur compounds, the compound must be converted to SO. Collection of SO in methanol containing a small amount of water converts the SO into ionic species. Although water is a satisfactory solvent for the sulfur or halogen modes, water containing an organic solvent is preferred for the nitrogen mode [9].

Mass Spectrometer and Fourier Transform Infrared Detectors

The mass spectrometer is a very widely used detector in gas chromatography. More detailed information about the use and capabilities of the mass spectrometer are presented in the chapter on mass spectrometry. The Fourier transform infrared (FT-IR) detector is also used in GC to provide structural information about components eluting from the GC column. The FT-IR detector is nondestructive and is often used in line with the mass spectrometer. This combination of detectors provide two very powerful complementary pieces of information for structural elucidation of analytes being eluted from the GC column. The FT-IR can also be used as a single detector and in many instances is also used off-line to identify components eluting from the GC column. In this instance, the components are trapped in a cool medium to prevent thermal losses and then analyzed off-line by FT-IR [10–13].

Other Gas Chromatographic Detectors

The Chemiluminescence-Redox Detector

The chemiluminescence-redox detector (CRD) is based on specific redox reactions coupled with chemiluminescence measurement. An attractive feature of this detector is that it responds to compounds such as ammonia, hydrogen sulfide, carbon disulfide, and sulfur dioxide. Hydrogen peroxide, hydrogen, carbon monoxide, sulfides, and thiols that are not sensitively detected by flame ionization detection can be detected with the CRD detector. Compounds that typically constitute a large portion of the matrix of many industrial samples are not detected, thus simplifying matrix effects and sample cleanup procedures for some applications.

The Helium Ionization Detector

The helium ionization detector (HID) is often used for the detection of inert gases.

The Surface Ionization Detector

Over the past few years, surface ionization detectors (SIDs) have been given considerable attention for the determination of organic compounds with low ionization potentials. Recently, however, a novel design based on hyperthermal positive surface ionization has been available. The primary requirement for the operation of this detector is the use of a supersonic free jet nozzle to introduce the sample to a high-work-function surface of rhenium oxide. The primary advantage of this new SID is that it produces a higher sensitivity for all organic compounds, providing a universal GC response.

The Ion Mobility Detector

The number of applications of ion mobility spectrometry (IMS) as a detector for gas chromatography continues to grow. In recent years, much of the emphasis on ion mobility detection after gas chromatography has been in the area of portable analytical instruments. IMS has been used as a detection method for pyrolysis GC.

Isotope Ratio Mass Spectrometry

Isotope ratio mass spectrometry (IR-MS) is used with GC for the high-precision measurement of isotopic ratios of D/H, $^{13}C/^{12}C$, and $^{15}N/^{14}N$ from organic mixtures. In general, gas chromatography is coupled to isotope ratio mass spectrometry via a combustion furnace [9].

Inductively Coupled Plasma Ionization

A powerful technique for the separation and speciation of volatile organometalic compounds is capillary gas chromatography coupled to inductively coupled plasma mass spectrometry (ICPMS) The main advantage of GC-ICPMS is that the total analyte is transferred into the ICPMS without loss due to nebulization. In some instances the interface does not require any changes in the ICP and can be completed in a relatively short time. Applications of GC/ICPMS include analysis of organometallic components.

Combination Detectors

The ability to combine GC detectors in a single analysis is a powerful approach for the investigation of complex mixtures and the identification of unknown compounds. Combinations of various detectors have also been used in novel ways to solve analytical problems during the past few years.

THE GC OVEN: TEMPERATURE CONTROL

The purpose of the oven is to ensure that the column is either kept at a constant temperature (isothermal GC) or programmed during the run. The temperature should be monitored, adjusted, and regulated at the injection port, in the oven surrounding the column, and at the detector. The temperature of the injection port must be sufficiently high to vaporize instantly the sample, yet not so high that thermal decomposition or molecular rearrangement can occur. The column temperature need not exceed the boiling point of the sample in order to keep the analytes in their vapor phases. Actually, the column will produce better separations if the temperature is below the sample's boiling point (above its condensation point), in order to increase the interaction with the stationary phase. The smaller the amount of the stationary phase, the lower is the temperature at which the column can operate: open tubular columns are usually required to run at lower temperatures than packed columns. The temperature of the detector housing should be sufficiently high so that no condensation of the effluent occurs, yet not so high that the detector malfunctions. In general, the detector must be kept at a higher temperature than the injector and higher or equal to the highest temperature of the CG column temperature program.

Isothermal Operation

Selecting the column temperature for isothermal operation is a complex problem. Although most often temperature-programmed analysis is preferred, in some instances isothermal analysis is convenient (i.e., when uniform flow rate due to detector response is required). A sample whose components have a wide range of boiling points cannot be satisfactorily chromatographed in a single isothermal run. A scouting run at a moderate column temperature may provide good resolution of the lower-boiling compounds but requires a lengthy period for the elution of high-boiling materials. One solution is to raise the column temperature to a higher value at some point during the chromatogram so that the higher-boiling components will be eluted more rapidly and with narrower peaks. A better solution to this problem is to change the band migration rates during the course of separation by using temperature programming.

Temperature Programming

In temperature programming, the sample is injected into the chromatographic system when the column temperature is below the lowest-boiling-point component of the sample, preferably 90°C or below. Then the column temperature is raised at some preselected heating rate. As a general rule, the retention time is halved for a 20–300°C increase in temperature. The final column temperature should be near the boiling point of the final solute but should not exceed the upper temperature limit of the stationary phase. Heating rates of 3–50°C/minute should be tried initially and then fine-tuned to achieve optimum separation.

DATA RECORDING AND PROCESSING

Chromatography data acquisition, processing, and archiving can be performed with the aid of a simple integrator or strip-chart recorder, with a highly sophisticated computer data system, or with any system in between. There are a great many choices for the analyst today. Depending on the type of analysis, the requirements may vary. For example, for high-speed GC, fast data acquisition systems are a prerequisite.

For complex operations with numerous instruments, a chromatographic network data system with the power needed to organize, store, and retrieve chromatography data quickly and easily is highly recommenced. Most commercially available systems offer the flexibility of working with a single system and then going to multiple instruments while still conserving the simplicity of the client personal computer. One of the most important and unique design capabilities of these systems is the fact that a single workstation will have exactly the same look and feel as very large client/server systems. Users have the ability to grow from a small single user systems to large client/server systems without the need to learn a new system protocol each time they expand.

The specific capabilities of simple or complex data acquisition and processing devices can vary significantly from source to source and is beyond the scope of this chapter.

SAMPLE PREPARATION

A key to successful chromatographic analysis lies in proper sample preparation. Ideally, it is preferred to dissolve the sample in a suitable solvent and analyze that solution directly, provided the presence of dissolved polymer does not complicate the chromatographic analysis. These cases are indeed rare. Normally, filtration or precipitation followed by final filtration is desirable to remove interferences in sample components (polymers) and higher-molecular-weight components. This approach works well when the polymers are, first, soluble, and second, can be precipitated with an antisolvent. Less soluble polymers, such as highly crystalline resins, require extraction to remove the components of interest from the resin matrix. Numerous extraction techniques (supercritical fluid extraction, solvent extraction, resin dissolution followed by antisolvent precipitation, etc.) are also available [14].

In many instances the dissolved polymer matrix (along with the components of interest in the solution) can be injected in the GC instrument directly (Fig. 7). This requires that the polymer be retained in the injector liner and that it is thermally stable at the injector temperatures of the analysis to minimize undesirable matrix interference. In reality this approach can work relatively well for some polymer systems if the injector liner is replaced or cleaned on a routine basis to prevent or minimize column contamination or degradation. A relatively new device which permits injection of polymer solutions for trace analysis of residual solvents or additives is the high-volume injector. This is further described at end of this chapter. In theory, any type of sample can be introduced into a GC for analysis. All that is required is the right accessory to provide the right conditions for sample introduction to make the sample amenable to gas chromatographic analysis. This includes GC-pyrolysis (for polymers and high-molecular-weight additives), headspace (to separate volatile components from nonvolatile matrix), purge-and-trap, thermal desorption devices (to extract volatile components from polymer matrix), etc. These techniques are presented in more detail later in the chapter.

A good approach for sample cleanup for GC analysis is solid-phase extraction (SPE). SPE is used primarily to clean up samples for analysis

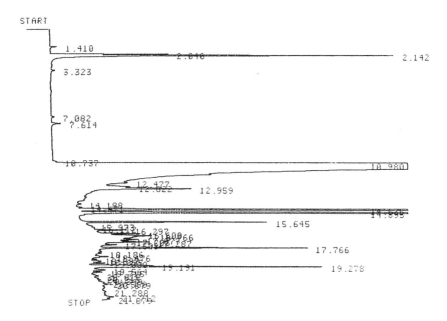

Figure 7 Analysis of trace monomers by GC by injection of whole polymer solution. Frequent change of the injector liner is required for this technique.

and/or concentrate samples to improve detection limits. SPE techniques usually provide better sample cleanup and recovery than liquid–liquid extraction techniques. SPE uses small volumes of common solvents, does not require the use of highly specialized laboratory equipment, and allows rapid sample preparation. A liquid or solid sample is dissolved in a proper solvent and poured into the conditioned SPE cartridge. Vacuum or pressure is used to force the sample through the sorbent in the cartridge. An SPE vacuum manifold can be used to process multiple cartridges simultaneously. Usually, SPE methods are designed to retain the analytes of interest; other sample components similar to the analytes may also be retained. A properly designed SPE method will minimize the amount of unwanted sample components retained by the sorbent. Weakly retained sample components are rinsed from the sorbent using a solvent. The analytes of interest are then eluted from the sorbent using a second solvent. This solvent is collected for analysis. In some cases, the analytes of interest are allowed to pass through the sorbent without being retained and are collected as they exit the sorbent. Most of the sample interferences are retained by the sorbent and thus are isolated from the analytes of interest. In most cases, retaining the analytes of interest followed by elution results in better sample cleanup [15].

SPE cartridges are commercially available from several manufacturers or can be packed by the analyst using commercially available resins to solve individual problems.

The cartridge body usually is a syringe-like barrel made of serological-grade polypropylene. Some manufacturers may also offer glass barrels. The barrel normally terminates in a male luer tip for ease of use. Different sizes of barrels are available to accommodate the various amounts of stationary phase used in SPE cartridges.

The stationary phase (sorbent) is the most important part of an SPE cartridge. The most common SPE phases are bonded silica-based materials. Various silanes are used to attach functional groups to the accessible areas of the silica particle. The functional group determines the identity and chromatographic characteristics of the phase. In addition, several non-silica-based phases are also available. Samples that are dirty, complex, or highly concentrated will require larger amounts of phase for sufficient sample cleanup.

GC ANALYSIS

Quantitative Analysis

Most chromatographic detectors respond to the concentration of the solute and yield a signal that is proportional to the solute concentration that passes through the detector. For these detectors the peak area is proportional to the mass of the component. This is true if the analyte concentration is located in the detector's working range. Peak quantitation is nornmally based on peak height or peak area.

Modern integrators or computers equipped with signal processing software make peak integration and quantitation a relatively easy task. These devices also provide the flexibility required to process complex chromatograms automatically or manually under different sets of conditions. In the case of overlapping peaks and complex chromatography, special algorithms allot areas to each component. During isothermal runs the software can automatically alter the slope sensitivity with time. This allows both sharp, narrow peaks and low, broad peaks to be measured with equal precision. If the peak height or peak areas are measured, there are four main methods that can be used to translate these numbers to the amounts of solute.

Area Normalization

For application of the area normalization method, the entire sample must elute from the column and all components must be separated. This is not a requirement if the composition of components not eluting from the GC is known or can be calculated by a different technique. The area under each peak is measured and corrected (if necessary) by a response factor. All the peak areas are added together. The percentage of individual components is obtained by multiplying each individual calculated area by 100 and then dividing by the total calculated area. Results are not correct if a sample component does not elute from the column or does not give a signal with the detector used. This type of report can be automatically obtained from a GC integrator or a computer acquisition system.

Internal Standard

A known quantity of an internal standard can be injected along with known amounts of the compound of interest. The area versus concentration is calculated to crate a calibration curve. Then an appropriate quantity of the internal standard is added to the raw sample prior to any sample analysis. The peak area of the standard in the sample run is compared with the peak area when the standard is run separately. This ratio serves as a correction factor for variation in sample size, losses during sample preparation, or incomplete elution of the sample. The internal standard must be completely resolved from adjacent sample components, should not interfere with the sample components, and should never be present in material to be analyzed.

External Standard

Calibration curves for each component can also be prepared from pure standards using identical injection volumes and operating conditions for standards and samples. The concentration of solute is read from its calibration curve. In this approach, only the areas of the peaks of interest need to be measured. This method is highly operator-dependent and requires good laboratory technique. The sample amount (volume) to be injected for this approach is critical. The variation in injection volume will compromise results, so it is recommended to use an auto-injector whenever possible for this technique.

Standard Addition

The standard addition approach is useful when only a few samples are to be analyzed. The chromatogram of the unknown is recorded. Then a known amount of the analyte(s) is added, and the sample is reanalyzed using the same reagents, instrument parameters, and procedures. From the increase in the peak area (or peak height), the original concentration can be computed by interpolation. The detector response should be linear for the analyte. Sufficient time should be allowed between addition of the standard and the analysis to allow equilibrium of the added standard with any matrix interferant.

The following equation can be used to calculate analyte concentration:

$$X = \frac{(A_1)(a)}{(A_2 - A_1)}$$

where A_1 = area count of analyte in original chromatogram
$\quad\quad A_2$ = area count of analyte in chromatogram of spiked sample
$\quad\quad X$ = original concentration
$\quad\quad a$ = known (added concentration)

The amount to spike should be around the expected concentration of the analyte in the original sample. Example: If expected concentration is 1%, spike sample to a known concentration (a) of 1% in sample.

A correction for dilution must be made if the amount of standard added changes the total sample volume significantly. Additions of analyte ranging from twice to one-half the amount of analyte present in the original sample are optimum conditions.

Qualitive Analysis (Identification of Unknowns)

Retention Time

Retention time is widely used in GC analysis as a way to identify components of mixtures. This is due mainly to the fact that retention time under fixed operating conditions is constant for a particular component. This is readily accomplished by comparing the retention times of the sample components with the retention times of pure standards. Additionally, retention index can be calculated for different analytes. This can be done by using isothermal (GC) elution times. Retention times usually vary in a regular and predictable fashion with repeated substitution groups (homologous series). One must be cautious, however, in interpretation of results because of the possibility of co-elution of an unknown with the standard.

Identification by Ancillary Techniques

Under normal circumstances, retention time is a good tool to identify components by GC. Often, however, retention time alone is not definitive because many compounds have similar retention times. Structural information can be obtained independently from several spectroscopic techniques. This has led to hyphenated techniques, such as gas chromatography-mass spectroscopy, gas chromatography-infrared spectroscopy and others. If spectroscopic, reference spectra are available, confirmation of analyte identity is very likely.

Enhancing GC Analysis

Derivatization Reactions

The majority of analytical derivatization reactions used for chromatography fall into three general reaction types: silation, alklylation, and acylation. These reactions are used to:

> Improve the thermal stability of compounds, particularly compounds that contain certain polar functional groups
> Enhance the volatility of (nonvolatile or polar) compounds
> Improve sensitivity by tagging a molecule that gives higher response with certain detectors

For analysis by gas chromatography, additives containing functional groups such as $-COOH$, $-OH$, $-NH$, and $-SH$ are of primary concern because of their tendency to hydrogen-bond. This affects the inherent volatility and thermal stability. Many derivatization methods are also intended to enhance detectabiliry by special detectors such as NPDs or ECDs.

Silylation

Silyl derivatives, which are widely used for gas chromatographic applications, are formed by the replacement of active hydrogens atoms from acids, alcohols, thiols, amines, amides, enolizable ketones, and aldehydes with trimethylsilyl groups. A wide variety of reagents are available. These reagents differ in their reactivity, selectivity, side reactions, and the character of the reaction by-products from the silylation reagent. The trifluoro group is commonly used for sensitizing substances to detection by electron capture. This derivatization method makes feasible the quantitative and qualitative analysis of amino acids.

Alkylation

Alkylation is used in the derivatization of additives and analytes containing active hydrogen atoms, aliphatic or aliphatic-aromatic substituents. This method is also used to modify compounds containing acidic hydrogen, such as carboxylic acids and phenols, which are converted into esters. Alkylation reactions can also be used to prepare ethers, thioethers and thioesters, N-alkylamines, amides, and sulfonamides. Although silyl derivatives of carboxylic acids are easily formed, these compounds suffer from low stability.

Acylation

Acylation is used to convert compounds containing active hydrogens into esters, thioesters, and amides through the action of carboxylic acid or a carboxylic acid derivative. The use of deuterated derivatives provides critical information for interpreting the mass spectra of silylated compounds.

Reagents to Enhance Detectability

When the detection sensitivity of certain analytes is low, the detection can be improved by adding (tagging) a component to which a specific detector is highly sensitive. Addition of halogen-containing tags enhances detectability and specificity for analysis by ECD. Addition of nitrogen-containing tags enhances detectability and specificity for analysis by NPD.

SAMPLE INTRODUCTION/SAMPLE CONCENTRATION

Pyrolysis Gas Chromatography

Pyrolysis gas chromatography has been used extensively to determine compositional analysis of polymeric materials. It is a technique that has long been used in a variety of investigative fields because it produces volatile compounds from macromolecules that are neither volatile nor soluble. Examples are polymers, rubbers, paint films, and resins. Volatile fragments are formed and introduced into the chromatographic column for analysis. A polymer can be used as an example to illustrate the use of this technique, which consists of two steps. The injection port is heated to about 250C;

when the sample is injected, the volatile ingredients that are driven off provide a fingerprint of the polymer formulation. Then the pyrolysis step develops the fingerprint of the nonvolatile components. Known monomers of suspected polymers can be injected with a microsyringe for identification of the peaks of an unknown pyrogram [15–18]. Specific identification of the peaks appearing in pyrograms is most effectively carried out by directly coupling gas chromatography-mass spectrometry together with the retention data of the reference samples. Gas chromatography-Fourier transform infrared spectrometry also can provide effective and provides complementary information.

Pyrolyzers can be classified into three groups:

1. Resistively heated electrical filament
2. High-frequency induction (Curie point)
3. Furnace types

The resistively heated pyrolyzer uses either a metal foil or a coil as the sample holder. The heat energy is supplied to the sample holder in pulses by an electric current. This permits stepwise pyrolysis at fixed or increasing pyrolysis temperatures. This feature makes it possible to perform discriminative analysis of volatile formulations and high polymers in a given compound without any preliminary sample treatment.

The Curie-point pyrolyzer uses the Curie points of ferromagnetic sample holders to achieve precisely controlled temperatures when the holder containing the sample is subjected to high-frequency induction heating. Foils of various ferromagnetic materials enable the analyst to select pyrolysis temperatures from 150 to 1000°C.

In the furnace pyrolyzer type, the sample is introduced into the center of a tubular furnace held at a fixed temperature. Temperature is controlled by a proportioning controller that utilizes a thermocouple feedback loop.

Pyrolysis GC has been used in the determination of compositional analysis and microstructure of chlorinated polyethylene (CPE). This method utilized specific aromatic compounds which were formed through dehydrochlorination of trimers after pyrolysis of CPE polymers at elevated temperatures. The composition and microstructure calculation was based on the difference between the levels of ethylene and vinyl chloride trimers formed [19–22].

A pyrolysis gas chromatography has also been used to study the composition and microstructure of styrene/methyl methacrylate (STY/MMA) copolymers. The composition was quantified by pyrolysis-GC using monomer peak intensity. Because of the poor stability of methyl methacrylate oligomers, neither MMA dimer nor MMA trimers were detected under normal pyrolysis conditions. The number-average sequence length for STY was determined by pure and hybrid trimer peak intensities. The number-average sequence length for MMA was determined by using formulas that incorporate composition and the number-average sequence length of STY [23].

Thermal degradation mechanisms under pyrolysis conditions have also been studied using a variation of this technique. It has been shown that the enhanced monomer production of poly(alkylvinylcarboxylate) during pyrolysis makes possible qualitative analysis of poly(vinylcarboxylic acid) through its derivatized alkyl ester. In this study, copolymers containing poly (methacrylic acid) and methacrylic acid were used to demonstrate the pyrolysis gas chromatography qualitative analysis of derivatized polymer [18].

Purge-and-Trap

The analysis of low levels of volatile organic compounds in samples is commonly performed using the technique of purge-and-trap gas chromatography. Using this technique it is possible to detect and identify various flavors, fragrances, off odors, and manufacturing by-products in a wide diversity of products and chemical formulations, colloidal suspensions, and liquid pastes, including perfumes, food products, blood, and latex paint. In the purge-and-trap method, samples are contained in a gas-tight glass vessel. With a sparging needle dipping nearly to the bottom of the sample vessel, the sample is purged by bubbling high-purity helium or nitrogen through the sample. If desired, the sample can be heated for the removal of higher boilers, but the temperature should be kept well below the boiling point of the sample. Volatile compounds are swept out of the sample and carried into and trapped in an adsorbent tube packed with Tenax or other adsorbent material such as activated carbon. Tenax has a low affinity for water. Also, a dry-gas purge is added just before the adsorption tube. Volatile organic materials can be efficiently collected from a relatively large sample, producing a concentration factor that is typically 500- to 1000-fold greater than the original and resulting in detection limits of parts per billion or parts per trillion (Fig. 8). The trapped material on the adsorbent is heated to release the sample and then backflushed using the GC carrier gas. This sweeps the sample directly onto the GC column for separation and detection by normal GC procedures. Trapped samples can easily be stored or shipped to another site for analysis.

Headspace Sampling

Headspace sampling is another technique that is widely used in the plastics industry for the analysis of volatile components in plastic resins. A major advantage of headspace GC is the relative cleanliness of the sample entering

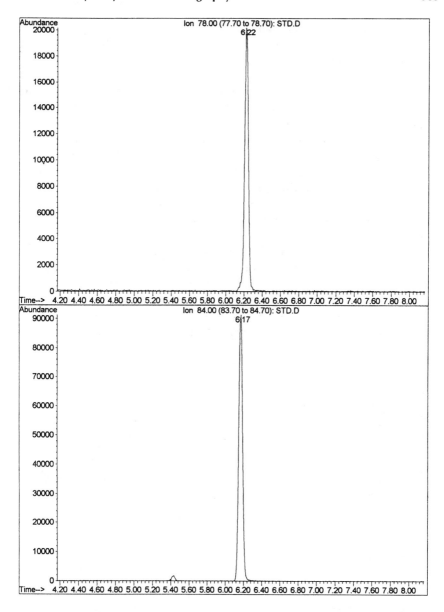

Figure 8 GC chromatogram: trace analysis of benzene in polymer by purge-and-trap GC-MS. The single-ion signal for benzene (m/z, 78) is displayed on top, while the single-ion signal for deuterated benzene (m/z, 84) is displayed below.

the GC (gas phase), free from nonvolatile, matrix interferences. This in turn results in chromatography of excellent signal-to-noise ratio. Headspace sampling can be done when only the vapor above the sample is of interest and the partition coefficient allows a sufficient amount of analyte into the gaseous phase. This is normally not a severe limitation, since temperature or matrix manipulation can be used to drive the compounds of interest into the vapor phase. Samples may be solid or liquid.

In a typical analysis a measured amount of sample, and often an internal standard, are placed in a vial, and then the septum and cap are crimped in place. The vials, contained in a carousel, are immersed in a heated oven or silicone oil bath operating from ambient temperature to up to 300C. A heated flexible tube that terminates in a needle samples each vial in turn. A gas-sampling valve provides a fixed volume of vapor sample to transfer into the GC injection port [19].

Headspace sampling is an excellent technique for the analysis of compounds that produce, odors in many commercial products, such as plastics, rubbers, paints, resins, etc. Interfacing of a headspace GC with a mass spectrometer provides valuable information for the identification of such components. Figure 9 shows the output from head space-GC analysis of residual benzene and toluene in polymeric material.

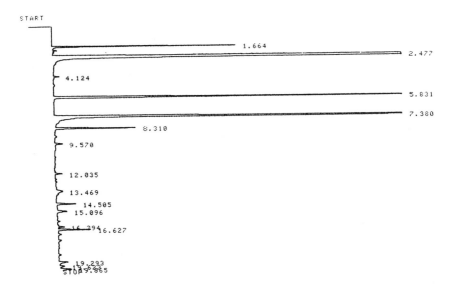

Figure 9 GC chromatogram: trace analysis of benzene and toluene in polymer by headspace GC-FID.

Thermal Desorption

Thermal desorption permits the analysis of plastic resins and other solid material samples without any prior solvent extraction or other sample preparation. Solid samples between 1 and 500 mg are placed inside a glass-lined, stainless steel desorption tube between two glass-wool plugs. After attaching a syringe needle to the tube and then placing the tube in the thermal desorption system, the desorption tube is purged with carrier gas to remove all traces of air or oxygen. The preheated heater blocks are closed around the adsorption tube to desorb samples at temperatures from 20 to 350°C and for program desorption times from 1 s up to 5 min or more. The procedure permits the thermal extraction of volatiles and semivolatiles from the sample directly into the GC injection port. The GC column is maintained at subambient temperatures or at a low enough temperature to retain any samples at the front of the GC column during the desorption step. This enables the desired components to be collected in a narrow band on the front of the GC column [20]. As an alternative to cryofocusing, a thick-film capillary column or a packed column with a high loading capacity may be used. After desorption is complete, the needle is removed and the GC gas turned on. The components trapped on the front of the GC column are separated and eluted via a temperature program in the GC oven. By selection of the desorption temperature, the number and molecular-weight distributions of components in the samples can be selected. Adsorbent tubes with samples collected during either dynamic headspace purging or purge-and-trap of liquids can also be analyzed with a thermal desorption unit [20,21].

Solid-Phase Microextraction

Solid-phase microextraction (SPME) is a straitforward sample extraction process. As pointed out above, the extraction of organic compounds from a sample matrix usually consists of purge-and-trap, thermal desorption, or headspace methods for concentrating volatiles; and liquid–liquid extraction, solid-phase extraction, or supercritical fluid extraction for semivolatiles and nonvolatiles. These methods have various drawbacks, including high cost and long sample preparation times. This relatively new sample preparation technique eliminates this to extract organic compounds from various matrices. SPME requires no solvents or complicated apparatus. It can concentrate volatile and nonvolatile compounds, in both liquid and gaseous samples, for analysis by GC or GC-MS. An SPME unit consists of a length of fused silica fiber coated with a phase material. The phase can be mixed with solid adsorbents, e.g., divinylbenzene polymers, templated resins, or porous carbons. The fiber is attached to a stainless steel plunger in a protective holder. In a relatively short step (a few minutes), the organic compounds of interest are concentrated by direct exposure to the fiber and subsequently injected directly into the instrument for characterization.

SPME has been applied to the analysis of various volatile components in plastic resins with gas chromatography with great success. Recently, it has also been demonstrated that SPME can be conveniently used with high-speed GC for quick sample analysis. While SPME offers a rapid sampling procedure for volatile analytes, the gas chromatographic analysis becomes the limiting factor in the speed of analysis. High-speed GC has been used to allow for the analysis of volatile components a factor of 10 times more rapidly than conventional GC. SPME allows the selectivity necessary to make these determinations, while rapid gas chromatography provides great speed of analysis, making this combination an excellent candidate for analysis of various products in the laboratory.

MULTIDIMENSIONAL GAS CHROMATOGRAPHY

Multidimensional gas chromatography has enjoyed a dramatic increase in usage recently to solve problems of materials characterization. Much activity has taken place in the coupling of liquid chromatography (LC) with GC, despite a first impression of apparent incompatibility of LC and GC for the analysis of large nonvolatile molecules. This field has further benefitted by new developments in high-volume injectors and sophisticated switching devices. However, in LC-GC, the liquid chromatograph has been used principally for prefractionation of complex mixtures to isolate a targeted GC-compatible substance. The availability of high-temperature GC should provide a brighter future to this approach.

In multidimensional gas chromatography, the components of a sample are separated by using series-connected columns of different capacities or selectivities. Two common multicolumn configurations are the packed-column and capillary-column combination and two capillary columns in series [21,22,24]. Two independently controlled ovens may be needed, and such a configuration is available commercially. In addition to decreased analysis time, this arrangement provides an effective way of handling samples containing components that vary widely in concentration, volatility, and/or polarity. Used in conjunction with techniques such as heart-cutting, backflushing, and peak switching, useful chromatographic data have been obtained for a variety of complex mixtures.

NEW TECHNOLOGIES IN GAS CHROMATOGRAPHY

Recent developments in gas chromatography have given this technique an additional boost in terms of power and attractiveness to solve analytical problems. New technical developments deserving special attention are high-temperature gas chromatography, high-speed GC (fast GC), retention-time locking, and high-volume injectors.

High-Temperature Gas Chromatography

High-temperature gas chromatography has extended the capability of GC analyses to analyses of samples previously not possible by this technique. Oven temperatures up to 450°C can be reached with this new equipment. A cool on-column injector, temperature programmable, is normally required for these applications, as well as high-temperature, aluminum-clad capillary columns that can take these temperatures without column bleeding. Most conventional GC detectors are available for this applications. While analytes of molecular weight of about 800 amu represent the upper limit for conventional CG, the upper limit for high-temperature GC is about 1200–1400 amu, depending on the nature of the analytes of interest. This means that many of chemical compounds previously analyzed by liquid chromatography can now be easily analyzed by high-temperature GC. Figure 10 illustrates the chromatographic and temperature capabilities of high-temperature GC.

Aluminum-Clad Columns

High-temperature capillary columns which are capable of operating at temperatures up to 420°C are coated with a thin (approximately 20 μm) uniform layer of aluminum replacing the conventional polyimide outer coating (typical maximum temperature of 370°C). Aluminum-clad columns offer excellent heat transfer and the same flexibility and inert fused silica surface as the polyimide coated columns. These columns are ideal for the analysis of a wide variety of high-molecular-weight compounds such as crude oils, waxes, polymers, etc. In order to obtain the best results with any of these high-boiling compounds, the injection technique is very important. Either cold on-column, movable needle, or programmable temperature injectors must be utilized to prevent component discrimination.

Stainless Steel Columns

The inertness of stainless steel columns matches that of fused silica and is derived from a multistep process which utilizes a multilayer pretreatment of the inner surface of the stainless steel. Each layer is chemically stable at elevated temperatures and has the same or higher mechanical properties as the steel tubing. The layers are chemically bonded together. Stationary phases are easily bonded to this stable inert surface, resulting in high-performance columns.

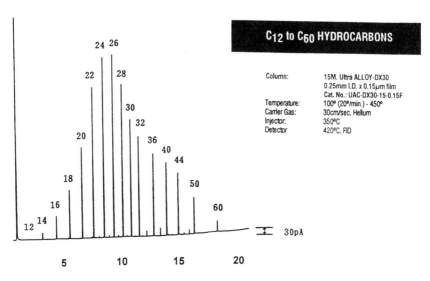

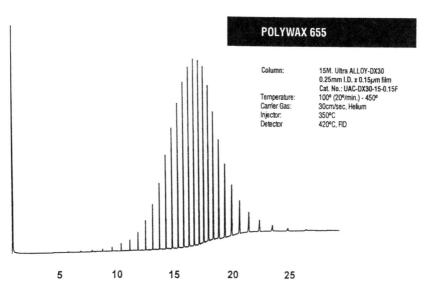

Figure 10 Analysis of high-molecular-weight waxes, illustrating the power of high-temperature GC columns. Note the high temperature limit for the column and the relatively short retention time. (Courtesy of Quadrex Corporation.)

Stainless steel columns are not lined with fused silica, which can crack or flake off when flexed or bent, thus exposing active sites. As a result, these columns can be tightly coiled to fit small GC ovens. Stainless steel capillary columns are available in three internal diameters, the standard 0.25-mm, 0.50-mm, and 0.80-mm I.D., and in lengths up to 60 m.

High-Speed Gas Chromatography

In recent years, the need for rapid GC analysis has led to the development of gas chromatographs with fast separation times. Several recent technologies have been developed to decrease analysis times in GC, such as using short narrow-bore columns and optimizing the flow rates, temperature rates, and sample focusing parameters. Two major requirements for successful fast GC are fast data acquisition rates and fast detector response.

High-speed GC offers several benefits:

Quicker results for timely decisions about sample or product fate
Faster sample turnaround times
Lower operating costs per sample analysis
Ability to handle more samples with fewer pieces of equipment

There are three major commercially available systems providing high-speed GC. They use different approaches to accomplish high speed of analysis. Their principles and capabilities are briefly described below.

High-Speed GC Using a Standard Instrument

High-speed GC analysis is now possible with the recent development of GC equipment that allows rapid heating of the CG oven as well as precise control of the carrier-gas pressure. Although this is not yet a widely used approach in chromatography, it has already been demonstrated that this new technique can reduce the analyses time by a factor of 5 or better, compared with conventional GC analyses. The advent of high-frequency sampling devices has been crucial in the development of this new technology.

One of the many remarkable features of modern GC instruments is their ability to perform fast GC without special modifications or expensive accessories. High-speed GC can be achieved using short, narrow-bore columns, resulting in analyses that are 5 times faster than with traditional methods run on conventional laboratory GCs. These GCs offer the capability to carry out fast GC without the need for cryofocusing or thermal desorption devices which may limit the flexibility or performance of the instrument. Properly configured for fast GC, the system can perform all types of analyses using existing detectors, injectors, and flow controllers.

Minimal system requirements are a GC equipped with electronic pneumatic control (EPC), off-the-shelf capillary columns, split/splitless or on-column inlets, standard detectors optimized for capillary columns, and a fast acquisition data system. At any time, users can switch from fast GC back to the original method without major difficulties, or optimize new methods to meet new analytical demands.

Flash GC

Flash temperature programming is a new technique for rapidly heating capillary GC columns. The technique utilizes resistive heating of a small-bore metal sheath that contains the GC column. This technology is based on the flash GC system, an innovative chromatographic system that accomplishes in 1–2 min what takes a conventional GC from 30 min to a 1 h or more. The flash GC can be over 20 times faster than conventional GC (Fig. 11) and is also more sensitive and far more versatile. Additionally, the flash GC can be used to concentrate samples as part of its purge-and-trap capabilities. On-column cryotrapping procedures can be accomplished within the system and incorporated as part of the very fast analysis procedure.

Development of the flash GC began in the 1980s. The first application was for the detection of explosives. The basic concepts, the proof of principle, and the initial designs were initially classified by the U.S. government. The patents were declassified in the early 1990s. and are now been issued to produce this equipment for public use.

Flash GC systems feature the following.

1. Reduction of total system cycle time exceeds other forms of temperature programmed "high-speed gas chromatography." This is possible thanks to rapid cool-down capabilities.
2. Increased sensitivity over conventional GC. The chromatographic peaks produced by the flash GC are typically only a fraction of a second wide. With mass-sensitive detectors, such as the FID, narrowing the peak widths of the same sample will give the user an additional signal-to-noise advantage.
3. Conventional GC columns 6 m or 12 m long are used in the flash column assemblies; (standard fused silica columns with typical column diameters and coatings are compatible with the flash GC). The flash column consists of a standard column inserted into a metal sheath designed for both rapid heating and cooling.
4. Sample introduction can be by direct on-column injection or isolated from the flash columns using "cold spots." Either method can be optimized for maximum performance.

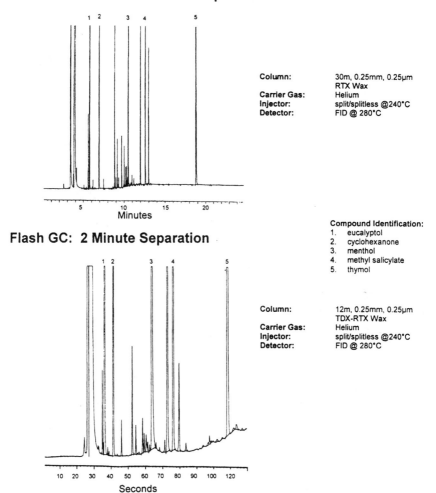

Conventional GC: 20 Minute Separation

Minutes

Column:	30m, 0.25mm, 0.25µm RTX Wax
Carrier Gas:	Helium
Injector:	split/splitless @240°C
Detector:	FID @ 280°C

Flash GC: 2 Minute Separation

Compound Identification:
1. eucalyptol
2. cyclohexanone
3. menthol
4. methyl salicylate
5. thymol

Column:	12m, 0.25mm, 0.25µm TDX-RTX Wax
Carrier Gas:	Helium
Injector:	split/splitless @240°C
Detector:	FID @ 280°C

Seconds

Figure 11 GC chromatograms comparing conventional GC versus high speed flash GC. Notice the reduction in run time by 10-fold by the use of the flash GC versus conventional GC. (Courtesy of Thermedics Detection.)

Samples can be introduced onto the cold spots, which serve to concentrate and focus the sample prior to introduction onto the flash columns. Samples can also be directly injected on to cooled flash columns.

5. The flash-GC comes equipped with up to four cold spots. The cold spots are computer-controlled independently of each other. These cold spots can be heated at rates from 4°C/s to 3000°C per second. Since standard column materials are used in the cold spots, their maximum upper temperatures are limited by the temperature rating of the column selected.

6. Because the sample compounds are typically in the hot zone of the flash columns for only a few seconds, thermally labile compounds, those compounds which are not normally amenable to GC, can often be analyzed.

7. The flash GC comes equipped with exceptionally fast electronics and software designed to handle the high-speed chromatographic data. The on-board A/D can operate at up to 200 Hz, providing 5-ms chromatographic resolution.

In summary, the flash GC provides great speed and flexibility to the analyst for the charaterization of a great variety of chemical compounds. This technology has been proven to work in conjunction with mass spectrometry.

There is also an upgrade kit available to easily convert a conventional GC into a flash GC. It is called the EZ Flash. This kit converts a conventional GC to a flash GC, providing the benefits from the speed and accuracy of flash GC technology with minimal investment. EZ Flash columns mount inside the oven of a conventional GC, replacing the existing column. The system offers column heating rates up to 20°C/s and a temperature control range of ambient to 400°C.

Cryofocusing Technology

Cryofocusing technology permits high-speed GC on already-existing GC equipment. ChromatoFast's proprietary Fast-GC technology enhances conventional capillary gas chromatographs with a novel sample inlet system to allow very rapid separations to take place (Fig. 12). It is ideal for use in a wide range of applications including plastic materials, industrial chemicals, and environmental applications. Using a unique cryofocusing inlet system, samples can be preconcentrated and are subsequently desorbed onto the analytical column in a very narrow band, thus eliminating the band broadening that occurs with conventional inlet systems. It can also be interfaced to other automated sample introduction devices such as autosamplers,

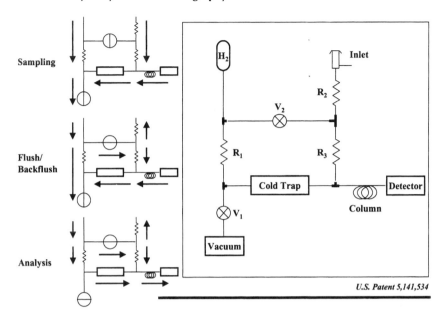

Figure 12 General diagram of a ChromatoFast GC. (Courtesy of ChromatoFast.)

purge-and-trap, headspace, and SPME. The Fast-GC is extremely sensitive, capable of subparts-per-billion detection This approach allows high-speed separations to take place using short lengths of conventional 0.25-mm columns, which provide increased sample handling capacity over microbore columns. Figure 13 shows a typical chromatogram using this technique.

The main advantages and applications of cryofocusing are

Affordable, can be installed on existing GC equipment
At least 10 times faster than conventional GC
Liquid and air analysis capabilities
Compatible with automated sample preparation/introduction devices
Enhanced sensitivity: sub-ppb limit of detection for air analysis
Utilizes conventional 0.25-mm columns and a wide range of stationary phases
Multiple detector options

Figure 12 is a diagram of a ChromatoFast system. This high-speed inlet system can deliver injection bandwidths of 5 ms. The main components

Fuels
Diesel Range Organics

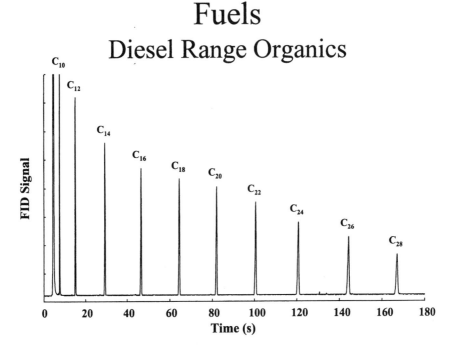

Figure 13 GC chromatogram illustrating the high speed of analysis using ChromatoFast technology. (Courtesy of ChromatoFast.)

include restrictor columns, solenoid valves, source of hydrogen or helium carrier (CG), vacuum pump, detector, and split injector for sample introduction. Short lengths of 0.25-mm capillary columns are typically used at high linear velocities. This provides the maximum rate of plate production as opposed to number of plates. The cold trap consists of a metal tube with inert coating that is cooled using liquid nitrogen and rapidly heated using a capacitive discharge power supply. This technology contains a high level of sophistication with several advantages. Three sequential stages of operation are dictated by the states of valves V1 and V2. All stages are controlled through an automated process for ease of use. Here are summarized the three main stages of operation.

1. *Sampling*. Sample is pulled through R2 and R3 and cryofocused at the right side of the trap. The solenoid valves used in the system are not in the sample flow path. This eliminates problems of sample carryover or decomposition from sample contact with valve surfaces. Since the sample is being pulled onto the trap by the vacuum pump, this system is available as

a separate model designed to sample directly from air or other sample containers. Large amounts of preconcentration can be obtained by varying the duration of the sampling stage.

2. *Flush.* A clean purge flow of carrier gas travels through V2 and splits between R2 and R3. Residual sample in R3 is swept onto the trap and purge flow through R2 eliminates any memory effect or sample carry-over from the split inlet.

3. *Analysis.* The flow reverses in the cold trap and the trap tube is rapidly heated at a rate of up to $100,000°C/s$. Because the sample is trapped and injected from the same end of the trap, it is in the heated metal trap tube for a minimal period of time. This reduces the possibility of decomposition of thermally labile samples. Broadening from the dead volume of the tube itself is also reduced.

High-Volume Injection

The use of high-volume injectors is a specialized option in gas chromatography which provides additional power to standard sample introduction techniques. This technique permits introduction of injection volumes of up to 1000 µl (versus 1–5 µl in conventional injectors). The major advantage provided by this technique is increased sensitivity, i.e., lower detection limits. Although there are various suppliers of high-volume injectors, essentially the technology is based on a sophisticated combination of cryogenic or chemical sample concentration and automatic valving [23,25].

A large-volume injector permits a speed-programmed injection of sample volumes of 5 µl to up to 1000 µl, with simultaneous venting of the solvent. These devices function as a cryotrap, focusing and concentrating the components to be determined, and then transferring them to the capillary column. Large-volume injectors are available as a single-shot system and as a fully automatic systems for processing up to 100 samples by using an autosampler tray. Specialized software packages are also commercially available to optimize the injection parameters. Critical parameters such as the sample injection speed as a function of initial injection temperature, and solvent split and split flow, can be easily determined. The programmed values can be transferred to a method and from instrument to instrument with a few keystrokes [26,27].

The major advantages provided by this technique are

Lower detection limits by injecting sample volumes of 5 µl to 1000 l
Reduced sample preparation time by eliminating solvent evaporation or preconcentration step
Solvent evaporation step is performed at optimal injection speed

Adjustable sampling height allows for automated injection of
solvent layer for "in vial" liquid/liquid extraction
Protection of the capillary column and the detector through solvent
venting

Some attractive applications for the analyst are

Lowering of detection limits for almost any analysis
Detection of impurities in ultrapure solvents
Elimination of solvent evaporation step when combined with solid-
phase extraction and liquid–liquid extraction techniques

Retention-Time Locking

Retention-time locking (RTL) is a unique feature available in some GC
equipment which has revolutionized gas chromatography by reproducing
retention times within hundredths of a minute from one instrument to
another. This capability is possible thanks to highly reliable electronic pneumatic control (EPC), (pressure and flow control), good temperature control;
and high reproducibility of capillary columns. The result is ultimately retention time stability between any GC systems with EPC.

Once a method is developed, a compound is selected, which becomes the
locking compound in the usual standard to establish the pressure-versus-
retention time relationship using the RTL software. Once developed, this
information can be used to lock the method on another GC system with EPC.

To lock a method on another GC system, you only have to make a
single run and enter the retention time of the locking compound into the
software. The software then calculates the new inlet pressure to match chromatograms exactly. Furthermore, the software automatically updates the
method with the new pressure and records the change in the instrument
log file that is saved with the method.

Retention-time locking provides the immediate benefits of increased
sample throughput, greater confidence in results, easier analysis for compliance, and lower costs of sample analysis. All you need is a GC equipped
with the RTL software, EPC, and a GC ChemStation. With RTL, all peaks
match and elution order is constant. Retention-time locking eliminates the
need to update calibration tables, timed events tables, and integration events
tables when a method is transferred, a new column is installed, or routine
maintenance is performed.

Pattern Recognition and Artificial Intelligence

In recent years, multivariate data analysis has been used powerfully in the
analysis of complex mixtures. This is an excellent tool for coping with the

large information density presented by high-resolution chromatography. Data processing has thus become an element of foundational significance in handling chromatographic data, extending and expanding the meaning of chromatographic resolution. Although the best examples come from complex petroleum or environmental samples, this tool can be valuable in the analysis of plastic materials, particularly for the analysis of trends in product quality when many samples must be compared to assess process improvement or product variability. Complex processes can be successfully modeled by the use of advanced artificial intelligence methods and chromatography data.

REFERENCES

1. Hyver, K. J. (Ed.), *High Resolution Gas Chromatography*, 3rd ed. Hewlett-Packard, Avondale, PA, 1989.
2. Gardea-Torresdey, J., Gas Chromatography, *Anal. Chem.* 1998; 70(12):321–340.
3. MacNamara, K., New Horizons in Capillary Gas Chromatography, *Irish Chem. News*, 1996; X(V):27–33.
4. Wang, C.-Y., Analysis of Synthetic Polymers and Rubbers, *Anal Chem.* 1997; 69(12):95-122.
5. Anderson, D. G., Coatings, *Anal. Chem.* 1997; 69(12):15–28.
6. Cacho, J., Quantitative Analysis of Pesticides in Postconsumer Recycled Plastics Using Off-Line Supercritical Fluid Extraction/GC-ECD, *Anal Chem.* 1997; 69(16):3304–3313.
7. Smith, P. B., Composition and Microstructure Analysis of Chlorinated Polyethylene by Pyrolysis Gas Chromatography and Pyrolysis Gas Chromatography/Mass Spectrometry, *Anal. Chem.* 1997; 69(4):618–622.
8. Niemax, K., GC Analysis of Chlorinated Hydrocarbons in Oil and Chlorophenols in Plant Extracts Applying Element-Selective Diode Laser Plasma Detection, *Anal. Chem.* 1997; 69(4):755–757.
9. Meier-Augenstein, W., On-Line Recording of 13C/12C Ratios and Mass Spectra in One Gas Chromatographic Analysis, *HRC* 1995; 18(1):28–32.
10. Hoffmann, A., R. Bremer, and J. A. Rijks, Applications of Multi Column Switching Capillary GC-MS in Identification of Trace Impurities in Industrial Products, *Proc. 15th Int. Symp. Capillary Chromatography*, Riva del Garda, Italy, P. Sandra (ed.), 1993, pp 830–836.
11. Mac Namara, K., and A. Hoffmann, Simultaneous Nitrogen, Sulfur, and Mass Spectrometric Analysis After Multi-Column Switching of Complex Whiskey Extracts, *Proc. 15th Int. Symp. on Capillary Chromatography*, Riva del Garda, Italy, P. Sandra (ed.), 1993, pp 877–884.
12. Wahl, H. G., H. M. Liebich, and A. Hoffmann, Identification of Fatty Acid Methyl Esters as Minor Components of Fish Oil by Multidimensional GC-MSD: New Furan Fatty Acids, *HRC* 1994; 17(5), 308–311.

13. Wahl, H. G., A. Chrzanowski, N. Ottawa, H.-U. Häring, and A. Hoffmann, Analysis of Plasticizers in Medical Products by GC-MS with a Combined Thermodesorption-Cooled Injection System, *Proc. 18th Int. Symp on Capillary Chromatography*, Riva del Garda, Italy, P. Sandra and G. Devos (eds.), 1996, pp. 988–991.

14. Brinen, J., Depth Distribution of Light Stabilizers in Coatings Analyzed by Supercritical Fluid Extraction-Gas Chromatography and Time-of-Flight Secondary Ion Mass Spectrometry, *Anal. Chem.* 1998; 70(18):3762–3765.

15. Baltussen, E., H.-G. Janssen, P. Sandra, and C.A. Cramers, A New Method for Sorptive Enrichment of Gaseous Samples: Application in Air Analysis and Natural Gas Characterisation, *HRC* 1994; 17(5):312–321.

16. Smith, P. B., Composition and Microstructure Analysis of Chlorinated Polyethylene by Pyrolysis Gas Chromatography and Pyrolysis Gas chromatography/Mass Spectrometry, *Anal. Chem.* 1997; 69(4):618–622.

17. Smith, P. B., Quantitative Analysis and Structure Determination of Styrene/ Methyl Methacrylate Copolymers by Pyrolysis Gas Chromatography, *Anal. Chem.*; 1996; 68(17):3033–3037.

18. Wang, F. C.-Y., Polymer Degradation Mechanism Reselection Through Derivatization for Qualitative Pyrolysis Gas Chromatography Analysis, *Anal. Chem.*; 1998; 70(17):3642–3648.

19. Brewer, W. E., A Sample Concentrator for Sensitivity Enhancement in Chromatographic Analyses, *Anal. Chem.* 1998; 70(10):2191–2195.

20. Raynie, E., Supercritical Fluid Chromatography and Extraction, *Anal. Chem.* 1996; 68(12):487–514.

21. Hoffmann, A., and R. Bremer, Design, Performance and Applicability of a Multi-functional Thermal Desorption System for Trace Analysis in Capillary GC, *Proc. 16th Int. Symp. on Capillary Chromatography*, Riva del Garda, Italy, P. Sandra and G. Devos (eds.), 1994, pp. 1165–1175.

22. Rogies, F., A. Hoffmann, and J. A. Rijks, Design and Evaluation of a Multi Column Switching Capillary GC Analyzer for Automated Simultaneous Analysis of Permanent Gases and Light Hydrocarbons in Natural Gas, *Proc. 16th Int. Symp. Capillary Chromatography*, Riva del Garda, Italy, P. Sandra and G. Devos (eds.), 1994, pp 1403–1410.

23. Sandra, P., F. David, R. Bremer, and A. Hoffmann, A Fully Automated and Robust LC-CGC Combination for Application in Environmental Analysis, *Int. Environ. Technol. 1997; 7(6):26–27.*

24. Teske, J., J. Efer, and W. Engewald, Large-Volume PTV Injection: New Results on Direct Injection of Water Samples in GC Analyis, *Chromatographia* 1997; 46(11/12):580–586.

25. Teske, J., J. Efer, and W.Engewald, Large Volume PTV Injection: Comparison of Direct Water Injection and In-Vial Extraction for GC Analysis of Triazines, *Chromatographia* 1998; 47(1/2):35–41.

26. Staniewski J., and J. A. Rijks, Solvent Elimination Rate in Temperature Programmed Injections of Large Sample Volumes in Capillary GC, *J. Chromatogr.* 1992; 623: 105–113.

27. Staniewski, J. and J. A. Rijks, Potentials and Limitations of the Liner Design for Cold Temperature Programmed Large Volume Injection in Capillary GC and for LC-GC Interfacing, *J. HRC*, 16(3) 1993; 182–187.

9

Nuclear Magnetic Resonance of Polymeric Materials

Anita J. Brandolini

William Paterson University of New Jersey, Wayne, New Jersey, USA

Nuclear magnetic resonance (NMR) spectroscopy is one of the most versatile techniques available to the analyst of plastics and other polymeric materials. Atomic nuclei respond to external magnetic fields and radiofrequency excitation in complex, but well-understood, ways, and the spectroscopist can exploit these interactions in many different ways. Dipolar and quadrupolar couplings extend along chemical bonds or through space, chemical shifts are extremely sensitive to the identities and positions of neighboring atoms, and scalar couplings reveal details of chemical structure. Each of these features could make a viable analytical tool on its own, but the real power of NMR arises from its ability to express, suppress, or correlate these interactions through experimental manipulations. One-, two-, and higher-dimensional spectroscopies provide the most detailed descriptions of polymer microstructure available from any characterization method, whether the sample is a solution or a solid. Measurements of nuclear relaxation, chemical shielding, and spin–spin interactions reveal details of the morphology and dynamics of polymer systems. Advances in medical imaging are beginning to be applied to materials, and NMR is becoming an important technique for monitoring resin properties on-line.

This chapter presents a mostly qualitative description of these different facets of NMR spectroscopy, summarizing diverse applications of NMR spectroscopy to polymeric materials. After a description of the basic phenomena that generate nuclear magnetic resonances, basic and more sophisticated instrumentation are covered. Execution of several different experiments, signal processing, and interpretation of the resulting data are discussed for both simple and more complex techniques. Common abbreviations used throughout the text are listed at the end of the chapter. The described applications focus on the characterization of polymer structure (both chemical and physical) and dynamics.

I. OVERVIEW OF NUCLEAR MAGNETIC RESONANCE

The origin of nuclear magnetic resonance lies, simply put, in the response of certain atomic nuclei to a magnetic field, B_0. This relationship governs the resonance frequency of the nucleus and the observed linewidths, positions, and splitting (if any) of the resonances in the NMR spectrum. The various interactions that give rise to these spectral features are discussed below in order of decreasing magnitude. The overview is not mathematical; classical and quantum-mechanical treatments are given in many general texts [1–16].

I.A. Zeeman Interaction

Only nuclides with nuclear spin, $I, \geq \frac{1}{2}$ give rise to NMR signals. For some elements, such as hydrogen, fluorine, and phosphorus, the most common naturally occurring isotopes (^1H, ^{19}F, and ^{31}P, respectively) are also NMR-active; they all happen to be spin-$\frac{1}{2}$ nuclei. Spectra of such species are relatively easy to observe because of the high concentration of NMR-active isotopes. However, for other chemically important elements, such as carbon and silicon, the most abundant forms (^{12}C and ^{28}Si) cannot be detected by NMR. Instead, their less abundant, or *rare-spin*, isotopes, ^{13}C (1.1%) and ^{29}Si (4.7%), are the NMR-observable species. Historically, these elements had been difficult to study, but current instrumentation makes them routine. Unfortunately, NMR-active forms of other interesting elements, such as oxygen (^{17}O, 0.04%) are even rarer. Acquisition of an ^{17}O spectrum in natural abundance may be time-consuming, but, for sufficiently concentrated samples, this task is well within the capabilities of modern spectrometers. Table 1 lists natural abundances for several nuclides relevant to polymer systems [17,18].

When an NMR-active nucleus is placed in a static magnetic field, B_0, its magnetic moment, m_i, begins to precess about the field, as illustrated in Fig. 1a. This *Zeeman interaction* depends on B_0, the magnitude of B_0, and occurs at a characteristic frequency, ν_0, called the *Larmor frequency*:

Table 1 Properties of Selected Nuclides

Isotope	Spin	Natural abundance (%)	Magnetogyric ratio, γ^a	Frequency (MHz) ν_0^b	Quadrupole moment, Q^c	Chemical-shift range (ppm)
^{13}C	1/2	1.1	6.73	101.8	N/A	250
^{1}H	1/2	99.99	27.75	400.0	N/A	12
^{2}H	1	0.015	4.11	61.4	2.73×10^{-3}	12
^{19}F	1/2	100	27.17	376.4	N/A	1200
^{29}Si	1/2	4.7	−5.31	79.5	N/A	300
^{31}P	1/2	100	10.83	161.9	N/A	1200
^{14}N	1	99.6	1.93	28.9	1.6×10^{-2}	1000
^{15}N	1/2	0.37	−2.71	40.6	N/A	1000
^{129}Xe	1/2	26.4	−7.40	110.6	N/A	4000
^{17}O	5/2	0.04	−3.63	54.2	-2.6×10^{-2}	600
^{27}Al	5/2	100	6.97	104.2	0.149	450
^{35}Cl	3/2	75.5	2.62	39.2	-7.89×10^{-2}	1200
^{81}Br	3/2	49.5	7.22	108.0	0.28	—[d]
^{127}I	5/2	100	5.35	80.0	−0.69	—[d]
^{23}Na	3/2	100	7.08	105.8	0.12	100
^{11}B	3/2	80.4	8.58	128.3	3.55×10^{-2}	200
^{119}Sn	1/2	8.58	−9.97	149.2	N/A	600

[a] $\times 10^7$ rad/T-s
[b] In 9.4-T magnetic field
[c] $\times 10^{-28}$ m^2.
[d] Insufficient data are available for these isotopes to assess the full range.
(*Source*: Refs. 17 and 18.)

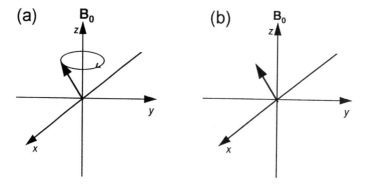

Figure 1 Behavior of a spin in a magnetic field at: (a) the Larmor frequency ω_0; (b) in a frame of reference rotating at ν_0.

$$\nu_0 = \frac{\gamma}{2\pi} B_0 \tag{1}$$

where the proportionality constant γ, the *magnetogyric ratio*, is a property of the nuclide (Table 1). Thus, a ^1H nucleus (usually referred to as a "proton," not as "hydrogen") that resonates at 400 MHz in a 9.4-T (tesla) magnetic field precesses at 200 MHz in a 4.7-T field. It can be demonstrated [1–16] that the magnitude of the observed NMR signal increases as γ^4, so high-γ (such as ^1H or ^{19}F) nuclei are intrinsically more sensitive than those with low γ's (such as like ^{14}N or ^{35}Cl). The behavior of magnetization during an NMR experiment is typically illustrated in the *rotating frame,* a coordinate system precessing at the Larmor frequency. In this frame of reference, m_i is stationary, as depicted in Fig. 1b. Because this representation simplifies the visualization of complex spin dynamics considerably, it is assumed throughout this discussion.

The Larmor frequencies in Table 1 are widely separated (by at least a few megahertz). As will be seen, the frequency range of spectral features is typically several orders of magnitude less than the Larmor frequency separation. The fortunate consequence of this is that the NMR spectra of these nuclides do not overlap. For example, a ^{29}Si spectrum displays resonances *only* from silicon-containing groups. Chemical entities that do not contain silicon do not contribute to the spectrum. In fact, with rare exceptions, the Larmor frequencies of all NMR-active species are isolated [17,18]. If a species of interest contains a unique element, its NMR spectrum can be observed without a distracting background signal. For example, the ^{31}P spectrum of polyethylene with low ($\mu g/g$) levels of phosphite antioxidants reflects only the phosphorus species; the carbon- and hydrogen-containing polymer background is not observed [19]. In many other spectroscopies, the polymer-matrix signal would dominate the spectrum. Second, probe molecules or isotopic labels can produce distinctive signals that allow study of specific features of the system. Xenon gas which is allowed to diffuse into a sample highlights its internal structure, which can be observed through the [129]Xe resonance [20]. Or, a polymer can be synthesized with ^{2}H at specific sites in order to isolate the behavior of a particular functional group [21].

The Zeeman interaction is highly specific, but it is, unfortunately, rather weak. When a sample is placed within a magnetic field, the magnetizations of the individual nuclei, m_i, align either with (low-energy) or against (high-energy), the direction of B_0. The vector sum of all m_i is m, the system's magnetization. At room temperature, there is only a slight excess of spins (on the order of 1 in 10^5 to 1 in 10^6) in the low-energy state. Since the relative populations depend on B_0^2, high magnetic fields ($\gg 1$ T) are generally preferred for NMR systems. While NMR is an extremely powerful tool, it is a relatively insensitive one, because of this small population difference, and spectral accumulation times may last many hours, especially for rare spins at natural abundance.

I.B. Dipolar and Quadrupolar Coupling

Each of the individual spins, m_i, in a sample has its own magnetic dipole, and the energy of dipolar interaction, at any particular moment, between spins A and B is

$$E_{\text{dipolar}} = \frac{\mu_A \cdot \mu_B}{r^3} - \frac{3(\mu_A \cdot \mathbf{r})(\mu_B \cdot \mathbf{r})}{r^5} \tag{2}$$

where μ_A and μ_B are the magnetic moments ($\mu = \gamma h I / 2\pi$; h is Planck's constant), $\mathbf{r}$ is the position vector between A and B, and r is the magnitude of $\mathbf{r}$. E_{dipolar} depends on the identity and relative positions of A and B, and it falls off rapidly with distance. In other words, the *dipolar coupling* experienced by a spin is strongly influenced by the nature and relative motion of its neighbors, and it is fairly short-range. Dipolar interactions can be *homonuclear*, between like spins such as ^{19}F and ^{19}F, or *heteronuclear*, between dissimilar spins such as [29]Si and [1]H. Most real samples are not assemblies of isolated A–B dipole pairs, and E_{dipolar}, when summed over all possible couplings (A–B, A–C, A–D, A–E, ...) can be quite large. In rigid solid materials, these interactions can dominate the spectrum, resulting in broad, apparently featureless spectra, often referred to as *wide-line* or *broad-line* spectra. Useful information can be extracted from such data, but since dipolar interactions are generally perceived as obscuring more interesting details, they are usually suppressed experimentally (Sec. II.E.2). In nonviscous fluids, the intrinsically isotropic and rapid molecular motion averages the dipolar coupling to zero (or nearly so) over the time scale of the NMR experiment. Because of this, dipolar interactions do not generally affect the spectra of small molecules in solution. They may contribute some line broadening to the spectra of viscous polymer solutions, in which molecular motion is slowed and somewhat restricted. On the other hand, some polymeric "solids," such as elastomers, have enough inherent molecular motion that the dipolar interactions are greatly attenuated, and a nearly liquid-like spectrum is obtained without any specialized experimental techniques.

Within nuclei with $I > \frac{1}{2}$, electrical charge is nonspherically distributed, and they possess a quadrupole, as well as a dipole, moment. Quadrupolar nuclei include ^2H, ^{17}O, ^{27}Al, ^{14}N, and ^{23}Na. The *quadrupole coupling constant*, χ, is given by

$$\chi = \frac{e^2 Q q_{zz}}{h} \tag{3}$$

in which e is the charge on the electron, Q is the quadrupole moment (Table 1), and q_{zz} is the electric field gradient at the nucleus. Quadrupolar interactions broaden resonances, more or less severely (depending on the magnitude of Q), in both solutions and solids. They can make the resonances of such nuclei difficult to observe, or can lead to peak overlap which obscures chemical information. However, χ can be used as a probe of polymer structure and dynamics, because q_{zz} is modulated by molecular motion [21].

I.C. Chemical Shift

The goal of most NMR spectroscopy is a chemical-shift spectrum, because this interaction is the source of NMR's remarkable utility as a molecular characterization tool. While a given nuclide is said to precess at the Larmor frequency, the exact precession frequency for a particular nucleus depends on its chemical environment. *Chemical-shift* frequencies scale with $\mathbf{B_0}$, so field-independent ppm units are generally preferred in order to facilitate comparison among spectra recorded at different fields. Because chemical shifts (commonly designated by δ) are usually of only a few hertz, relative to the megahertz-scale Larmor frequency, they are usually expressed as *parts per million,* or *ppm*. In a 4.7-T magnet, ^{13}C will resonate at 50 MHz. A peak separation of 50 Hz corresponds to a chemical shift of 1 ppm (50 Hz/50 MHz). In a 9.4-T field ($^{13}C \approx 100$ MHz), these same peaks will be 100 Hz apart. While the difference, in frequency units, has doubled, the chemical shift is still 1 ppm (100 Hz/100 MHz). The chemical-shift δ scale is referenced relative to the frequency of a standard material. For example, in ^{13}C, ^1H, and ^{29}Si NMR, the reference is tetramethylsilane (TMS), whose resonance is defined to be 0 ppm. In a 9.4-T magnet, the chemical shift of a ^{13}C peak resonating at a frequency 6500 Hz higher than TMS will be $\delta = 65$ ppm. In older terminology, shifts to higher δ are said to be *downfield*; those to higher δ, *upfield*, [22]. Since, in modern experiments, the field is held constant, the preferred terms are *deshielded* and *shielded*, respectively, as this usage more clearly describes the effects of electrons. The range over which observed chemical shifts appear varies from nuclide to nuclide. Commonly encountered ^{1}H resonances, for example, cover just 12 ppm, while ^{31}P resonances span 1200 ppm. Chemical-shift ranges are listed in Table 1.

Chemical shifts originate in the interactions among a nucleus, its surrounding electrons, and $\mathbf{B_0}$. In organic materials such as polymers, the major contributor to δ is the electronic shielding around the nucleus (called the *diamagnetic* contribution). For example, carbons bonded to oxygen resonate at higher values of δ because the electron-withdrawing character of oxygen decreases the electron density around the ^{13}C; the carbon in question is said to be *deshielded*. Other contributions to the chemical shift may include: (1) *paramagnetic* effects, which originate in asymmetry caused by the electric fields of neighboring nuclei and which dominate for heavy elements; (2) *magnetic anisotropy* and *ring currents*, which arise from anisotropic molecular features such as multiple bonds or aromatic groups; (3) *unpaired-electron shifts*, which are caused by the presence of a paramagnetic center; and (4) *solvent effects*. The relative strengths of these contributions vary with from nuclide to nuclide, but it is possible, for a particular isotope, to generalize about chemical-shift trends (Secs. II.G.1–II.G.7). Chemical-shift theory is well developed, but a more detailed discussion is beyond the scope of this overview. Simple calculational schemes can estimate shifts for some systems; several ^{13}C models will be discussed in Sec. II.G.1.

Because chemical shift depends so strongly on the spatial arrangement of nearby electrons, it has a directional character i.e., it is a *tensor* quantity. The observed frequency ω_i for a particular nucleus depends on the orientation of its tensor axes relative to $\mathbf{B_0}$:

$$\omega_{ref} - \omega_i \propto \sigma_{xx} \cos^2 \alpha \sin^2 \beta + \sigma_{yy} \sin^2 \alpha \sin^2 \beta + \sigma_{zz} \cos^2 \beta \tag{4}$$

where σ_{xx}, σ_{yy}, and σ_{zz} are the principal values of the chemical-shielding tensor, and the angles α and β specify the position of the tensor axes relative to $\mathbf{B_0}$. In solutions, this *chemical-shielding anisotropy* (CSA) is averaged out by rapid, omnidirectional molecular motion, as are the dipolar interactions (Sec. I.B), and the peak appears at the *isotropic shift* value:

$$\delta_{iso} = \frac{\omega_{iso} - \omega_{ref}}{\omega_{ref}} = \sigma_{ref} - \sigma_{iso} \tag{5}$$

where $\sigma_{iso} = (\sigma_{xx} + \sigma_{yy} + \sigma_{zz})/3$. The CSA is, however, an important feature of solid-state NMR spectra, because of the comparative molecular immobility of the nuclei in the sample. The CSA is reflected in a nonsymmetric lineshape (Sec. I.H.2), which depends on the symmetry of the tensor and on the dynamics and geometry of molecular structure. For example, the lineshapes of materials which are structurally anisotropic (such as oriented liquid crystals or deformed polymers) are convolutions of the CSA and the orientation distribution function of the material [23]. These contributions can be separated to derive orientation information from the observed spectrum.

I.D. Scalar Coupling

In addition to being affected by the electronic environment induced by neighboring atoms, a nucleus is also influenced by the spin states of some of those atoms, through an interaction called *scalar*, or *J*, *coupling*. For example, a ^{13}C nucleus in a methylene group is, most likely, bonded to two ^1H nuclei (a *very* small number will be adjacent to naturally occurring ^2H). Relative to $\mathbf{B_0}$, the possible states of the two ^1H spins are, in order of increasing energy,

$$\uparrow\uparrow < \uparrow\downarrow = \downarrow\uparrow < \downarrow\downarrow \tag{6}$$

Because the energy differences among these states are extremely small, they are all essentially equally populated. As a result of this interaction, a methylene ^{13}C resonance appears as three lines (with the center one twice the intensity of the outer two), corresponding to (1) $\uparrow\uparrow$, (2) $\uparrow\downarrow + \downarrow\uparrow$ (hence the doubled intensity), and (3) $\downarrow\downarrow$. $^{13}C-^1H$ coupling patterns are fairly simple, with CH_n producing $n + 1$ peaks (Fig. 2). The distance, in frequency units, between the peaks in a *doublet, triplet, quartet,* or higher *multiplet* is called the *J-coupling constant*, J_{XY}, where X and Y indicate the coupled spins. Unlike chemical shift, J_{XY} does not scale with magnetic field. The example above describes *first-order coupling*, which involves the nearest coupled spins. Higher-order patterns are sometimes observed, although this is seldom the case for polymer solutions.

Heteronuclear scalar coupling is observed between dissimilar spin pairs: $^{13}C-^1H$, $^{13}C-^{19}F$, $^{29}Si-^1H$, and $^{31}P-^1H$, to name a few. Homonuclear *J* coupling between abundant spins, such as $^1H-^1H$, $^{19}F-^{19}F$, or $^{19}F-^1H$ can produce more intricate patterns, which are often obscured by residual dipolar broadening in spectra of polymer solutions (Sec. II.B). In solid-state NMR spectra, scalar couplings are not detected

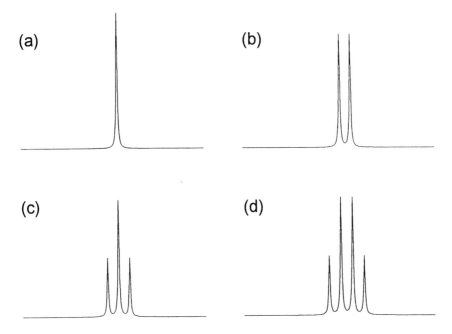

Figure 2 $^{13}C-^1H$ J-coupling patterns in the ^{13}C spectrum for: (a) $>C<$; (b) $>CH-$; (c) $-CH_2-$; (d) $-CH_3$.

because much stronger interactions dominate. In principle, rare-spin het-ero- or homonuclear scalar interactions arise for directly bonded spins (such as $^{13}C-^{15}N$ or $^{13}C-^{13}C$), but because of their intrinsically low natural abundance, the probability of this juxtaposition is extremely small (0.005% and 0.012%, respectively). Such interactions are, therefore, not readily observed. J-coupling patterns that involve a quadrupolar nucleus (such as $^{13}C-^2H$) are more complex, because a quadrupolar spin has more than two possible energy states.

   In practice, solution spectra (other than ^1H) are usually run in *J-decoupled mode*, as will be discussed in Sec. III.D.3. This removes the sca-lar-coupling multiplicity, so that each resonance appears as a single peak. While this results in a less complicated spectrum, valuable clues to the molecular structure may be lost. In an attempt to regain this information, experiments have been devised which allow the J-couplings to be completely or partially restored without undue complication of the spectrum (Secs II.E.3 and II.E.4).

II. EQUIPMENT, DATA ACQUISITION, AND SPECTRAL INTERPRETATION

Producing an interpretable spectrum utilizing the physical phenomena discussed in Sec. I is the ultimate goal of NMR spectroscopy. Instrumentation and experimental procedures have been developed that allow manipulation of nuclear interactions to provide useful information.

II.A. Basic Instrumentation

The basic components of a research-grade, high-resolution (i.e., solution-state) NMR spectrometer are shown schematically in Fig. 3. They include magnet, probe, gating circuitry, radiofrequency (rf) synthesizer or oscillator, rf transmitter, amplifiers, lock circuit, decoupler, rf receiver, phase-sensitive detector, analog-to-digital converter (ADC), and data system [24]. Some instruments have accessories such as additional rf channels, gradient circuits, high-power amplifiers, temperature controllers, and/or automatic sample changers.

Most modern NMR spectrometers employ large superconducting magnets, or *supercons*, as they are called. Very stable high fields, up to ~21 T, or 900 MHz for the ^{1}H resonance, are commercially available at this time. A superconducting magnet can be pictured as a set of concentric

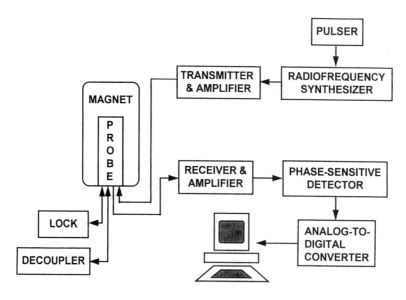

Figure 3   Schematic of basic NMR spectrometer.

cans, with a vertical hole, called the *bore*, through the center. The inner-most can holds a large coil of Nb–Ti or Nb–Sn wire immersed in liquid helium (LHe) at 4.2 K. At this temperature, the niobium alloy is super-conducting, so that no continuous power source is needed to maintain the field. The outer can contains liquid nitrogen (LN_2) at 77 K, which miti-gates the liquid helium boiloff. Evacuated space and highly effective insu-lating materials around the cans also help to preserve the LHe. Both cryogens need to be refilled regularly; the schedule depends on the design and condition of the magnet. Typically, liquid helium fills are performed one to 12 times per year; liquid nitrogen fills, one to six times per month. In many parts of the United States, commercial cryogen fill services are available at reasonable cost. Between fills, cryogen levels (particularly LHe) must to be monitored to ensure that the system is stable. Because these superconducting magnets are so strong, *fringe fields* can extend sev-eral feet from the center of the magnet bore. The layout of the room in which the instrument is placed must take this into account, as fringe fields can erase magnetic storage media and distort computer monitor displays that are positioned too close. One instrument manufacturer has recently introduced *ultrashielded* magnets which reduce the fringe fields substan-tially.

The cylindrical *probe*, which slides into the magnet bore, houses the electronics needed to transmit and receive pulses of one or more frequencies. The main part of the probe is a tuned circuit which includes, among other components, a resistor, a capacitor, and an inductor [24]. The inductor is called the *coil* (not to be confused with the magnet coil), even though its shape is usually not a spiral. Probes may be either *fixed-frequency* or *tunable*. Fixed-frequency probes are intended for observation of only one nucleus, usually ^{13}C, ^{1}H, or ^{19}F. They may also allow for ^{1}H (or possibly ^{19}F) decoupling, and it may also be possible to observe this second nuclide using the decoupler coil. The resonant frequency of a tunable probe can be adjusted over a particular range, so that the same unit can be used to observe many different nuclides. Probes also have air bearings that cause the sample to spin at several hertz (for solutions) or kilohertz (for solids) during the experiment; the air system may also be used for temperature control. A solution NMR probe accepts a specific-diameter glass tube, usually either 5 or 10 mm. The 10-mm size is generally more useful for polymer studies, because samples are often in the form of pellets or a finished product, and because the resulting solutions are viscous. It is difficult to cut pieces small enough to fit in a 5-mm tube, and air bubbles that may develop in the solution tend to become trapped in the smaller size. Samples for solid-state NMR are packed in plastic or ceramic rotors; the rotor design must be appropriate for the specific probe used.

As will be described more fully in Sec. II.D, an NMR signal is acquired after imposition of one or more very short (typically, several microsecond) radiofrequency (rf) pulses, produced by an array of rf synthesizers and/or oscillators, amplifiers, and control circuits (called *gates*). The exact sequence of pulses employed during the experiment can be quite complex, and the pulse programmer, or pulser, controls their timing. Common experiments are usually preprogrammed into the pulser by the manufacturer, but custom, user-defined sequences can also be developed. Most NMR systems have at least three distinct transmitter channels: (1) a "low-frequency" system for observing nuclei such as ^{13}C, ^{29}Si, or ^{31}P; (2) a "high-frequency" system for ^1H (and possibly ^{19}F) decoupling and observation; and (3) a ^2H system for stabilizing the signal and optimizing the field homogeneity. The frequency of the receiver is often shifted by a few kilohertz from that of the transmitter; this *frequency offset* changes the center point of the spectrum. The received signals are digitized at kilohertz rates; this sampling rate governs the width of the observed spectrum.

Data processing for modern NMR spectroscopy requires significant computing power, and most systems employ Unix-based workstations. The computer system controls the experiment, performs data analysis, and provides file storage. Each NMR vendor provides its own software, and third-party programs are also available.

II.B. Special-Purpose Instrumentation

II.B.1. Solids

While any commercial spectrometer is capable of acquiring spectra of solution samples, solid-state NMR experiments necessitate the use of special equipment [8,9,12,14,25–27]. Such instrumentation is readily available, but is not part of a "basic" spectrometer configuration. To generate a high-resolution solid-state spectrum, i.e., one with resolved chemical shifts, high-power amplifiers and a probe capable of spinning the sample very rapidly at a specific angle are needed, in order to overcome the strong dipolar couplings and large chemical-shift anisotropies found in solid materials. However, not all solid-state NMR experiments aim for chemical-shift resolution. Spectra of abundant-spin or quadrupolar nuclei can be extremely broad (many kilohertz) because of the dominant dipolar or quadrupolar coupling. The width of a spectrum is inversely proportional to the duration of the original signal (Sec. II.H.1), so these signals decay rapidly. Observing spectra of such nuclei requires a very fast (megahertz) digitizer. Abundant-spin experiments which remove the dipolar interaction but leave the CSA [28] employ very short, intense pulses; if such studies are to be done, the system must be capable of generating, transmitting, and receiving such pulses.

II.B.2. Diffusion Measurements and Imaging

Spatially resolved NMR can be used to follow diffusional processes, or to generate images; this capability is also an upgrade to a basic spectrometer system. In either case, a system that includes controllable field gradients superimposes a frame of reference on the sample; the positions of spins can then be pinpointed relative to this coordinate system (Sec. II.E.5) [29–33].

II.B.3. Flow Techniques: On-Line, Rheo-, and LC-NMR

NMR is also becoming a viable tool for on-line analysis [34] (this application is sometimes called *industrial magnetic resonance*, IMR) and a useful detector for liquid chromatography [35]. In both cases, material flows through the magnet continuously, rather than being analyzed as a succession of discrete samples. In on-line applications, the analyses must be completed rapidly enough to give useful time resolution. On-line spectrometers are usually sited in manufacturing facilities, not in laboratory settings. Most are built around small, low-field permanent magnets in which the ^{1}H resonance frequency is 20–60 MHz. In most cases, ^{1}H is detected because of this nuclide's high sensitivity. Liquid chromatography-NMR (LC-NMR) systems employ NMR as the detector. Separation is effected on a standard LC column, and the eluent is pumped through a specially constructed probe, in which the spectrum is recorded. In both on-line and LC-NMR, the sampling system's reliability is often the greatest instrumental concern. Rheo-NMR probes house a mechanical arrangement for keeping a polymer sample under stress while the spectrum is recorded [36]; such probes are custom-built, and are not yet commercially available.

II.C. Sample Preparation

Polymeric materials come as liquids, pellets, powders, films, molded parts, melt-index strands, or other form. For solution NMR, the sample must be cut or ground into pieces small enough to fit into a 5- or 10-mm-O.D. (outer diameter) glass tube. Samples for solid-state NMR need to fit tightly into small cylindrical rotors (outer diameter of a few millimeters). Powders and films can often be packed as is, but larger pieces must first be ground or cut into plugs that fit into the rotor.

In solution NMR, the quality of the final spectrum depends strongly on the homogeneity of the sample. Because polymer solutions tend to be viscous, care must be taken to minimize concentration gradients and to eliminate air bubbles. NMR spectra are usually run in deuterated solvents to provide a ^{2}H signal for the lock system. These solvents, which are available from the larger chemical supply houses and from specialty isotope vendors, can be quite expensive, with prices ranging from \$0.50 to \$50.00 per gram for commonly used materials. In many cases, it is possible to dilute the deuterated solvent with its protonated analog, or with another miscible solvent, to help minimize costs. Experiments can be run unlocked, but the quality of the spectrum will likely be somewhat degraded.

Obviously, a good solvent for the sample is needed [37], but the selection of an appropriate solvent for an NMR analysis is governed by other factors as well. First, the deuterated analog must be available; this is not necessarily the case for uncommon solvents. Second, the solvent's spectrum should not exhibit resonances that overlap significantly with those of the sample. For example, neither benzene-d_6 nor toluene-d_8 would be a good choice for the ^{13}C spectrum of polystyrene. Although they dissolve the polymer effectively, their aromatic resonances would obscure that region of the polymer's spectrum. Table 2 suggests some appropriate solvents for common types of plastics. Some polymers, such as the acrylates, dissolve readily in a wide variety of solvents. Others, such as poly(ethylene terephthalate), are only sparingly soluble, even at high temperature. Still other materials, such as polyethylene and polypropylene, can be analyzed in the melt with a miscible deuterated solvent added for lock. Heating a sample for a few minutes to several hours before the experiment can promote a homogeneous mixture of sample and solvent, and help to drive air bubbles out of viscous solutions. Some polymer additives can interfere with the quality of the NMR signal. The presence of insoluble fillers, such as glass or talc, may broaden the peaks because the sample is microscopically inhomogeneous. Most orange, tan, or brown pigments, which contain iron oxide, make it virtually impossible to obtain high-quality spectra because of the colorant's magnetic character. In general, a high concentration of any paramagnetic or ferromagnetic species will degrade the quality of the spectrum significantly.

Additional substances are sometimes added to the sample solution in order to facilitate measurements or to provide an internal reference for chemical shift or quantitation. Certain experiments benefit from the use of a small amount of a paramagnetic relaxation agent such as chromium(III) acetylacetonate (Secs. II.D.4 and II.G). A chemical-shift reference standard may be added if exact shifts are required. Some quantitative analyses employ an internal *spin-counting* intensity reference; such a standard must not interfere with the sample spectrum. If the reference material is inert, it can be added directly to the sample. If it might react with the sample, the

Table 2 Recommended NMR Solvents for Various Polymer Types

Polymer type	Suggested deuterated solvent[a]
Polyolefins	1,2-Dichlorobenzene-d_4[b]; 1,1,2,2,-tetrachloroethane-d_2[b]; decalin-d_{18}
Polystyrenes	Chloroform-d; dichloromethane-d_2; tetrahydrofuran-d_8
Polyacrylates	Benzene-d_6; toluene-d_8; chloroform-d; dichloromethane-d_2
Polymethacrylates	Benzene-d_6; toluene-d_8; chloroform-d; dichloromethane-d_2
Polyacrylonitriles	Chloroform-d; dichloromethane-d_2; acetone-d_6; n,n-dimethylformamide-d_7; dimethylsulfoxide-d_6
Polyacrylamides	Water-d_2; methanol-d_4; acetone-d_6
Poly(vinyl halides)	Toluene-d_8; chloroform-d; dichloromethane-d_2; n,n-dimethylformamide-d_7
Poly(vinylidene halides)	Tetrahydrofuran-d_8; acetone-d_6
Poly(vinyl ethers)	Benzene-d_6; toluene-d_8; chloroform-d; dichloromethane-d_2
Poly(vinyl ketones)	Chloroform-d; dichloromethane-d_2; acetone-d_6
Poly(vinyl esters)	Benzene-d_6; toluene-d_8; chloroform-d; dichloromethane-d_2; acetone-d_6
Poly(vinyl pyridines)	Chloroform-d; Dichloromethane-d_2; acetone-d_6
Polydienes	Chloroform-d; dichloromethane-d_2
Polyethers	Benzene-d_6; toluene-d_8; chloroform-d; dichloromethane-d_2; methanol-d_4; acetone-d_6
Polyesters, aliphatic	Benzene-d_6; toluene-d_8; chloroform-d; dichloromethane-d_2
Polyesters, aromatic	Dimethylsulfoxide-d_6[b]
Polyamides	N,N-Dimethylformamide-d_7[b]; dimethylsulfoxide-d_6[b]
Polysiloxanes	Chloroform-d; dichloromethane-d_2
Polyphosphazenes	Chloroform-d; dichloromethane-d_2
Polysulfones	Chloroform-d; dichloromethane-d_2

[a] Suggestions intended as guidelines only; solvent may not be appropriate for all members of a class.
[b] At elevated temperature.
(*Source:* Ref. 37.)

standard should be placed in a special capillary insert that slides into the NMR tube.

II.D. Basic Experiments

Virtually all modern spectrometers operate in *pulse*, or *Fourier transform* (FT), mode, in which the experiment is performed by discrete pulses of radiofrequency. The resulting signal is analyzed using the Fourier transform, which sorts out the component frequencies. This simple description, however, belies the complexity of many modern NMR experiments, which employ carefully crafted sequences of pulses, as will be discussed in Sec. II.E.

II.D.1. Overview of Pulse NMR

Any NMR experiment begins when a sample is placed within the magnetic field by manual loading or by an automatic sample changer. The sample is usually set spinning at an appropriate rate, and its temperature may be adjusted. If the nuclide to be observed is different from the previous experiment, the probe is tuned to the new resonant frequency. Some systems require slight retuning if the sample temperature has been changed, even if the observation frequency has not. During solution experiments, the solvent's ^2H signal is monitored by the lock system. The system can then compensate for any slight magnetic field drift during the experiment, although, with modern superconducting magnets, this is not a serious problem over the course of a typical experiment. More important, the ^2H lock signal is also used for *shimming*, or optimization of the magnetic field homogeneity. Small corrections to the static field are made by varying the electrical current in a set of *shim coils* located within the magnet housing. The shim quality, or the degree to which the homogeneity is optimized, has a significant impact on the width and symmetry of the peaks in the spectrum. Polymer solutions can be difficult to shim because of their relatively high viscosity, so it is important to ensure, through careful sample preparation, that the sample has no strong concentration gradients, and that it is bubble-free. NMR systems have automated shimming routines, but the procedure can also be performed manually.

In theory, a rectangular rf pulse of duration t contains substantial frequency components from 0 to $1/t$ Hz. Thus, an ideal rectangular 20-μs pulse produces frequencies over a 50-kHz range, with the high- and low-frequency components somewhat attenuated. After the sample's spins are excited by an rf pulse, they produce a signal called a *free-induction decay* (FID), which diminishes with time. While the FID may be simply a decreasing exponential, a more typical signal is composed of more than one oscillating component (Fig. 4a), the rate of decay being characterized by a time constant, T_2^*. The Fourier transform, when applied to the *time-domain* FID, identifies the component frequencies and weights them appropriately. The resulting *frequency-domain* spectrum is displayed on chemical shift-intensity (or, less commonly, frequency-intensity) axes (Fig. 4b). Experimental parameters must be chosen so that the FID is not truncated along either the y (intensity) or the x (time) direction, or there will be baseline distortions in the final spectrum (Fig. 5). Truncation in the y direction is avoided by assuring that the receiver gain is not set too high; modern instruments can do this automatically. If x truncation is the problem, the data-collection time axis should be extended by increasing the number of data points or by decreasing the sampling rate, although the latter approach also changes the chemical-shift range of the transformed spectrum. If the spectral width, sometimes called the *sweep width* (an anachronistic term from the early days of NMR [21]), is too narrow, resonances which are beyond the observed range will be *aliased*. Such peaks appear at incorrect chemical shifts, and are usually distorted in shape (Fig. 6).

In the idealized Fig. 4a, data sampling starts right after the rf pulse, when the FID is at its maximum. In reality, residual coil noise, called *ringdown*, makes it impossible to acquire meaningful data immediately after an rf pulse. Consequently, a short (few microsecond) delay, called the *dead time*, during which the ringdown abates, is inserted into the pulse sequence. This interval is not normally depicted in pulse-sequence diagrams, but it is understood to be present. As a result of the delayed onset of signal acquisition, a phase shift is introduced; this effect can be easily corrected during data processing and will be discussed further in Sec. II.F.

While a satisfactory spectrum can sometimes be acquired with just one pulse cycle, it is common practice to repeat the sequence many times because

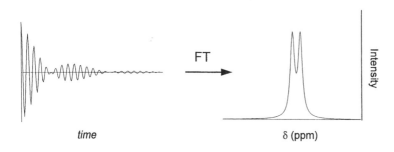

Figure 4 Fourier transformation (FT) of a time-domain free-induction decay (FID) into a frequency-domain spectrum.

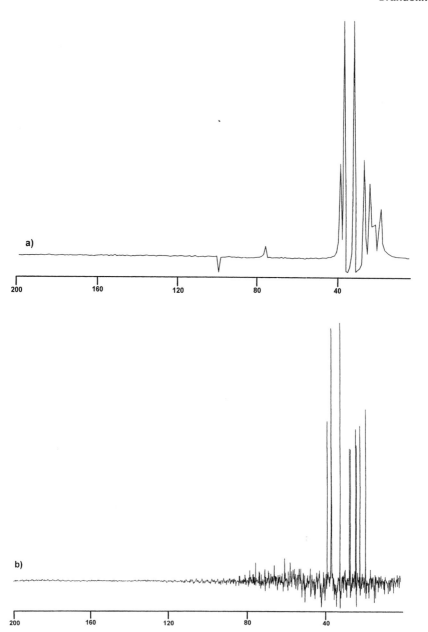

Figure 5 Effect of free-induction decay (FID) truncation in the (a) time or (b) intensity dimension. Truncation effects have been exaggerated to make the distortions more obvious. The sample is squalane in CDCl₃.

the *signal-to-noise ratio* (S/N) improves with the number of scans, N. This ratio increases as $N^{1/2}$, so the number of transients must be quadrupled in order to double S/N. It is not unusual to record several thousand transients for a spectrum, particularly if the sample's concentration is low, if a rare spin (such as ^{13}C or ^{15}N) is being observed, or if the peaks of interest are small. The length of time between pulse cycles, called the *relaxation delay*, can vary from less than 1 s to several minutes.

II.D.2. Single-Pulse Experiment (without Decoupling)

The most straightforward NMR experiment, represented schematically in Fig. 7, consists of a single pulse; this sequence is used to acquire a simple ^1H spectrum. Once the sample has been placed in the magnetic field, the net magnetization, **m**, aligns along $\mathbf{B_0}$ (by convention, the $+z$ axis), as shown in Fig. 8. A short (few microsecond) burst of radiofrequency is sent into the probe circuitry. Transmitting the pulse along the $+x$ axis causes **m** to tip toward the $+y$ axis, where it is detected. The tip angle, α, through which **m** is tilted is proportional to the length of the pulse. If the length of a 90° pulse (Fig. 8a) is 30 μs, then the magnetization can be tipped through 30° by applying a 10-μs pulse (Fig. 8b), or through 180° with a 60-μs pulse (Fig. 8c), assuming that the power level is kept constant. In each case, the intensity of the detected signal is proportional to $\sin \alpha$, the projection of the resulting vector onto the x–y plane. The maximum signal, I_{max}, is observed after a 90° pulse. For a 30° pulse, the signal is reduced to $0.5I_{max}$; no signal is seen after a 180° pulse. The *phase* of the pulse, the axis along which it is applied, can also be varied. The direction in which **m** rotates can be predicted by a right-hand rule (when the thumb is held parallel to the pulse axis, the fingers curl along the path of **m**). While Fig. 8 portrays detection only along the y axis, modern instruments employ *phase-sensitive* or *quadrature detection*, which senses the signal along both the x and y axes.

After a pulse, **m** eventually returns to its original alignment with $\mathbf{B_0}$. This process occurs exponentially, with a characteristic time constant, T_1, the *spin-lattice relaxation time*. As discussed in Sec. I.A, the magnetization vector **m** is the sum of the individual $\mathbf{m_i}$ from all spins in the sample. These individual $\mathbf{m_i}$ are exposed to slightly different local magnetic fields, due to their motion relative to nearby spins, to variations within the sample, or to inhomogeneities in $\mathbf{B_0}$. As a consequence of these differing effective fields, some $\mathbf{m_i}$ precess faster than the rotating frame during the FID; others, slower. In other words, the $\mathbf{m_i}$ lose *phase coherence,* and they spread out in the x–y plane. This dephasing occurs with an exponential time constant, T_2, the *spin–spin relaxation time*. Both T_1 and T_2 processes may be reflected in the final spectrum, and they may provide useful information about mole-

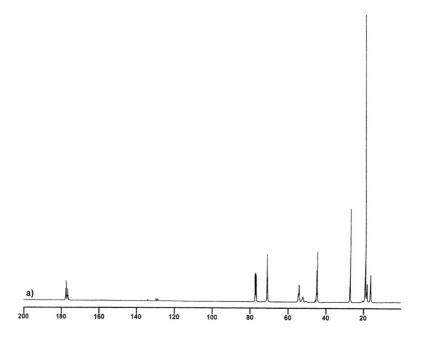

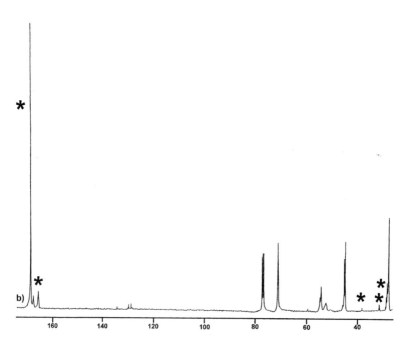

90°

Figure 7 Pulse sequence and response for a single-pulse experiment (without decoupling).

cular motion; these subjects will be discussed in detail in Secs II.D.4 and II.H.1, respectively.

II.D.3. Single-Pulse Experiment (with Decoupling)

A slightly more complicated experiment, comprised of a single pulse with decoupling, is diagrammed in Fig. 9a. This sequence, or a minor variant thereof, is commonly used for one-dimensional spectra of nuclei other than ^1H. The X pulse sequence (where X is the observed nucleus, such as ^{13}C, ^{31}P, or ^{29}Si) consists of a pulse which tips the magnetization through an angle α, just as described in Sec. II.D.2. If the Y spins (usually ^1H) are to be decoupled from the X, a stream of Y-radiofrequency is simultaneously applied (this stream may be continuous or a series of pulses). The decoupling field saturates the Y-spin population, equalizing the populations in the low- and high-energy states. This suppresses the scalar coupling of X and Y, reducing the J multiplets to single lines.

A side effect of leaving the decoupling rf on during the relaxation delay is the *nuclear Overhauser effect* (NOE), which results in signal enhancements up to threefold for a $^{13}C-^1H$ pair (the maximum NOE for other nuclide pairs is related to the ratio of the magnetogyric ratios, γ) [16]. As explained in Sec. I.A, nuclear spins in a magnetic field exhibit an equilibrium population of high- and low-energy states, with a slight

Figure 6 Aliasing due to an improper sweep-width setting: (a) normal spectrum recorded with a sweep width encompassing all resonances; (b) spectrum with aliased peaks (asterisk) taken with a sweep width narrower than the sample's chemical-shift range. Note the change in position of resonances at either end of the spectrum and the uncorrectable baseline distortions observed for peaks in the middle. The sample is poly(isobutyl methacrylate) in $CDCl_3$.

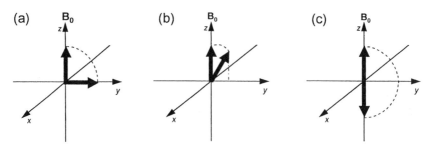

Figure 8 Behavior of nuclear magnetization during pulses of varying length: (a) 90°; (b) 30°; (c) 180°.

excess aligned with $\mathbf{B_0}$ (low energy); these give rise to the NMR signal. When Y-rf is applied for a sufficiently long time, the Y-spin population is *saturated*, i.e., the two states become equalized. Magnetization (or polarization) is then transferred between the Y spins and the X, with the X-spin distribution usually becoming skewed toward lower energy. In such cases, when an X-pulse is applied, more of these spins are able to contribute to the observed signal and the NOE significantly improves S/N. However, it can lead to distorted relative peak intensities, because the effective enhancement is not necessarily equivalent for all X resonances. For example, ^{13}C with no directly bonded protons (such as quaternary carbons or carbonyls) experiences less enhancement than protonated carbons (such as methyls or methylenes). This problem can be circumvented by *gating* the decoupling; this pulse scheme is depicted in Fig. 9b. In this experiment, the decoupling field is turned on for only a brief period during the experiment cycle. This time is not sufficiently long for a significant NOE to develop, but it is long enough to suppress the *J* couplings. Spectra taken with decoupler gating exhibit *J*-decoupled peaks with correct relative areas, provided that the other conditions for quantitation, to be discussed in Sec. II.D.4, are met. Some nuclides, which have negative γ's (Table 1), such as ^{29}Si and ^{15}N, can actually produce negative NOEs, leading to inverted, or even null, signals; spectra of these isotopes should be recorded with decoupler gating. The NOE can be calculated as the ratio of the integrated peak area recorded under continuous decoupling (with NOE) to that under gated decoupling (without NOE). Measurement of the NOE can provide useful information on the relative distance between X and Y spins, because the enhancement is mediated by the heteronuclear dipolar interaction which, as seen in Sec. II.B., depends on geometry. Hetero-

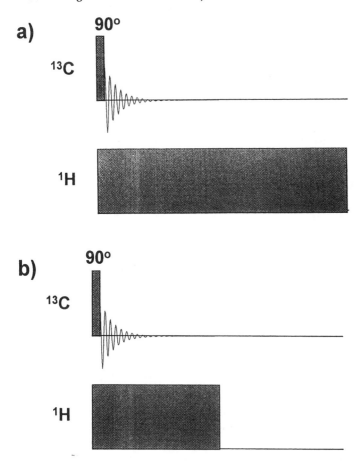

Figure 9 Pulse sequence and response for a single-pulse experiment: (a) with decoupling; (b) with decoupler gating.

nuclear distances (such as C−H) are usually governed by the bond length, but homonuclear NOEs can arise from through-space interactions.

II.D.4. Quantitation

While qualitative NMR spectra can be readily obtained, acquisition of spectra suitable for quantitation requires that careful attention be paid to several experimental parameters. First, the entire spin system must be allowed to return to equilibrium between scans, i.e., **m** must revert to its original position, aligned with **B₀**. The rate at which transients can be acquired is, therefore, governed by the longest T_1 in the system. In fact, because T_1 is an exponential decay constant, a relaxation delay of $5 \times T_1$ will effect recovery of 99% of the original magnetization after a 90° pulse; an interval of $10 \times T_1$ is needed to regain 99.99%. Since T_1 depends on, among other variables, the identity of the sample, the solution concentration, and the temperature, this constant should be measured (Sec. II.E.1) as part of any method-development procedure. T_1 can vary from a few milliseconds to many minutes; it is longest for rigid solids and nonviscous liquids. The requirement of waiting at least $5 \times T_1$ between scans can, obviously, lead to prohibitively time-consuming experiments for long-T_1 samples. One strategy for circumventing this problem is to use a tip angle less than 90°:

$$\cos(\alpha') = \exp\left(\frac{-t}{T_1}\right) \tag{7}$$

where α' is the *Ernst angle*, or optimum tip angle for an experiment with a desired relaxation delay t on a system whose longest relaxation time is T_1. Fortunately, most polymers, whether in solution or as solids, have relatively short T_1's, and so fairly brief pulse delays can be used (typically a few seconds to 1 min). Since T_1 processes are facilitated by dipolar interactions, plentiful spins such as ^1H and ^{19}F tend to relax much more quickly than rare spins. Faster recycle times can, therefore, generally be used for abundant spins. For slowly relaxing systems, it is sometimes advisable to add a paramagnetic relaxation agent, such as chromium(III) acetylacetonate or $Cr(acac)_3$, to a solution sample. Such paramagnetic species dramatically curtail the system's T_1's, thereby shortening experiment run-times.

As mentioned in Sec. II.D.3, the NOE is also a consideration for quantitative analysis. The use of decoupler gating ensures that the NOE will be suppressed, but it is not always necessary, or even desirable, to eliminate the S/N enhancement that the NOE affords. If the molecules in the sample move relatively slowly (as do polymers in solution), and if all or most of the X nuclei are directly bonded to Y's (e.g., if most carbons are protonated), then most NOE's tend to be equivalent. This happens because of *spin diffusion*, a process by which magnetization migrates to neighboring spins. Correct relative peak areas are maintained for most peaks, with a significant decrease in the experiment time. Some analogous concerns apply to quantitation in solid-state NMR; these will be discussed in Sec. II.E.2.

II.E. Advanced Experiments

The experiments described in Secs. II.D.1 and II.D.2 will yield qualitative or quantitative one-dimensional spectra of solutions, which can provide a wealth of information. Such spectra exhibit isotropic chemical shifts, with or without scalar coupling. This simplicity has been achieved, however, by destroying any solid-state structure present in the sample, and by discarding the information inherent in the dipolar interactions, chemical-shift anisotropy, and (in appropriate cases) quadrupolar coupling. Also, any correlations between these discrete spectra must be inferred from knowledge of chemical shifts or J coupling. NMR spectroscopists have devised many sophisticated methods for recovering this lost information, some of which will be discussed briefly in this section.

II.E.1. Relaxation Measurements

T_1 is usually measured by the *inversion-recovery* experiment, which is illustrated in Fig. 10. This sequence, unlike the two covered previously (Figs 7 and 9; single pulse with and without decoupling), involves more than one X pulse. The behavior of **m** during this sequence is shown in Fig. 11. In the first step, a $180°$ pulse inverts the magnetization, as shown in Fig. 11a. At this point, a detector placed along the $+y$ axis would sense no signal because there is no intensity along that axis. After a very short time, τ_1, the magnetization has relaxed only slightly (Fig. 11b). Next, a $90°$ pulse along the $+x$ axis tips **m** onto the $-y$ axis, producing a signal with negative intensity (Fig. 11c). Once the system has relaxed back to equilibrium (**m** is once again aligned with $\mathbf{B_0}$), the experiment can be repeated, beginning with the $180°$ pulse (Fig. 11d). If the system is now allowed to evolve for a longer time, τ_2, between the two pulses (Fig. 11e), the $90°$ pulse generates a large positive signal (Fig. 11f). If τ is made very long, **m** will have reverted to its equilibrium position before the second pulse, and the $90°_{+x}$ pulse results in a positive peak with the maximum possible intensity (this is equivalent to a one-pulse experiment). In practice, the inversion-recovery experiment is performed many times, as a function of τ, and the exponentially varying signal intensities are analyzed as described in Sec. II.H.1 [38].

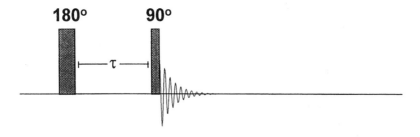

Figure 10 Pulse sequence and response for the inversion-recovery experiment used to measure the spin–lattice relaxation time, T_1.

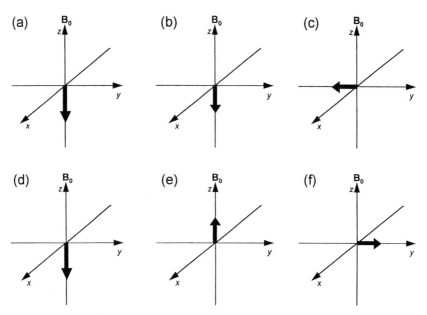

Figure 11   Behavior of nuclear magnetization during the inversion-recovery experiment: (a) inversion by 180° pulse; (b) evolution during delay time, τ; (c) 90° pulse and detection; (d) inversion by 180° pulse; (e) evolution during longer delay time, τ'; (f) 90° pulse and detection.

One experiment that can be used to measure T_2 is illustrated in Fig. 12; it is called the *Carr-Purcell-Meiboom-Gill* (CPMG) sequence. It consists of a $90°_{+x}$ pulse, a short time delay τ', and a train of $180°_{+y}$ pulses. As mentioned in Sec. II.D.2, the components of **m** dephase after the initial pulse (Fig. 13a) with a time constant T_2, because their precession frequencies are slightly different. Thus, some spins move "ahead" of others (shown by the white and black arrows in Fig. 13b). A $180°_{+y}$ pulse after τ' flips the entire set of spins, so that the faster spins end up "behind" the slower ones (Fig. 13c). After another τ' interval, the faster spins catch up to the slower, refocusing the signal and generating a *spin echo* at a time of $2\tau'$ (Fig. 13d). If another $180°_{+y}$ pulse is applied (again after τ'), a second, smaller echo will form at $4\tau'$. Yet another pulse generates a third, even less intense echo at $6\tau'$, and so on. If data are recorded at intervals of $2n\tau'$ (where $n = 1, 2, 3, \ldots$), the set of transformed spectra exhibits peak intensities which decrease with $n\tau'$; the rate of decay yields T_2. T_2 is related to T_2^* (Sec. II.D.1), the FID decay constant; in fact, $T_2 = T_2^*$ in an ideal magnetic field. Since perfect field homogeneity cannot be attained, $T_2^* < T_2$ because

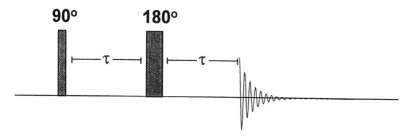

Figure 12 Pulse sequence and detected signal for the Carr-Purcell-Meiboom-Gill (CPMG) experiment used to measure the spin–spin relaxation time, T_2.

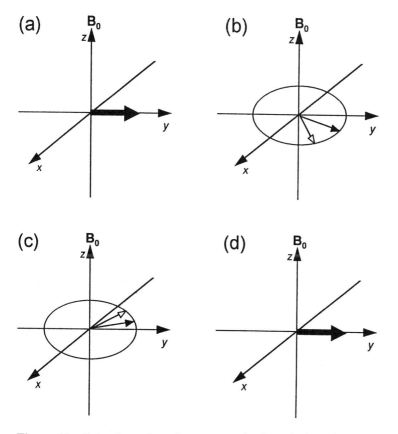

Figure 13 Behavior of nuclear magnetization during the CPMG experiment: (a) 90° pulse; (b) dephasing of m_i with "faster" spins shown in white; (c) 180° pulse; (d) refocusing of magnetization.

the inhomogeneities induce additional relaxation processes. The use of even-numbered echoes in the CPMG experiment obviates these inhomogeneity effects, and so yields a true T_2 [7].

II.E.2. Solid-State NMR

As mentioned in Sec. I.A, interactions other than isotropic chemical shifts and scalar couplings dominate the NMR of solids. If a "high-resolution" spectrum is desired, these other effects must be deliberately removed; molecular motion alone will not average them during the course of the measurement. Alternatively, experiments have been designed which isolate these interactions—dipolar coupling, quadrupolar coupling (in appropriate nuclei), and chemical-shift anisotropy—and exploit their useful features to provide information about the system under study.

In rare spin-$\frac{1}{2}$ (e.g., ^{13}C, ^{29}Si, or ^{15}N) NMR of solids, the heteronuclear dipolar interaction can be removed by high-power, abundant-spin decoupling ($\sim 10\times$ more intense than that used to remove scalar coupling in solutions). The use of a lower-power decoupling field makes only the more mobile parts of a system observable; this approach can, for example, be used to isolate the amorphous domains in a semicrystalline polymer. Rapid (several kilohertz) rotation of the sample about an angle 54.7° (the *magic angle*) to $\mathbf{B_0}$ removes the chemical-shift anisotropy (CSA). This experiment, in which a pulse is applied to the rare-spin system, under dipolar decoupling while the sample is being spun rapidly about the magic angle, is called the *Bloch decay*. When the sample is spun, the shift in Eq. (4) becomes dependent on both the spinning speed and the angle θ between the axis of rotation and $\mathbf{B_0}$. This expression contains a term multiplied by $(3\cos^2\theta - 1)$, which goes to zero when θ is the magic angle [8,9,12]. Under these circumstances, a peak appears at the isotropic shift, which is the center of gravity of the CSA. The sample must be spun at a rate faster than the width of the anisotropy (expressed in frequency units) in order to completely suppress the effects of this interaction. If the spinning speed is too slow, extraneous peaks called *spinning sidebands* appear (special experimental approaches can reduce or eliminate them [39]). If the relaxation delay (Sect. II.D.4) is chosen appropriately, Bloch-decay spectra are quantitative.

However, because of the relatively restricted motion in solids, rare-spin T_1's tend to be extremely long, which necessitates the use of long relaxation delays (sometimes more than several minutes); this situation can be ameliorated by *cross-polarization* (CP), in which fast-relaxing abundant spins govern the behavior of the system. Cross-polarization is similar to the NOE, involving an enhancement of magnetization by interacting spins. Because of this magnetization exchange, the system's spin-lattice relaxation becomes dominated by the faster-relaxing abundant spins. Cross-polarization can induce a fourfold enhancement in rare-spin signal intensity (compared to threefold for the NOE), but relative peaks intensities may become skewed, making quantitation unreliable. In particular, because of the geometric dependence of cross-polarization, rare spins physically close to abundant spins are affected more strongly than those far away. The length of time the system is *spin-locked* (i.e., the length of time during which magnetization can be exchanged) is called the *contact time*.

When optimal polarization-transfer efficiency is attained, the system is said to be in the *Hartmann-Hahn condition*. The exponential decrease of signal intensity at long contact times is governed by *rotating-frame spin-lattice relaxation time*, $T_{1\rho}$. The combination of dipolar decoupling, magic-angle spinning, and cross-polarization is referred to as CP/MAS NMR.

Because of their appearance, dipolar-coupled abundant spin-$\frac{1}{2}$ (usually ^{1}H or ^{19}F) spectra are described as "broad-line" or "wide-line." The FIDs of such signals decay very quickly, often within microseconds, rather than milliseconds, as is typical for liquids. Broad-line spectra can be obtained by a simple one-pulse sequence as discussed in Sec. II.D.2, although basic spectrometers are not capable of capturing these extremely brief FIDs. The homonuclear dipolar interaction can be removed by application of a "multiple-pulse" sequence such as WAHUHA-4 or MREV-8 [28]; this leaves the next largest interaction, the chemical shift and its associated anisotropy (CSA), as the dominant features of the spectrum. High-resolution (i.e., chemical-shift-resolved) spectra of abundant spins in solids can be recorded with the CRAMPS experiment (*c*ombined *r*otation *a*nd *m*ultiple-*p*ulse *s*pectroscopy) [40], which employs both a multiple-pulse sequence to remove the dipolar interaction, and magic-angle spinning to obviate the CSA. This experiment requires very short, intense rf pulses which a standard instrument cannot deliver; such capabilities are available on specially designed solid-state NMR instruments or as upgrades to some spectrometer systems.

NMR spectra of quadrupolar nuclei, such as ^{2}H, are, in general, even broader than abundant-spin wide-line spectra. It is generally not possible to detect their signals directly, as the signal decay is much too rapid (often shorter than the instrument deadtime) . To circumvent this difficulty, the *quadrupolar-echo* experiment is used (Fig. 14) [21]. This sequence generates an echo at a time t, which is longer than the deadtime. Like the wide-line and multipulse techniques, this experiment requires specialized solid-state NMR equipment. Recently introduced innovative spinning techniques can narrow quadrupolar lines [41], but they have been mostly applied to inorganic solids. While this new approach results in better chemical-shift resolution, it obscures the geometric and dynamic information that is often the goal of quadrupolar NMR studies of polymers.

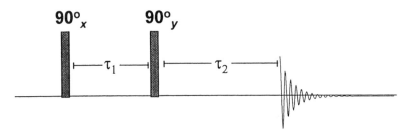

Figure 14 Pulse sequence and detected signal for the quadrupolar-echo experiment used to record solid-state ^{2}H spectra.

II.E.3. Spectral editing

The J-coupling patterns in spectra can help to elucidate chemical structure, but they also complicate the spectrum. *Spectral-editing experiments* allow such interactions to be observed in a controlled way. One such technique, DEPT (*d*istortionless *e*nhancement by *p*olarization *t*ransfer), edits the spectrum based on scalar coupling; the pulse sequence is depicted in Fig. 15. DEPT is actually a set of three experiments, with tip angles, ψ, of 45°, 90°, and 135°. In ^{13}C NMR, appropriate addition and subtraction of the resulting subspectra separate the contributions from methyl, methylene, and methine carbons; quaternaries are not observed.

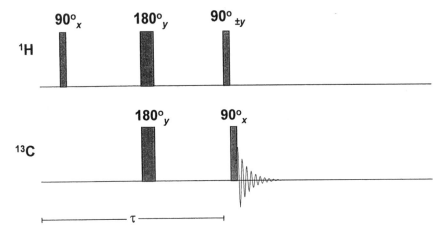

Figure 15 Pulse sequence and detected signal for the distortionless enhancement through polarization transfer (DEPT) experiment used to identify peak multiplicity.

In *selective decoupling,* when a particular resonance is decoupled, the multiplets of all peaks coupled to it collapse to singlets; uncoupled resonances are unaffected. In *off-resonance decoupling,* the coupling constants are decreased, thereby reducing overlap and simplifying identification of the multiplets. While still useful, these experiments have been largely supplanted by some of the multidimensional experiments to be covered in Sec. II.E.4. There is a similar experiment for solid-state ^{13}C NMR, known as *interrupted decoupling* or *dipolar dephasing,* which causes the selective attenuation of protonated and relatively immobile carbons [42].

II.E.4. Multidimensional NMR

Two- and higher-dimensional NMR experiments can identify coupling patterns, establish correlations between the spectra of different nuclei, and provide estimation of intra- and intermolecular distances [3,13,15,43–48]. They have become standard experiments in the NMR arsenal, and automated versions are incorporated into most standard software packages. While they are more often applied to biological macromolecules than to synthetic ones, they are still useful adjuncts to the other experiments discussed in this chapter [43,44,46] for the analysis of plastic materials.

The simple two-pulse sequence shown in Fig. 16 [11] can help in explaining the general concept of multidimensional NMR. The basic approach—multiple pulses with variable delay times—is similar to that used to measure relaxation times (Sec. II.E.1.). Even though the FID is only detected at the end of the sequence, the magnetization evolves throughout. If, as shown in Figure 16a, τ sufficiently is long that the first FID decays completely (i.e., **m** returns to equilibrium), then the sequence is equivalent to a simple one-pulse experiment in which every other FID is collected. In Fig. 16b, the time between pulses has been significantly decreased, and the second pulse operates on a perturbed, or nonequilibrium, magnetization; the resulting signal will be altered from that in Fig. 16a. Use of a different τ value obviously leads to yet another state at the start of the second pulse. The exact character of this state depends on the behavior of the magnetization between the two pulses. Even if τ is kept constant, the final FID will be affected if the component spins do not behave in the same way. For example, if one component oscillates at a lower frequency, as shown in Fig. 16c, the second pulse operates on a state different from that depicted in Fig. 16b, and a distinguishable signal results. Since FIDs of real samples are generally composed of many superimposed frequencies, the array of data collected as a function of τ characterizes the behavior of each peak's magnetization over the course of the experiment. All multidimensional experiments include: (1) the *preparation state* (analogous to the first pulse); (2) at least one *evolution period* (such as the time period τ); (3) at least one *mixing time* (such as the

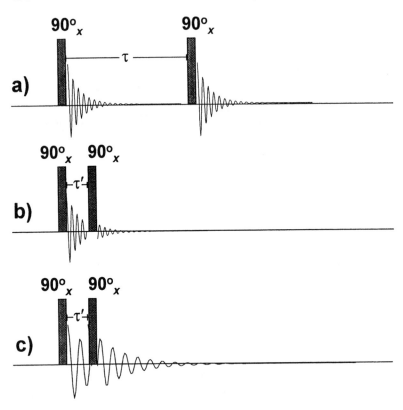

Figure 16 Model double-pulse sequences for a two-dimensional experiment: (a) long delay time τ′; (b) short delay time τ′; (c) short delay time τ′ with lower-frequency signal.

second pulse); and (4) the *detection period*. More sophisticated experiments vary the number, length, phase, and timing of the pulses; an astonishing number of multidimensional NMR experiments has been developed.

The most common two-dimensional NMR experiment is COSY (pronounced "cozy"), which is an abbreviation for *co*rrelated *s*pectroscop*y*. This experiment exploits internuclear coupling to establish relationships among peaks in the spectrum; an example will be discussed in Sec. II.H.4. In homonuclear COSY, the pulse sequence is essentially the same as that depicted in Fig. 16; the heteronuclear version utilizes a somewhat more complex scheme. The whimsically named INADEQUATE (*i*ncredible *n*atural-*a*bundance *d*ouble-*qua*ntum *t*ransfer *e*xperiment) reveals a molecule's carbon skeleton. Similarly, *h*eteronuclear *m*ultiple-*b*ond *c*orrelation (HMBC) establishes connectivities over two to three chemical bonds. *Double-quantum filtered correlated spectroscopy* (DQFCOSY) simplifies 2-D spectra by taking advantage of "forbidden" (i.e., allowed, but unobservable) transitions to eliminate solvent peaks and produce correlated spectra with both positive and negative peaks. A combination of *heteronuclear multiple-quantum correlation* and COSY (HMQC-COSY) provides simultaneous ^{1}H–^{1}H and ^{13}C–^{1}H correlations. The *homonuclear Hartmann-Hahn* (HOHAHA) experiment employs cross-polarization (as discussed in Sec. II.E.2.) to establish correlations through a spin-locking mechanism.

Other basic 2-D experiments include *J spectroscopy*, which displays scalar-coupled multiplets along the second dimension, and *nuclear Overhauser effect spectroscopy* (NOESY, pronounced "nosy"), which exploits through-space dipolar couplings to provide conformational information. In the NOESY experiment, protons are selectively irradiated as data are acquired, and the effect of neighboring nuclei is monitored by their mutual enhancement. NOESY is one example of an *exchange* experiment, in which magnetization is transferred from spin to spin.

There are also multidimensional solid-state NMR experiments, even though, because of the more stringent instrumental requirements, they are more difficult to perform than solution experiments [49–51]. Correlations can be established between the isotropic chemical shift and chemical-shift anisotropy, and between isotropic shift and dipolar coupling. Solid-state exchange NMR provides information about the geometry of molecular motion in the sample, and spin-diffusion measurements are useful for probing domain size [52,53].

II.E.5. Diffusion Measurements and Imaging

Diffusion and imaging experiments rely on the imposition of controlled field gradients to provide a coordinate system. Take, for example, the idealized one-dimensional system shown in Fig. 17, composed of two identical stationary spins. During a "normal" NMR experiment, the field is homogeneous across the sample, and the black and white spins precess at the same frequency, appearing at the same chemical shift (Fig. 17a). If, on the other hand (Fig. 17b), the field is varied linearly across the sample, the two spins do not experience the same magnetic field, and so resonate at different frequencies. They therefore appear as distinct peaks in the spectrum.

The *pulsed (field) gradient-spin echo* (PFGSE or PGSE) experiment is used to measure translational diffusion constants [29] (which must not be confused with spin diffusion). It is based on the two-pulse spin-echo experiment (Fig. 12; Sec.II.E.1), with the addition of two field-gradient periods to create a spatial frame of reference in the x and y dimensions. The first gradient is applied between the two pulses; the second, between the 180°

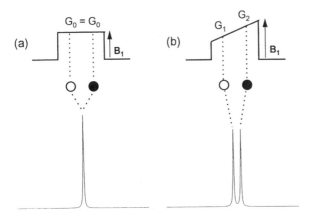

Figure 17   Use of a linear, one-dimensional field gradient, G, to encode spin locations: (a) without gradient; (b) with gradient.

pulse and acquisition of the FID. At its simplest, the imaging experiment is a three-dimensional analog of the PGSE experiment; a third gradient, along the z axis, is applied to provide a coordinate in that dimension. Depending on the type of visual image desired, overall signal intensity, chemical shift, T_1, or other NMR parameter is measured and displayed.

II.E.6.   Flow Techniques: On-Line, Rheo, and LC-NMR

Most flow techniques are based on simple abundant-spin, one-pulse experiments such as those discussed in Sec. II.D.2, because they are both rapid and sensitive. In general, signals must be acquired every few seconds to few minutes in order to provide sufficient information to be useful.

II.F.   Data Processing

In Sec. II.D. and Fig. 4, the FID was modeled as a two-component, damped cosine function. While a few real NMR signals resemble this, most are much more complex. An example of actual data is shown in Fig. 18a. If this signal is expanded $100\times$ in the time (x) domain, it is seen to be composed of many overlapping frequency components of varying intensity (Fig. 18b). Prior to Fourier transformation of data such as this, some manipulations are usually performed to improve the quality of the ultimate spectrum. Any DC (direct current) offset is subtracted to ensure that the FID is centered around zero on the intensity (y) axis. Failure to do this can result in a spike at the center point of the spectrum; this artifact might be mistaken for a peak. Note that the FID of Fig. 18a has decayed completely before the end of the data collection period; the right third contains nothing but noise. If the FID is multiplied by a decreasing-exponential *window* or *broadening function*, the relative contribution of this noise to the spectrum is diminished, enhancing the final signal-to-noise ratio. This procedure does, however, broaden the peaks slightly, as shown in Fig. 19. Other window functions affect the spectrum differently; a rising exponential, for example, increases the resolution at the expense of S/N.

The Fourier transform's function is to identify the various frequency components in the FID, and to weight them appropriately, so that they can be displayed as a frequency (or chemical-shift) spectrum. As mentioned in Sec. II.D.1, due to the unavoidable receiver deadtime, signal acquisition cannot commence immediately after a pulse. Therefore, the first point of the digitized FID is probably not at a maximum in its oscillation, but rather, is phase-shifted. Because the distinct components of the FID precess at different frequencies, this phase shift is not equivalent across the spectrum. Phasing errors are reflected as a "twisted" baseline in the transformed spectrum, as depicted in Fig. 20a. These distortions are eliminated by *zero-order* (Fig. 20b) and *first-order* (Fig. 20c) *phase corrections*, which may be performed automatically and/or manually. Some experiments, such as certain relaxation measurements and multidimensional spectra, intentionally produce inverted peaks; this feature must be preserved during phase correction. *Baseline roll* is sometimes observed, particularly when the FID has decayed rapidly; it may only be apparent at high vertical expansions ($100\times$ in Fig. 21a). Software correction routines will flatten the baseline, but they may distort peak shapes somewhat (Fig. 21b). *Chemical-shift referencing* is accomplished by establishing the position of an internal standard, or other peak whose shift is well known (such as a solvent), at the appropriate value. Many spectrometer systems can perform this operation automatically. These are the most common types of data manipulation performed, but this list is certainly not exhaustive.

When NMR is used as a quantitative tool, peak areas are evaluated by numerical *integration*. Assuming that the correct experimental conditions are chosen (Sec. II.D.4), the major sources of error in peak integration are (1) a poor baseline; (2) insufficient digital resolution, (3) poor S/N; and (4) incomplete peak resolution. Proper phasing and baseline correction minimize errors due to baseline distortions. Because polymer spectra tend to be broad, even in solution, digital resolution (seconds/point in the FID, or ppm(Hz)/point in the spectrum) is seldom a limitation unless a very small data set is collected. The breadth of polymer resonances can, however, adversely affect S/N and peak resolution. In a broader peak, the area is spread over a wider chemical-shift range, lowering the maximum intensity and increasing the relative contribution of noise to the integral value.

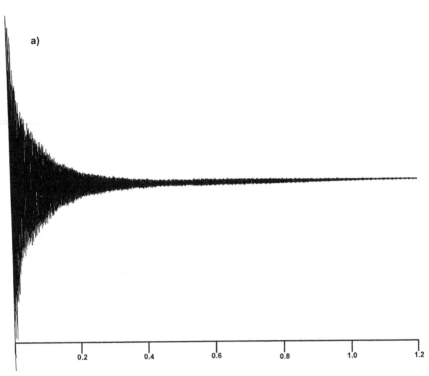

Figure 18 Experimental free-induction decay (FID): (a) full scale; (b) first 1% of points; the x axes are in seconds.

Obviously, broader peaks are more likely to overlap, and any procedure used to resolve them introduces additional uncertainties into the area calculations. Attempts to simultaneously optimize S/N and resolution often necessitate compromises for polymer systems. S/N can be improved by increasing the sample concentration and/or the number of scans. However, many polymers are inherently only sparingly soluble, and significant improvements in S/N may, therefore, require extremely long experiments. Peak resolution is best enhanced by decreasing the sample concentration, but this results, of course, in a loss of signal intensity. The options for improving S/N and resolution are even more limited in solid-state NMR because the sample concentration is fixed, and because additional broadening mechanisms prevail.

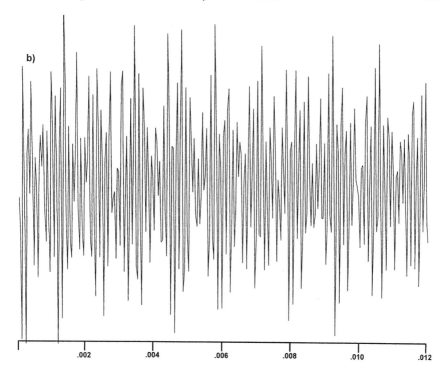

b)

| | | | | | |
|.002|.004|.006|.008|.010|.012|

Figure 18 Continued

II.G. Basic Spectral Interpretation

While NMR can provide valuable insights into the physical properties of polymers (as will be seen in Secs III.C–III.E), it is the technique's ability to clarify details of molecular structure that is most frequently exploited. The goal of most NMR experiments is a chemical-shift-resolved spectrum, so correlating these shifts with chemical structure is important. The nuclei most commonly studied in polymer analysis are ^{13}C and ^{1}H, although ^{2}H, ^{19}F, ^{29}Si, ^{31}P, ^{15}N, ^{129}Xe, and others find specific applications. If the system contains more than one NMR-active nuclide, complementary multinuclear studies often provide valuable information [17,18]. The next seven sections (II.G.1–II.G.7) highlight chemical-shift trends for the more commonly studied nuclei. These generalizations apply to chemical-shift-resolved spectra in both solutions and solids.

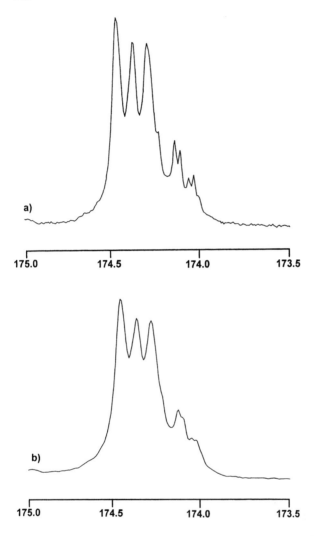

Figure 19 Effect of a single decreasing exponential window function, $\exp(-\kappa t)$: (a) $\kappa = 1\,\text{Hz}$; (b) $\kappa = 5\,\text{Hz}$. Note the poorer peak resolution and improved signal-to-noise in (b). These ^{13}C spectra depict the carbonyl region of poly(isobutyl methacrylate) in CDCl_3; the x axes are in ppm.

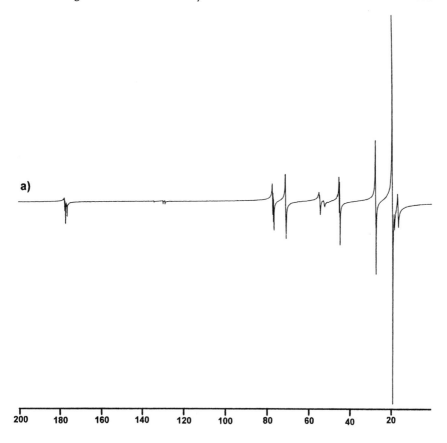

a)

| 200 | 180 | 160 | 140 | 120 | 100 | 80 | 60 | 40 | 20 |

Figure 20 Effect of phase correction procedure: (a) as-transformed data; (b) after a zero-order correction; (c) after both zero- and first-order corrections on a ^{13}C spectrum. The sample is poly(isobutyl methacrylate) in CDCl$_3$; the x axes are in ppm.

II.G.1. ^{13}C NMR

Figure 22 shows a chemical-shift correlation chart for ^{13}C, illustrating the δ_C ranges for functional groups frequently found in polymers [2,16]; all shifts are cited relative to the usual reference standard, tetramethylsilane (TMS), at 0 ppm. Aliphatic carbons bonded to silicon (**C**–Si) also resonate near this point. **C**–**C** aliphatic carbons appear at somewhat higher δ_C's, between 5 and 50 ppm. As a rule of thumb, δ_C increases as the number of protons decreases:

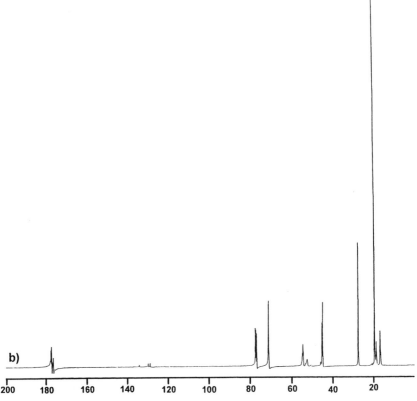

Figure 20   Continued

$$\delta_{methyl} < \delta_{methylene} < \delta_{methine} < \delta_{quaternary} \qquad (8)$$

Carbons singly bonded to oxygen ($C-O$) typically resonate between 40 and 80 ppm, as do aliphatic $C-Cl$ and $C-N$. $C-F$ covers a wide range, from 70 to 120 ppm. Aromatic and olefinic carbons overlap from 100 to 150 ppm, and carbonyls appear between 150 and 250 ppm. ^{13}C NMR spectra are usually recorded under ^{1}H decoupling. While this procedure simplifies resonances arising from [13]C–[1]H pairs, it does not affect coupling of carbon to other nuclides, such as ^{19}F or [2]H. Deuterated carbons exhibit slight isotope shifts relative to the protonated form (-0.25 ppm/deuteron for the α carbon, and -0.1 ppm/deuteron for the β carbon) [16].

Several schemes have been devised for the approximation of ^{13}C chemical shifts. Two which are particularly useful for polymers are the

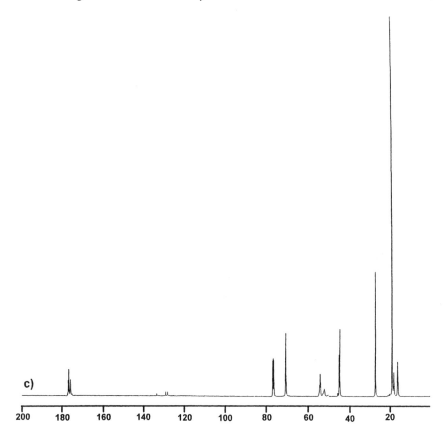

c)

| | | | | | | | | | |
|200|180|160|140|120|100|80|60|40|20|

Figure 20 Continued

Lindeman-Adams [54] and Grant-Paul [55,56] models. In both, the effects of various α, β, γ, and δ substituents on the chemical shielding of a particular carbon are summed. The Lindeman-Adams scheme focuses on saturated hydrocarbons and is therefore most applicable to polyolefins. The Grant-Paul rules apply to saturated carbons, but include corrections for other functional groups. Models which approximate chemical shifts for substituted aromatic [16] and olefinic [57] carbons have also been developed. An example of the use of these calculational schemes is given in Sec. III.B.1.

II.G.2. ^1H and ^2H NMR

The correlations between ^1H chemical shift and structure are shown in Fig. 23. It can be seen that δ_H's follow the same general trends as carbon reso-

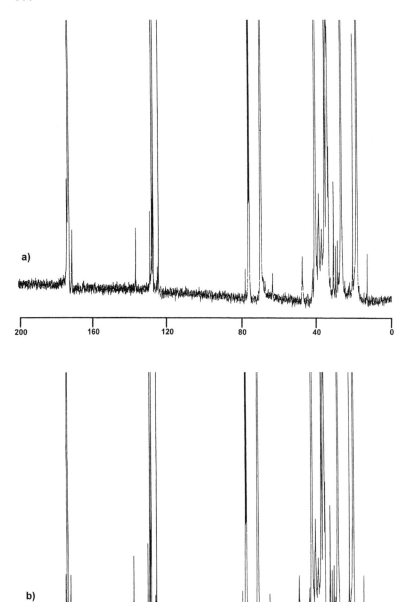

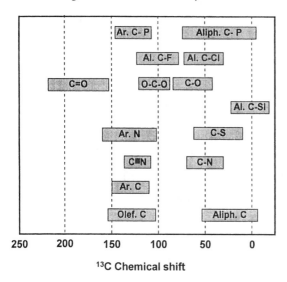

Figure 22   Chemical-shift correlation chart for ^{13}C.

nances, except that aromatics and olefinics do not overlap. Once again, the shift reference is TMS at 0 ppm. ^{1}H NMR spectra of small molecules in solution are usually complicated by scalar-coupling patterns, but in polymer solutions these may be partly or completely obscured by the linewidth. Chemical-shift trends for ^{2}H parallel those for ^{1}H [17,18]; in fact, the shifts are essentially the same. Most ^{2}H NMR studies utilize isotopically enriched samples because this nuclide is difficult to observe in its low natural abundance (0.015%).

II.G.3. ^{19}F NMR

^{19}F NMR is used primarily for the characterization of fluoropolymers; the relevant chemical-shift ranges are depicted in Fig. 24 [17,18]. The reference is CFCl$_3$ (0 ppm). The ^{19}F nucleus is an extremely sensitive probe of copolymer structure, because of its wide chemical-shift range and high natural abundance [58,59]. Since many fluorine-containing materials also contain hydrogen, ^{19}F spectra are usually acquired with ^{1}H decoupling. However,

---

Figure 21   Effect of baseline correction: (a) before correction; (b) after correction. Note the curved baseline in (a) and the baseline distortions in (b). The sample is poly(isobutyl methacrylate) in CDCl$_3$; x axes are in ppm.

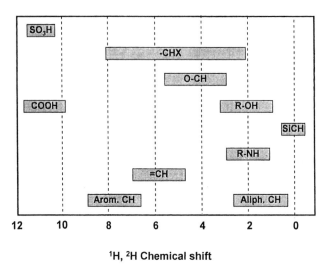

¹H, ²H Chemical shift

Figure 23   Chemical-shift correlation chart for ¹H and ²H.

¹⁹F–¹⁹F scalar interactions remain; the splitting patterns are similar to those encountered for ¹H. A broad background resonance, due to fluoropolymer components in the probe, is often seen in ¹⁹F spectra. Baseline correction can eliminate this interference, but the spectral resonances may become distorted.

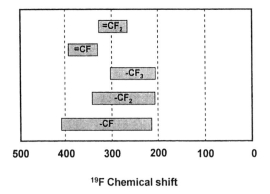

¹⁹F Chemical shift

Figure 24   Chemical-shift correlation chart for ¹⁹F.

II.G.4. ^{29}Si NMR

Not surprisingly, the principal use of ^{29}Si NMR in polymer analysis is to study the polysiloxanes (silicones), although the technique is equally service-able for other Si-containing materials, such as polysilanes and polysilicates. The chemical-shift correlation chart [17,18] in Fig. 25 reveals that the primary predictor of δ_{Si} is the number of oxygen molecules bonded to silicon: (1) for SiO_4 (silicate, also designated Q), the peaks appear between -70 and -110 ppm (relative to TMS); (2) for SiO_3 (T), -70 to -30 ppm; (3) for $> SiO_2$ (siloxane, D), -20 to $+20$ ppm; (4) for $\equiv SiO$ (M), 0 to $+50$ ppm; and (5) for $> Si <$ (silane), -30 to $+10$ ppm. While ^{29}Si spectra are usually acquired under ^{1}H decoupling, the decoupler should be gated because this nuclide has a negative magnetogyric ratio, which can lead to inverted or null peaks. Since ^{29}Si T_1 relaxation times tend to be very long, it is recommended that $Cr(acac)_3$ relaxation agent be added to solutions. A broad resonance centered about -110 ppm often appears in the ^{29}Si spectrum of a solution; this arises from the glass sample tube. Baseline correction may remove the interference, but this procedure can adversely affect nearby sample resonances.

II.G.5. ^{31}P NMR

While phosphorus is not a common element in polymeric materials, other than in the polyphosphazenes, it does occur in a number of additives, such as secondary antioxidants and some plasticizers. The chemical-shift correlation chart of Fig. 26 illustrates that δ_P depends largely on the oxidation state of the phosphorus, with P(III) species (such as phosphites) resonating between -450 and $+250$ and P(V) species (such as phosphates), between -50 and $+100$ ppm [17,18,60]. These shifts are relative to the ^{31}P reference

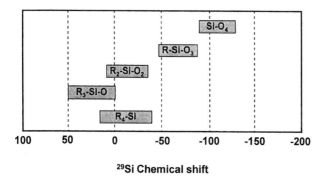

Figure 25 Chemical-shift correlation chart for ^{29}Si.

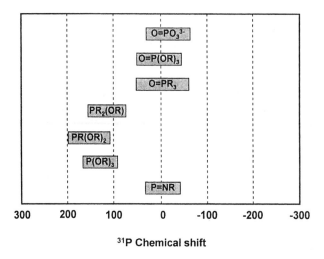

Figure 26   Chemical-shift correlation chart for ^{31}P.

standard, 85% H_3PO_4, at 0 ppm. Since this standard is highly corrosive, it is best contained in an internal capillary tube. ^{31}P spectra are usually recorded with ^{1}H decoupling.

II.G.6. ^{15}N and ^{14}N NMR

Both ^{15}N and ^{14}N are NMR-active, but their spectra look quite different, even though the chemical-shift ranges are similar (Fig. 27); shifts are cited relative to neat nitromethane [17,18]. Some of the N-containing groups commonly found in polymers include amides (−200 to −400 ppm) and nitriles (−200 to −100 ppm). ^{14}N is quadrupolar, and its resonances tend to be broad, limiting its utility for chemical studies. While the spin-$\frac{1}{2}$ ^{15}N nucleus can produce high-resolution solution- and solid-state spectra, its low natural abundance makes it difficult to study unless the sample is artificially enriched. Because of its negative magnetogyric ratio and generally long T_1, Cr(acac)$_3$ should be added to the sample, and the decoupler should be gated.

II.G.7. ^{129}Xe NMR

While there are no commercial xenon-containing polymers, this nucleus finds applications as a probe molecule [17,18,20]. It has been successfully employed to investigate the pore structure of zeolites, and its use in polymer characterization is increasing. The chemical shift of ^{129}Xe gas depends on the pressure or, more specifically, on the number of collisions per second,

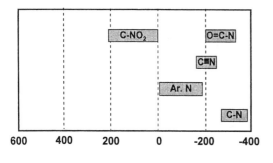

^{14}N, ^{15}N Chemical shift

Figure 27 Chemical-shift correlation chart for ^{14}N and ^{15}N.

that a nucleus undergoes. Since motionally restricted xenon atoms experience a higher collision frequently than the unconstrained gas, it can be is a sensitive probe of morphology and bulk structure.

II.H. Advanced Spectral Interpretation

Not surprisingly, the advanced experiments outlined in Sec. II.E employ more sophisticated data interpretation than the basic techniques, in order to extract the more detailed chemical and physical information discarded in the simpler experiments.

I.H.1. Relaxation Measurements

If the inversion-recovery experiment (Sec. II.F.1) is repeated as a function of τ, the signal intensity (A) is found to increase exponentially:

$$\frac{A(\tau)}{A(\infty)} = B \exp\left(\frac{-\tau}{T_1}\right) + \frac{A(0)}{A(\infty)} \tag{9}$$

where A_∞ is the peak area at very long τ and B is a constant. Obviously, T_1 can be easily extracted from a semilogarithmic plot of a function of intensity versus τ. Since an entire spectrum is recorded at each τ, one set of data can be used to determine T_1 for each peak. Samples in which there is a distribution of molecular mobilities (such as in an inhomogeneous solution or a multiphase system) exhibit behavior more complicated than Eq. (9) predicts; a plot of intensity versus τ will be composed of multiple components [38]. T_2 is calculated from a semilogarithmic plot of peak areas versus $2n\tau'$, the total time of each echo train (see Sec. II.E.1). As was the case for T_1, each peak in the spectrum will likely have a different T_2, and inhomogeneous systems will exhibit multiexponential decays. T_2^* characterizes the decay of the FID, and includes both relaxation from true spin–spin (T_2) processes and those arising from magnetic-field inhomogeneities; it is related to the linewidth:

$$\Delta v_{1/2} = \frac{1}{\pi T_2^*} \tag{10}$$

where $\Delta v_{1/2}$ is the full width of the resonance at half-height, expressed in hertz. $T_{1\rho}$ is measured from the decay of cross-polarization intensity with contact time, or from experiments with variable spin-lock conditions.

Each of these parameters (T_1, T_2, T_2^*, and $T_{1\rho}$) relates to molecular mobility. In general, T_2^*, T_2, and $T_{1\rho}$ are of the order of milliseconds for polymer solutions and solids, and they relate to molecular motions such as short segmental reorientations, that occur at kilohertz rates. T_1, which is typically somewhat longer, is most influenced by fast (megahertz) motions, such as methyl group rotations. Figure 28 illustrates the effect of temperature on the FID and ^{13}C spectrum of polyethylene. As expected, increased

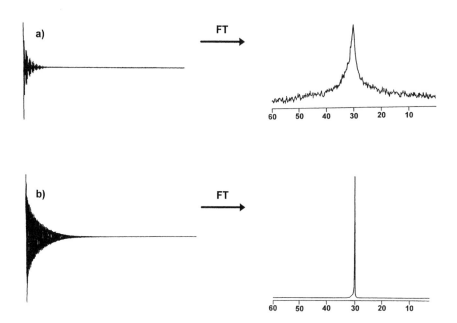

Figure 28 Free-induction decay (FID) and resulting ^{13}C spectra for polyethylene recorded at (a) 50°C and (b) 130°C. The sample had previously been dissolved in 1,2-dichlorobenzene-d$_4$/ 1,2,4-trichlorobenzene (1/3 v/v); x axes are in ppm.

thermal energy at the higher temperature induces more rapid polymer-chain reorientations, which are reflected in the longer T_2^* (i.e., longer decay) and narrower linewidth. The temperature dependence of relaxation times generally follows an Arrhenius relationship:

$$\frac{1}{T_x} = A \exp\left(\frac{-E_a}{kT}\right) \tag{11}$$

where E_a is an activation energy and k is Boltzmann's constant. Relaxation-time measurements have been used to determine E_a for particular types of molecular motions.

II.H.2. Solid-State NMR

High-resolution, solid-state CP/MAS and CRAMPS spectra bear a superficial resemblance to the corresponding solution spectra, but the resonances are generally broader. Peak positions may be interpreted according to the chemical-shift trends summarized in Figs 22–27. However, certain high-resolution solid-state spectra also reflect features that are lost when the material is dissolved. Chemical shift, for example, may be morphology-dependent because the electronic environment around a nucleus can vary with the spatial arrangement of nearby atoms [61–64].

Chemical-shift anisotropy is very sensitive to molecular structure and dynamics. Each nucleus can be pictured as being surrounded by an ellipsoidal chemical-shift field, $\tilde{A}$, arising from the influences of neighboring spins, as described by Eq. (4). If the molecules in the sample have no preferred orientational order, these tensors will be randomly distributed, and the lineshape is predictable. If the shielding is equivalent in all directions ($\sigma_{xx} = \sigma_{yy} = \sigma_{zz}$; $\tilde{A}$ is spherical), a symmetric peak, like shown that in Fig. 29a, will be observed at σ_{iso}, which is defined in Eq. (5). Axial symmetry ($\sigma_{xx} = \sigma_{yy} \neq \sigma_{zz}$; $\tilde{A}$ is, more or less, football-shaped) results in a *powder pattern* like that shown in Fig. 29b. In this case, the tensor elements may be labeled $\sigma_{\parallel}$ (σ_{zz}) and $\sigma_{\perp}$ (σ_{xx} and σ_{yy}). If there is no symmetry in the chemical-shift field ($\sigma_{xx} \neq \sigma_{yy} \neq \sigma_{zz}$; $\tilde{A}$ is a flattened football), then the spectrum assumes the form depicted in Fig. 29c.

If the rate of molecular motion in the sample increases, any of these lineshapes will narrow, but the exact shape will depend on the geometry of motion. If the sample is spun at the magic angle, the resonance collapses to a symmetric peak at σ_{iso}. If some preferred orientation is induced in the sample, e.g., by mechanical deformation, the resulting spectrum is a convolution of the original powder pattern with the imposed distribution. CSA spectra of oriented materials have been used to derive order parameters and to propose models for the orientation process [23]. Multiple-pulse experiments are needed to observe the CSA of abundant spins, but not of rare

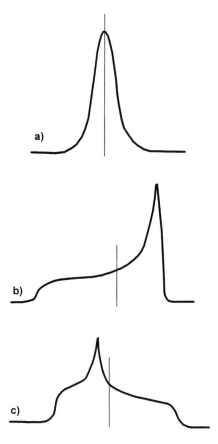

Figure 29 Powder patterns for shielding tensor Ã with: (a) spherical symmetry; (b) axial symmetry; (c) no symmetry.

spins. In the latter case, there are no significant homonuclear dipolar inter-actions to overcome; the CSA can be isolated by heteronuclear dipolar decoupling with no spinning. The many overlapping powder patterns (one for each peak) may make such spectra difficult to interpret, however. There are alternative ways of extracting the CSA lineshapes from slow-spinning MAS spectra [65] and from 2D NMR [66].

Homonuclear dipolar-coupled, abundant-spin, or broad-line spectra may appear to be featureless and uninformative, but they contain a great deal of information about molecular structure and motion. For example, the wideline ^1H spectrum of polyethylene, which might look like the simulation depicted in Fig. 30, exhibits two overlapping resonances. The broader arises

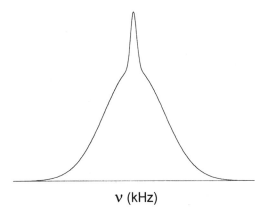

ν (kHz)

Figure 30 Simulated ^1H broadline spectrum for a two-phase material such as polyethylene. The motionally narrowed component arises from the amorphous domain; the broad, from the crystalline.

from the crystalline phase of the polymer; the narrower, from the amorphous. The intensity ratio of these two peaks gives the degree of crystallinity. The details of such a spectrum could change with variables that affect molecular motion and morphology, such as branching and molecular weight.

Quadrupolar-echo experiments yield lineshapes related to those obtained from the CSA. For example, for an $I = 1$ nucleus such as ^2H, the spectrum of a powder material with axially symmetric quadrupole tensors resembles overlapping mirror-image CSA lineshapes (Fig. 31). The

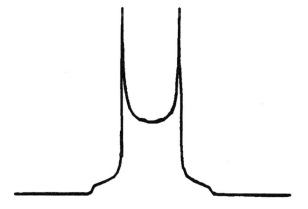

Figure 31 Simulated ^2H powder pattern for an unoriented solid sample.

width and form of these quadrupolar resonances change in ways that are characteristic of molecular motion and imposed order. In fact, these effects are sufficiently specific that dynamic models can be tested by comparing experimental data with computer simulations. Solid-state ^{2}H NMR studies are performed on materials that have been synthesized with ^{2}H labels in particular positions within the molecule, making it possible to isolate and study very localized motions [21,67–69].

II.H.3. Spectral Editing

The tip angle (ψ)-dependent DEPT subspectra (S_{45}, S_{90}, and S_{135}) (Sec. II.D.3) are combinations of CH, CH$_2$, and CH$_3$ resonances; an example is shown in Fig. 32. The pure contributions can be calculated, using standard spectral addition and subtraction routines:

$$S(\text{CH}) = S_{90} \tag{12a}$$
$$S(\text{CH}_2) = S_{45} - S_{135} \tag{12b}$$
$$S(\text{CH}_3) = S_{45} + S_{135} - 0.707 S_{90} \tag{12c}$$

Note that peaks that occur in the normal spectrum, but do not appear in any subspectrum, arise from quaternary carbons.

II.H.4. Multidimensional NMR

The interpretation of multidimensional spectra can be quite involved, so only a simple overview will be presented here. Two-dimensional spectra are usually represented as *contour plots* with two frequency axes forming the base plane (e.g., δ_H–δ_H, δ_C–δ_C, or δ_C–J_{CH}), and intensity depicted by curves spaced according to the steepness of the ascent; these plots resemble geological survey maps. The *stacked plot* tilts the frequency–frequency plane, showing the peaks as vertical intensity; they look like perspective drawings of mountainous terrain. Three-dimensional spectra may be portrayed as false-color (to show intensity) spots in a cube formed by three frequency axes.

Figure 33 shows a schematic homonuclear ^{1}H–^{1}H COSY contour plot for ethyl acetate. This experiment identifies interacting pairs of spins. Each peak is correlated to itself; these uninformative peaks appear on the diagonal (depicted by white circles). The off-diagonal peaks, called *cross peaks* (black circles), disclose *J*-coupling relationships among the peaks. The *projection* of the contour plot onto either axis gives the ^{1}H spectrum, and a *slice* of the plot at a particular δ_H highlights all nuclei that are coupled to the ^{1}H resonating at that shift. Heteronuclear COSY spectra are, in principle, even simpler to interpret, as shown in Fig. 34 (also for ethyl acetate). One axis is ^{13}C chemical shift; the other is δ_H. Correlations appear at the intersections

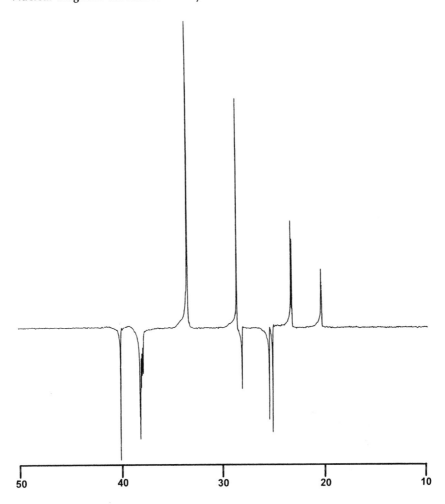

Figure 32   DEPT subspectrum ($\psi = 135°$) for squalane, $[(CH_3)_2CH(CH_2)_3$ $(CH_3)CH(CH_2)_3(CH_3)(CH_2)_2\cdot]_2$, with CH_3 and CH resonances pointing up; CH_2, down. The x axis is in ppm.

(δ_C, δ_H), of the peak positions. The projection onto the δ_C axis gives the ^{13}C spectrum, and correspondingly for the δ_H axis. The absence of a cross peak at a particular ^{13}C or ^1H resonance indicates a carbon with no attached hydrogen (such as $C{=}O$) or a hydrogen with no carbon attached (such as OH).

Figure 35 portrays a schematic J-resolved spectrum for ethyl acetate. The scalar splitting of each carbon is expanded vertically; the coupling

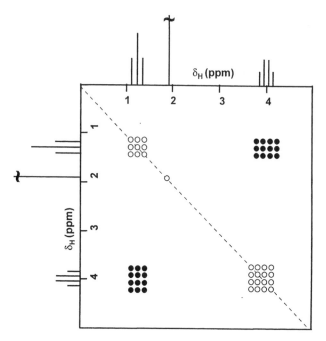

Figure 33 Schematic homonuclear $^1H^1H$ COSY spectrum with diagonal peaks (o) and cross peaks (•).

constant can be read directly from the J_{CH} axis. A projection on the δ_C axis yields the ^1H-decoupled spectrum. In a NOESY spectrum, cross-peak intensities reflect the degree of nuclear Overhauser enhancement experienced by that nucleus, which can be related to interatomic distances. This calculation is complicated by the fact that a given nucleus is simultaneously affected by all neighboring spins; numerical-analysis software helps deal with this problem.

II.H.5.   Diffusion Measurements and Imaging

The result of a PGSE experiment is a train of echoes that decays with a diffusion constant D:

$$A = A_0 \exp(-\gamma^2 G^2 \beta D) \tag{13}$$

where A is the echo amplitude, A_0 is a preexponential factor, γ is the magnetogyric ratio, G is the magnitude of the field gradient, β is a timing variable, and D is the translational diffusion constant. The timing variable β is defined as $\xi^2(\Delta - \xi/3)$, where ξ is the length of the field-gradient pulse and Δ

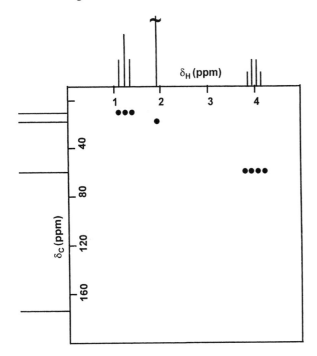

Figure 34   Schematic heteronuclear ^{13}C–^{1}H COSY spectrum.

is the time between pulses. Since γ and G are known, D can be deduced from the slope of a semilogarithmic plot of A versus β [29]. Alternatively, D can be evaluated from a semilogarithmic plot of A versus G, if β is held constant.

Because imaging experiments produce a pictorial representation of the experimental measurements, their interpretation is primarily visual. The image may be a map of signal intensity, molecular motion, or occurrence of a specific functional group [70–74]. Standard image-analysis software can provide quantitative information about the size and distribution of the observed features.

II.H.6.   Flow NMR: On-Line, Rheo, and LC-NMR

Since on-line instrumentation is generally used to monitor process stability, rather than to investigate molecular details, such systems do not necessarily display intermediate FIDs or spectra. The output is often a control chart tracking the property to be monitored. The unseen correlation between NMR signal and control chart is based on a chemometric model previously derived from an extensive calibration set. The parameters used for the fit might include relative signal intensities and/or relaxation times.

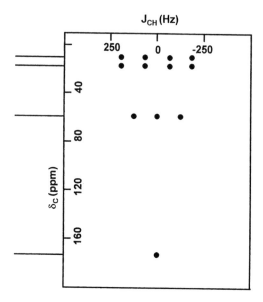

Figure 35 Schematic ^{13}C J-resolved spectrum.

Because NMR is so closely related to molecular motion, there has been much effort to tie spectroscopic parameters, such as relaxation times, to the known rheological and mechanical properties of materials. While there has been some success, as will be seen in Sec. III.D, a general correlation has yet to be established. One difficulty is that NMR is most sensitive to relatively rapid, local motions, while it is primarily slower, longer-range motions that are responsible for a polymer's bulk properties. Rheo-NMR experiments have attempted to provide the connection between micro- and macroscopic observations, and they will likely become a very useful tool in the NMR of polymers if the equipment becomes widely available [36].

In LC-NMR experiments, NMR spectra are recorded as a function of elution time; these spectra can be interpreted in the normal way to identify the chemical structure of the eluent. By proper choice of experimental parameters, these spectra can be made quantitative.

III. APPLICATIONS TO POLYMER SYSTEMS

This section reviews some uses of NMR for the investigation of polymer systems [25,26,75–81]. These citations are intended to show the diversity of such applications, not to present a comprehensive review. The subjects of these studies may be either solutions or solids, and the experiments run the

gamut from the simple to the most complex. The discussion will focus on:
(1) polymerization reactions; (2) polymer-chain structure; (3) bulk properties; (4) polymer-chain dynamics; and (5) multicomponent polymer systems.

III.A. Polymerization Reactions

While NMR does not easily lend itself to the direct analysis of many polymerization processes, the technique has been used to probe catalysis and
mechanisms. NMR also provides information about reaction kinetics, as
will be discussed more extensively in Secs III.B.2 and III.B.3.

III.A.1. Catalysis

Commercially important catalyst systens have been the subject of numerous
NMR studies. Many Ziegler-Natta complexes of titanium alkoxides and
magnesium halides are used as catalysts in solution polymerizations, and
NMR allows them to be investigated in this state [83].

Similar systems, prepared on silica supports, were examined by solid-
state ^{13}C NMR [84–86]; the method of preparation was found to lead to
compositional differences in the catalyst. At present, there is much interest
in metallocene-based catalysts, which give rise to some novel materials [87];
they are composed of a metallocene (usually a zirconocene) and an alkyla-
luminoxane (often methylaluminoxane, or MAO). The formation of
"cation-like" alkylzirconocene upon exposure to MAO has been detected
by ^{13}C CP/MAS NMR[88]. Isotopically enriched $Cp_2Zr(^{13}CH_3)_2$ was used to
make the metallocene component more readily observable, even though it is
present in low concentration (Al/Zr ranged from 5/1 to 20/1) [88]. Isotopic
labeling also enabled comparison of the early stages of ethylene polymerization catalyzed by a metallocene and by an older Ziegler-Natta system.
Differences in the coordination of the monomer to the metal centers were
indicated by the different end groups formed in poly(ethylene-d$_4$) [89,90].
Figure 36 compares the ^{13}C methyl-group region for polymers prepared
with Ziegler-Natta and metallocene (specifically, Cp_2ZrCl_2/MAO) catalysts.
Two resonances are observed. One is a singlet at 14.1 ppm due to $^{13}C^1H_3$
ends; the other is a triplet at 13.7 ppm from $^{13}C^1H_2{}^2H$ ends ($^{13}C–^2H$ scalar
coupling has not been removed). As pointed out in Sec. II.G.1, an isotope
effect causes an upfield shift (to lower δ) for deuterated carbons, allowing
the two end groups to be differentiated.

III.A.2. Mechanisms

NMR has been used to monitor polymerization reactions in real time.
During the emulsion formation of poly(butyl acrylate), solid-state ^{1}H spectra were recorded every minute, and ^{13}C spectra, every 8.5 min [91]. Not

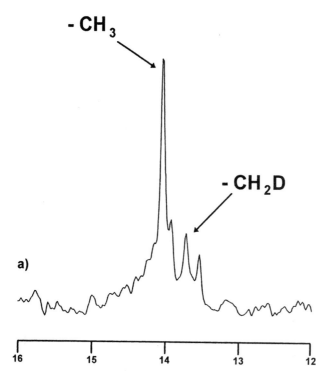

Figure 36 Methyl region of ^{13}C spectra, showing the partially deuterated end groups (CH$_2$D) observed in polyethylene prepared with (a) Ziegler-Natta and (b) metallocene catalyst and ethylene-d. The x axes are in ppm.

surprisingly, it was found that ^{1}H NMR offered higher S/N and more facile quantitation, while ^{13}C NMR exhibited improved peak resolution and improved structural detail. The living cationic reaction of isobutyl vinyl ether was followed by in-situ monitoring of the growing species [92,93], NMR imaging has provided a visual representation of the polymerization of methyl methacrylate [94]. The viscosity of the medium, the degree of polymerization, and the reaction kinetics could all be determined from measurements based on these images. Comonomer sequence distributions, which will be discussed in detail in Sec. II.B.5, have been used to deduce details of copolymerization mechanisms [95].

III.B. Polymer-Chain Structure

The most straightforward, and probably the most common, application of NMR is the identification of a material and/or characterization of its com-

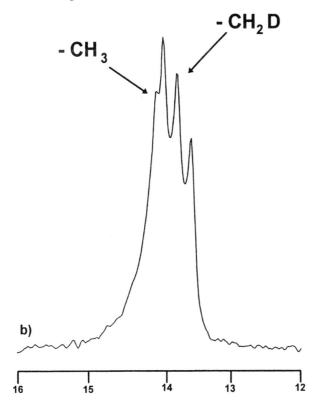

Figure 36 Continued

position. Once a spectrum has been recorded under the appropriate experimental conditions, the sample's identity can be inferred from comparison to the spectrum of a reference material, to published compilations of spectra [96–98], to calculated chemical-shift approximations, or to data for model compounds. More subtle structural features, such as stereoregularity, branching, cross-linking, and end groups, are also observable. In particular, spectra of copolymers provide detailed information about the microstructure of the polymer chains.

III.B.1. Material Identification

The starting point for identification of an unknown material by NMR is almost always a simple one-dimensional, high-resolution ^{13}C spectrum. Once a suitable solvent has been found and a good-quality spectrum acquired, normal peak-assignment strategies can be applied. Figure 37 shows the ^{13}C chemical-shift ranges observed for many common polymer

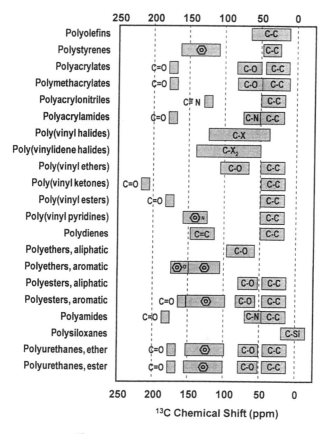

Figure 37 ^{13}C chemical-shift ranges for common polymer types.

types. Experienced spectroscopists can recognize many materials by inspection, and several compilations provide large collections of reference spectra of polymers [96–98]. It must be kept in mind, however, that many "real-world" materials are blends of several polymers, or contain fairly high levels of additives. Interpretation is often complicated by the contributions of these other components to the spectrum. If the sample proves to be intractable, or if an appropriate solvent is not available, it may be necessary to record a solid-state spectrum. If these approaches do not provide sufficient information, more sophisticated experiments may be in order.

As discussed in Sec. II.G.1, there are several calculational schemes which can be used to estimate chemical shifts, particularly for ^{13}C. The following example demonstrates the use of the Grant-Paul rules, to approximate the chemical shift of the *ualpha*-methyl group in poly(methyl methacrylate) (PMMA):

$$
\begin{array}{c}
\text{CH}_3 \\
| \\
-(-(\text{CH}_2-\text{C}-)_n- \\
| \\
\text{O}=\text{C}-\text{OCH}_3
\end{array}
\qquad (14)
$$

The procedure, which is summarized in Tables 3 and 4, is as follows.

Table 3 ^{13}C Chemical-Shift Substituent Parameters ($\delta_C = -2.3 + \Sigma\alpha + \Sigma\beta + \Sigma\gamma + \Sigma + \gamma + \Omega^a$)

Substituent	α	β	γ	δ
$>$C$<$*	9.1	9.4	-2.5	0.3
$-$O$-$*	49.0	10.1	-6.0	0.3
$>$N$-$*	28.3	11.3	-5.1	0.3
$-$S$-$*	11.0	12.0	-3.0	-0.5
$-$C$_6$H$_5$	22.1	9.3	-2.6	0.3
$-$F	66.0–70.1^b	7.8	-6.8	0.0
$-$Cl	31.1^c	10.0	-5.1	-0.5
$-$C$\equiv$N	3.1	2.4	-3.3	-0.5
$\equiv$SO$_3$H	38.9	0.5	-3.7	0.2
$-$C(O)$-^d$	22.5	3.0	-3.0	0.0
$-$COOH	20.1	2.0	-2.8	0.0
$-$COO$^-$	24.5	3.5	-2.5	0.0
$-$C(O)O$-^d$	22.6	2.0	-2.8	0.0
$-$OC(O)$-^d$	54.5–62.5^e	6.5	-6.0	0.0
$-$C(O)NH$-^d$	22.0	2.6	-3.2	-0.4
$-$NHC(O)$-^d$	28.0	6.8	-5.1	0.0
$-$C$=$C$-$*	21.5	6.9	-2.1	0.4

*When in α position, use steric correction factors, Ω in Table 4. If there is more than one α substituent, consider the most substituted.
^{a}See Table 4.
$^b\alpha$ contribution is 69.1 for CF$_3$; 70.1 for CF$_2$ and CF.
$^c\alpha$ contribution is 35.0 for CCl$_3$; 31.1 for CCl$_2$ and CCl.
^{d}C(O) denotes a carbonyl, C$=$O.
$^e\alpha$ contribution is 62.5 for quaternary carbon; 54.5 for others.
(*Source*: Refs. 55 and 56.)

Table 4 ^{13}C Steric Correction
Parameters

Type [a]	$i = 1$ [b]	$i = 2$ [b]	$i = 3$ [b]	$i = 4$ [b]
Primary	0.0	0.0	−1.1	−3.4
Secondary	0.0	0.0	−2.5	−7.5
Tertiary	0.0	−3.7	−9.5	−15
Quaternary	−1.5	−8.4	−15.0	—

[a] Carbon for which shift is being estimated.
[b] Number of nonproton substituents on α-substituent.
(*Source*: Refs. 55 and 56.)

1. Begin with the basis shift of −2.3 ppm (not coincidentally, the shift for methane).
2. Add 9.1 ppm as the α-contribution of the quaternary carbon, 2×9.4 ppm for the two backbone β-carbons, and 2.0 ppm for the ester group in the β-position. The estimated δ is now 27.6 ppm.
3. Subtract 2×-2.5 for the two γ-carbons in the backbone. No correction is needed for the γ-oxygens; they were included with the ester group. The estimated δ is 22.6 ppm.
4. Add 5×0.3 for the five δ-carbons: two in the backbone, two α-methyls in adjacent methyl methacrylate units, and one methoxy in the pendant group. The estimated shift is now 24.1 ppm.
5. Use a steric correction parameter of −3.4, for a primary carbon with a quaternary α-substituent, for a final result of 20.7 ppm.

This approximation compares favorably with the observed value of 18.8 ppm. Such calculations are usually within 1–3 ppm of the experimental value, although the error for highly substituted carbons may be somewhat larger. Table 5 compares the shifts estimated in this way for the noncarbonyl resonances of PMMA with the experimentally observed values. The approximated shifts for three of the four alkyl carbons agree with the experimental values to within 2 ppm; the highly substituted quaternary carbon deviates by ∼4 ppm.

Tabulations of small-molecule shifts can also offer clues to a polymer's structure. A cursory examination of the ^{13}C spectrum shown in Fig. 38 suggests that it is a styrenic polymer because of the presence of both aliphatic (30–40 ppm) and (probable) aromatic resonances (110–150 ppm), and the

Table 5 Calculated ^{13}C Chemical Shifts for
Poly(methyl methacrylate)

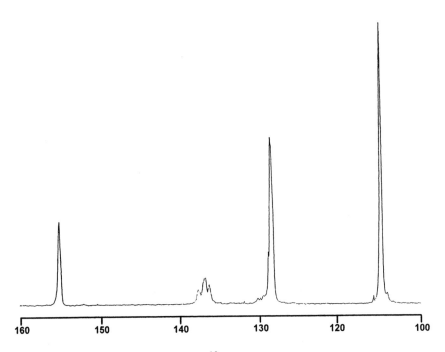

Carbon	δ_C, ppm (calculated)	δ_C, ppm (experimental) [96]
A	45.6	44.9
B	40.5	44.6
C	20.7	18.8
D	50.6	51.6

(*Source*: Refs. 55 and 56.)

Figure 38 Aromatic region of the ^{13}C spectrum of poly(4-hydroxystyrene).
The sample was dissolved in acetone-d$_6$; the x axis is in ppm.

simple pattern of the aromatic resonances indicates that the pendant ring is para-substituted. The presence of a peak at 114.8 ppm is, however, somewhat unusual, as aromatic resonances typically occur at higher δ's. A comparison of these data to published shifts for disubstituted benzenes reveals that phenolic species exhibit a peak around 115 ppm, leading to the reasonable deduction that the material in question is poly(4-hydroxystyrene). Most of the other observed aromatic shifts are, in fact, reasonable matches to those for *p*-cresol (Table 6). The large discrepancy at the *ipso*-carbon (A) is expected, due to the presence of β-, γ-, and δ-backbone carbons in the polymer. If published chemical shifts from appropriate models are not available, such compounds may be purchased or synthesized [99].

If a one-dimensional spectrum does not provide sufficient information to determine the polymer structure, two-dimensional techniques, such as INADEQUATE, may be employed (see Sec. II.E.4). For example, this experiment was used to measure the sizes of the cyclic materials, poly(ethyl α-[(allyloxy)methyl] acrylate) and poly(allyl α-[hydroxymethyl] acrylate); carbon–carbon connectivities could be traced around the rings [100]. Such advanced techniques are particularly useful for characterizing the *regioregularity* of some polymers. Certain monomers, such as vinyl fluoride, can add to a growing chain in either a "head-to-tail,"

$$-CH_2-CHF-CH_2-CHF- \qquad\qquad (15a)$$

Table 6 Aromatic^{13}C Chemical Shifts for Poly(4-hydroxystyrene)

Carbon	δ_C (ppm), experimental [96]	δ_C (ppm), *p*-cresol [2]
A	155.0	152.6
B	114.8	115.3
C	128.3	130.2
D	136.7	130.5

or "head-to-head" fashion,

$$-CH_2-CHF-CHF-CH_2-$$ (15b)

Such arrangement can be identified and quantified by 2-D ^{19}F studies of poly(vinyl fluoride) [58,59].

III.B.2. Branching and Cross-Linking

Nuclei located at or near branch points are chemically shifted relative to other atoms in the polymer chain. It is usually possible to identify pendant groups and determine their lengths. NMR is, for example, the only technique which can easily and unambiguously distinguish copolymers of ethylene with propylene, 1-butene, 1-hexene, 4-methyl-1-pentene, or 1-octene (the linear, low-density polyethylenes), regardless of comonomer content. Some materials, such as low-density (high-pressure) polyethylene, exhibit complex branching distributions, with varying levels of many different alkyl side chains—methyl, ethyl, butyl, and so on. ^{13}C NMR can quantify the levels of each of these branch types [101] (Fig. 39). Other branched polymers, such as dechlorinated poly(vinyl chloride) [102], have also been extensively characterized by NMR. Other NMR studies reveal that the level of branching which arises from chain transfer during the polymerization of *n*-butyl acrylate varies with molecular weight and degree of conversion [103], and that the emulsion polymerization of vinyl acetate leads to more highly branched materials than the bulk reaction [104].

Cross-linked polymers and other network materials are difficult to study in detail because of their insolubility and generally irregular structure [105]. Solid-state NMR can be effectively used to characterize such systems, since cross-linking reactions produce increasingly insoluble materials. For example, ^{13}C NMR reveals the connections between the structure of cross-linked unsaturated polyesters and their mechanical properties [106]. ^{15}N-enriched epoxy resins have been studied by two-dimensional NMR, which provides direct measurements of the extent of cure and crosslinking [107]. Natural-abundance ^{15}N CP/MAS spectra of epoxies have been obtained with a special high-volume spinner; this approach reveals information not available from ^{13}C data [108]. Cross-linked polystyrene (PS) beads, which are often used as catalyst supports, have been swollen and observed by both solution [109] and solid-state NMR techniques [110]. The static (nonspinning) ^{13}C spectrum of the solid beads, shown in Fig. 40a, is broad, and peak resolution is poor, while spectra taken under magic-angle spinning (Fig. 40b) demonstrate the line-narrowing effected by this technique. The small peaks marked by an asterisk (*) are spinning sidebands (Sec. II.E.2). Interestingly, this sample does not require multiple-pulse heteronuclear dipolar decoupling, because, in the gel state, the PS molecules are sufficiently

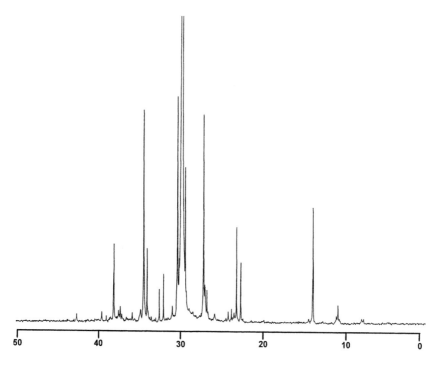

Figure 39 ^{13}C NMR spectrum of low-density polyethylene (LDPE). Sample was dissolved in 1,2-dichlorobenzene-d_4/ 1,2,4-trichlorobenzene (1/3 v/v); the x axis is in ppm.

mobile to average the coupling. Polydimethylsiloxane gels have been studied in a similar way [111]. The use of high temperatures ($>200°$C) can also improve the spectral resolution, as for epoxy resins, by imparting increased molecular mobility [112]. Network formation in interpenetrating polymers systems has also been monitored by ^{13}C NMR [113], and the decrease in molecular motion that accompanies the sulfur vulcanization of rubber has been followed by imaging experiments [114].

III.B.3. End Groups

NMR is a useful tool for identifying and quantifying polymer-chain end groups. Even a familiar polymer may be found to incorporate new types of terminal groups if a different catalyst system is being tested, or if the polymerization conditions have been altered [115]. End groups can be difficult to observe because they are present at such low levels, but selective isotopic

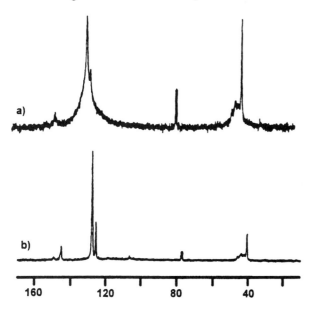

Figure 40 Effect of magic-angle spinning on ^{13}C spectra of solid, cross-linked polystyrene beads: (a) no spinning; (b) spinning at 1.8 kHz. The x axis is in ppm. (From Ref. 111, 1991 American Chemical Society.)

labeling can make them more readily observable [116]. For example, the introduction of ^{13}C-enriched methylaluminoxane cocatalyst during polymerization results in poly(3-methyl-1-pentene) which is labeled in the terminal positions.

Each of the two quantitative ^{1}H spectra of poly(ethylene glycol) (PEG) depicted in Fig. 41 exhibits an end-group (**OH**) resonance at 3.2 ppm and a main $-O-CH_2-CH_2O-$ peak at 3.9 ppm. The molecular weights of these two products can easily be calculated by NMR, based on the relative intensity ratio of the end to main peaks. Rare spins (^{13}C and ^{29}Si) can also be used to determine molecular weight, as demonstrated for several polysiloxanes [117]. A commercially available on-line system tracks both density and melt index (which is related to molecular weight) based on wide-line ^{1}H experiments [118]. GPC/NMR, a variant of LC/NMR, has been used to track the molecular-weight distribution of poly(methyl methacrylate) [119].

III.B.4. Configurational Isomerism

NMR is uniquely suited to the analysis of structural details such as configurational isomerism, whether geometric or stereochemical in origin, which

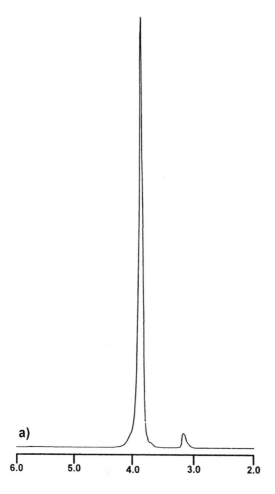

Figure 41   ¹H NMR spectra of poly(ethylene glycol)'s (PEG) of different molecular weights: (a) MW ≈ 1000; (b) MW ≈ 5000. The samples were dissolved in CDCl₃; the *x* axes are in ppm

have a significant impact on the bulk properties of a material. ¹³C spectra of polymers with in-chain unsaturation, such as polyisoprene,

$$-(-CH=\overset{\overset{\displaystyle CH_3}{|}}{C}-)_n-\qquad\qquad(16)$$

clearly show resonances attributable to the *cis* and/or the *trans* form (Fig. 42). Similarly, the configuration of the polymer backbone through the cyclo-

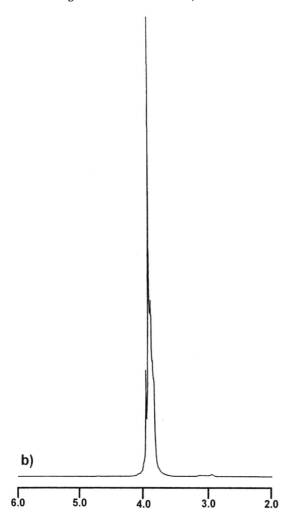

b)

6.0   5.0   4.0   3.0   2.0

Figure 41 Continued

hexyl ring in poly(1,4-cyclohexanedimethylene terephthalate) (Fig. 43) is clearly reflected in its ^{13}C spectrum.

NMR's ability to characterize the tacticity, or stereoisomerism, of polymers, such as polypropylene, polystyrene, or poly(vinyl chloride), is also significant. The tacticity of these, or of any other stereoirregular polymer, is analyzed as sequences of stereo pairs, or *dyads*. Each pair is designated as either *meso* (m) or *racemic* (r), depending on the stereostructure. In *isotactic*

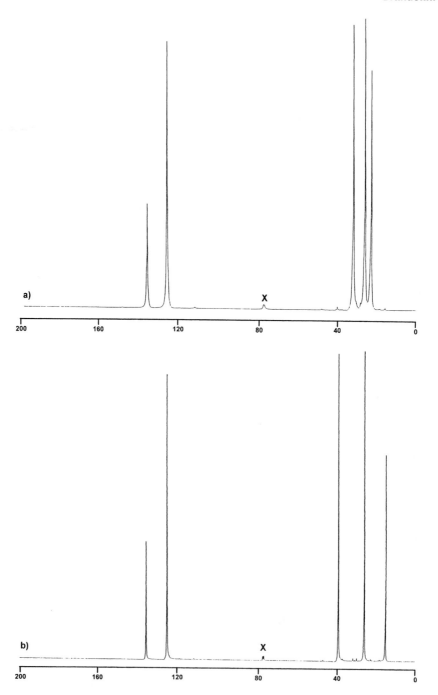

polymers, the pendant groups are all arranged on one side of the backbone (Fig. 44a) in an all-meso sequence; in the *syndiotactic* form, they alternate (Fig. 44b), and all are racemic. No definitive structure can be drawn for *atactic* materials; their stereosequences are randomly distributed (such as in Figure 44c). Figure 45 compares methyl regions from the ^{13}C spectra of: (a) isotactic; (b) syndiotactic; and (c) atactic polypropylene solutions; peak assignments are also shown. As expected, the predominantly isotactic form contains mostly meso configurations (*mm*), the syndiotactic consists mainly of racemic (*rr*), and the atactic, of nearly equivalent amounts of *mm*, *mr*, and *rr*. Resonances due to longer sequences, such as *mmm, mmr, mrm, rmr, rrm,* and *rrr* are seen as fine structure within each of the dyad groupings [76]. No technique can fully characterize the stereosequence of every polymer chain in a sample, but statistical analyses provide an overall picture of the structure. An isotactic chain with only one configurational shift, such as the one shown in Fig. 44d, is 75% *mm*, 25% *mr*, and 0% *rr*. On the other hand, a chain with many shifts, such as Fig. 44c, is described as 12.5% *mm*, 62.5% *mr*, and 25% *rr*. As can be seen from Fig. 45, resonances arising from sequences longer than triads (four monomer units) may overlap severely; the analysis of higher *n*-ads is a good candidate for 2-D NMR study. Heteronuclear correlated spectroscopy has aided in the detailed assignment of the ^{1}H spectrum of poly(pentyl acrylate), and a homonuclear COSY experiment definec the ^{1}H–^{1}H J-couplings [120]. Tacticity effects have even been studied in the solid state [121], although they are not always observed, and LC/NMR has been used to study the tacticity of oligostyrene species [122].

Tacticity measurements can be correlated with reaction mechanisms and physical properties. For example, the incorporation of an electron donor into the polymerization catalyst formulation has been found to increase isotacticity in a propylene-1-butene copolymer [123], and the distribution of propylene and 1-butene contents as a function of molecular weight varied, depending on donor type. External donors, such as dimethoxysilane, decrease the butene content more than internal electron donors (in this case, di-*n*-butyl phthalate). Mechanisms of new polymerization reactions, such as the group-transfer copolymerization of methyl methacrylate and lauryl methacrylate, can be elucidated by comparing NMR-derived structural details [124]. The presence of unanticipated peaks in the spectrum of poly(ethylene-co-norbornene) suggest the occurrence of epimerization

---

Figure 42 ^{13}C NMR spectra of (a) *cis-* and (b) *trans*-polyisoprene. The samples were dissolved in CDCl$_3$ (marked by "X"); the *x* axes are in ppm.

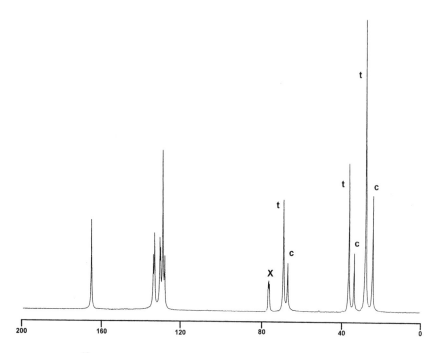

Figure 43 ^{13}C NMR spectrum of poly(cyclohexanedimethylene terephthalate) showing cis (c) and trans (t) configurations of the cyclohexyl ring. The sample was dissolved in CDCl$_3$ (marked by "X"); the x axis is in ppm

[125]. Computer simulations of polymer-chain growth can be compared to experimental data in order to propose or support models of polymerization, as has been done for polylactide [126]. Physical properties, such as the dielectric constant of poly(N-vinyl carbazole), can often be correlated with structural features such as tacticity [127].

III.B.5. Copolymer Structure, Composition, and Sequence
          Distribution

Many commercial plastic materials are actually copolymers, formed by the co-reaction of one or more different monomers. Copolymers can be blocky, alternating, or random. The spectrum of a blocky $-$AAAAAABBBBBB$-$ copolymer, such as poly(styrene-co-butadiene) (Fig. 46a) strongly resembles the weighted sum of the spectra of the corresponding homopolymers (Figs 46b and 46c). If the blocks are not very long, resonances arising from $-$ AB$-$ junctions between the $-$AAAAA and BBBBB$-$ sequences may be

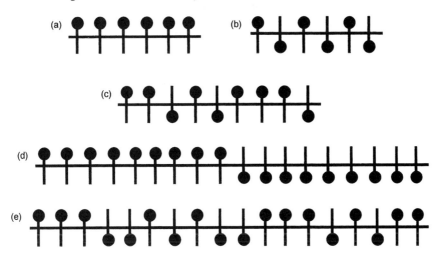

Figure 44 Schematic depictions of tacticity: (a) isotactic; (b) syndiotactic; (c) atactic; (d) primarily isotactic chain, with one stereo-alteration; (e) primarily atactic chain.

observed, but in most copolymers, they are very small [128]. If the comonomers are randomly distributed, such as in poly(vinyl acetate-co-ethylene) (Fig. 47a), peaks appear that are not present in either homopolymer spectrum (Fig. 47b; the main resonance for polyethylene is a single peak at 30 ppm). Where appropriate, nuclides other than ^{1}H and ^{13}C can be used for analysis of copolymers, as demonstrated by [29]Si studies of several copolysiloxanes [129]. In certain cases, two-dimensional approaches such as NOESY can clarify ambiguous assignments [130]. For some copolymers, such as poly(propylene-co-1-hexene), sequence and tacticity effects coexist, making detailed interpretation of the spectra very involved [131].

As discussed previously (Sec. II.D.4), acquisition of a quantitative NMR spectrum presupposes careful planning. NMR offers several advantages over other techniques, such as infrared spectroscopy, for the determination of copolymer composition. First, in a truly quantitative NMR spectrum, the relative areas of all peaks are self-consistent, so that if the intensity of peak A is twice that of peak B, it is certain that the number of nuclei contributing to A is double that of B. In other words, there is no NMR equivalent of an extinction coefficient, and precharacterized standards are not needed for method development, if the experiment is properly designed and executed. Second, because of NMR's great sensitivity to subtle structural changes, even chemically similar species can be distinguished. It is, for example, possible to calculate the concentration of each pendant-

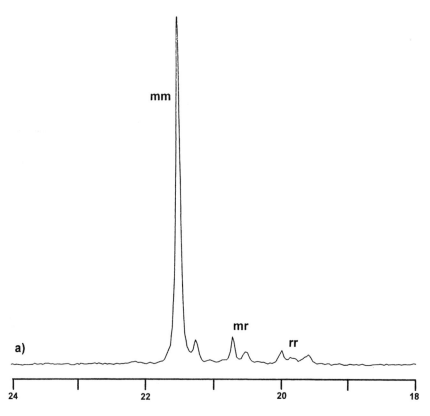

Figure 45 Methyl regions of the ^{13}C NMR spectra of polypropylene: (a) isotactic; (b) syndiotactic; (c) atactic. The samples were dissolved in 1,2-dichlorobenzene-d$_4$/1,2,4-trichlorobenzene (1/3 v/v); the x axes are in ppm.

group type in low-density polyethylene, whose spectrum is depicted in Fig. 39, even though the branches differ only by a few carbons. Third, NMR spectra often contain redundant peaks, as shown in Fig. 48, the ^{13}C spectrum of poly(ethylene-co-1-hexene). To shorten the experiment time, the relaxation delay for this spectrum was set to 15 s, which is too short to allow the methyl-group resonance (14.1 ppm) to relax fully. This peak is not needed for the calculation, however, because any or all of the well-resolved peaks due to side-chain carbons (23.4 and 34.9 ppm), the methine branch point (38.1 ppm), or the backbone carbons near the branch point (34.5 and 27.3 ppm) can, in principle, be used to determine the comonomer content [132].

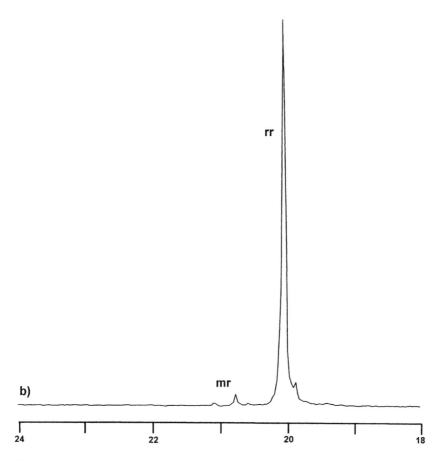

Figure 45 Continued

The procedure for calculating comonomer concentration directly from an NMR spectrum is straightforward; it will be demonstrated for the ^{13}C spectrum of poly(propylene-co-ethylene-co-1-butene), shown in Fig. 49. In the first step, at least one peak attributable to each component must be identified; ideally, this peak should not overlap with resonances from any other spectral features. If more than one such peak is available, the areas can be averaged. In some cases, it is necessary to calculate peak contributions. For example, if the spectrum has two peaks, one assignable to component A and the other to A + B, then the contribution for B can be obtained by subtracting the first area from the second. In the case of poly (propylene-co-ethylene-co-1-butene), one can use the averages of the peaks

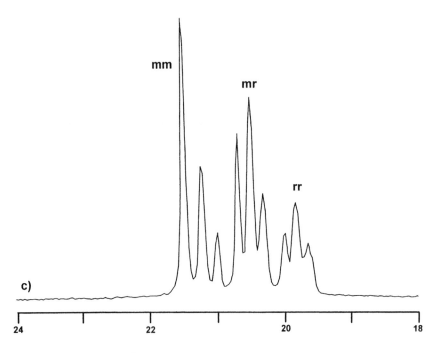

Figure 45 Continued

at: (1) 46 and 22 ppm (one carbon each) as the propylene contribution, $\langle P \rangle$; (2) 48 ppm (two carbons), 31 ppm (two carbons), and 24 ppm (one carbon) for ethylene, $\langle E \rangle$; and (3) 43.5 ppm (two carbons) and 35 ppm (one carbon) ppm for 1-butene, $\langle B \rangle$. The mole fraction, $[X]$, for each component can be calculated from:

$$[X] = \frac{\langle X \rangle}{\langle P \rangle + \langle E \rangle + \langle B \rangle} \tag{17}$$

where $[X]$ is $[P]$, $[B]$, or $[E]$. The material whose spectrum is shown in Fig. 49 is 79 mol% P, 17 mol% B, and 4 mol% E, or, as more commonly expressed, 75 wt% P, 22 wt% B, and 3 wt% E.

It has already been shown (Figs 46 and 47) how easily NMR distinguishes block from random copolymers. Many copolymers appear to be irregular, but they are not truly random in a statistical sense; there may be some slight tendency to blockiness that is not readily apparent from the spectrum. NMR can characterize the extent of this blockiness, through an

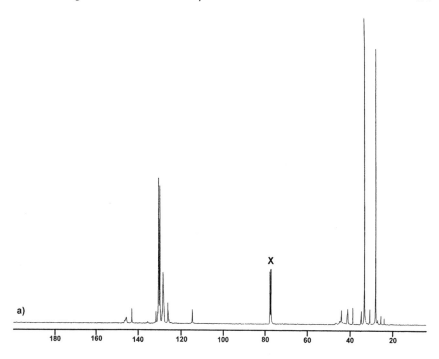

Figure 46 ^{13}C NMR spectra of: (a) styrene/butadiene block copolymer; (b) polystyrene; (c) polybutadiene (*cis-* and *trans-*). The samples were dissolved in CDCl$_3$; the x axes are in ppm.

analysis similar to that used to characterize tacticity (Sec. II.B.4). A (very short) hypothetical copolymer,

$$\text{AAAABBBAAABAABBBBABB} \tag{18}$$

can be analyzed as a series of triads, i.e., groups of three comonomer units. The triad sequence for structure (18) is: AAA (first three units); AAA (second through fourth units); AAB (third through fifth units); ABB; BBB; BBA; BAA; AAA; and so on. Analysis of this distribution reveals that there are 17% AAA triads, 28% AAB, 5% ABA, 5% BAB, 28% BBA, and 17% BBB. The mole fractions, [A] and [B], are given by [78]

$$[A] = [AAA] + [BAA] + [BAB] \tag{19a}$$

$$[B] = [BBB] + [ABB] + [ABA] \tag{19b}$$

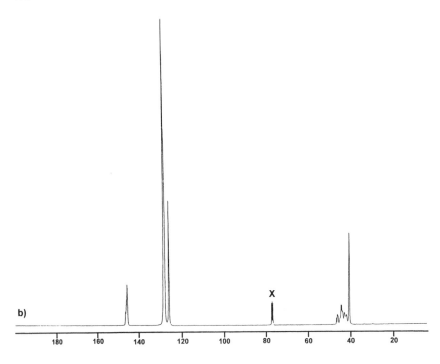

Figure 46   Continued

and both equal 0.50. If structure (18) is compared to a block material which is also 50/50 A/B,

AAAAAAAAAABBBBBBBBBB   (20)

the sequence distribution is seen to be quite different from structure (18). This block material contains 44% each AAA and BBB, 6% each AAB and BBA, and no ABA or BAB sequences. A strictly alternating copolymer, on the other hand,

ABABABABABABABABABAB   (21)

has 50% ABA and 50% BAB; no other sequences exist. In the spectrum of poly(ethylene-co-1-hexene) depicted in Fig. 48, small resonances due to sequences are observed at 36.7 and 24.3 ppm; they are due to HHE and HEH, respectively [132]. While ^{13}C NMR is usually the nuclide of choice for comonomer-sequence analyses, others are sometimes appropriate, such as for poly(ethylene terephthalate-co-hydroquinonephenylphosphate); in this case both ^{1}H and ^{31}P NMR spectra were studied [133].

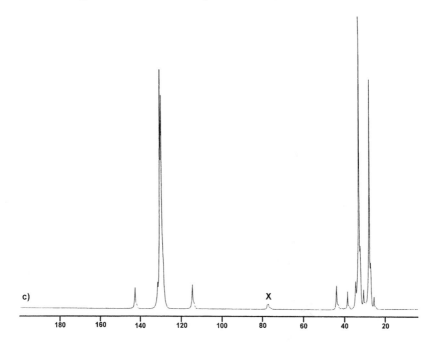

c)

180   160   140   120   100   80   60   40   20

Figure 46 Continued

Derived sequence distributions can be compared to those predicted by statistical copolymerization models, such as the Bernoullian and the *n*th-order Markovian models [78]. Bernoullian systems are statistically random. As a polymer chain grows, the probability that an A unit will be subsequently added is simply the probability that the next comonomer to approach the active site is A, i.e., the fraction of A in the comonomer mixture. Copolymerization of ethylene and styrene, when initiated by metallocene catalysts, produces a mostly isotactic alternating copolymer, for which, as seen above, ABA and BAB sequences predominate [134]. The fact that the copolymerization is truly random (Bernoullian) was confirmed by an analysis of the NMR data. In systems governed by Markovian statistics, the likelihood that A will add to a growing chain depends on the structure of the end of the polymer. In an *n*th-order model, the probability of reacting A varies with the identity and sequence of the last *n* comonomer units. In the first-order Markovian model, the likelihood of adding A to the polymer chain might be 95% if the chain ends with an A (P_{AA}), and only 5% if it ends with a B (P_{AB}), and $P_{AA} \gg P_{AB}$. If a similar situation occurs for addition of B ($P_{BB} \gg P_{BA}$), the resulting polymer will be blocky, as this

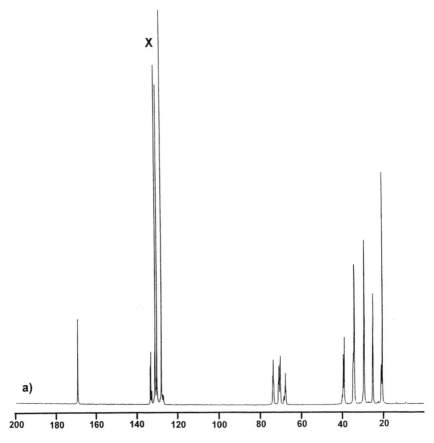

Figure 47 ^{13}C NMR spectra of: (a) poly(vinyl acetate-co-ethylene) (70/30) random copolymer in 1,2-dichlorobenzene-d$_4$/1,2,4-trichlorobenzene (1/3 v/v); and (b) poly(vinyl acetate) in CDCl$_3$. The x axes are in ppm.

model predicts mostly continuing addition of the same comonomer to the chain end. These probabilities are related to reactivity ratios, r_1 and r_2, which are often used to characterize copolymerization kinetics, and NMR has been used to determine r_1 and r_2, as for the coreaction of acrylonitrile and vinyl acetate [135].

Even more sophisticated data analyses have been applied to comonomer sequences in order to extract even more specific information from the data [136–142]. These detailed reaction models allow derivation of mechanistic and kinetic parameters from sequence-distribution data. While most of these

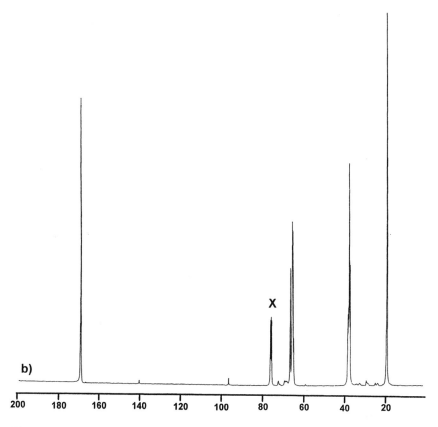

Figure 47 Continued

analyses have been directed at vinyl polymers, including polyolefins, they are generally applicable.

III.B.6.   Degradation Mechanisms

As a material undergoes degradation, whether from heat, light, chemical reaction, irradiation, or other source, alterations in the polymer's chemical structure, which can be characterized by NMR, occur. Upon thermal degradation, poly(ethylene oxide) and poly(propylene oxide) form ester and/or hydroxyl end groups, and decrease in molecular weight; a mechanism has been postulated based on the spectroscopic insights [143]. Mechanical shearing during extrusion decreases the molecular weight of poly(ethylene-co-propylene) [144]; 2-D ^{13}C–^{1}H correlation has facilitated assignment of the ^{1}H end-group resonances. Heat and UV aging of polyhexaneamide (Nylon

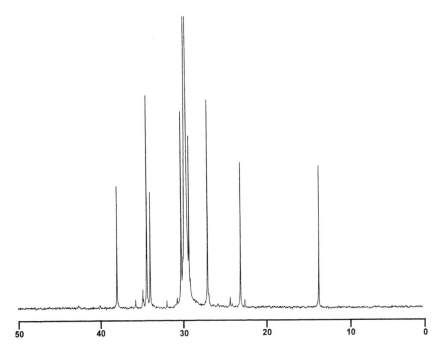

Figure 48 ^{13}C NMR spectrum of poly(ethylene-co-1-hexene). The sample was dissolved in 1,2-dichlorobenzene-d$_4$/1,2,4-trichlorobenzene (1/3 v/v); the x axis is in ppm.

6) have been studied by 1-D and 2-D NMR (^{13}C and ^{15}N) experiments which reveal the occurrence of chain scission and the formation of hydroxyl end groups [145]. In irradiated polycarbonate, the developement of a yellow color has been attributed to an unstable quinone species not directly detected. Reaction with a phosphorus-containing reagent stabilizes the quinone, allowing its existence to be confirmed by ^{31}P NMR [146]. Irradiation of poly(tetrafluoroethylene-co-perfluoromethyl vinyl ether) leads to a decrease in its solubility, so solid-state ^{13}C and ^{19}F NMR have been used to identify the end groups resulting from chain scission and cross-linking [147]. The process of biodegradation has recently become a subject of significant interest in polymer science. The enzymatic degradation of siloxane oligomers has been studied by a variety of 2D techniques [148], and NMR studies have improved the understanding of the biodegradation of poly(lactic acid) [149].

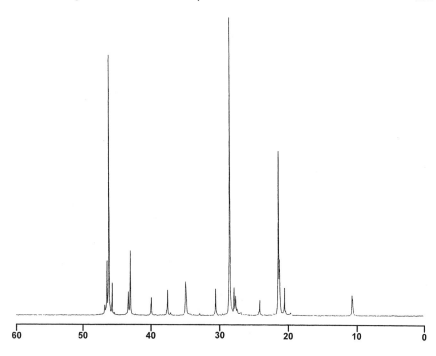

Figure 49 ^{13}C NMR spectrum of poly(propylene-co-ethylene-co-1-butene). The sample was dissolved in 1,2-dichlorobenzene-d$_4$/1,2,4-trichlorobenzene (1/3 v/v); the x axis is in ppm.

III.C. Bulk Physical Properties

To this point, the applications of NMR to polymer analysis have focused on the description of molecular structures. The section discusses supramolecular features such as morphology and orientation. These uses of NMR are mostly solid-state studies. Since they are more specialized, they are not covered in as much detail. The references will, however, provide an overview of the technique's capabilities.

III.C.1. Morphology

In the solid state, polymer chains can be (more or less) randomly coiled or arrayed in an ordered state, and these amorphous and crystalline domains give rise to different signals in the solid-state NMR spectrum of a bulk material. The broad-line ^{1}H NMR spectrum of a semicrystalline polymer such as polyethylene has already been introduced (Sec. II.H.2 and Fig. 30). The linewidth difference is caused by the more restrictive molecular motion in the crystalline phase. In high-resolution solid-state NMR, morphological differences can result in chemical-shielding variations which are reflected in a CP/MAS or CRAMPS spectrum. Chemical shift originates in the electronic environment created by the spatial arrangement of neighboring nuclei, so it follows that diverse crystalline forms will give rise to different shielding values. The relationship between shielding and morphology can be calculated, as was done for poly(phenylene sulfide) [150]. The various morphologies of poly(4-methyl-1-pentene) have been compared [151,152], and several are depicted in Fig. 50. These data lead to a detailed description of molecular conformations; the results agree with the conclusions from X-ray diffraction studies. For example, in all cases, the methyl side-chain ends are inequivalent by ^{13}C NMR, which limits the possible polymorphs to those with asymmetric positions for these carbons. The existence of two crystalline polymorphs (orthorhombic and monoclinic) are detected in ultrahigh-molecular-weight polyethylene fibers by ^{13}C NMR [153]. The amorphous component in some polyethylenes has been found to be comprised of multiple components; they are assigned to an interfacial region and a rubbery domain [154]. Polyethylene morphology has been studied as a function of reactor type (slurry or gas phase) and reaction temperature; the relative amounts of orthorhombic and monoclinic phases varied with the process used to produce the resin [155]. A new type of ordered nematic state has been proposed for a random, liquid-crystalline copolymer of 4-hydroxybenzoic acid and 6-hydroxy-2-naphthoic acid [156]. The crystal forms of ^{15}N-labeled polylauramide (Nylon 12) are identifiable by ^{13}C and ^{15}N CP/MAS experiments [157], and the fact that the two components of poly(3-hydroxybutyrate-co-3-hydroxyvalerate) co-crystallize has been confirmed by ^{13}C NMR [158]. The crystallization process of a liquid-crystalline polyurethane could be followed through slow cooling [159]. Hydroxyl groups in poly(ethylene-co-vinyl alcohol) are partitioned proportionately between the crystalline and amorphous phases [160]. Other thermodynamic changes that have been correlated with NMR include the freezing of a poly(vinyl alcohol) solution [161] and the glass transition of poly(vinyl chloride) [162]. Manipulating the spin-lock time before cross-polarization allows the spectra of specific phases to be recorded. Mixtures of [13]C- and [2]H-labeled polycarbonates has been analyzed by a 2-D technique called *rotational-echo double resonance* (REDOR), in which [13]C–[2]H couplings are interpreted as intermolecular distances [163]. The location of a counter-ion in an ionomer could be pinpointed based on observed morphological changes, evidenced by ^{23}Na NMR [164]. The effect of radiation on the morphology of linear low-density polyethylene has been studied by solid-state ^{13}C NMR; peaks attributable to cross-links have even been observed [165].

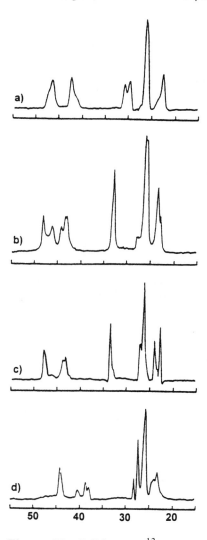

Figure 50 Solid-state ^{13}C NMR spectra of poly(4-methyl-1-pentene) (P4MP) showing different crystalline forms: (a) form I, isotactic P4MP; (b) form II, isotactic P4MP; (c) form III, isotactic P4MP; (d) form IV, isotactic P4MP. The x axes are in ppm (From Ref. 151, © 1997 American Chemical Society.)

Probe molecules such as ^{129}Xe can also be used to investigate phase structure. In general, the gas diffuses into the amorphous regions of the polymer, where the chemical shift is thought to be governed by the van der Waals interaction [166]. Using this approach, a microporous structure has been found in polyethylene [167], and the presence of more than one subphase has been detected in the amorphous region of this polymer [168]. Similar studies have been performed on ethylene-propylene rubber (EPDM) [169] and impact polypropylene [170]. Figure 51 depicts the ^{129}Xe spectra of EPDM before and after curing. The least shielded peaks (> 200 ppm) correspond to the smallest voids in the amorphous phase; the most shielded (< 200 ppm), to the largest open spaces. Cross-linking decreases the average void size, which is reflected in the tendency toward higher chemical shifts after curing.

Analysis of multicomponent exponential curves from relaxation measurements can offer insights into the microphase structure of polymers, such as poly(vinyl chloride), polyester, and nitrile rubber [171]. Spin-diffusion measurements are useful for estimating domain sizes [52,53], as for poly(ethylene terephthalate), poly(ethylene isophthalate), and poly(ethylene naphthenoate) [172].

Materials imaging is not a highly developed technique, partly because of some of the difficulties inherent in attempting to view the small to microscopic features in solids [70–72]. Simply put, observation of microscopic structures requires very steep field gradients that are hard to generate.

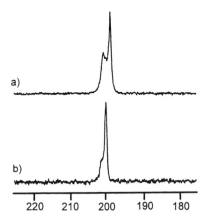

Figure 51 ^{129}Xe NMR spectra of gas adsorbed into ethylene/propylene rubber before (a) and after (b) curing. (Reprinted with permission from Ref. 169, © 1990, Springer-Verlag.)

Also, since dipolar interactions and chemical-shift anisotropy are operative, they need to be suppressed. It is easier to image mobile molecules, such as those found in a polymer's amorphous regions. This fact has been exploited to study the disordered phase preferentially during the enzymatic biodegradation of poly(3-hydroxybutyrate) and poly(3-hydroxybutyrate-co-3-hydroxyvalerate) [173]. Images of a poly(methacrylic acid) gel reveal the motion of the Mn^{2+} ions in this material under an applied electric field [174]. The effect of tensile stress on the molecular dynamics of poly(Bisphenol A carbonate) has been investigated [175]. Images produced of aging-related phase separation of natural rubber and poly(styrene-co-isoprene) demonstrate the utility of these experiments for studying changes in materials [176].

III.C.2. Orientation

Polymers that undergo mechanical deformation experience morphological changes which can be observed in certain types of NMR spectra. Not surprisingly, poly(ethylene terephthalate) (PET) has been the subject of many NMR studies of oriented fibers and films. For example, PET yarns were found to be almost completely amorphous by XRD, but ^{13}C NMR reveals a region with relatively restricted motion within the disordered phase [177]. Orientation distribution functions for PET materials have been derived from multidimensional exchange NMR, particularly an experiment called *DECODER* (*D*irection *e*xchange with *c*orrelation for *o*rientation-*d*istribution *e*valuation and *r*econstruction) [178]. During this experiment, the physical orientation of the sample, relative to B_0, is changed, altering the projection of the chemical-shift anisotropy tensor axes onto the magnetic field. This leads to very different spectra when the sample's position is changed, even in the 1-D spectrum (Fig. 52). In DECODER, contour plots are similar to the pole diagrams produced by XRD, in that they provide direct evidence of the distribution of orientations, as shown in Fig. 53. Unoriented samples, as portrayed in Fig. 53a, generate an elliptical contour plot with equivalent intensity around the perimeter. Biaxial distributions, on the other hand, cause a decrease in signal at certain points, depending on the features of the sample's orientation (Figs 53b and 53c). Similar information has been obtained from ^2H NMR studies of deuterated PET through the use of specific labels, in either the glycol or the aromatic part of the polymer [179]. Phase changes which accompany the drawing of poly(ethylene naphthenoate) and poly(butylene naphthenoate) have been observed by ^{13}C CP/MAS NMR [180].

A simpler approach is to dissolve a small amount of some probe molecule into the material [181], where it accumulates in the amorphous regions. Images of filled elastomers under mechanical stress have been produced. Inhomogeneities which appear in silica-filled polydimethylsiloxane have

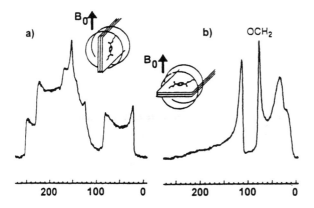

Figure 52 ^{13}C spectra of biaxially oriented poly(ethylene terephthalate) film at two orientations (a) and (b) relative to $\mathbf{B_0}$. (From Ref. 178, © 1993 American Chemical Society.)

been clearly seen [182], and deformation-induced heating in carbon black-filled styrene-butadiene rubber (SBR) has been mapped [183].

III.D.   Polymer-Chain Dynamics

Since the time dependence of variations in magnetization, dipolar or quadrupolar coupling, and chemical-shift anisotropy are inherent parts of the NMR phenomenon, it is a natural tool for studying molecular motions. These relationships are usually well understood, and NMR data can therefore be used to test dynamic models. Conformations, free volume, diffusion, and mechanical properties have all been correlated with NMR observations. This section will first review mobile systems (solution, melts, and elastomers), and then discuss solids.

III.D.1.   Solutions, Melts, and Elastomers

NMR's ability to resolve chemical features makes the technique very powerful for characterizing the dynamics of specific groups within a molecule. For example, the thermodynamically preferred conformation of the 1,3-dioxane rings in poly(vinyl formal) and poly(vinyl butyral) has been confirmed to be a cis-chair form, but the second conformation is a twisted form rather than trans-chair [184]. Multidimensional NOE and J-resolved experiments have provided conformational descriptions based on analysis of scalar couplings and through-space dipolar interactions [185]. ^{1}H T_1 measurements indicate that the dissolution of polystyrene in cyclohexane, a θ solvent, occurs in two

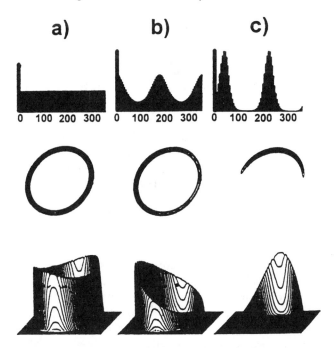

Figure 53 DECODER spectra for: (a) no orientation; (b) some biaxiality; (c) pronounced biaxiality. The top row shows the orientation distribution; the middle row, the contour plot of the two-dimensional spectrum; the bottom row, the stacked plot. (From Ref. 178, © 1993 American Chemical Society.)

discrete steps [186]. Terminal ionic groups in polytetrahydrofuran have been found to affect the motion of the polymer chains only in low-dielectric-constant solvents [187]. The binding constants of polyvinylpyrrolidone with halide anions have been determined by ^{35}Cl, ^{81}Br, and ^{127}I NMR [188]. While these nuclides are not commonly observed, because of their resonances' inherent breadth, the excess linewidth is found to relate to the strength of the binding.

Many NMR dynamics studies attempt to correlate the observed data with models of molecular motion. Some of the more successful are the Hall-Helfand [189,190], the Dejean-Lauprêtre-Monnerie (DLM) [189,191, 192], and the Williams-Watts [189,194,195] models. They predict the relaxation times and NOEs expected from specific types of motion such as unrestricted diffusion or discrete jumps. Very often, distinct parts of the molecule are be best modeled by different functions. For a poly(isobutyl methacrylate) solution, the DLM model, with its conformational jumps and bond librations, describes backbone motion well [196]. The isobutyl side-group dynamics, on the other hand, most closely follows a three-site jump scheme. The DLM model has often been found to be in good agreement with relaxation data for vinyl polymers, such as poly(vinyl alcohol) in water [197], poly(vinyl pyridine) in chloroform [198], and poly(vinyl chloride) in dibutyl phthalate and tetrachloroethane [199]. Similar measurements can be compared to well-known descriptions of fluid dynamics, such as the Doi-Edwards tube-disentanglement model [200,201] and reptation [202–204]. NMR parameters can also be correlated with the well-known Williams-Landel-Ferry (WLF) equation [205], which describes the temperature dependence of many thermodynamic properties, generically designated by φ:

$$\log\left(\frac{\varphi}{\varphi_0}\right) = \frac{-C_1(T - T_0)}{C_2 + (T - T_0)} \tag{22}$$

where C_1 and C_2 are constants, T is temperature, and the subscript 0 indicates a reference state. The mobility inferred from the deuterium quadrupolar splitting of ^2H_2O dissolved in a block polyamide-polyether copolymer follows Eq. (20) and does not depend on the type of polyamide [206]. This type of behavior has been found to be applicable to a broad range of polymers [207]. More generically, free-volume approaches have been helpful in understanding NMR results [208]. Many attempts have been made to correlate NMR with rheological measurements. During the cross-linking of unsaturated polyesters, ^1H NMR has been found to indicate gel formation earlier than rheology during the crosslinking of unsaturated polyesters [209]. Rheo-NMR, discussed in Secs II.B.3, II.E.6, and II.H.6, has been demonstrated for polyisobutylene, polydimethylsiloxane (PDMS), and poly(vinyl methyl ether) [36]. As shown in Fig. 54, the linewidth of PDMS increases in a shear field, and gradually decreases once the stress is relieved. Furthermore, the magnitude of the $\Delta v_{1/2}$ increase is greater at higher shear rate. NMR imaging, which is amenable to the study of polymer fluids, generates profiles of flowing aqueous solutions of poly(ethylene oxide) [210].

The investigation of diffusional processes by the PGSE (pulsed gradient-spin echo) experiment has been discussed in Sections II.B.2., II.E.5., and II.H.5. The distance scale over which measurements can be made depends on the strength of the imposed gradient. With very steep, high-magnitude gradients, it is possible to reduce the effective range to < 100 nm [211]. Modified poly(ethylene oxide) polymers, which are used as water thickeners, form clusters in solution, and their diffusion has been characterized [212]. The PGSE experiment can be combined with rheo-NMR to study entanglements in a polystyrene/cyclohexane system [213]. Diffusion coefficients for poly(ethylene glycol) (PEG) in several acrylamide/acrylic acid gels have

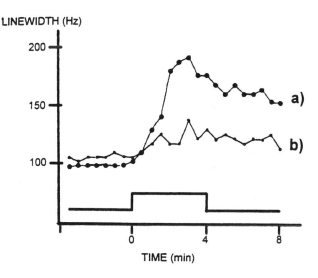

Figure 54 Evolution of the ^{1}H NMR signal linewidths for polydimethylsiloxane at different shear rates: (a) 4.35 s^{-1}; (b) 1.16 s^{-1}. (From Ref. 35. 1990 American Chemical Society.)

recently been measured; a hydrogen-bonded complex forms between the PEG and the gel [214]. Like the simpler measurements of T_1 and NOE, diffusion results can be analyzed in terms of free-volume and dynamic models, as has been done for the curing of an epoxy resin [215] and for short-chain diffusion in PDMS [216].

III.D.2. Solids

Variable-temperature NMR experiments provide information about the dynamic processes that occur in solids. It is well known that, for a semicrystalline polymer such as polyethylene (PE), a higher rate of cooling from the melt leads to lower crystallinity; ^{13}C CP/MAS NMR establishes that the cooling rate also affects the structure and motion of the amorphous domain [217]. Many of the dynamic properties of PE, such as jump rates and activation energy, can be explained in terms of chain diffusion between the phases, as detected by 2-D exchange ^{13}C NMR [218]. Very slow motions in the crystalline α-form of deuterated poly(vinylidene fluoride) have also been detected by exchange NMR [219]. Double-quantum transitions can occur in materials in which ^{13}C–^{13}C spin pairs are only a few bonds apart; this experiment can yield torsion angles for enriched PE [220]. ^{19}F multiple-quantum NMR has been used to differentiate polytetrafluoroethylene samples with different crystallinities and thermal histories [221]. The structure and dynamics of polycaprolactone have been comprehensively studied. The relative fractions of rigid crystalline, mobile crystalline, interfacial, and amorphous material have all been calculated [222]. Solid-state ^{2}H NMR of phenyl-deuterated poly(ethylene terephthalate) reveals that the rings undergo discrete 180° flips [223]. As shown in Fig. 55, computer simulations suggest that the flip rates exhibit a bimodal distribution, with a narrow component centered at 10 Hz and a broader one around 10 kHz.

The mobility of the soft phase in a two-segment thermoplastic elastomer polyurethane, as derived from ^{2}H NMR, correlates well with mechanical properties [224,225]. The soft poly(tetramethylene oxide) segments consist of both crystalline and amorphous phases. The observed 100-fold decrease in storage modulus, G', between 200 and 280 K corresponds to the transition of the amorphous part of the soft blocks from a glassy to a rubbery

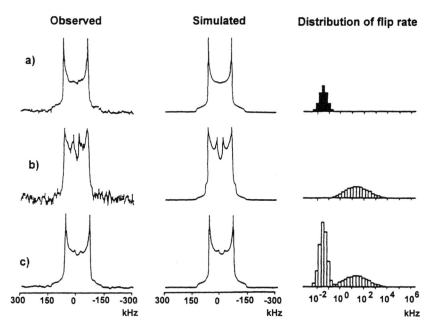

Figure 55   Experimental ^{2}H powder patterns, simulated spectra, and model phenyl-ring flip-rate distributions; for quenched, phenylene-labeled poly (ethylene terephthalate) for: (a) crystalline phase; (b) mobile noncrystalline; (c) total noncrystalline. (From Ref. 223, © 1998 Elsevier Science Ltd.)

state, and of the crystalline part, to a molten form. A further decrease in G' above 280 K is attributed to a softening in hard piperazine segments. This relationship between storage modulus and NMR parameters has been calculated for a set of 16 diverse materials, including three polyurethanes, polystyrene, poly(2,6-dimethyl-1,4-phenylene oxide), poly(vinyl formal), poly(vinyl butyral), plasticized poly(vinyl butyral), low-density polyethylene, and plasticized poly(vinyl chloride) [226].

III.E. Multicomponent Polymer Systems

Most plastic products contain additives or other components, whose purpose is to modify the properties of the base resin, change its appearance, or lower the cost. Blends of more than one polymer are common, and composites contain inorganic fillers. Additives are incorporated into polymers for many reasons, such as altering transition temperatures, improving stability, or enhancing surface properties. NMR is very useful for characterizing the composition of blends, and solid-state techniques provide insight into compatibility and phase structure. Since NMR's sensitivity is limited, its utility for filler and additive analysis is limited, as they are often used at very low ($\mu g/g$) levels. In specific cases, however, NMR can be useful even for these components.

III.E.1. Blends

Only an elementary knowledge of chemical shifts is needed to predict that blends of very different polymers, such as polystyrene and poly(methyl vinyl ether) are easy to distinguish based on their NMR spectra. In many cases, the most difficult part of the analysis is finding a mutually appropriate solvent. But NMR can be useful even for the characterization of materials with only slight structural variations. Figure 56, for example, shows ^{13}C spectra of an octene linear low-density PE (LLDPE), low-density polyethylene (LDPE), and an 80/20 blend of the two. In octene-LLDPE, the only side chains present are six carbons long, which give rise to a limited number of nonbackbone resonances (Fig. 56a). As discussed earlier (Sec. III.B.2), LDPE has branches of many different lengths, and so it exhibits a complex spectrum (Fig. 56b). No unique resonances appear in the spectrum of the blend (Fig. 56c), but close inspection reveals that the ratio of the peaks at 23.4 and 22.9 ppm has decreased relative to that in LDPE due to the contribution of the poly(ethylene-1-octene)'s resonance at 22.9 ppm. In a case where the peaks overlap even more severely, a combination of DEPT and heteronuclear COSY discloses peak assignments for blends of polypropylene and ethylene-propylene rubber [227].

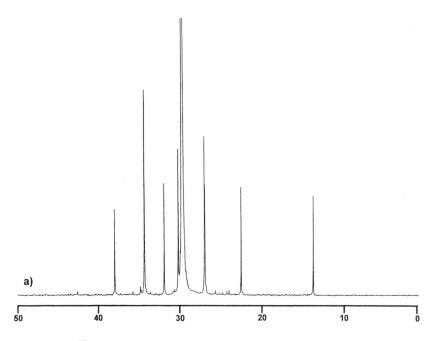

Figure 56 ^{13}C NMR spectra of: (a) poly(ethylene-co-1-octene); (b) low-density polyethylene; (c) 20/80 blend of (a) and (b). The samples were dissolved in 1,2-dichlorobenzene-d$_4$/1,2,4-trichlorobenzene (1/3 v/v).

Polymer blends exhibit varying degrees of compatibility, and NMR provides an effective tool for determining the degree of miscibility. Poly(methyl methacrylate) (PMMA) and poly(vinyl chloride) (PVC), for example, can form compatible blends, and NOESY indicates that hydrogen bonds form between the methine carbon of PVC and the carbonyl of PMMA in concentrated tetrahydrofuran solutions (Fig. 57) [228]. At relatively low concentrations (22 wt%), only intramolecular crosspeaks are observed (Fig. 57a), while at higher concentrations (38 wt%), correlations between the two polymers are observed (marked by *). This intensity indicates that these protons are within 5 Å of each other. In a similar way, NOESY demonstrates molecular-level mixing in mixed polyurethanes [229] and in blends of poly(N,N-dimethylacrylamide) with a variety of styrenic polymers and copolymers [230]. One-dimensional ^{13}C CP/MAS NMR can be used to investigate solid-state miscibility. Strong interactions between blend components can even lead to changes in chemical shifts. For completely miscible blends, all resonances exhibit essentially the same intermediate relaxation time (T_1 or $T_{1\rho}$); this is analogous to the behavior of the glass-

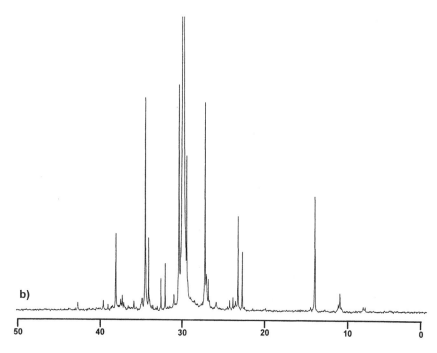

Figure 56 Continued

transition temperature, T_g. Finally, in a mixture of a protonated and a deuterated polymer, the appearance of peaks from the labeled component under cross-polarization is evidence of proximity. As discussed in Sec. II.E.2, only carbons near protons are enhanced by this process. If a carbon is not directly bonded to a proton, it will only be cross-polarized if it is close enough to experience a through-space dipolar interaction. Blends of poly(-ethylene terephthalate) and poly(ethylene naphthenoate) have been found to be miscible according to this criterion [231], as have mixtures of poly(buty-lene terephthalate) and a polyarylate [232]. Other approaches for establish-ing compatibility include measurement of spin diffusion [52,53,233–237], 2D exchange [238], and infusion of ^{129}Xe gas [239–241].

These procedures for investigating blends can be combined with the chain-dynamics techniques presented in Sec. III.D. Addition of polycarbo-nate to a poly(hexaneamide)/poly(propylene oxide) blend hardens the mate-rial; this has been attributed to restrictions in the mobility of the amine nitrogen in the polyamide caused by interfacial interactions among the other blend components [242]. The solid-state heteronuclear *WISE* (*wi*deline *se*paration) experiment can be tailored to selectively highlight the interface

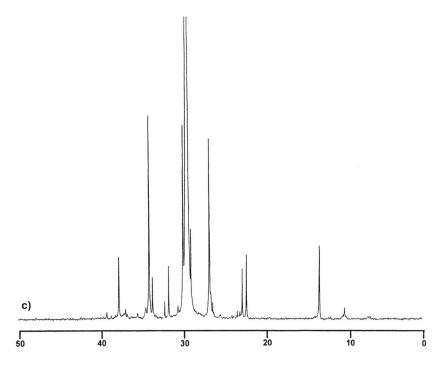

Figure 56 Continued

regions selectively [243,244]. It produces a chemical-shift resolved ^{13}C spectrum along one axis; a broadline ^{1}H signal along the other. The phase separation of a polystyrene/poly(vinyl methyl ether) blend, as monitored by this technique, has been found to correlate well with rheological measurements [244].

III.E.2. Composites

In polymer composites, the presence of an inorganic filler such as SiO_2 (often as glass fibers) or calcium carbonate significantly modifies the properties of the base resin. The efficiency of a particular composite depends heavily on the strength of the interface between polymer and filler. Coupling agents are often used to promote this adhesion. An (aminopropyl)triethoxysilane coupler mixed with bismaleimide resin has been studied by ^{13}C and ^{2}H NMR [245]. Based on comparison to the starting material, Fig. 58 clearly illustrates that a reaction has occurred. The dipolar-dephased spectrum of Fig. 58b shows the simplification effected by this sequence. In general, the dipolar- dephasing process accentuates interfacial regions [246].

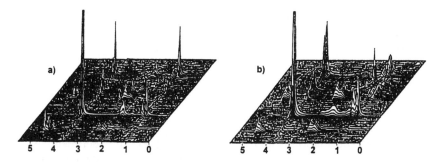

Figure 57 NOESY spectra of a 40/60 blend of poly(vinyl chloride) and syndiotactic poly(methyl methacrylate) dissolved in tetrahydrofuran-d$_8$ at concentrations of (a) 22 wt% and (b) 38 wt%. (From Ref. 228, © 1992 American Chemical Society.)

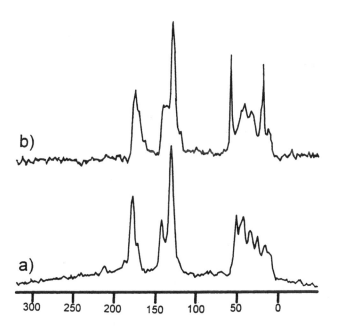

Figure 58 ^{13}C CP/MAS spectra of (aminopropyl)triethoxysilane coupling agent and bismaleimide resin: (a) with dipolar decoupling; (b) with dipolar dephasing. (From Ref. 245, © 1992 American Chemical Society.)

The molecular dynamics of deuterated poly(methyl acrylate) (PMA) change when the polymer is adsorbed onto silica. Two discrete species can be observed: (1) the bulk, which was identical to silica-free PMA; and (2) the interface, which exhibits evidence of an interaction [247]. Rather than studying the amount of polymer deposited onto a substrate, the solution from which adsorption has taken place can be analyzed. By difference, this spectrum indicates the polymer level in the model composite [248]. Images of polydimethylsiloxane which had been reinforced with *in situ*-generated silica show that SiO_2 formation is inhomogeneous [249]. PGSE experiments on a similar system indicate that the sorption of PDMS depends strongly on the molecular weight of the polymer chains [250].

III.E.3. Additives

Additives, which are commonly used to modify polymer properties, can be identified and quantified by NMR in favorable cases, such as the quantitation of poly(ethylene glycol) (PEG) [251] and other additives [252] in polyethylene (PE) or the characterization of the degradation of phosphorus-containing antioxidants in polyethylene and high-impact polystyrene (HIPS) [19]. Even though the additives are used in low amounts ($\sim$0.1–0.5 wt%), particular features of these systems make the analysis possible. Nearly all of the protons in PEG are equivalent, and its ^1H resonance is well resolved from that of PE. PE and HIPS typically contain no phosphorus-containing components other than secondary antioxidants, so a ^{31}P spectrum isolates these species. The molecular dynamics of the slip additive, erucamide, in polypropylene has been studied by several solid-state NMR techniques; erucamide microdomains were detected [253]. Plasticizers and antiplasticizers affect the glass-transition temperature, T_g, by changing the molecular motion in a material. It is not surprising, then, that their effect can be investigated by NMR. ^{13}C CP/MAS experiments have been carried out under a high pressure of CO_2 to compare its plasticization effect with that of temperature [254]. The rotational diffusion of a phosphorus-containing plasticizer, tetraxylyl hydroquinone diphosphate, in a blend of polystyrene and poly(2,6-dimethyl-1,4-phenylene oxide), has been found to occur at a rate which corresponds to a loss peak in the dynamic mechanical spectrum; these two phenomena, therefore, appear to be related [255].

IV. CONCLUDING REMARKS

As can be inferred from the previous sections, there is scarcely an area of polymer science that has not been the object of NMR investigations. From relatively straightforward material identification to sophisticated, multidimensional studies of dynamic models, the technique is continually being advanced and refined to address more complex questions and materials. Nearly 40 years ago, an NMR spectroscopist wrote, "What then does the future of the method hold . . . ? At present the possibilities seem unlimited" [22]. Since that time, spectroscopists have learned how to acquire spectra of rare spins, to extend experiments into two or more dimensions, to observe or suppress couplings, almost at will, and to generate detailed images based on nuclear interactions. The future still seems unbounded, particularly when applied to polymer science, a field with broad horizons of its own.

ACKNOWLEDGMENTS

Special thanks are due to Marie V. Brezina, who acquired some of the spectra used as illustrations, to Eileen A. Gurney, who assisted with compilation of the references, and to Cecil Dybowski for his helpful comments on this manuscript.

ABBREVIATIONS

ADC	analog-to-digital converter
COSY	correlated spectroscopy
CP	cross-polarization
CP/MAS	cross-polarization/magic-angle spinning
CPMG	Carr-Purcell-Meiboom-Gill experiment
CRAMPS	combined rotation and multiple-pulse spectroscopy
CSA	chemical-shift anisotropy
DECODER	direction exchange with correlation for orientation-distribution evaluation and reconstruction
DEPT	distortionless enhancement by polarization transfer
DQFCOSY	double-quantum filtered correlated spectroscopy
FID	free-induction decay
FT	Fourier transform
HMBC	heteronuclear multiple-bond correlation
HMQC-COSY	heteronuclear multiple-quantum correlation-correlation spectroscopy
HOHAHA	homonuclear Hartmann-Hahn
IMR	industrial magnetic resonance
INADEQUATE	incredible natural-abundance double-quantrum transfer experiment
LC-NMR	liquid chromatography-NMR
MAS	magic-angle spinning
NMR	nuclear magnetic resonance
NOE	nuclear Overhauser effect

NOESY	nuclear Overhauser effect spectroscopy
PGSE	pulsed gradient-spin echo
REDOR	rotational-echo double resonance
rf	radiofrequency
S/N	signal-to-noise ratio
T_1	spin-lattice relaxation time
T_2	spin-spin relaxation time
TMS	tetramethylsilane

REFERENCES

1. A. Abragam, *The Principles of Nuclear Magnetism*, Oxford, UK: Clarendon Press (1985).

2. B. Breitmaier, W. Voelter, *Carbon-13 NMR Spectroscopy*, New York: VCH (1987).

3. W. S. Brey, *Pulse Methods in ID and 2D Liquid-Phase NMR*, San Diego, CA: Academic (1988).

4. M. D. Bruch (ed.), *NMR Spectroscopy Techniques*, New York: Marcel Dekker (1996).

5. C. Dybowski, R. L. Lichter (eds.), *NMR Spectroscopy Techniques*, New York: Marcel Dekker (1987).

6. R. Ernst, G. Bodenhausen, A. Wokaun, *Principles of Nuclear Magnetic Resonance in One and Two Dimensions*, Oxford, UK: Clarendon Press (1987).

7. T. C. Farrar, E. D. Becker, *Pulse and Fourier Transform NMR*, New York: Academic (1971).

8. C. A. Fyfe, *Solid-State NMR for Chemists*, Guelph, ON: CFC (1983).

9. B. C. Gerstein, C. R. Dybowski, *Transient Techniques in NMR of Solids*, New York: Academic Press (1985).

10. R. K. Harris, *Nuclear Magnetic Resonance: A Physicochemical View*, Essex, UK: Longman (1986).

11. S. W. Homans, *A Dictionary of Concepts in NMR*, Oxford, UK: Clarendon Press (1995).

12. M. Mehring, *Principles of High-Resolution NMR in Solids*, Berlin: Springer-Verlag (1983).

13. J. Schraml, J. M. Bellama, *Two-Dimensional NMR Spectroscopy*, New York: Wiley (1988).

14. E. O. Stejskal, J. D. Memory, *High-Resolution NMR in the Solid State*, New York: Oxford University Press (1994).

15. F. J. M. van de Ven, *Multidimensional NMR in Liquids*, New York: VCH (1995).

16. F. W. Wehrli, T. Wirthlin, *Interpretation of Carbon-13 NMR Spectra*, London, UK: Heyden (1980).

17. R. K. Harris, B. E. Mann, *NMR and the Periodic Table*, London: Academic (1978).

18. J. Mason (ed.), *Multinuclear NMR*, New York: Plenum (1987).

19. A. J. Brandolini, J. M. Garcia, R. E. Truitt, *Spectroscopy* 7: 34 (1992).
20. P.J. Barrie, J. Klinowski, *Progr. NMR Spectrosc.* 24: 91 (1992).
21. R. F. Colletti, L. J. Mathias, Solid State ^{2}H NMR: Overview with Specific Examples. In *Solid State NMR of Polymers*, L. J. Mathias (ed.), New York: Plenum (1991).
22. L. M. Jackman, *Applications of Nuclear Magnetic Resonance Spectroscopy in Organic Chemistry*, New York: Pergamon (1959).
23. A. J. Brandolini, C. Dybowski, High-Resolution NMR Studies of Oriented Polymers. In *High-Resolution NMR Spectroscopy of Synthetic Polymers in Bulk*, R. A. Komoroski (ed.), Deerfield Beach, FL: VCH (1986).
24. E. Fukushima, S. B. W. Roeder, *Experimental Pulse NMR: A Nuts and Bolts Approach*, Reading, MA: Addison-Wesley (1981).
25. R. A. Komoroski (ed.), *High-Resolution NMR Spectroscopy of Synthetic Polymers in Bulk*, Deerfield Beach, FL: VCH (1986).
26. L. J. Mathias (ed.), *Solid State NMR of Polymers*, New York: Plenum (1991).
27. C. Dybowski, *Anal. Chem.* 70: 1R (1998).
28. C. Dybowski, Multi-Pulse ^{1}H and [19F Techniques. In *Solid State NMR of Polymers*, L. J. Mathias (ed.), New York: Plenum (1991).
29. F. D. Blum, *Spectroscopy* 1: 32 (1986).
30. P. Stilbs, *Progr. NMR Spectrosc.* 19: 1 (1987).
31. R. A. Komoroski, *Anal. Chem.* 65: 1069A (1993).
32. J. M. Listerud, S. W. Sinton, G. P. Drobny, *Anal. Chem.* 61: 23A (1989).
33. M. S. Went, L. W. Jelinski, Chemical Aspects of NMR Imaging. In *Microscopic and Spectroscopic Imaging of the Chemical State*, M. D. Morris (ed.), New York: Marcel Dekker (1993).
34. I. M. Pykett, Applications of Magnetic Resonance to Industrial Process Control. In *High Magnetic Fields: Applications, Generation, Materials*, H. J. Schneider-Muntau (ed.), Singapore: World Scientific (1996).
35. K. Albert, E. Bayer, *Trends Anal. Chem.* 7: 288 (1988).
36. A. I. Nakatani, M. D. Poliks, E. T. Samulski, *Macromolecules* 23: 2686 (1990).
37. J. Brandrup, E. H. Immergut, *Polymer Handbook*, New York: Wiley (1989).
38. D. M. Grant, C. L. Mayne, F. Liu, T.-X. Xiang, *Chem. Rev.* 91: 1591 (1991).
39. W. T. Dixon, *J. Chem. Phys.* 77: 1800 (1982).
40. B. Gerstein, R. Pembleton, R. Wilson, L. Ryan, *J. Chem. Phys.* 66: 361 (1997).
41. J. Virlet, *J. Chem. Phys.* 89: 359 (1992).
42. S. J. Opella, M. H. Frey, *J. Am. Chem. Soc.* 101: 5854 (1979).
43. A. Bax, *Two-Dimensional Nuclear Magnetic Resonance in Liquids*, Dordrecht: Delft University Press (1982).
44. A. Bax, L. Lerner, *Science* 232: 960 (1986).
45. F. A. Bovey, P. A. Mirau, *Acc. Chem. Res.* 21: 37 (1988).
46. F. A. Bovey, P. A. Mirau, *Makromol. Chem. Macromol. Symp.* 34: 1 (1990).
47. Kessler, M. Gehrke, C. Griesinger, *Angew. Chem. Int. Ed. Engl.*, 27: 490 (1988).
48. T. Saito, P. L. Rinaldi, *J. Magn. Res.* 132: 41 (1998).
49. H. W. Spiess, *J. Non-Cryst. Solids* 131–133: 766 (1991).
50. H. W. Spiess, *Chem. Rev.* 91: 1321 (1991).

51. W. S. Veeman, A. P. M. Kentgens, R. Janssen, *Fresenius Z. Anal. Chem.* 327: 61 (1987).
52. J. Clauss, K. Schmidt-Rohr, H. W. Spiess, *Acta Polymer* 44: 1 (1993).
53. A. Natansohn, *Polymer Eng. Sci.* 32: 1711 (1992).
54. L. P. Lindeman, J. Q. Adams, *Anal. Chem.* 43: 1245 (1971).
55. D. M. Grant, E. G. Paul, *J. Am. Chem. Soc.* 86: 2984 (1964).
56. H. N. Cheng, S. J. Ellingsen, *J. Chem. Inf. Comput. Sci.* 23: 197 (1983).
57. D. E. Derman, M. Jautelat, J. D. Roberts, *J. Org. Chem.* 36: 2757 (1971).
58. M. D. Bruch, *Macromolecules* 22: 151 (1989).
59. M. D. Bruch, F. A. Bovey, R. E. Cais, *Macromolecules* 17:2547 (1984).
60. J. C. Tebby, *Handbook of Phosphorus-31 Nuclear Magnetic Resonance Data,* Boca Raton, FL: CRC Press (1991).
61. B. Blümich, A. Hagemeyer, D. Schaefer, K. Schmidt-Rohr, H. W. Spiess, *Adv. Mater.* 2: 72 (1990).
62. F. Lauprêtre, *Progr. Polymer Sci.* 15: 425 (1990).
63. V. J. McBrierty, NMR Spectroscopy of Polymers in the Solid State. In *Comprehensive Polymer Science*, C. Booth, C. Price (eds.), Oxford, UK: Pergamon (1989).
64. A. L. Segre, D. Capitani, *Trends Polymers* 1: 280 (1990).
65. J. Herzfeld, J. E. Roberts, R. G. Griffin, *J. Chem. Phys.* 86: 597 (1986).
66. A. C. Kolbert, R G. Griffin, *Chem. Phys. Lett.* 166: 87 (1990).
67. E.Günther, B. Blümich, H. W. Spiess, *Molec. Phys.* 71: 477 (1990).
68. S. Wefing, S. Kaufman, H. W. Spiess, *J. Chem. Phys.* 89: 1234 (1988).
69. S. Wefing, H. W. Spiess, *J. Chem. Phys.* 89: 1219 (1988).
70. B. Blümich, P. Blümler, E. Günther, G. Schauss, H. W. Spiess, *Makromol. Chem. Macromol. Symp.* 44: 37 (1991).
71. B. Blümich, *Makromol. Chem.* 194: 2133 (1993).
72. D. G. Cory, A. M. Reichwein, J. W. M. Van Os, W. S. Veeman, *Chem. Phys. Lett.* 143: 467 (1988).
73. M. A. Hepp, J. B. Miller, *Macro. Symp.* 86: 271 (1994).
74. H. N. Cheng, NMR Characterization of Polymers. In *Modern Methods of Polymer Characterization*, H. G. Barth, J. W. Mays (eds.), New York: Wiley (1991).
75. I. Ando, T. Yamanobe, T. Asakura, *Progr. NMR Spectrosc.* 22: 349 (1990).
76. J. C. Randall, *J. Macro. Sci.—Rev. Macro. Chem. Phys.* C29: 201 (1989).
77. J. C. Randall, *Polymer Sequence Determination: Carbon-13 Method,* New York: Academic (1977).
78. J. C. Randall (ed.), *NMR and Macromolecules: Sequence, Dynamic, and Domain Structure,* Washington, DC: American Chemical Society (1984).
79. F. A. Bovey, *Chain Structure and Conformation of Molecules,* New York: Academic (1982).
80. A. E. Tonelli, *NMR Spectroscopy and Polymer Microstructure,* Deerfield Beach, FL: VCH (1989).
81. C. Dybowski, A. J. Brandolini, NMR Spectroscopy of Solid Polymer Systems. In *Characterization of Solid Polymers*, S. J. Spells (ed.), London, UK: Chapman & Hall (1994).

82. Deleted in proof.
83. L. Abis, G. Bacchilega, S. Spera, U. Zucchini, T. Dall'Occo, *Makromol. Chem.* 192: 981 (1991).
84. L. Abis, E. Albizzati, U. Giannini, G. Giunchi, E. Santoro, L. Noristi, *Makromol. Chem.* 189: 1595 (1988).
85. J. C. Vizzini, J.-F. Shi, J. C. W. Chien, *J. Appl. Polymer Sci.* 48: 2173 (1993).
86. P. Sormunen, T. Hjertberg, E. Iiskola, *Makromol. Chem.* 191: 2663 (1990).
87. W. Kaminsky, *Adv. Polymer Sci.* 127: 143 (1997).
88. C. Sishta, R. M. Hatborn, T. J. Marks, *J. Am. Chem. Soc.* 114: 1112 (1992).
89. Y. V. Kissin, T. E. Nowlin, R. I. Mink, A. J. Brandolini, *Top. Catal.* 7:69 (1999).
90. Y. V. Kissin, A. J. Brandolini, *J. Polymer Sci. Polymer Chem.* 37:4273 (1999).
91. K. Landfester, S. Spiegel, R. Born, H. W. Spiess, *Colloid Polymer Sci.* 276: 356 (1998).
92. M. Kamigaito, Y. Maeda, M. Sawamoto, T. Higashimura, *Macromolecules* 26: 1643 (1993).
93. H. Katayama, M. Kamigaito, M. Sawamoto, T. Higashimura, *Macromolecules* 28: 3747 (1995).
94. P. J. Jackson, N. J. Clayden, N. J. Walton, T. A. Carpenter, L. D. Hall, P. Jezzard, *Polymer Int.* 24: 139 (1991).
95. M. T. Roland, H. N. Cheng, *Macromolecules* 24: 2015 (1991).
96. A. J. Brandolini, D. D. Hills, *NMR Spectra of Polymers and Polymer Additives*, New York: Marcel Dekker (2000).
97. Q. T. Pham, R. Petiaud, H. Waton, *Proton and Carbon NMR Spectra of Polymers*, New York: Wiley Interscience (1983).
98. L. J. Mathias, R. F. Colletti, R. J. Halley, W. L. Jarrett, C. G. Johnson, D. G. Powell, S. C. Warren, *Solid-State NMR Polymer Spectra: Collected Volume 1*, Hattiesburg, MS: MRG (1990).
99. M. Perez, J. C. Ronda, J. A. Reina, A. Serra, *Polymer* 39: 3885 (1998).
100. R. D. Thompson, W. L. Jarrett, L. J. Mathias, *Macromolecules* 25: 6455 (1992).
101. E. F. McCord, W. H. Shaw, Jr., R. A. Hutchinson, *Macromolecules* 30: 246 (1997).
102. W. H. Starnes, Jr, B. J. Wojciechowski, A. Velazquez, G. M. Benedikt, *Macromolecules* 25: 3638 (1992).
103. N. M. Ahmad, F. Heatley, P. A. Lovell, *Macromolecules* 31: 2822 (1998).
104. D. Britton, F. Heatley, P. A. Lovell, *Macromolecules* 31: 2828 (1998).
105. G. I. Sandakov, I. P. Smirnov, A. I. Sosikov, K. T. Summanen, N. Volkova, *J. Polym. Sci. B: Polymer Phys.* 32: 1585 (1994).
106. J. Grobelny, *Eur. Polymer J.* 34: 235 (1998).
107. M. E. Merritt, L. Heux, J. L Halary, J. Schaefer, *Macromolecules* 30: 6760 (1997).
108. I.-S. Chuang, G. E. Maciel, *Polymer* 35: 1621 (1994).
109. W. T. Ford, M. Peryasamy, H. O. Spivey, *Macromolecules* 17: 2881 (1984).
110. H. D. H. Stöver, J. M. J. Frechet, *Macromolecules* 24: 883 (1991).
111. J. P. Cohen Addad, H. Montes, *Macromolecules* 30: 3678 (1997).
112. R. K. Harris, R. R. Yeung, P. Johncock, D. A. Jones, *Polymer* 37: 721 (1996).

113. D. W. Crick, S. D. Alexandratos, *Macromolecules* 26: 3267 (1993).
114. S. R. Smith, J. L. Koenig, *Macromolecules* 24: 3496 (1991).
115. A. Rossi, G. Odian, J. Zhang, *Macromolecules* 28: 1739 (1995).
116. L. Oliva, P. Longo, A. Zambelli, *Macromolecules* 29: 6383 (1996).
117. D. A. Laude, Jr., C. L. Wilkins, *Macromolecules* 19: 2295 (1986).
118. C. I. Tanzer, A. K. Roy, *Proc. ANTEC* 2700 (1995).
119. K. Ute, R. Miimi, S.-Y. Hongo, K. Hatada, *Polymer J.* 30: 439 (1998).
120. A. S. Brar, K. Dutta, *Polymer J.* 30: 304 (1998).
121. S. F. Dec, R. A. Wind, G. E. Maciel, *Macromolecules* 20: 2754 (1987).
122. H. Pasch, W. Hiller, R. Haner, *Polymer* 39: 1515 (1998).
123. J. Xu, L. Feng, S. Yang, Y. Yang, X. Kong, *Macromolecules* 30: 7655 (1997).
124. B. Sannigrahi, B. Garnaik, *Polymer J.* 30: 340 (1998).
125. C. H. Bergström, B. R. Sperlich, J. Ruotoistemäki, J. V. Seppälä, *J. Polymer Sci. A: Polymer Chem.* 36: 1633 (1998).
126. K. A. M. Thakur, R. T. Kean, E. S. Hall, J. J. Kolstad, E. J. Munson, *Macromolecules* 31: 1487 (1998).
127. S. J. Mumby, M. S. Beevers, *Polymer* 26: 2014 (1985).
128. M. Tokles, P. A. Keifer, P. L. Rinaldi, *Macromolecules* 28: 3944 (1995).
129. E. A. Williams, J. H. Wengrovius, V. M. VanValkenburgh, J. F. Smith, *Macromolecules* 24: 1445 (1991).
130. A. M. Aerdts, J. W. de Haan, A. L. German, G. P. M. van der Velden, *Macromolecules* 24: 1473 (1991).
131. Y. V. Kissin, A. J. Brandolini, *Macromolecules* 24: 2632 (1991).
132. E. T. Hsieh, J. C. Randall, *Macromolecules* 15: 1401 (1982).
133. M. Murano, *Polymer J.* 30: 281 (1998).
134. T. Arai, T. Ohtsu, S. Suzuki, *Macromol. Rapid Commun.* 19: 327 (1998).
135. P. F. Cheetham, T. N. Huckerby, B. J. Tabiner, *Eur. Polymer. J.* 30: 581 (1994).
136. H. N. Cheng, *Macromolecules* 30: 4117 (1997).
137. H. N. Cheng, P. C. Gillette, *Polymer Bull.* 38: 555 (1997).
138. H. N. Cheng, Macromol. *Theory Simul.* 3: 979 (1994).
139. H. N. Cheng, L. J. Kasehagen, *Macromolecules* 26: 4774 (1993).
140. H. N. Cheng, G. N. Babu, R. A. Newmark, J. C. W. Chien, *Macromolecules* 25: 6980 (1992).
141. H. N. Cheng, S. B. Tam, L. J. Kasehagen, *Macromolecules* 25: 3779 (1992).
142. H. N. Cheng, *Macromolecules* 25: 2351 (1992).
143. L. Yang, F. Heatley, T. G. Blease, R. I. G. Thompson, *Eur. Polymer. J.* 32: 535 (1996).
144. M. R. Krejsa, K. Udipi, J. C. Middleton, *Macromolecules* 30: 4695 (1997).
145. A. Factor, P. E. Donahue, *Polymer Degrad. Stabil.* 57: 83 (1997).
146. J. S. Forsythe, D. J. T. Hill, A. L. Logothetis, T. Seguchi, A. K. Whittaker, *Macromolecules* 30: 8101 (1997).
147. A. C. Kolbert, J. G. Didier, J. Polymer. *Sci. Part B: Polymer Sci. B: Polymer Phys.* 35: 1955 (1997).
148. P. L. Jackson, S. Rubinsztajn, W. K. Fife, M. Zeldin, D. G. Gorenstein, *Macromolecules* 25: 7078 (1992).

149. J. L. Espartero, I. Rashkov, S. M. Li, N. Manolova, M. Vert, *Macromolecules* 29: 3535 (1996).
150. A. E. Tonelli, D. B. Chestnut, *Macromolecules* 29: 2537 (1996).
151. C. DeRosa, A. Grassi, D. Capitani, *Macromolecules* 31: 3163 (1998).
152. C. DeRosa, D. Capitani, S. Cosco, *Macromolecules* 30: 8322 (1997).
153. A. Kaji, Y. Ohta, H. Yasuda, M. Murano, *Polymer J.* 22: 455 (1990).
154. Y. Shimizu, Y. Harashina, Y. Sugiura, M. Matsuo, *Macromolecules* 28: 6889 (1995).
155. W. L. Jarrett, L. J. Mathias, R. S. Porter, *Macromolecules* 23: 5164 (1990).
156. L. J. Mathias, C. G. Johnson, *Macromolecules* 24: 6114 (1991).
157. D. L. VanderHart, W. J. Orts, R. H. Marchessault, *Macromolecules* 28: 6394 (1995).
158. D. L. Vander Hart, S. Simmons, J. W. Gilman, *Polymer* 36: 4223 (1995).
159. M. Gentzler, J. A. Reimer, *Macromolecules* 39: 8365 (1997).
160. H. Ishida, H. Kaji, F. Horii, *Macromolecules* 30: 5799 (1997).
161. F. Horii, K. Masuda, H. Kaji, *Macromolecules* 30: 2519 (1997).
162. R. A. Komoroski, M. H. Lehr, J. H. Goldstein, R. C. Long, *Macromolecules* 25: 3381 (1992).
163. A. Schmidt, T. Kowalewski, J. Schaefer, *Macromolecules* 26: 1729 (1993).
164. E. O'Connell, D. G. Peiffer, T. W. Root, S. L. Cooper, *Macromolecules* 29: 2124 (1996).
165. J. H. O'Donnell, A. K. Whittaker, *Radiat. Phys. Chem.* 39: 209 (1992).
166. J. B. Miller, J. H. Walton, C. M. Roland, *Macromolecules* 26: 5602 (1993).
167. M. A. Ferrero, S. W. Webb, W. C. Conner, Jr., J. L. Bonardet, J. Fraissard, *Langmuir* 8: 2269 (1992).
168. A. P. M. Kentgens, H. A. van Boxtel, R.-J. Verweel, W. S. Veeman, *Macromolecules* 24: 3712 (1991).
169. G. J. Kennedy, *Polymer Bull.* 23: 605 (1990).
170. F. M. Mirabella, D. C. McFaddin, *Polymer* 37: 931 (1996).
171. S.-Y. Kwak, S. Y. Kim, *Polymer* 39: 4099 (1998).
172. L. Abis, G. Floridi, E. Merlo, R. Po, C. Zannoni, *J. Polymer Sci. B: Polymer Phys.* 36: 1557 (1998).
173. A. Spyros, R. Kimmich, B. H. Briese, D. Jendrossek, *Macromolecules* 30: 8218 (1997).
174. A. Yamazaki, Y. Hotta, H. Kurosu, I. Ando, *Polymer* 39: 511 (1998).
175. P. Weigand, H. W. Spiess, *Macromolecules* 28: 6361 (1995).
176. P. Blümler, B. Blümich, *Macromolecules* 24: 2183 (1991).
177. W. Gabriëlse, H. A. Gaur, F. C. Feyen, W. S. Veeman, *Macromolecules* 27: 5811 (1994).
178. B. F. Chmelka, K. Schmidt-Rohr, H. W. Spiess, *Macromolecules* 26: 2282 (1993).
179. S. Röber, H. G. Zachmann, *Polymer* 33: 2061 (1992).
180. T. Yamanobe, H. Matsuda, K. Imai, A. Hirata, S. Mori, T. Komoto, *Polymer J.* 28: 177 (1996).
181. M. G. Brereton, M. E. Ries, *Macromolecules* 29: 2644 (1996).

182. P. Blümler, B. Blümich, *Acta Polymer* 44: 125 (1993).
183. D. Hauck, P. Blümler, B. Blümich, *Macro. Chem.* 198: 2729 (1997).
184. P. A. Berger, E. E. Remsen, G. C. Leo, D. J. David, *Macromolecules* 24: 2189 (1991).
185. P. A. Mirau, S. A. Heffner, F. A. Bovey, *Macromolecules* 23: 4482 (1990).
186. S.-Z. Mao, H.-Q. Feng, *Colloid Polymer Sci.* 276: 247 (1998).
187. S. A. M. Ali, D. J. Hourston, T. N. Huckerby, *Eur. Polymer J.* 29: 137 (1993).
188. J. D. Song, R. Ryoo, M. S. Jhon, *Macromolecules* 24: 1727 (1991).
189. A. Bandis, W. Y. Wen, E. B. Jones, P. Kaskan, Y. Zhu, A. A. Jones, P. T. Inglefield, J. T. Bendler, *J. Polymer Sci. B: Polymer Phys.* 32: 1707 (1994).
190. C. K. Hall, E. Helfand, *J. Chem. Phys.* 77: 3275 (1982).
191. R. Dejean de la Batie, F. Lauprêtre, L. Monnerie, *Macromolecules* 21: 2045 (1988).
192. R. Dejean de la Batie, F. Lauprêtre, L. Monnerie, *Macromolecules* 22: 122 (1989).
193. Deleted in proof.
194. G. Williams, D. C. Watts, *Trans. Faraday Soc.* 66: 1323 (1971).
195. G. Williams, D. C. Watts, S. B. Dev, A. M. North, *J. Chem. Phys.* 73: 3348 (1980).
196. S. Ravindranathan, D. N. Sathyanarayana, *Macromolecules* 29: 3525 (1996).
197. J.-M. Petit, X. X. Zhu, *Macromolecules* 29: 2075 (1996).
198. S. Ravindranathan, D. N. Sathyanarayana, *Macromolecules* 28: 2396 (1995).
199. T. Radiotis, G. R. Brown, P. Dais, *Macromolecules* 26: 1445 (1993).
200. P. G. deGennes, *J. Chem. Phys.* 55: 572 (1971).
201. M. G. Brereton, I. M. Ward, N. Boden, P. Wright, *Macromolecules* 24: 2068 (1991).
202. T. Cosgrove, M. J. Turner, P. C. Griffiths, J. Hollinghurst, M. J. Shenton, J. A. Semiyen, *Polymer* 37: 1535 (1996).
203. M. Doi, S. F. Edwards, *The Theory of Polymer Dynamics*, Oxford, UK: Clarendon Press (1986).
204. H. W. Weber, R. Kimmich, *Macromolecules* 26: 2597 (1993).
205. J. D. Ferry, *The Viscoelastic Properties of Polymers*, New York: Wiley (1980).
206. J. Rault, C. Mace, P. Judeinstein, J. Courtieu, *J. Macromol. Sci.—Phys.* B35: 115 (1996).
207. F. Lauprêtre, Makromol. *Chem. Macromol. Symp.* 34: 113 (1990).
208. S.-U. Hong, A. J. Benesi, J. L. Duda, *Polymer Int.* 39: 243 (1996).
209. K. Hietalahti, M. Skrifvars, A. Root, F. Sundholm, *J. Appl. Polymer Sci.* 68: 671 (1998).
210. Y. Xia, P. T. Callaghan, *Macromolecules* 24: 4777 (1991).
211. M. Appel, G. Fleischer, F. Fujara, *Polymer Sci.* 35: 1618 (1993).
212. H. Walderhaug, F. K. Hansen, S. Abrahmsén, K. Persson, P. Stilbs, *J. Phys. Chem.* 97: 8336 (1993).
213. B. Manz, P. T. Callaghan, *Macromolecules* 30: 3309 (1997).
214. S. Matsukawa, I. Ando, *Macromolecules* 30: 8310 (1997).
215. W. Yu, E. D. von Meerwall, *Macromolecules* 23: 882 (1990).

216. T. Kubo, T. Nose, *Polymer J.* 24: 1351 (1992).
217. Q. Chen, T. Yamada, H. Kurosu, I. Ando, T. Shiono, Y. Doi, *J. Polymer Sci. B: Polymer Phys.* 30: 591 (1992).
218. K. Schmidt-Rohr, H. W. Spiess, *Macromolecules* 24: 5288 (1991).
219. J. Hirschinger, D. Schaefer, H. W. Spiess, A. J. Lovinger, *Macromolecules* 24: 2428 (1991).
220. K. Schmidt-Rohr, *Macromolecules* 29: 3975 (1996).
221. D. A. Lathrop, K. K. Gleason, *Macromolecules* 26: 4652 (1993).
222. H. Kaji, F. Horii, *Macromolecules* 30: 5791 (1997).
223. T. Kawaguchi, A. Mamada, Y. Hosokawa, F. Horii, *Polymer* 39: 2725 (1998).
224. A. D. Meltzer, H. W. Spiess, C. D. Eisenbach, H. Hayen, *Makromol. Chem. Rapid Commun.* 12: 261 (1991).
225. J. A. Kornfeld, H. W. Spiess, H. Nefzger, H. Hayen, C. D. Eisenbach, *Macromolecules* 24: 4787 (1991).
226. A. A. Parker, J. J. Marcinko, P. Rinaldi, D. P. Hedrick, W. M. Ritchey, *J. Appl. Polymer Sci.* 48: 67 (1993).
227. N. M. Da Silva, M. I. B. Tavares, S. M. C. De Menezes, *J. Appl. Polymer Sci.* 60: 1419 (1996).
228. G. Kögler, P. A. Mirau, *Macromolecules* 25: 598 (1992).
229. X. Lu, M. Hou, X. Gao, S. Chen, *Polymer* 35: 2510 (1994).
230. L. W. Kelts, C. J. T. Landry, D. M Teegarden, *Macromolecules* 26: 2941 (1993).
231. M. Guo, H. G. Zachmann, *Polymer* 34: 3503 (1993).
232. P. P. Huo, P. Cebe, *Macromolecules* 25: 5561 (1993).
233. G. C. Campbell, D. L. vanderHart, *J. Magn. Res.* 96: 69 (1992).
234. D. L. vanderHart, *Makromol. Chem. Macromol. Symp.* 34: 125 (1990).
235. J. H. Walton, J. B. Miller, C. M. Roland, *J. Polymer Sci. B: Polymer Phys.* 30: 527 (1992).
236. S. Kaplan, *Macromolecules* 26: 1060 (1993).
237. P. Wang, A. A. Jones, P. T. Inglefield, D. M. White, J. T. Bendler, *New Polymer. Mater.* 2: 221 (1990).
238. A. Asano, K. Takegoshi, K. Hikichi, *Polymer* 35: 5630 (1994).
239. K. Hikichi, A. Tezuka, K. Takegoshi, *Polymer J.* 30: 356 (1998).
240. T. Miyoshi, K. Takegoshi, T. Terao, *Polymer* 38: 5475 (1997).
241. S. Schantz, W. S. Veeman, *J. Polymer Sci. B: Polymer Phys.* 35: 2681 (1997).
242. D. A. Costa, C. M. F. Oliveira, M. I. B. Tavares, *J. Appl. Polymer Sci.* 69: 129 (1998).
243. K. Schmidt-Rohr, J. Clauss, H. W. Spiess, *Macromolecules* 25: 3273 (1992).
244. I. S. Polios, M. Soliman, C. Lee, S. P. Gido, K. Schmidt-Rohr, H. H. Winter, *Macromolecules* 30: 4470 (1997).
245. J. E. Gambogi, F. D. Blum, *Macromolecules* 25: 4526 (1992).
246. W. S. Veeman, *Makromol. Chem. Macromol. Symp.* 69:149 (1993).
247. W.-Y. Lin, F. D. Blum, *Macromolecules* 30: 5331 (1997).
248. F. Bossé, H. P. Schreiber, A. Eisenberg, *Macromolecules* 26: 6447 (1993).
249. L. Garrido, J. E. Mark, C. C. Sun, J. L. Ackerman, C. Chang, *Macromolecules* 24: 4067 (1991).

250. E. D. von Meerwall, T. Pryor, V. Galiatsatos, *Macromolecules* 31: 669 (1998).
251. T. A. Kestner, R. A. Newmark, C. Bardeau, *Polymer Preprints* 37: 232 (1996).
252. F. C. Schilling, V. J. Kuck, *Polymer Deg. Stab.* 31: 141 (1991).
253. I. Quijada-Garrido, M. Wilhelm, H. W. Spiess, J. M. Barrales-Rienda, *Macromol. Chem. Phys.* 199: 985 (1998).
254. T. Miyoshi, K. Takegoshi, T. Terao, *Macromolecules* 30: 6582 (1997).
255. C. Zhang, P. Wang, A. A. Jones, P. T. Inglefield, R. P. Kambour, *Macromolecules* 24: 338 (1991).

10

Inorganic Analyses of Polymers

John Lemmon

GE Corporate Research, Schenectady, New York, USA

Galina Georgieva

GE Medical Systems, Waukesha, Wisconsin, USA

INTRODUCTION AND OVERVIEW

The vast majority of plastic materials are based on organic polymers. Inorganic elements/compounds are part of the plastic's composition, either as additives (stabilizers, fillers, etc.) or as trace element contaminants. In the case of additives, one may expect significant amounts of inorganic elements (up to percent levels) when analyzing such materials (silicone levels in glass-filled polymers, for example). However, trace-level metal (element) contaminants are present at much lower levels of concentration (ppm to ppt), which imposes certain requirements related to handling/preparation.

Several different analytical techniques are currently used to quantify levels of inorganics in plastic materials:

> *Atomic spectroscopy.* Atomic spectroscopic methods includes atmoci absorption (AA), inductively coupled plasma atomic emission (ICP-AES), and inductively coupled plasma mass spectroscopy (ICP-MS). These methods are based on emission and absorption of electromagnetic radiation by atoms and provide information about levels of different elements in the sample (except some light elements such as H, C, N, O, etc.)

511

X-ray methods. Wavelength dispersion X-ray analysis (WDXRF) is based on emission of element characteristic secondary X-ray radiation using wavelengths shorter than the absorption edge of the spectral lines desired. The method is applicable only for analyses of major constituents (percent Si in glass-filled plastics) and is not suitable for trace-level analyses because of high detection limits for most elements. Matrix effects tend to be very significant in this type of analysis, which makes the choice of standards extremely critical.

Energy-dispersion X-ray analysis (EDXRF) is similar in nature to WDXRF, but the X-rays emitted from the sample are absorbed (simultaneously for multielement composites) by a solid-state detector and pluse-height analyzer. Detection limits are usually in the parts-per-million range, but quantification limits are about 50 times less than in WDXRF.

Electrochemical methods (ISE). Ion-selective electrode measurements provide information about a number of anions/cations through potentiometric measurements against a suitable reference electrode in water solutions (requires polymer digestion/ashing). Detection limits vary for different ions but are usually in the upper parts-per-million range. Interference by similar ions (Br^-, F^- when measuring Cl^-) is one of the major drawbacks of this techique, which limits its applicability.

Neutron activation analysis. Neutron activation is another technique for elemental analysis based on selective radioactivity measurements for certain elements induced by sample bombarment with nuclear particles. Thermal (slow) neutrons generated in a nuclear reaction or fast neutrons (generated by an acceleraor) are usually used to bombard the sample. The method's detection limits are usually in the range of 50 to 1000 ppm.

Ion chromatography. Ion chromatography is a powerful technique for analysis of both anions and cations. Detection limits usually range from low parts-per-billion to low parts-per-million. For application in plastic materials analysis, the method provides two type of information: (a) free ionic content and (2) total content of some of the inorganic elements (for example, chloride, sulfur, etc.) after suitable sample mineralization.

Capillary electrophoresis. Capillary electrophoresis provides in general the same information as ion chromatography and is in a sense an alternative technique for this type of application. However, the principle is different because the analyte separation is based on application of a high-voltage electrical field across a capillary column.

The material presented through the rest of this chapter focuses on several of the methods for inorganic analysis such as atomic spectroscopy (ICP in particular) and chromatography. These methods proide high versaility, low element quantification limits, and high degree of automation, which make them indispensable in modern analysis of plastics.

ATOMIC SPECTROSCOPY

A family of techniques based on atomic spectroscopy is currently the most widely used approach for analysis of trace-element concentrations in samples. These techniques are based on the different energies required by each element to result in atomic transitions. The basic processes taking place (Fig. 1) are the following.

Excitation: when the energy added to the atom in a specific energy state is absorbed by the atom causing electron transition to higher energy state orbital

Ionization: when the energy absorbed by system is sufficient to completely dissociate an electron leaving an ion with net positive charge

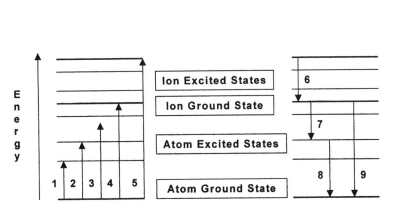

Figure 1 Atomic energy-level transitions diagram 1–3 are excitation transitions; 4 is ionization; 5 represents ionization/excitation; 6 is ion emission; 7–9 are atom emission.

Emission: when an atom or ion in its excited state decays to its ground state via emission of a discrete quantity of electromagnetic radiation called a "photon"

Energy dissipation: when energy is transferred to the surrounding atoms/molecules through collision.

The specific pathways for energy transitions are shown in Figure 1 and can be derived by Plank's equation,

$$E = h\nu$$

where E is the energy difference between two energy levels (specific for each element), h is Plank's constant, and ν is the radiation frequency. Another way of expressing Plank's equation is to substitute for n with c/l, where c is the speed of light and l is the wavelength:

$$E = \frac{hc}{l}$$

Each element has a specific set of energy levels, which translates to a unique set of absorption/emission wavelengths. This is the basis for element-specific analysis in atomic spectroscopy.

The major strengths of atomic spectroscopy techniques over other methods is that they are relatively inexpensive, and they provide outstanding flexibility in terms of automation and multielement analysis capabilities (almost the whole Periodic Table). These advantages, coupled with high precision and accuracy, make atomic spectoscopy a preferred method of analysis.

Analytical Techniques in Atomic Spectroscopy

The two primary approaches in atomic spectroscopy utilize atomic absorption (FAAS, GFAAS) and atomic emission (FAES, ICP-AES) for analyte quantification. The principal arrangement for the spectrometry system of these two techniques is shown in Fig. 2.

Another method included in this chapter is ICP-MS, which, although not based on atomic spectroscopy in a true sense, utilizes the same instrumental approach as ICP-AES for sample introduction. The difference is that analyte quantification takes place using a mass spectroscopy detector.

Table 1 provides side-by-side comparison of the most important characteristic features for the technqiues in Fig. 2. A graphical representation of element detections for different atomic spectroscopy techniques is given in Fig. 3.

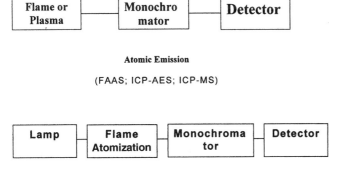

Figure 2 Block scheme diagrams of atomic emission/absorption spectrometry assembly.

Description of the Individual Techniques

Graphite Furnace Atomic Absorption Spectroscopy (GFAAS)

Background

The absorption process utilized in GFAAS refers to transitions between atomic ground-state and atomic excited states of a given element. For a transition to take place, the energy (wavelength) of the source (lamp) must be equal to the energy of the specific transition. A portion of the source radiation (energy) is absorbed by the analyte present in the sample. The radiation transmitted through the sample, T, is given by

$$T = T_0 e^{-(K_n l)}$$

where T_0 is the incident radiation; K_n is an absorption coefficient specific for the transition of the analyte, and l is the sample path length.

The radiation sources used in atomic absorption are hollow cathode lamps or electrodeless discharge lamps. Although in most cases a lamp needs to be changed for each element of interest, some are designed to be used for several elements (up to seven).

The analyte is quantified by the difference between the radiation energy at a given wavelength in the absence of a sample, and the radiation energy in the presence of a sample in the atomization chamber. The amount of the analyte in the sample is thus proportional to the absorbed energy. To obtain

Table 1 Atomic Spectroscopy Methods Comparison

Technique	Sample Introduction	Atomization	Detection	Detection limits	Advantages	Disadvantages
FAAS	Nebulization	Flame: 2000–2800°C	Absorption of radiation measured by spectrophotometer	ppm–ppb	• Inexpensive	• High DL • Matrix/flame interferences • Very limited multielement capabilities • Low throughput
GFAAS	• Liquid • Slurry • Solids	Electrothermal—graphite furnace	Absorption of radiation measured by spectrophotometer	ppb–ppt	• High selectivity, better than FAAS and ICP-AES • High dissolved solids tolerance	• Matrix effects • Very limited multielement capabilities • Low throughput
FAES	Nebulization	Flame: 2000–2800°C	Emission measured by spectrophotometer	ppm—most ppb—alkaline earth elements	• Inexpensive • Simple • Sensitive for alkali and alkaline earth elements	• Spectral interferences • Molecular species emission
ICP-AES	Nebulization: pneumatic or ultrasonic	High-temperature argon plasma	Emission measured by spectrophotometer	ppm–ppb	• Multielemental • High throughput • Low interferences • Refractory elem. • High dissolved solids tolerance	• Relatively expensive • Limited sensitivity • Spectral interferences
ICP-MS	Nebulization: pneumatic or ultrasonic	High-temperature argon plasma	Ion separation, m/z measured by mass spectrometer	ppt with few exceptions, ppq for some elements	• High sensitivity—better than GFAAS • Multielemental • High sample throughput	• Expensive • Matrix interferences • Low dissolved solids tolerance

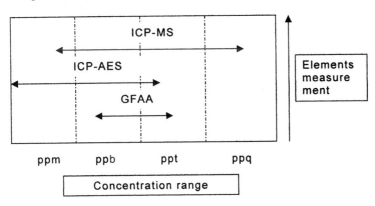

Figure 3 Atomic spectroscopy techniques capability.

the analyte concentration, the energy absorbed in the sample is compared to a calibration curve of absorption data for the same analyte. The calibration curve is constructed by using certified standards to prepare solutions of known analyte concentration.

Sample Introduction

Samples in GFAAS are introduced onto a platform in the furnace chamber as liquids (solutions), solids, or slurries. Only small amounts (total 10^{-8} to 10^{-11} g) of the analyte are usually required, due to the high method sensitivity. The amount of liquid samples required is usually in the range of 5–10 ml.

Atomization

The heating of the sample in the electrothermal atomizer is programmable to fit the variety of sample types and analytes. The atomization process involves the following steps:

Drying includes heating of the sample for a short period of time (20–50 s) at 110°C to remove solvents or other volatiles.

Ashing or charring involves heating to medium temperatures (400–900°C, depending on the analytes) to decompose organic matrices and volatize higher boilers. Prolonged ashing or too-high temperature settings in this step may result in loss of analyte. Addition of matrix modifiers [Palladium(II) nitrate in particular] is often required to decrease analyte volatility. This allows ashing to be carried out at higher temperatures for removal of matrix constituents, which results in better analysis consistency.

Recommended ash temperatures for selected elements are listed in Table 2.

Atomization is accomplished by quickly raising the temperature of the graphite furnace to the maximum allowable level (or a pre-selected level) after completion of the ashing step.

Detection

Analyte detection and quantification is accomplished using spectrophotometer devices. These are usually photoamplifier tubes or diode array assemblies. A computerized control system allows for continuous signal monitoring. This system can be interfaced with an autosampler/diluter system which uses a feedback loop based on signal intensity to automatically dilute overrange samples until they reach calibration range.

Data Analysis

Data generated by the spectrophotometer device are stored and processed in the computer system of the instrument. GFAAS raw data usually require background correction, especially for analyses with resonance frequencies in

Table 2 Ash Temperatures with a Graphite Furnace

Element	Recommended ash temperature (°C)	Ash temperature with Pd modifier (°C)
Antimony	800	1400
Arsenic	800[a]	1500
Bismuth	500	1100
Cadmium	300	550
Chromium	1300	1300
Cobalt	900	1200
Copper	900	1100
Iron	800	1300
Lead	400	1000
Manganese	800	1200
Mercury	120	450
Nickel	900	1200
Tin	800	1300
Thallium	400	1500
Zinc	400	900

[a] Nickel modifier added.
Source: Ref. 21.

the far-UV region. Background correction is also necessary to achieve high accuracy for trace-level analyses in complex matrices. Different background correction approaches are available, most of which are usually included in the equipment software.

ICP

Background

The ICP methods (both ICP-AES and ICP-MS) utilize an argon plasma as a high-temperature/high-energy atomization and ionization source. This results in quick, simultaneous excitation of the analytes in the sample. Emissions of elements (ICP-AES) includes the atomic ground-state transitions, but also first and second ionization-state transitions.

The high plasma temperature ensures complete atomization of the sample (even in the case of refractory compounds) and also prevents the formation of di- or polyatomic species. Therefore, most matrix effects common for flame and electrothermal atomization techniques are eliminated in ICP-AES.

Sample Introduction

The sample is usually introduced into the ICP torch in the form of an aerosol obtained from a liquid sample in a process called nebulization. This process is critical for the sensitivity of the ICP analyses. Because only small droplets are efficiently atomized in the ICP plasma, the type and capabilities of the particular nebulizer equipment relate directly to the sensitivity capability of the instrument.

Two types of nebulizers used with current ICP equipment are pneumatic and ultrasonic nebulizers. Pneumatic nebulizers include concentric, cross-flow, Babington, and v-groove types. All these are based on dispersing the liquid sample in a jet of argon gas to form an aerosol. One of the requirements which has driven nebulizer design is the need to nebulize liquids of different viscosity and dissolved solids content without clogging the equipment. For example, the Babington nebulizer is the least susceptible to clogging by viscous liquids, while the V-groove nebulizer has the highest dissolved solids tolerance.

Ultrasonic nebulizers utilize high-frequency mechanical waves to generate an aerosol. A distinct advantage of this approach is the ability to very efficiently generate small droplets in the aerosol. This results in

Greater sensitivity—up to 20 times greater
Lower polyatomic ion interference
Better flexibility for sample matrices—capable of introducing samples in many different organic solvents.

The nebulization efficiency of the ultrasonic nebulizer is so high that a desolvator unit must be added to the system to prevent overloading the plasma with solvent from the sample.

Another integral part of the ultrasonic (as well as some pneumatic) nebulizer systems is a peristaltic pump needed to deliver constant sample solution flow. These pumps use a series of rollers which push the sample solution through plastic tubing without contacting the sample directly (thus preventing sample contamination). The principle utilized by these pumps is called *peristalsis*.

A typical layout of an ultrasonic nebulization and membrane desolvation system is shown in Fig. 4.

Although not part of the standard ICP equipment, ultrasonic nebulizers are increasingly becoming a necessity for a high-performance ICP systems. This is particularly true for trace-level element analyses.

Alternative Sample Introduction Techniques

Several alternatives of sample delivery have been explored or are currently under development. These include:

Hydride generation based on conversion of certain elements (Hg, As, Pb, Sb, Bi, Te, Sn, and Ge) to the corresponding volatile hydrides, which leads to significant improvement in detection limits

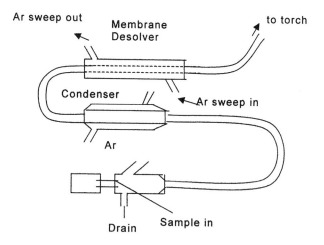

Figure 4   CETAC U-6000 AT+ ultrasonic nebulizer—schematic representation.

Direct injection of liquid sample into the plasma (requires special torch construction)
Graphite furnace for sample atomization
Direct insertion of solids
Laser ablation of solids

Most of these techniques have either limited applicability or suffer from inconsistent precision and accuracy and therefore have not been adopted as routine approaches. Laser ablation is probably one of the most promising methods in the above list, with high potential to provide an alternative sample introduction route for the different atomic spectroscopy techniques.

Atomization

The atomization/ionization in ICP spectroscopy is accomplished via the high-temperature plasma source. The plasma is obtained by applying a high-frequency magnetic field to an ionized argon gas stream in the ICP torch device (Fig. 5).

Radiofrequency (rf) power is applied to the load coil (positioned at the top end of the torch), causing an alternating current oscillation at a rate corresponding to the rf frequency (usually 27 or 40 MHz). These oscillations induce an electromagnetic field at the top of the coil. When a spark is applied to the argon gas swirling through the torch, partial ionization of

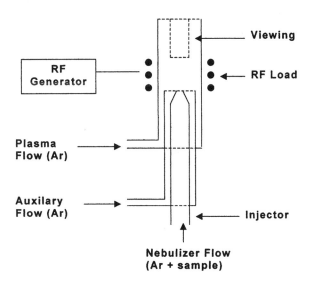

Figure 5 Schematic representation of an ICP torch.

some of the argon atoms occurs. The electrons are then accelerated by the magnetic field, causing additional collisional ionization of the argon gas in a chain-reaction fashion. This type of energy transfer from the coil to the argon gas is known as *inductive coupling*. The argon plasma obtained is known as *inductively coupled plasma* (ICP). The ICP discharge is maintained by continuous transfer of rf energy to the plasma through the load coil.

The torch positioning in the ICP instrument can be either radial or axial. Some instruments allow for switching between these two positions. In general, axial torch positioning allows for lower analyte detection limits. However, matrix effects are much more pronounced in this torch geometry.

The sample solution is introduced into the plasma as fine droplets from the nebulizer. The following transformation steps occur due to the high plasma temperature (1000–6000°K):

1. Desolvation to produce solid particles.
2. Vaporization/cracking to produce gaseous species.
3. Atomization to completely dissociate any polyatomic species; excitation/emission takes place.
4. Ionization to produce ionic species; excitation/emission also takes place.

The ICP discharge consists of different zones, each of which belongs to a different temperature region (see Fig. 6). The induction region (bottom portion of the discharge), where the energy transfer from the rf source to the plasma takes place has the highest temperature (10,000°K) and is toroidal. This particular plasma region is shaped like a doughnut because the nebu-

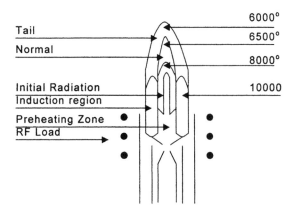

Figure 6 Different zones and temperature regions of the ICP discharge.

lizer flow literally punches a hole through the center of the discharge. Of significant importance is the *normal analytical zone*, because this region of the plasma is used for analyte emission measurements (this is where excitation/ionization occur).

Detection

ICP-AES: The emission radiation of the excited analytes from the normal analytical zone of the plasma is transferred to the spectrophotometer using an optical device called "transfer optics," which focuses the incoming polychromatic radiation onto the entrance slit of the wavelength-dispersing device.

The emission of the excited analytes in the plasma is polychromatic (several different wavelengths per analyte). The spectrophotometers used to detect and quantify the emitted radiation contain a monochromator (polychromator for simultaneous measurements at several different wavelengths) and a detector (photomultiplier tube) to quantify the amount of radiation for each specific wavelength.

The signals generated by the spectrophotometer are further processed by the instrument software and ultimately stored as a set of wavelengths with corresponding intensities. A calibration curve obtained through calibration standard measurements is necessay in order to quantify analyte concentration in the sample.

ICP-MS: In the ICP-MS systems, the analytes are detected as ions formed in the ICP. These ions are first transported through an ICP-MS interface and ion optics to the detector, which in this case is a mass spectrometer. The ions are separated according to their mass/charge (m/z) ratio and then measured in the mass spectrometer to provide qualitative and quantitative information.

Selecting the Proper Atomic Spectroscopy Technique

Selection between the different atomic spectroscopy techniques (ICP-AES, ICP-MS, GFAAS, FAAS) depends largely on the types of analyses performed in each particular laboratory. For an application requiring single-element analysis, relatively low sample throughput, or low initial cost, FAAS could be the technique of choice. If the application requires very low detection limits, then GFAAS will be better suited to meet the analysis requirements.

For an application dealing with multielement analysis of samples in complex matrices and requiring high throughput rates with moderate sensitivity, ICP-AES would be the best choice. If, however, low detection limits are required in addition to the high throughput/multielement analyses features of a particular application, then ICP-MS becomes the preferred candidate.

In reality, many laboratories are equipped with multiple instruments (GFAAS, ICP-AES, ICP-MS), each with its own set of applications. The selection criteria are generally more complex than described in the above examples. The maturing of these analytical techniques, along with continuous advancement and decreasing cost of the computing equipment, make most of this instrumentation more commonly available in analytical laboratories.

IONIC SEPARATION AND SEPARATION TECHNIQUES

Overview

As described in a later section of this chapter, there are several methods available for the determination of metal cations, and in some cases anions, in polymer materials (e.g., FAAS, ICP-AES). Although these techniques are capable of measuring total elemental concentrations, they do not differentiate between various ionic species of the same element. In the case of anionic analytes, methods such as ion chromatography (IC) and capillary electrophoresis (CE) often represent the only suitable determination methods. This section will describe the two prevalent methods, IC and CE, used for the determination of metal cations, elemental anions, oxyanions, and the speciation of metal cations. Most commonly, samples are prepared in the liquid state as described later. The preparation method used will determine the type of analyte to be measured in the polymer. In the case of liquid extraction, only soluble ionic species will be quantified. However, using wet or dry ashing, oxygen bomb, or acid dissolution techniques, the total amount of analyte (soluble and insoluble) can be determined. In general, the use of organic dissolution methods for sample preparation should be avoided, to minimize matrix effects and adverse conditions on equipment.

Description of the Individual Techniques

Ion Chromatography

Background

Ion chromatography is an analytical technique that separates ionic species by combining chromatographic and ion equilibrium theory into one application. In theory, IC can be used to analyze any ionic species. There are several ion equilibrium mechanisms utilizing different solid stationary phases and mobile phases (eluents) which allow for the analysis of a variety of organic and inorganic ions. A brief summary of these methods can be found in Table 3.

Ion exchange is probably the most widely used IC method for the elemental analysis of polymers. Although the other IC techniques are useful for several different applications, they are not covered in the scope of this section.

Most common ion-exchange systems used today combine high-performance liquid chromatography (HPLC) technology with ion-exchange capability (Fig. 7). A stationary phase of small-diameter uniform particles is packed into a column. Eluent is passed through the column with a high-pressure pump. A fixed-volume injection loop is placed between the pump and column for sample introduction. The components of the sample are separated on the column and flow to a detector. Data from the detector are collected on a data station, where quantification of the analyte is calculated and stored. Modern IC systems generally include an integrated package of pumps, detectors, autosampler, column oven, and are completely computer automated.

Table 3 Chromatographic Methods Utilizing Ion-Exchange Mechanisms

Method	Stationary-phase functional group	Eluent	Analytes
Ion exchange	Cations: sulfonic carboxylic acid; Anions: functionalized amines	Aqueous buffer, (may contain organic solvent)	Metal ions, halides, oxyanions, transistion metals
Ion pairing	Reverse phase C18 Functionalized silica	Aqueous buffer with ion-paring reagent	Wake acids/bases, large hydrophobic ions
Ion exclusion	Sulfonic acid	Water or dilute acid	Hydrophilic organic, weak acids
Chelation	Dicarboxylates	Strong acids	Main-group metals

Source: Refs. 1 and 2.

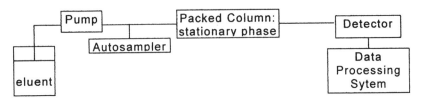

Figure 7   Diagram of a typical IC instrument. Separation of analytes takes place in the column.

Typically, the analysis of elemental or ionic species in polymers by ion exchange first requires the elimination of the polymer matrix as described in the previous section. Once the analyte of interest is in solution, ion exchange has proven to be a versatile and robust technique. Ion exchange is capable of identification and quantification of many ions, including halides, alkali metals, alkaline earths, transition metals, and oxyanions. Depending on the detector response of the analytes, ion-exchange detection limits for most common ions are 1–5 ppb. For routine analysis, accuracy is around 3% relative standard deviation (RSD), and with the use of an internal standard RSDs of less than 1% can be achieved [2].

Ion-Exchange Theory

Ion exchange is an analytical technique in which ions are separated chromatographically based on their interaction with a stationary phase and eluent. The stationary phase (ion-exchange resin) is commonly comprised of cross-linked polystyrene derivitized with a fixed ion. To maintain charge balance, a counterion must be present. The concentration of the counterion is maintained in the eluent. It is this counterion that is replaced during the ion-exchange process by the analyte during an analysis (Fig. 8). The ion-exchange resin is classified as a cation-exchange resin when negatively charged, and anionic when positively charged.

The competition for charged sites on the stationary phase (S) between the eluent ion (counterion, H) and the analyte ion (A) can be expressed by a simple equilibrium mechanism. The equilibrium for an ion-exchange process involving a cation-exchange resin is shown below [3,4].

$$S^-H^+ + A^+ \leftrightarrow S^-A^+ + H^+ \tag{1}$$

Competition between the analyte ion and the eluent ion is stoichiometric, with both concentration and valency influencing the equilibrium. Equation (1) can be generalized for y moles of A^{x+} exchanging with x moles of H^{y+} as demonstration in Eq. (2):

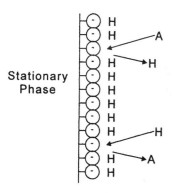

Figure 8 Schematic of ionic exchange between a cation analyte (A^+) and acid eluent on a cation-exchange column.

$$yA^{x+}_{\text{soln}} + xH^{y+}_{\text{res}} \leftrightarrow yA^{x+}_{\text{res}} + xH^{y+}_{\text{soln}} \tag{2}$$

where the subscript "res" refers to resin-bound ions. Two important effects of these interactions are: (1) ions present in higher concentrations will take a greater portion of the ion exchange sites, and (2) a higher-valency eluent ion is a stronger eluent than a lower-valency eluent ion. The equilibrium constant for the reaction shown in Eq. (2) is

$$K_{\text{AH}} = [A^{x+}_{\text{res}}]^y [H^{y-}_{\text{soln}}]^x / [A^{x-}_{\text{res}}]^y [H^{y-}_{\text{res}}]^x \tag{3}$$

Concentrations can be used because the activity of the ion is assumed to be unity. This constant is also referred to as the *selectivity coefficient*. The selectivity coefficient derives its name from the probability of an exchange occurring between the analyte ion and the counter ion. Therefore if $K_{\text{AH}} = 1$, then the ratios of the concentrations of ions in the resin and solution are equal and the resin will show no selectivity for A^{x+} over H^{y+}. If $K_{\text{AH}} > 1$, then the resin will contain a higher concentration of ion A^{x+} than the solution phase and will select A^{x+} preferentially over H^{y+}. The reverse situation applies for $K_{\text{AH}} < 1$.

Selectivity

In general, selectivity coefficients can be used to approximate elution order of cations or anions. However, in most cases simple ion exchange may not be the only mechanism influencing selectivity. In addition to selectivity coefficients, the conditions listed in Table 4 can help predict the elution order of a group of ions from an ion exchanger. A typical anion-exchange

Table 4 Selectivity Factors

Variable	Selectivity effect
Charge on analyte ion	The higher the charge, more coulombic interaction with resin, long elution time
Solvation size of analyte ion	The more strongly hydrated ion is bound most weakly, faster elution time
Resin cross-linking	The higher the degree of cross-linking in the resin, the greater the preference for small analyte ions.
Polarizability of analyte ion	In general, the more polarizability, the more affinity to the resin, longer elution time.

Source: Ref. 5

chromatogram is shown in Fig. 9 and demonstrates some of the mentioned selectivity rules.

Selectivity of transition metal cations is not as straightforward as for the alkali and alkaline earth metals. Metal ions which have different valencies, or which have the same valency but are in different rows, can be separated by cation exchange. However, many transition metal cations have the same valency and therefore cannot be separated by cation exchange. One common solution to increase selectivity among transition metal cations is to form coordination complexes which form strong ionic species. In aqueous solutions some transition metals can hydrolize to form coordination complexes in which water is covalently bound. The water can be replaced by an ion or molecule which can donate one or more electron pairs to the metal. Such molecules are referred to as ligands and are classified by how many electron pairs they donate. Several anionic chelating molecules are shown in Table 5 which coordinate with many of the transition metal cations to form coordination complexes.

To be viable, the formation of such complexes must be quick and efficient, thus the stability constant, K, should be high. Other factors can influence complexation; these include pH, ionic strength, and competition with other ligands. Once the complexes are formed, elution order is influenced by charge, molecular geometry, and hydrophobic characteristics [5].

Methods of Detection

Several methods of detection are available for ion chromatography; the most common are listed in Table 6. However, the scope of this section will be limited to the detection of inorganic ions, thus detection by conduc-

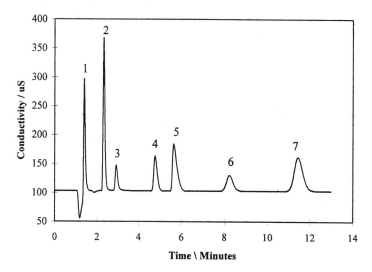

Figure 9 Typical ion-exchange chromatogram for a mixture of anions, concentration 1 μg/ml. Chromatographic conditions: Dionex AS4a column, carbonate/bicarbonate eluent buffer, 1.0 ml/min flow rate, supressed conductivity detection. Peak identifications: 1, F^-; 2, Cl^-, 3, NO_2^-; 4, Br^-; 5, NO_3^-; 6, PO_4^{3-}; 7, SO_4^{2-}. (Chromatogram provided by Ann Young, GE-CRD.)

tivity and postcolumn derivatization followed by UV-Vis absorption spectroscopy will be discussed.

Conductivity: The nature of ion chromatography places several constraints on the detection of ions once they are separated. Unlike HPLC, in which most analyte species are UV-Vis absorbers, few ions demonstrate this property. A common strategy applied toward the detection of ions,

Table 5 Anionic Ligands and Chelating Molecules Used to Form Coordination Complexes with Transition Metals for IC

Coordination	Ligand
Monodentate	NH_3, CN^-, Cl^-
Bidentate	Oxalate, ethylenediamine
Tridentate	Pyridine-2,6-dicarboxylate (PDCA)
Hexadentate	Ethylenediaminetetra-acetate (EDTA)

Table 6 Detection Methods

Detection method	Application
Conductivity suppressed	Good for strong to moderate ionic species, good sensitivity
Postcolumn derivatization spectroscopy	Transition metals, high sensitive
Direct	Used for iodide, nitrate, sulfide
Indirect	Universal application, lacks sensitivity
Electrochemical	
Amperometry	Good for phenols, catecholamines
Pulsed amperometry	Good for alcohols, amines, carbohydrates

Source: Ref. 6.

and in many cases the only method available, is conductivity. Conductivity detection has the major advantage in that all ions are conductive in solution, thus the response is universal. In general, conductivity is the detection mode of choice for the alkali metals, alkaline earth metals, transition metal oxyanions, and most halogen and other nonmetal oxyanions. Table 7 shows the limiting ionic conductivities for several ions in aqueous solution.

Two different modes of conductivity can be practiced, direct or indirect detection. Direct detection can be defined as when the limiting ionic conductance of an analyte species is much greater than that of the eluent and an

Table 7 Limiting Ionic Conductivities in Aqueous Solutions at 25°C ($\lambda = 10^{-4}\,\text{S-m}^2/\text{mol}$)

Cation	λ	Anions	λ
H_3O^+	350	OH^-	198
Li^+	3.87	F^-	55
Na^+	50	Cl^-	76
K^+	74	Br^-	78
NH_4^+	73	I^-	77
Mg^{2+}	53	NO_3^-	71
Ca^{2+}	59	SO_4^{2-}	80

Source: Ref. 7.

increase in conductance is observed when the analyte passes through the detector. Alternatively, the indirect detection involves an eluent with a high conductance, and a decrease in the eluent conductance is observed when the analyte is detected.

Although indirect conductivity detection has been applied to anions using hydroxide eluents [8] and cations using mineral acids [9], direct conductivity remains the most common approach. The key to direct conductivity measurements has been the development of the suppressor, which results in an increase in signal-to-noise ratio between the eluent and analyte. This is accomplished by passing the eluent through an additional ion-exchange membrane and replacing highly conductive ions with less conductive species. Another benefit that occurs is that the measured conductivity of an ion is the sum of the equivalent conductance of the anion and cation. Since either hydronium or hydroxide, possessing the highest limiting equivalent conductance, is exchanged in the suppressor, the overall signal is greatly enhanced. The limits of detection using suppressed conductivity detection for several ions in water are shown in Table 8.

Postcolumn derivatization with absorbance spectroscopy: Postcolumn derivatization reactions involve the chemical reaction of the analyte with a chromophore, after the analyte has passed through the column, followed by the detection of the analyte by UV-Vis absorbance spectroscopy. This type of detection system is typically used for transition and lanthanide metal cations due to the incompatibilities between the transition metal complexing eluent and the conductivity supressor. The properties of the postcolumn derivatization species should include high molar absorptivity of complexes, reactivity with most metals, formation of stable metal com-

Table 8 Practical Limits of Detection for Several Ions in Water Using Supressed Conductivity

Cation	LOD (µg/l)	Anion	LOD (µg/l)
Li^+	0.05	F^-	0.6
Na^+	0.08	Cl^-	1
K^+	0.25	Br^-	8
NH_4^+	0.18	PO_4^{3-}	5
Mg^{2+}	0.25	NO_3^-	1
CA^{2+}	0.5	SO_4^{2-}	3
Fe^{3+}	2.0		

Source: Ref. 10.

plexes, and rapid formation kinetics [11]. The most common derivatization agent in use is 4-(2-pyridylazo) resorcinol (PAR) to give complexes that absorb at 520 nm.

When designing a postcolumn reactor, care must be taken in the delivery of the postcolumn reagent. Detector noise caused by the column reactor pump can lower the sensitivity of the analysis. Pneumatic devices have been shown to deliver a pulseless flow of derivatization eluent, thereby decreasing detector noise. Several vendors of IC instruments have developed postcolumn reaction technology, and a reactor system can be purchased as an accessory.

Capillary Electrophoresis

Background

Capillary electrophoresis (CE) is an analytical technique that separates ionic species in electrolytic solution by the application of an electric field. This technique can be used to quantitatively measure a number of inorganic components present in polymers. The analysis can be conducted to determine either the total concentration present following dissolution of the polymer by one of the techniques discussed in the sample preparation section, or the amount present as extractable ionic species.

Small inorganic ions of interest to the polymer chemist that can be separated and quantified by traditional CE systems include halides, alkali metals, alkaline earths, and oxyanions. Detection limits in solution are typically 50–100 ppb using indirect UV absorbance detection, but have been reported as low as 1–10 ppb for common inorganic anions, using supressed conductivity detection [12]. The detection limit in a polymer is determined by the amount of polymer used to prepare the solution as discussed in the sample preparation section. The observed precision can be less than 0.1% relative standard deviation [13].

Ion chromatography (IC) and CE are competing techniques in that both can be used to determine inorganic ions. As such, each is often used to validate results obtained by the other method. Major advantages of CE compared to IC include smaller sample requirements (nanoliters rather than microliters) and greater separation efficiency due to less zone spreading in the column, as discussed in the next section. Another advantage of CE is the speed of analysis. For example, 36 inorganic and small organic anions present in concentrations ranging from 0.7 to 3.3 ppm were separated in less than 3 min [14]. However, there are also some limitations in using CE compared to IC. Since the most common mode of detection in CE systems is UV absorbance, greater sensitivities, and therefore detection limits, can be achieved with IC.

Capillary Electrophoresis Theory

Capillary electrophoresis is an analytical technique which enables efficient and rapid separation of ionic compounds under the influence of an electric field. The analysis is conducted in a fused silica capillary that is filled with an electrolytic solution, or buffer. Each end of the capillary is placed in a buffer reservoir along with electrodes attached to a high-voltage power supply. The power supply is then used to apply an electric field, up to 30 kV, between the two reservoirs. A schematic of a typical CE instrument is shown in Fig. 10.

When a sample solution is introduced into the capillary, the direction and rate of solute migration is dependent on the charge-to-size ratio, which affects the electrophoretic mobility [15]. Cations will migrate toward the cathode, the negatively charged electrode, while anions will migrate toward the anode, the positively charged electrode. Ions in the buffer will also be affected by the applied voltage, and the movement of solvent in the capillary is called electroosmotic flow [16,17]. In normal operation, the direction of electroosmotic flow is from the anode to the cathode, as explained later. Most anions will be carried by the electroosmotic flow toward the anode. Neutral solutes travel down the column with the electroosmotic flow and therefore are not separated. The detector is then placed near the cathodic end of the capillary to detect the solute zones as they migrate toward the cathode. A schematic of capillary electrophoresis separation is shown in Fig. 11.

Solutes in capillary electrophoresis are separated in the capillary due to differences in electrophoretic mobility, or the rate of migration. In chromatography, separation is the result of interaction with the stationary phase, or the retention of the solutes. This leads to a difference in terminology for electrophoresis compared to chromatography. The identification of species is based on the migration time (t_m) rather than retention time (t_R), and

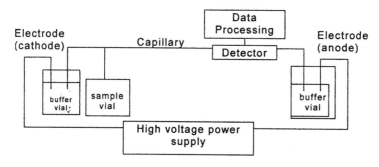

Figure 10   Diagram of a typical capillary electrophoresis instrument.

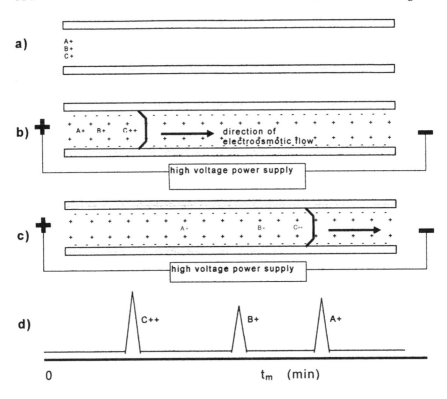

Figure 11 Schematic of capillary electrophoretic separation: (a) introduction of sample; (b) application of electric field; (c) separation of analytes into distinct zones; (d) resulting electropherogram. Note: the capillary surface charge does not interact directly with the analytes.

the detector response which is analogous to a chromatogram is called an electropherogram.

Capillary Zone Electrophoresis

There are several modes of capillary electrophoresis, some of which combine electropheretic and chromatographic techniques, as listed in Table 9.

The primary CE method used for separation of inorganic ions and thus discussed here is capillary zone electrophoresis (CZE). The movement of species in capillary electrophoresis is due to the combination of electrophoretic mobility and electroosmotic flow. In normal operation, the direction of solvent migration, or electroosmotic flow (EOF), is toward the cathode. At pH above about 3, the surface silonal groups (Si-OH) in the

Table 9 Modes of Capillary Electrophoresis

Method	Separate mechanism	Conditions	Analytes
Capillary zone electrophoresis (CZE)	Electrophoretic mobility	Open silica capillary column	Cations and anions
Capillary gel electrophoresis (CGE)	Molecular sieve and electrophoretic mobility	Capillary packed with polymer gel	Large biomolecules, i.e. DNA fragments
Micellar electrokinetic capillary chromatography (MECC or MEKC)	Partitioning of solutes between micellar phase and solution phase	Detergent added to buffer above critical micelle concentration	Neutral molecules
Capillary isoelectric focusing (CIEF)	Difference in isoelectric points	Electroosmotic flow is eliminated by treating the capillary	Amphoteric solutes, i.e., proteins

fused silica column exist as ionized silanoate (Si-O$^-$). The negatively charged surface attracts cations from the solvent to form an electric double layer. When a potential is applied, the loosely held cations in the diffuse layer are pulled toward the negatively charged cathode. The solvation of these cations causes the bulk of the solution to be pulled toward the cathode as well, resulting in the electroosmotic flow. The velocity of the EOF, ν_{EOF}, is due to the buffer characteristics and the applied potential, E, as shown:

$$\nu_{EOF} = \frac{\varepsilon \zeta E}{4\pi\eta} \qquad (4)$$

where ε is the dielectric constant of the buffer, ζ is the zeta potential (influenced by pH and concentration), and η is the buffer viscosity [18].

It is important for the electroosmotic flow to remain constant throughout an analysis to prevent variation in analyte migration times, and errors in peak area measurements. The EOF will increase with increased buffer pH, increased applied voltage, and increased temperature. Increasing the buffer concentration or ionic strength will result in a decrease in EOF. Addition of an organic solvent can either increase or decrease the flow, while modification of the capillary wall can increase, decrease, or even reverse the electro-osmotic flow [19].

The velocity at which charged solutes travel through an electrolytic solution under the influence of an electric field (v_{EP}) is due to the magnitude of the field strength (E) and the electrophoretic mobility (μ_{EP}):

$$v_{EP} = \mu_{EP}E \tag{5}$$

The field strength is given by

$$E = \frac{V}{L} \tag{6}$$

where V is the applied voltage and L is the capillary length. The electro-phoretic mobility of an ion is determined by the buffer viscosity (η) and the charge-to-size ratio as follows:

$$\mu_{EP} = \frac{q}{6\pi\eta r} \tag{7}$$

where q is the charge of the ionized solute and r is the ionic radius. The electrophoretic mobility of a solute is really analogous to the electroosmotic mobility of the solvent. As can be seen from the above equation, the rate of migration increases with increasing charge-to-size ratio. Thus, small highly charged ions migrate more quickly and reach the detector first. Smaller ions with less charge will travel more slowly down the capillary.

Cation Separation

As mentioned previously, solute migration in the capillary is due to both electrophoretic mobility and electroosmotic flow. In normal capillary elec-trophoresis, the direction of both electroosmotic flow and cation migration is toward the cathode, while anion migration is toward the anode. However, the electroosmotic flow is usually greater than the electrophoretic migration of anions. Neutral components are not affected by the electric field and thus are carried at the same rate as the electroosmotic flow. This results in the migration of all solutes toward the cathode, with cations migrating more quickly than the solvent and anions migrating more slowly. Therefore, the order of elution to the detector is small highly charged cations, then larger cations with less charge, followed by neutral molecules, then large and low-charged anions, and finally small, highly charged anions.

Anion Separation

When analyzing primarily for anions, the analysis time can be decreased by reversing the direction of electroosmotic flow. The flow is then toward the anode rather than the cathode, and the anions will elute first. To reverse the direction of the electroosmotic flow, a cationic surfactant such as cetyltrimethylammonium bromide (CTAB) or tetradecyltrimethylammonium bromide (TTAB) is added to the buffer [20]. The positively charged hydrophilic end of the surfactant interacts with the surface of the capillary (Si-O⁻). The hydrophobic tails align with those of surfactants in the solution. The result is a positively charged surface which attracts anions in the buffer and causes them to travel toward the anode. In addition to reversing the EOF, the polarity of the applied field must be reversed or the anions will migrate from the sample inlet to the anode and not pass through the detector (Fig. 10). A typical electropherogram from the separation of anions is shown in Fig. 12.

Instrumentation

A capillary electrophoresis instrument includes two buffer vials, sample vial, fused silica capillary column, high-voltage power supply, detector, and a

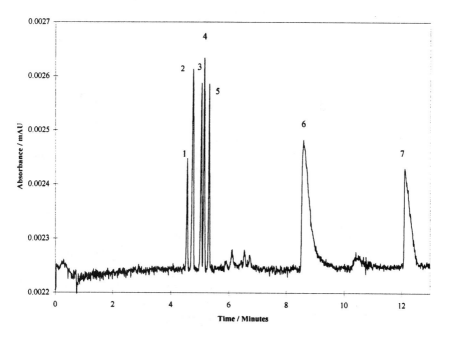

Figure 12 Electropherogram for a mixture of anions; concentration is 50 ng/ml. Separation carried out by CZE using an open silica capillary and detected by indirect absorbance. Peak identifications: 1, Br^-; 2, Cl^-; 3, SO_4^{2-}; 4, NO_2^-; 5, NO_3^-; 6, F^-; 7, PO_4^{3}. (Electropherogram provided by Joanne Smith, GE-CRD.

data processing system as shown in Fig. 10. This section will discuss the capillary, sample introduction system, and detectors in more detail.

The fused silica capillaries used in electrophoresis typically have an inner diameter (I.D.) between 50 and 75 µm and range from 30 to 100 cm in length. If absorbance or fluorescence detectors are used, the capillary can serve as the detector window because fused silica is transparent to ultraviolet and visible light. This is called *on-column* detection. However, the small diameter of the capillary also results in a short path length for the detector and thus low sensitivity compared to chromatography absorbance and fluorescent detectors.

Sample sizes in capillary electrophoresis are very small, 1–30 nl, which can be an advantage when sample volume is limited. The sample is introduced into the capillary by gravity, vacuum, pump, or electrokinetic injection. In all cases, the sample is introduced by momentarily moving the inlet of the capillary from the inlet buffer vial to the sample vial, and then back. When gravity introduction is used, the sample is siphoned onto the column by raising the sample vial. The volume injected (V_i) in nanoliters is calculated by

$$V_i = 2.84 \times 10^{-8} \frac{L}{Htd^4} \tag{8}$$

Where H is the height of the sample (mm), t is the time the sample is raised (s), d is the I.D. of the capillary (µm), and L is the length of the capillary.

Samples can also be introduced by creating a pressure difference across the capillary. The pressure difference can be created by pressurizing the sample vial or by pulling a vacuum on the outlet of the capillary. The sample volume (V_i) introduced by this method can be found by

$$V_i = \frac{Pd^4\pi t}{128\eta L} \tag{9}$$

where P is the pressure and η is the viscosity of the sample.

The third method of sample introduction is electrokinetic injection in which a potential is applied between the sample vial and the outlet end of the capillary. The sample is drawn into the capillary by the same forces that will cause it to migrate toward the detector, electrophoretic mobility and electroosmotic flow. The amount of each solute injected (Q) is determined by

$$Q = \frac{V\pi ctr^2(\mu_{EP} + \mu_{EOF})}{L} \tag{10}$$

where μ_{EP} and μ_{EOF} are the electrophoretic mobility and the electroosmotic mobility, respectively. Because electrokinetic injection is based on electrophoretic mobility, sampling bias can occur. For this reason, electrokinetic injection is not as precise as the other injection methods [19].

Detection

There have been many detection methods used in capillary electrophoresis systems, including mass spectrometry, amperometry, and conductivity. The most common detectors are absorbance and fluorescence. These detectors allow solutes to be detected while still on the column by using the capillary as the detector window. As discussed above, the short path length results in higher detection limits than similar chromatography detectors, which have cell path lengths of 50–75 µm.

When analyzing inorganic ions that are nonabsorbing, indirect absorbance detection is used. The only difference is that a chromophore is added to the buffer. The nonabsorbing analytes displace the chromophore in the detector, causing a dip in the baseline and giving rise to negative peaks. However, the output polarity of the detector is typically reversed to produce positive peaks on the integrator or computer. Some examples of chromophores are pyromelletic acid for detection of inorganic ions, and phthalate for analysis of small organic anions.

Capillary Electrophoresis versus Ion Chromatography

The efficiency of capillary electrophoresis is greater compared to chromatography because there is less zone spreading that occurs as the analytes migrate down the column. There are several reasons for the minimal zone spreading that occurs. First, the profile of electroosmotic flow is relatively flat (as depicted in Fig. 11) compared to pumped flow profiles in chromatographic techniques. With a flat flow profile, the solvent molecules travel at the same velocity regardless of their location in the column, which minimizes zone spreading of the solutes. In addition, the solutes migrate in a tighter plug because there is no interaction with the column which results in zone broadening. One factor in electrophoresis that increases solute dispersion compared to chromatography is the heat produced from the passing of current through the capillary, or Joule heating. Joule heating is minized by decreasing the diameter of the column, which led to the use of capillary columns for electrophoresis analysis. Capillary columns can also be cooled to further decrease Joule heating and therefore thermal convection.

Applications of Ionic Separation and Speciation Techniques

Small inorganic ions of interest to the polymer chemist that can be separated and quantified by traditional CE and IC systems include halides, alkali metals, alkaline earths, and oxyanions. The analysis can be conducted to determine either the total elemental concentration or the concentration of extractable ionic species, depending on the sample preparation method (see sample preparation section).

The detection method most often used for the analysis of small inorganic ions by CE is indirect absorbance. Pyromellitic acid is a typical chromophore and can be purchased as a buffer solution from instrument manufacturers. The most common method of detection in IC analysis is suppressed conductivity.

The detection limits of relevant ions in solution are typically 50–100 ppb using indirect UV absorbance CE and range from 0.1 to 50 ppb using suppressed conductivity IC. The amount of analyte present in the polymer is determined by the weight of polymer analyzed and the final solution volume, as explained in the section below.

METHODS FOR SAMPLE PREPARATION IN THE INORGANIC ANALYSIS OF PLASTICS

Introduction

Most of the commonly used methods of inorganic analysis of plastics, including inductively coupled plasma emission spectroscopy, atomic absorption spectroscopy, ionic chromatography, and capillary electrophoresis, require that the samples for analyses be introduced preferably as solutions with low viscosity and minimal salt content.

The choice of the particular approach used to prepare solutions containing the inorganic analytes from plastic samples depends on the nature of the sample (sample matrix), the analytical technique used, as well as the nature of the analytes and their concentration levels in the sample.

Most of the methods (except for direct dissolution in organic solvents) used to convert the analytes of interest in the plastics to solution involve sample mineralization which results in oxidation of its organic constituents to carbon dioxide and water. The oxidation process can be accomplished as sample combustion (dry ashing) whereby the sample is oxidized in a stream of oxygen or air, or alternatively, strong liquid oxidizers (mineral acids, peroxides, etc.) can be used for the same purpose (wet ashing).

Sample Digestion (Wet Ashing)

Plastic samples are generally dissolved in mineral acids in a process called *acid digestion*. This usually requires heating of the sample with acid(s) using a hot plate or microwave oven (*microwave acid digestion*). In special cases (if the sample contains silicates or certain types of mineral oxides which are difficult to dissolve in acids), a *fusion* approach can be applied in which the sample is reacted with inorganic salts (carbonates, hydroxides, peroxides, etc.) at high temperatures. The last approach has very limited applicability for plastic materials and will not be discussed in this chapter.

Some general considerations when sample digestion is applied involve the possibility of loss of some analytes due to their volatilization. Halogen compounds of certain elements, such as antimony, arsenic, boron, chromium, germanium, silicon, lead, mercury, tin, zirconium, zinc, and titanium, can be partially lost as volatiles if formed during the sample preparation. Oxides of some late transition metals in their highest oxidation state, such as ruthenium(VIII), osmium(VIII), and rhenium(VII), are also known to be volatile. Closed-vessel digestion technique can in some cases provide solution to these potential issues.

Chemical Reagents Used for Acid Digestion of Plastics

Since oxidation of the polymeric backbone is an integral part of the plastics dissolution process, strong oxidizing acids such as nitric, perchloric, and sulfuric, as well as acid mixtures, are most commonly used for this purpose.

Extra care must be exercised when handling concentrated mineral acids due to the high risk of chemical burns and sometimes explosion. Table 10 summarizes the physical properties of the most commonly used mineral acids:

Nitric acid in its concentrated form (65–69%) is probably the most widely used reagent for wet ashing because of its dual action as a strong oxidizer and strong acid. The nitrate salts formed are water-soluble and the nitrate anion is only weakly complexing. Nitric acid readily attacks the polymeric backbone of plastic samples, although complete destruction is rare, and it is often used in combination with other acids such as hydrochloric, hydrofluoric, and sulfuric acids.

Table 10 Physical Properties of Mineral Acids Commonly Used for Sample Digestion

Acid	Concentration (wt%)	Molarity (mol/l)	Boiling point (°C)	Specific gravity (g/ml)
Nitric	70	16	121 (67% azeotrope)	1.42
Sulfuric	96	18	330	1.84
Perchloric	70	11.6	203	1.66
Hydrochloric	35	11.3	109 (6 M/l azeotrope)	1.18
Hydrofluoric	46	26.5	111 (36% azeotrope)	1.15

In some cases, when the samples contains higher levels of metallic elements such as Al, Ta, Ti, Cr, or Nb, dissolution in HNO_3 is incomplete due to metal surface passivation as a result of a formation of nonsoluble metal oxide layer which inhibits further reaction. In addition, some metal nitrates, such as Sn, Sb, and W, can be hydrolyzed and precipitated as hydrated oxides.

Concentrated *sulfuric acid* (about 98%) is an effective oxidizer, especially hot, strong acid, and is also a very effective dehydrating agent, which often is a significant aid in decomposing organic materials. Its effectiveness in dissolving plastic samples can be attributed to its high boiling point (about 340°C), which is sufficient to decompose the organic content of most samples. Sulfuric acid is often used in combination with oxidizers such as nitric acid or hydrogen peroxide in order to prevent formation of carbon residue. Sulfate salts have generally low volitality, which reduces the risk of volatility losses for some of the analytes. Elements forming insoluble or low-soluble sulfates such as Ba, Sr, Pb, and Ca are not suitable for acid digestion including sulfuric acid treatment.

Perchloric acid is an extremely powerful oxidizing agent when hot and concentrated (60–72%). Its use is limited, however, and it is applied usually in mixture with other oxidizers such as nitric or sulfuric acid because it constitutes an exceptional explosion hazard, especially when overheated.

Hydrochloric acid is a very useful agent in dissolving a variety of inorganic compounds as well as electropositive metals, although it is not an oxidizing agent. Therefore hydrochloric acid is used only in conjunction with other oxidizers (nitric, sulfuric, or perchloric acid) for the digestion of plastic samples. Samples containing metals forming low-soluble chloride salts such as Ag and Tl as well as Pb, as well as in case when volatilization of some of the analytes of interest may be suspected, should not be digested in the presence of hydrochloric acid.

Although *hydrofluoric acid* is not an oxidizer or a weak acid, it is often used as an aid in digesting important refractory materials such as silicates (in glass-filled plastic materials), titanates, niobates, etc., with which it forms stable complexes. Silicon is usually lost in open digestion as SiF_4 and boron as BF_3. Traces of SiF_4 can sometimes be complexed with boric acid. Hydrofluoric acid is usually used in mixtures with nitric or sulfuric acid for digestion of plastic materials.

Oxidizing mixtures comprised mainly of combinations of different acids and sometimes addition of oxidizers to mineral acids are most commonly used in wet digestion procedures for plastic samples. Such an approach is used to incorporate the individual actions of the constituting mixture components in order to achieve effective sample solubilization. A typical example in this respect is a HNO_3/HF mixture used for digestion of a plastic sample containing glass fill, where nitric acid is used to dissolve the polymeric organic backbone of the sample while hydrofluoric acid is used to dissolve the glass filler additive.

Instrumental Techniques for West Ashing of Plastic Samples

Three general approaches are currently used for sample dissolution by mineral acids:

1. Open-vessel acid digestion
2. Acid digestion bombs
3. Microwave acid digestion

Open-Vessel Acid Digestion

Acid digestion in open vessels is one of the oldest techniques used for mineralization of solid samples. It is a relatively simple—requiring only the use of a hot plate in terms of equipment—but time- and resources-consuming technique. It needs to be carried out in specially designated fume hoods and requires constant attention (full protective equipment as well as protective shields required), due to the potentially violent character of the reactions. Extra care and precaution need to be taken if formation of explosive mixtures (for example, when working with perchloric acid) may be possible.

Acid Digestion in Closed Vessels

The acid digestion bomb is another powerful technique for solubilization of plastic samples, in which the digestive process is carried out in a sealed pressurized vessel called a digestion bomb and heated in a muffle furnace. The advantage of this approach is that the temperature of the process may reach well above the boiling point of the mixtures in the normal process, thus making possible the complete solubilization of some slow or incompletely dissolving components under open-vessel conditions. Volatilization of some of the analytes as well as sample contamination from airborne particulates can be also minimized.

This approach has, however, certain limitation in terms of the amount of organic sample (most plastics materials) which can be digested while avoiding significant pressure buildup. Acid digestion bomb loading limits for some typical bomb sizes are shown in Table 11.

Nitric acid is the preferred medium for acid bomb digestion. Special care must be exercised when a mix with sulfuric acid is to be applied, to avoid violent reactions with the samples. Operating limits on most acid digestion bombs are usually set at around 250°C and 1800 psi. Perchloric

Table 11 Recommended Loading Limits for Acid Digestion Bombs

Vessel capacity (ml)	Maximum organic sample (g)	Maximum inorganic sample (g)	Minimum and maximum volume of HNO_3 for use with organics (ml)
125	0.5	5	12–15
45	0.2	2	5–6
23	0.1	1	2.5–3

acid is not recommended for use in acid bombs, due to its unpredictable behavior when heated.

Microwave Acid Digestion

The use of microwave radiation provides an alternative heating source for acid digestion of solid samples. It has distinctive advantages over conventional heating systems (hot plate or muffle furnace), including:

Faster, complete and reproducible digestion
High throughput and rapid sample turnaround
Improved recoveries
Retention of volatile analytes
Elimination of cross-contaminants

Background: Microwave energy has a frequency in the range from 300 to 300,000 MHz (2450 MHz for home appliances). The power output ranges from 600 to 1500 W and may also be programmed to suit any particular application. Microwave energy is absorbed by the sample by two mechanisms:

Ionic conduction: migration of dissolved ions in the applied electrical field
Dipole rotation: alignment of polarized molecules to the poles of the electromagnetic field

Heat is generated by molecular collision taking place due to ion migration or molecular dipole alignment/misalignment in the changing electric field causing the molecules to rotate back and forth.

Closed-vessel microwave dissolution: Closed-vessel microwave dissolution utilizes several types of microwave digestion vessels designed for heating of liquids under pressures usually from 200 up to 1500 psi (100 bar).

These vessels are usually made from Teflon-based composite materials, which makes them transparent to the microwave energy and allows for it to be applied directly to the samples for complete and rapid digestion. The vessels are equipped with pressure and temperature sensors which also provide the means to control these parameters. The vessels are placed on a rotating turntable to assure equal distribution of the microwave energy.

Heavy-duty vessels capable of withstanding 100 psi pressure or more are recommended for plastic sample digestion, as these samples are expected to develop high internal pressures during sample digestion due to their organic nature. Another recommended approach for hard-to-digest samples (as plastic samples are usually classified) involves a microwave charring step using quartz digestion vessels prior to closed-vessel digestion.

CONTAMINATES IN POLYMERS: STRATEGIES OF A CASE STUDY

The need for contaminate-free polymers has been driven by the advent of plastics in the use of microelectronics, memory storage, and video storage, along with increasing demands for optical products. In most cases, the quantification and control of such species throughout the entire synthetic process is important to meet a product's demands. To satisfy these requirements, experimental methods must be designed to selectively identify and confidently quantify contaminates which affect the performance characteristics of a given product.

Inorganic contaminates are known to affect the product performance of several polymers. Through the use of sample preparation and the previously described characterization techniques, the identification, quantification, and speciation of such contaminates can be accomplished. The acquired information can lead to the source and the subsequent remedy of a contamination issue.

To approach such problem solving, the first and most important step is the sampling and storage of the specimen. Care must be taken in the choice of sampling container and the tools used to extract the sample. The same attention to specimen handling is required while the sample is in the possession of the analyst. Before the analysis begins, it is important to discuss with the appropriate people the specific details of what is needed and expected from the analysis. Such discussions can save a tremendous amount of time and eliminate any miscommunications.

After the specimens are obtained, a general strategy can be employed for the analysis. To acquire data on the total amount of contaminant in a sample, the specimen should be incinerated or acid-digested followed by one of the spectroscopic characterization methods described previously. To obtain data on speciation (cationic and anionic) and ion mobility in the polymer, the specimen should be manipulated as little as possible. A typical method employs the dissolution of the polymer in an organic solvent, followed by the addition of an aqueous solution. The sample is shaken or stirred and then the aqueous phase is separated from the organic phase. The aqueous phase can then be analyzed by ion-exchange chromatography or capillary electrophoresis for extractable ions. Incineration of the polymer in an oxygen bomb followed by IC or CE provides the total amount of anions, both polymer bond and ionic. Thus, the ability to separate contaminates as bound and unbound species allows the analyst to pinpoint contaminant processes as well as to explain the cause and effect of such contamination.

BIBLIOGRAPHY

ZB Alfassi, ed. *Determination of Trace Elements*, Weinheim: VCH, 1994.

KD Altria, ed. *Capillary Electrophoresis Guidebook: Principles, Operation, and Applications.* Totowa, NJ: Humana, 1996.

CB Boss, KJ Fredeen. *Concepts, Instrumentation and Techniques in Inductively Coupled Plasma Atomic Emission Spectrometry.* Perkin Elmer, 1989.

PWJM Boumans, ed. Inductively Coupled Plasma Emission Spectroscopy—Part 2, Application and Fundamentals, *Chemical Analysis, Volume 90.* New York: Wiley Interscience, 1987.

HM Kingston, LB Jassie, eds. *Introduction to Microwave Sample Preparation— Theory and Practice.* Washington, DC: American Chemical Society, 1988.

R Kuhn, S Hoffstetter-Kuhn. *Capillary Electrophoresis: Principles and Practice.* Berlin, New York: Springer-Verlag, 1993.

LHJ Lajunen. *Spectrochemical Analysis by Atomic Absorption and Emission.* Cambridge, UK: Royal Society of Chemistry, 1992.

A Montaser, DW Golightly, eds. *Inductively Coupled Plasma in Analytical Atomic Spectroscopy.* Weinheim: VCH, 1992.

J Ottaway, A Ure. *Practical Atomic Absorption Spectroscopy.* New York: Pergamon, 1983.

RD Reeves, RR Brooks. Trace Element Analysis of Geological Materials. *Chemical Analysis, Volume 51.* New York: Wiley, 1978.

H Shintani, J Polonsky, eds. *Handbook of Capillary Electrophoresis*, London, UK: Blackie, 1997.

Q Yang, K Hidajat, SFY Li. *J. Chromatog. Sci.* **35**: 358, 1997.

R Weinberger. *Practical Capillary Electrophoresis.* Boston: Academic Press, 1993.

REFERENCES

1. J Statler. Ion Chromatography. In: FA Settle, ed. *Handbook of Instrumental Techniques for Analytical Chemistry.* Upper Saddle River, NJ: Prentice Hall, 1997, pp. 199–220.

2. H Small. *Ion Chromatography*. New York: Plenum, 1989, pp. 1–58.
3. PR Haddad, PE Jackson. *Ion Chromatography: Principles and Applications*. Amsterdam: Elsevier, 1990, pp. 1–78.
4. JG Tarter, ed. *Ion Chromatography*. New York: Marcel Dekker, 1987.
5. J Weiss, ed. *Ion Chromatography*. New York: VCH, 1995.
6. RD Rocklin. Detection in Ion Chromatography. *J. Chromatogr.* 546:175, 1991.
7. OA Shpigun, UA Zolotov. *Ion Chromatography in Water Analysis*. Chichester, UK: Ellis Horwood, 1988, p. 85.
8. T Okada, T Kuwamoto. *Anal. Chem.* 55:1001, 1983.
9. PR Haddad, RC Foley. *Anal. Chem.* 61:1435, 1989.
10. E Prichard, GM MacKay, J Points, eds. *Trace Analysis: A Structured Approach to Obtaining Reliable Results*. Cambridge, UK: Royal Society of Chemistry, 1996, p. 151.
11. PR Haddad, PE Jackson. *Ion Chromatography: Principles and Applications*. Amsterdam: Elsevier, 1990, pp. 387–405.
12. N Avdalovic, CA Pohl, FD Rocklin, JR Stillian. *Anal. Chem.* 65:1993, 1970.
13. PD Grossman, JC Colburn, eds. *Capillary Electrophoresis: Theory and Practice*. San Diego, CA: Academic, 1992.
14. G Bondoux, P Jandik, WR Jones. *J. Chromatog.* 602:79, 1992.
15. JW Jorgenson, K Lukacs. *Anal. Chem.* 53:241, 1981.
16. JW Jorgenson, K Lukacs. *J. Chromatogr.* 218:209, 1981.
17. JW Jorgenson, K Lukacs. *Science* 222:266, 1983.
18. DR Baker. *Capillary Electrophoresis*. New York: Wiley, 1995, pp. 19–51.
19. D Baker. Capillary Electrophoresis: In: FA Settle, ed. *Handbook of Instrumental Techniques for Analytical Chemistry*. Upper Saddle River, NJ: Prentice Hall, 1997, pp. 165–182.
20. SFY Li. *Capillary Electrophoresis: Principles, Practice and Applications*. Amsterdam: Elsevier, 1992, pp. 1–28.
21. JA Dean. *Analytical Chemistry Handbook*. New York: McGraw-Hill, 1995.

11

Liquid Chromatography of Polymers

Gary J. Fallick and Rick Nielson

Waters Corporation, Milford, Massachusetts, USA

I. INTRODUCTION

A. Overview of Liquid-Phase Chromatography

Modern high-performance liquid chromatography (HPLC) is a versatile analytical technology that can provide a range of essential information about the soluble components of plastics, including the base polymer, the formulated resin, and the additives. All modes of HPLC use a column packed with materials to separate species in a complex mixture. In some modes of HPLC the separation is based on differences in chemical composition or selectivity. The mode of HPLC that is most familiar to polymer analysts is gel permeation chromatography (GPC). It is also termed size-exclusion chromatography (SEC).

As suggested by the names, the mechanism of separation in GPC/SEC is based on apparent size differences among the sample components in solution. This makes it especially useful for characterizing polymers. This chapter focuses primarily on the use of GPC and to a lesser extent on HPLC in plastics analysis. While gas chromatography (GC) is also used extensively for characterizing monomers and some polymers, it is beyond the scope of this review.

549

B. Molecular-Weight Distribution of Polymers

Monomers, the building blocks of polymers, have a single molecular weight. As low-molecular-weight organic compounds, they are monodisperse. Typical monomers include styrene, ethylene, and phenol. At the onset of polymerization these molecules combine to form oligomers such as dimers, trimers, tetramers, and so on. Ultimately, the growing chains of molecules reach sufficient size to be termed polymers. At this stage not every molecule contains the exact same number of monomer units. This results in a distribution of chain lengths or molecular weights. Depending on the type and conditions of polymerization, this molecular-weight distribution can be very narrow or quite broad. For example, a condensation, or step-growth polymer such as a polyester, poly(ethyleneterephthalate), will have a narrow distribution of molecular weights. Conversely, a free-radical polymerization of an olefin such as ethylene may produce a polyethylene polymer with a very broad distribution of chain lengths and molecular weights.

Both the molecular-weight distribution and the average molecular weight are among the most important determinants of the polymer performance. This encompasses the physical properties of the polymer as well as its processability [1]. Consequently, it is vital to be able to accurately measure molecular weights and molecular-weight distributions for many purposes. These uses include controlling the kinetics of the polymerization to achieve a desired distribution, ensuring the quality and consistency of incoming materials that will be formulated and fabricated, evaluating competitive materials and diagnosing an undesired change in performance.

The widespread use of gel permeation chromatography is due to its ability to accurately and reproducibly determine polymer molecular-weight distributions and to distinguish subtle differences between closely related samples. Figure 1 shows the slight shifts in molecular-weight distribution of a polypropylene sample at various points in the injection-molding process, especially before and after the gate. The profiles for the pellet and the sample taken from the runner are offset for ease of comparison.

C. Composition and Chemical Structure

While polymer molecular-weight average and distribution have a major influence on behavior, many other factors are also very significant. The structure of the polymer chain, polymer chemical composition, and the additive packages may all play a key role. For example, a polymer chain structure that is linear will have very different thermal and processability characteristics than one that is branched. Of course the chemical composition of the monomer or the use of two or more monomers to produce a copolymer or terpolymer will often account for greater differences than a

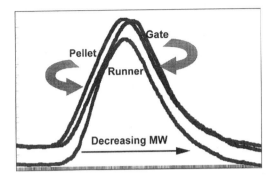

Figure 1 Injection molding effects. Solvent: 1,2,4-trichlorobenzene. Temperature: 140°C. Columns: Styragel HT6, HT5, HT4, and HT3. The shift in the molecular-weight distribution of the "Gate" sample may be due to melt shear-induced reduction of the largest molecules after passing through the mold gate.

slight shift in molecular-weight distribution between two samples of the sample polymer type. The vital role of additives is also well known.

In addition to molecular weight and distribution, modes of HPLC can also be used to characterize polymer structure, polymer composition, and additives. Sections II.D. and III.B describe the use of multiple-detector GPC for measuring branching and other aspects of polymer structure, while Secs. II.F and III.C contain brief examples of additive and copolymer analysis by HPLC.

D. Gel Permeation Chromatography

GPC is the most predictable mode of HPLC. The separation is based on the size of the sample in solution, not the molecular weight. There must not be any interaction with the column packing material (adsorption, partition, etc.), as there is in other modes of HPLC. GPC column packings are particles of cross-linked gel that contain surface pores. The sizes of these pores are controlled and vary from small to large. They act as a molecular filtration system. The most widely used gels are styrene/divinylbenzene copolymers for organic solvent-soluble polymers and acrylate gels for water-soluble polymers.

In practice, the polymer is dissolved in a suitable solvent and a small sample of the dilute polymer solution is injected into a solvent (usually the same as the polymer solvent) that is being pumped through the columns. For most applications, several columns of varying pore size are connected in series. When the column packing porosity range is chosen properly, the largest polymer molecules in the distribution will fit into only a few of the pores and will pass through the columns primarily in the interstitial volume between the packing particles. They will elute sooner than the smaller polymer molecules that can fit into more of the pores and therefore be retained in the columns longer. The solvent path out of the column is through one or more detectors that can sense the presence of the polymer fractions as they elute. The detector signal is then recorded as a chromatogram showing the distribution of the various size components in the sample, proceeding from largest to smallest. The amount of each fraction corresponds to the height of the peak.

The schematic separation shown in Fig. 2 represents the molecular-weight (MW) distribution of the polymeric gum base used in chewing gum, followed by the lower-MW components in decreasing order of size. A similar example could be a chromatogram of PVC with a mixture of plasticizers, antioxidants, and UV stabilizers.

One of the original techniques for using GPC was simply to superimpose the chromatograms of two or more samples and display regions of difference in the distribution. This was used to explain differences in behavior, Fig. 3, or to establish acceptance envelopes for incoming materials, Fig. 4. Note the detail of the oligomeric region of the uncured phenolic resin in Fig. 4. Also, recall that since GPC is performed with dilute solutions of the polymer sample, this type of analysis is only suitable for thermosetting resins and elastomers before they are cured.

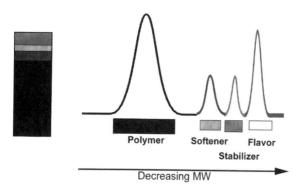

Figure 2 GPC example: chewing gum. The chromatogram is not to scale—the stabilizer and flavor peaks would probably be much smaller and would consist of one or more monodisperse peaks instead of the distribution shown to illustrate the size separation concept.

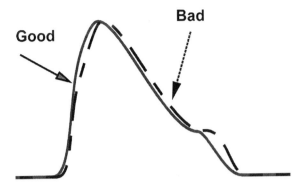

Figure 3 Comparison of results. Many differences in properties or processability of polymers can be explained by comparing molecular-weight regions. In this example the bad sample has slightly less high-molecular-weight and more low-molecular-weight content than the good sample.

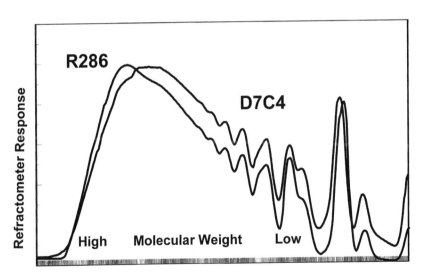

Figure 4 Phenolic resins from different vendors. GPC is useful for incoming quality control, technical support, and competitive analysis. Differences in the molecular-weight distributions of these two phenolic resins could correspond to acceptable and unacceptable processability or performance.

With an understanding of the influence specific molecular-weight regions have on behavior, this can be a very useful way to use GPC information. However, for many purposes it is desirable, if not essential, to be able to quantify the molecular-weight values.

E. Polymer Characterization: Molecular-Weight Values

In order to assign numerical values to the GPC chromatogram, it is divided into a number of vertical slices and treated statistically, Fig. 5. When a mass-sensitive detector is used in the GPC system, the height of each slice, H_i corresponds to the number of molecules present in that slice. The retention time (or volume) of that slice is assigned a molecular weight value, M_i, which is obtained from a calibration curve, such as that shown in Fig. 6. Calibration methods are discussed in Sec. II.B.

Once values for H_i and M_i are assigned to each slice, the summations indicated in Fig. 5 are performed to obtain the various molecular-weight averages that describe the molecular-weight distribution. The values are statistical moments of the distribution, while PD, the polydispersity, is an indication of the narrowness of the distribution.

There are other techniques for determining the various molecular-weight averages. Number average, Mn, can be obtained by membrane osmometry or end-group analysis. Weight average, Mw, is determined by

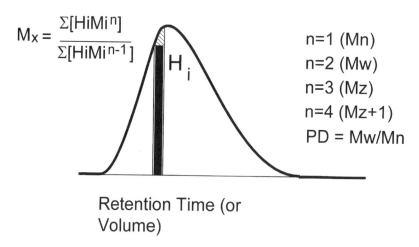

$$M_x = \frac{\Sigma[H_i M_i^n]}{\Sigma[H_i M_i^{n-1}]}$$

H_i

n=1 (Mn)
n=2 (Mw)
n=3 (Mz)
n=4 (Mz+1)
PD = Mw/Mn

Retention Time (or Volume)

Figure 5 A number of molecular-weight averages are calculated from the distribution curve to fully define the distribution.

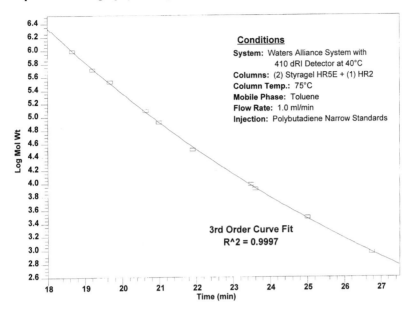

Figure 6 Polybutadiene calibration. The curve is developed for a specific column set with the same solvent and at the same conditions that will be used to characterize unknown samples.

light scattering, while Z and $Z + 1$ averages, M_Z and M_{Z+1} are measured by ultracentrifugation. GPC is unique in that all of these averages can be obtained in a single analysis.

The value in determining the different molecular-weight averages is related to the influence that various molecular-weight fractions have on properties and behavior. For example, additional high-molecular polymer chains may change the flexibility of the polymer to a much greater extent than the same number of low-molecular chains in the material. Raising the molecular weight to a higher power in the calculation of Mz than in the calculation of Mn reflects the greater influence of these longer chains. Mz is related to elongation and flexibility, while Mn is related to brittleness and flow properties. The region of the molecular-weight distribution where each of these molecular-weight averages occurs is shown in Fig. 7. MP is the only chromatographic value in the figure, corresponding to the peak molecular weight.

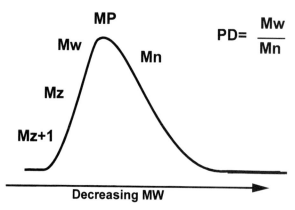

Figure 7 The calculated molecular-weight values correspond to regions of the full distribution. The peak molecular weight, Mp, is not calculated but read directly from the chromatogram. The polydispersity, PD, is the ratio of the weight-average to number-average molecular-weight, an indication of the broadness of the molecular-weight distribution.

F. Structure–Property Correlations

The relationship between certain molecular-weight averages and properties extends beyond the brief examples mentioned in the preceding section. Various studies and correlations have been published [2–5].

Weight-average molecular weight, Mw, is related to characteristics such as tensile strength and impact resistance. The actual relationship may be the consequence of molecular weight and structure contributing to the balance of amorphous and crystalline regions, which in turn influence physical behavior. This is shown in a study of three polypropylene samples that were characterized by GPC, differential scanning calorimetry (DSC), dynamic mechanical analysis (DMA), and melt rheometry. Molecular-weight values and intrinsic viscosity were determined by GPC. Enthalpy of melting, a measure of crystallinity, was determined by DSC. Tan δ, the ratio of loss modulus to storage modulus, a measure of amorphous content at glass transition, was measured by DMA, and zero-shear melt viscosity was measured by melt rheometer. The results appear in Tables 1 and 2.

The most apparent correlation is between the rheology and the GPC measurements. The samples had a broad range of melt flow index values, the amount of material that will flow through a standard orifice in 10 min under a fixed combination of temperature and pressure. Higher melt flow index values indicate a greater tendency to flow. The most frequently

Table 1 Correlating GPC and Rheology

Sample melt flow [g/10 min]	Mn × 10³	Mw × 10³	Intrinsic viscosity (solution) (η)	Zero shear viscosity (melt) [Pa.S]
30	26	201.4	1.26	2,960
10	44	275.5	1.47	3,990
1.8	69	328.6	1.76	41,800

When comparing the same type of polymer, molecular weight and flow properties may correlate very strongly, as shown for these polypropylene samples.

reported molecular-weight values, number average, Mn, and weight average, Mw, decreased as melt flow index increased. Lower-molecular-weight polymers contain shorter polymer chains that will entangle less often in flowing conditions and therefore have less resistance to flow. Hence the higher melt flow index samples had lower dilute solution viscosity and melt viscosity values with decreasing molecular weights. The intrinsic viscosity was determined with a GPC system containing a viscometer detector. More information is given in Secs. II.D and III.B.

As suggested above, the influence of molecular weight on various mechanical properties may, in some instances, actually be attributed to differences in the degree of crystallinity, which, in turn, is related to chain length. Longer polymer chains may be less mobile and therefore sterically inhibited from aligning as many chain segments into ordered, crystalline regions as the shorter polymer chains. This simplified mechanistic model is supported by the values listed in Table 2. Enthalpy of melting shows more

Table 2 Correlating GPC and Thermal Analysis

Sample melt flow [g/10 min]	Mn × 10³	Mw × 10³	Enthalpy of melting ΔHm[a] [J/g]	tan δ @ Tg[b]
30	26	201.4	99.5	0.053
10	44	275.5	96.7	0.061
1.8	69	328.6	95.4	0.072

Thermal and thermal mechanical properties are related to crystallinity, which in turn can be influenced by molecular weight.

[a] Increases with higher crystallinity (less amorphous).

[b] Increases with lower crystallinity (more amorphous).

crystallinity with the lower-molecular-weight samples, while tan δ, a measure of toughness contributed by the amorphous region of the polymer, also shows more crystallinity (less amorphous material) at the lower molecular weight.

Any such analysis is valid only for a closely related set of samples of the same polymer type. Otherwise, factors such as chemical composition of the polymer, differences in degree of branching, or presence of additive packages can overshadow the contributions of shifts in molecular weight. In addition, the history of the sample may also account for a difference in behavior. Recall in Fig. 1, the shift in the MW distribution toward the lower range due to the melt shear imparted to the resin during the injection-molding operation. This also suggests the utility of monitoring molecular-weight profiles to estimate the amount of regrind that can be used without impairing performance.

II. CHROMATOGRAPHY PRACTICES

A. System Configuration

All HPLC systems, including GPC, employ a pumping system to deliver a steady flow rate of solvent through the columns and detectors. There is also some means of introducing the sample solution into the flowing solvent before it enters the columns, as well as a data handling device. The components may be modular or incorporated into an integrated system. Manufacturers of GPC systems are listed in Appendix E. Whatever the configuration, the pumping system is the key to the performance of the system. It must be a sophisticated fluid manager capable of delivering flow accurately, reproducibly, and smoothly. The significance of flow precision is illustrated in Sec. III.A.

The most basic GPC system consists of a basic, single solvent delivery system, a sample injection device, columns, and a differential refractometer detector. The sample injection mechanism can be a simple, single-sample, manually activated valve or an autosampler for unattended processing of multiple samples. A basic modular system with autosampler is shown schematically in the top section of Fig. 8.

Numerous variations and enhancements are available for many of the system components. Solvent delivery systems that operate with multiple solvents to produce programmable changes in composition during the course of the analysis are used often. This enables the system to also perform HPLC analysis of additives or oligomeric resins such as phenolics and epoxies. For these applications it is also necessary to use additional detectors, such as the photodiode array (PDA) detector indicated in the middle

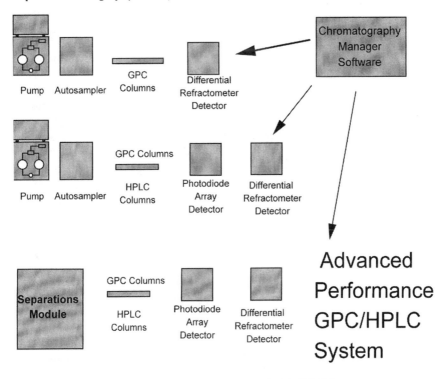

Figure 8   Primary components of modern GPC/HPLC instruments.

system of Fig. 8 (see Secs. II.D and F). Manufacturers of complete HPLC systems are listed in Appendix E.

Some of the other components commonly found in GPC systems are column heater compartments and solvent degassers. The column heater is often used, even for polymers that are soluble at room temperature, to reduce solvent viscosity and ensure uniform column operating conditions, despite any fluctuations in room temperature. The solvent degasser is located before the pump intake. Degassing prevents noisy signals caused by intermittent outgassing in the refractive index detector. In all systems the data management system is essential. It can range from a simple strip-chart recorder or integrator to a computer-based software package that can control the system as well as processing the data and reporting the results. See Sec. II.E.

The lower system schematic in Fig. 8 shows a module in which the pumping system and autosampler have been integrated into a single Separations Module. This unit, the 2690 Separations Module, is the basis of high-performance HPLC and GPC Alliance Systems that combine the high performance and capacity of the solvent and sample management with the flexibility of detector choices according to the ways the system is to be used. It can be used for single-solvent (isocratic) GPC work or for up to four solvent gradient separations of additives and oligomers. It offers the option of integral solvent degassing. A number of design enhancements incorporated into the 2690 significantly improve GPC performance associated with traditional gear-driven dual reciprocating-head pumps. The two heads are independently microprocessor controlled, producing extremely smooth, precise flow that contributes to exceptionally reproducible molecular-weight values. See Sec. III.A. Figure 9 is a photo of a basic Alliance HPLC/GPC System using a refractive index detector, while Fig. 10 depicts the major internal configuration of this system.

Figure 9 A versatile integrated solvent and sample management module configured with modular refractive index detector for performing basic GPC.

Figure 10 The primary components of the system shown in Fig. 9 include the control panel (upper left), injection mechanism (upper right), auto-sampler carousels (mid-level), independently driven dual reciprocating solvent delivery pistons (lower left), and degasser/solvent select module (lower right).

In order to perform GPC of polyolefins and other polymers that are only soluble at elevated temperatures, a completely integrated system is used to ensure that solvent, injector, columns, and detector are all maintained at temperature to preserve solubility and provide greatest reproducibility. Such systems are produced by Polymer Laboratories and Waters Corp. The Waters Alliance GPC 2000 System is shown in Fig. 11. It features the same innovative flow management design as the 2690 Separations Module, plus a precise dual-temperature-zone injector and the option of single detection with integral differential refractometer or dual detection with refractometer and viscometer. See Secs. III.B and III.D. The autosampler can agitate the samples by spinning to ensure complete solubility and homogeneity and also filter the samples. This is especially useful for samples prepared from filled materials or elastomers that may contain gels. The agitation and filtration operations take place in the pre-injection zone, which is thermally isolated and independently controlled relative to the holding positions in the autosampler. They can be at the same temperature or the pre-injection zone temperature can be set higher than the holding positions. This enables the samples to be injected at temperature without impairing sample stability while in the queue. The benefits of the integrated system and the sample processing capabilities can also be applied to ambient soluble polymers.

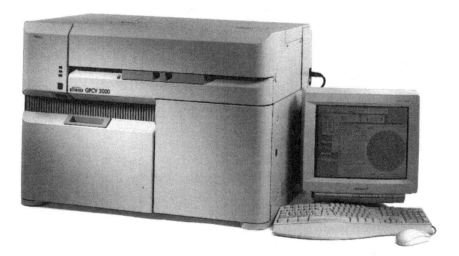

Figure 11 A completely integrated, automated GPC system capable of analyzing samples ranging from those soluble at ambient conditions to those which are only soluble at temperatures up to 180°C.

B. Column Conventions: Selection and Calibration

To ensure that a representative molecular-weight distribution is obtained, it is critical to choose the correct columns. This refers to the correct packing pore sizes to maximize the resolution in the molecular-weight range of concern. The solvent must also be considered. GPC is a separation technique based on the size of the polymer in solution, and the solvent will influence the conformation of the polymer in solution. The solvents that are most suitable for specific polymers are listed in the Appendix.

Column designations are related to the pore size of the packing, but the numerical grades are not the actual pore size. Instead, they are the nominal size (in solution) of the smallest polymer molecule that will be excluded by that column. That is, all polymers of that size and larger will not fit into even the largest pores and will be excluded by the column. So a 10^3 Å column will exclude all polymer chains in the sample that have an effective size in solution of 10^3 Å or greater.

For convenience, the organic soluble polymer molecular-weight ranges corresponding to the specific column designations have been tabulated in Table 3.

The breadth of molecular-weight ranges for any one column is a consequence of the variety of polymer types that are characterized by GPC. Furthermore, the process of producing the porous, cross-linked packings results in a narrow but finite distribution of pore sizes. Typically, three or more columns representing a range of pore sizes are used in series to provide adequate resolution and ensure that the full molecular-weight range of the samples is evaluated.

Table 3 Organic Soluble Polymer Molecular-Weight Ranges Corresponding to the Specific Column Designations

MW range	Column (Å)
100–1000	50
250–2500	100
1000–18K	500
5000–40K	10^3
10K–200K	10^4
50K–1M	10^5
200K–> 5M	10^6
500K–~ 20M	10^7

For example, to analyze a low-molecular-weight material such as an uncured epoxy resin, a suitable column set would be one each of 10^3, 500, 100, and perhaps a 50. For a medium-molecular-weight PVC sample a likely column set would contain a 10^3, 10^4, and 10^5. Choosing individual pore sizes targeted at the molecular-weight range of the polymer provides the highest resolution. If the molecular-weight range is not known or is very broad, there are also mixed-bed, or "linear" columns that contain blends of packings to provide an extended range of use, over several orders of magnitude of molecular weight. The trade-off is a loss of resolution within any specific polymer molecular-weight range, so multiple columns are also needed with these mixed bed types to achieve the necessary resolution.

For some applications mixed-bed columns are combined with individual-pore-size columns for optimum utility. Using two mixed-bed columns plus a 500 Å provides extra pore volume at the low-molecular-weight "tail" of the distribution and ensures that all polymer chains will be well resolved from any low-molecular-weight additive or impurity peaks that may appear at the end of the chromatogram.

There are other factors to consider when choosing a column, especially the particle size of the packing materials, the column dimensions, and the solvent in which the column is packed. For GPC of polymers in organic solvents, columns are commercially available prepacked in some of the most frequently used solvents. Particle size influences the ruggedness and resolving power of the columns, while column length can also affect resolving power and diameter relates directly to solvent consumption.

Once the column set has been chosen and assembled, it must be calibrated to enable calculation of the molecular-weight and polydispersity values. As noted in Sec. I.E, calibration enables assignment of molecular-weight values to each retention-time slice of the raw chromatogram. There are several ways to perform a calibration. The simplest is to use a relative calibration based on a set of well-characterized polymer standards, each with as narrow a molecular-weight distribution as possible.

The ideal standards are monodisperse, with $Mw/Mn = 1$. There are polymer standards that are polymerized specifically for this use, such as anionically polymerized polystyrene. They have dispersity of < 1.10 and cover molecular weights ranging from monomer to $> 10,000,000$. Other narrow standards available for organic solvent GPC include poly(methylmethacrylates), polyisoprenes, polybutadienes (Fig. 6), and poly(THF), while poly(ethylene oxides), poly(ethylene glycols), and pullulans (polysaccharides) are used for aqueous GPC.

Polystyrene is the most frequently used narrow standard for organic GPC analysis. A series of the narrow standards of known molecular weight are analyzed using the same conditions that will be employed with the samples, especially flow rate, injection volume, and concentration. Then a plot of log MW versus retention time (or volume) is generated with a polynomial fit, usually third (Fig. 6) or fifth order. The values are taken from this curve to calculate the molecular-weight averages of the samples.

This conventional narrow standard calibration procedure yields "relative" molecular-weight values because the averages obtained are relative to the calibration polymer. For example, if polyethylene molecular weights are determined with a calibration curve produced with polystyrene narrow standards, the results would be incorrect for polyethylene. For many purposes relative molecular-weight values are adequate, since the results obtained with an unknown sample may be compared to preestablished acceptable relative values. If narrow standards of the same polymer as the unknown sample are available, then it is possible to calculate "absolute" values.

Other calibration techniques that enable determination of "absolute" molecular-weight values include broad standard calibration [6] and universal calibration [7]. The broad standard approach uses a polymer of the same type as the sample for which the various molecular-weight averages have been characterized by alternative methods such as membrane osmometry, light scattering, and ultracentrifugation. These molecular-weight averages are entered into the software and the broad standard is chromatographed by the same conditions used for the samples. The software does a Simplex search routine, fitting the broad standard chromatogram to the given molecular-weight averages.

The resulting broad standard calibration curve will consist of the data points for each average. If only number- and weight-average molecular-weights are available, the resulting calibration curve will consist of these two points plus the peak molecular-weight, or a three-point calibration curve. This is in contrast to a typical nine-point minimum that is common with narrow standard calibration. However, for the QC lab regularly testing the same polymer in the same molecular-weight range as the broad standard, this is sufficient to produce absolute molecular-weight values.

Modern GPC systems incorporating viscometry detection together with the differential refractometer are able to generate universal calibration curves and from them determine absolute molecular-weight values. In 1967 Benoit and co-workers introduced the concept of universal calibration. Instead of plotting the log molecular-weight M versus retention time, the log of the product of intrinsic viscosity $[\eta]$ times molecular-weight M is plotted versus retention time. This product, $[\eta]M$, is related to the hydrodynamic volume of the polymer in solution.

The hydrodynamic volume is the reciprocal of density, so the plot of log (volume/mass) versus time becomes independent of the polymer type, enabling construction of the universal calibration curve, Fig. 12. This concept is applicable to all random-coil polymers, the most commonly used synthetic polymers. Other conformations, such as rods, spheres, or globular, may not comply with the universal calibration approach.

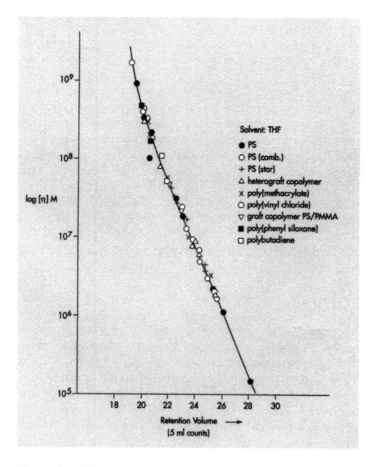

Figure 12   The universal calibration curve makes it possible to determine the absolute molecular-weight values for a polymer even if standards of the same chemical composition are not available.

C.   Sample Handling and Solvent Considerations

GPC analysis is a dilute solution procedure. Among the most frequently used solvents for characterizing synthetic organic polymers, tetrahydrofuran (THF) is probably the most common. Toluene is also popular, especially for elastomers. Appendix A–D contains an extensive list of solvent/polymer combinations plus information about the solvents. Depending on the effectiveness of the solvent that is being used, the molecular-weight range of the polymer, and its crystallinity, preparation of the standard and sample solutions will take time. This usually requires a few hours. Gentle stirring or shaking can be used, but high-speed mixing, ultrasonic, or microwave dissolution should not be used without first determining if it causes shear degradation or other damage to the polymer. Polyolefins and some other classes of polymers are soluble only at elevated temperatures. They are also listed in Appendix C.

The appropriate polymer concentration is somewhat a function of the molecular-weight. It must be high enough to produce a good detector signal but low enough to avoid column overload and resulting concentration-related viscous effects that can skew the results. The ranges shown in Table 4 are based on experience with typical analyses in which the sample injection size is 100 μL per 7.8 mm ID × 30 cm column in a set. Concentrations are mass/volume, so 0.10% is 1.0 mg/mL.

In GPC, the solvent in which the standards and sample are dissolved should be identical to the mobile-phase solvent in which the analysis will be performed. In most cases filtration is the only step needed to prepare the mobile phase. Organic solvents should be vacuum filtered through a 0.45-μm fluorocarbon filter, while acetate-type filters are used with aqueous mobile phases. In some cases mobile-phase additives are required. When polar solvents such as N,N-dimethyformamide, N,N-dimethylacetamide, and *n*-methyl pyrrolidone are used to analyze polar polymers such as poly-

Table 4   Polymer Concentrations

MW range	Concentration range (%)
> 1,000,000	0.007–0.02
500K–1,000,000	0.02–0.07
100K–500K	0.07–0.10
50K–100K	0.10–0.13
10K–50K	0.13–0.16
< 10K	0.16–0.20

urethanes or polyimides, a dipole interaction can occur, causing artificial shoulders to appear on the high-molecular-weight end of the distribution. This interaction is eliminated by adding 0.05 M lithium bromide to the solvent.

Salts are also used in aqueous GPC because the methacrylate-based gel packing has a net anionic charge. This can cause ion exclusion with anionic samples and ion adsorption with cationic samples instead of the simple size-exclusion separation mechanism that is essential for valid GPC. Sodium nitrate is the preferred salt for minimizing these ionic interferences

D. Detection Options

There are a number of detection options, some used primarily for GPC and others that have use for GPC as well as other modes of HPLC. The differential refractometer, viscometer, and light-scattering detectors are associated mostly with GPC, while absorbance detectors such as the UV/visible or photodiode array (PDA) are widely used in all HPLC modes, including GPC. The UV/visible and PDA are especially useful for characterizing polymers and oligomers with chromophoric groups and for HPLC analyses of additives. Mass spectrometry is also used for some analyses. This is described in Sec. II.F.

The most popular GPC detector is the differential refractometer. It is a concentration-sensitive detector that measures the difference in refractive index (ΔRI) between the eluent (the flowing solvent) and the sample solution. It is a universal detector that will respond to any polymer with a significant refractive index difference from the solvent. So another consideration when selecting a solvent, besides being a good solvent for the polymer, must be a refractive index that will provide a significant ΔRI. The solvent/polymer combinations listed in the Appendix fulfill this criterion.

Two additional detectors specific to GPC are the viscometer and the light-scattering detectors. They are used to complement the information provided by the differential refractometer, since they respond to molecular-weight and structural characteristics of the polymer, while the refractometer responds primarily to mass. The viscometer determines viscosity by measuring differences in the pressure drop as the solvent and sample solutions flow through capillaries. Determination of absolute molecular-weight information with the viscometer and refractometer was mentioned in Sec. II.B. Having the viscometer detector in line with the refractometer also makes it possible to calculate the intrinsic viscosity [η] across the molecular-weight distribution and to obtain information about long-chain branching. See Sec. III.B.

Light scattering, coupled with the refractometer, is another powerful mode of advanced GPC detection. A laser beam is scattered by the dissolved polymer molecules as they flow through a cell. The intensity of scattered light is proportional to the size of the scattering molecules. Measuring the intensity at various angles enables very accurate determination of weight-average molecular-weight, Mw. Light-scattering detection also provides information about the polymer radius of gyration and branching. See Sec. III.B.

Absorbance detection, either UV/visible or photodiode array, has broad but specific applicability, especially for styrenic polymers, epoxies, phenolics, polycarbonates, polyurethanes, aromatic polyesters, and many additives. When other HPLC modes are used, additional separating capability is sometimes achieved by changing the solvent composition during the analysis (gradient elution). For this work the UV or PDA detector is essential, since the RI detector would drift excessively as the composition, and therefore the refractive index, changes. See Sec. II.F.

The PDA can monitor the separation at hundreds of wavelengths simultaneously in the range from 190 to 800 nm. This makes it possible to generate the actual UV spectra of polymers and additives and evaluate the chemical composition distribution. Block SBR copolymers are distinguishable from random copolymers [8]. Spectral libraries can be created for known polymers and additives to assist identification and deformulation of unknowns.

E. Data Reduction

A wide variety of choices exist for processing the GPC data, extending from simple integrators to powerful computer-based systems. Most integrators will process only one sample chromatogram at a time and offer little or no flexibility in presenting information.

Using computers and commercially available GPC software, calibration, molecular-weight averages and distributions, and structural information are determined quickly and easily. These packages can process chromatograms from the various detectors and present the information in multiple, user-selectable formats. Multiple calibration procedures are supported. GPC software is available from Polymer Laboratories, Polymer Standards Service, Viscotek, and Waters Corp. Some packages have enough flexibility to enable the user to automate the entire procedure in a "run and report" mode or to integrate specific chromatograms manually if desired.

In addition, the most advanced software will also control the instrumentation and archive the raw data and results in a relational database that links instrumentation records with operating conditions, data processing, and reporting methods. User-designed reporting for standard or custom purposes is also integrated into the software. Statistical options permit tailored information retrieval for tracking and trending purposes. Workstations may be operated independently or as elements of a network within the laboratory or throughout the operations of multinational corporations.

F. HPLC of Polymers and Additives

Modes of HPLC other than GPC can be used to evaluate polymer mixtures, copolymer composition, and to analyze additives. As indicated in Fig. 8 and in Secs. II.A and II.D, for this work a number of changes are made to the system, especially the type of column and detection methods. The mechanism of separation is no longer exclusion, in which no adsorptive interaction between the sample components and the column packing material is tolerable. Instead, there is deliberate adsorptive interaction between the samples and the column packing in order to achieve separations based on differences in chemical composition of the sample components, independent of molecular size, Sec. III.C.

HPLC columns usually contain packings based on porous silica particles, often with organic chemical groups bonded to the surface. The solvents used with these bonded-phase packings range from nonpolar to aqueous phases, depending on the characteristics of the sample. Solvent composition is often programmed during the analysis to further regulate the separation.

In GPC instrumentation the pumping system delivers a single eluent throughout the analysis. It may be a single solvent, a blend of solvents, or a solvent plus an additive, Sec. II.C, but the composition remains constant. Solvent programming, also termed gradient elution, in HPLC requires more capability from the solvent delivery system to ensure precise flow and reproducible solvent composition profiles.

While RI detectors are used for isocratic HPLC, the other detectors, viscometer and light scattering, are not generally suitable for HPLC. As noted in Sec. II.D. absorbance detectors, either UV/visible or photodiode array (PDA), are much more useful. Table 5 and Fig. 13 show the HPLC of a polymer additive mixture. The reproducibility of 12 consecutive injections shown in Fig. 13 demonstrates exceptional reproducibility of the analysis, especially considering that both the solvent composition and solvent flow rates were programmed for this work.

Depending on the nature of the work, other detection techniques that can be employed in conjunction with absorbance detection, or alone, include mass spectrometry (MS) and evaporative light scattering, Sec. III.C. While MS has been used to characterize polymers for a number of years, the

Table 5 Chromatographic Conditions for Polymer Additive Analysis by
Gradient HPLC

System: Waters Alliance System with 996 PDA detector
Column: Symmetry C8, 3.9 × 150 mm
Column Temperature: 50°C
Solvent program: 70% acetonitrile (ACN)/water to 100% ACN in 5 min
Flow program: Initially 2.0 mL/min, at 6.0 min ramped to 3.0 mL/min in 0.2 min
Injection volume: 10 μL of standard additive mixture
Reproducibility study: 12 duplicate injections

The separation was performed using a packing material with a nonpolar C8 bonded phases.
The column temperature was set above ambient to ensure reproducible conditions. Analytical
HPLC injection volumes are typically an order of magnitude smaller than GPC injection
volumes.

utilization of MS as an HPLC detector is a more recent development. An
example of this use is shown in Table 6 and Figs. 14 and 15. The data system
is able to extract ion and adsorbance spectra from the total chromatogram,
Fig. 14. Closer examination of electron ionization spectrum in Fig. 15 shows
the correspondence of fragments with the parent molecule and likely pathways to the dominant ions, enabling definitive identification of the Irganox
1076 (Ciba Specialty Chemicals, Additives Division, Tarrytown, NY).

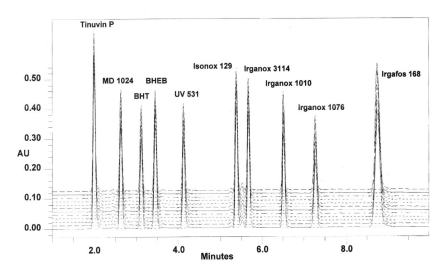

Figure 13 Polymer additive analysis by HPLC of a synthetic mixture of
antioxidants and UV stabilizers (Alliance reproducibility study). An analysis
of this type could be used to deformulate a competitive material.

Table 6 Chromatographic Conditions for Polymer
Additive Analysis by LC/MS

LC conditions
  Gradient conditions:
    Solvent A: water
    Solvent B: acetonitrile
    Gradient: 80% to 100% B in 5 min, holds for 10 min
    Column: Waters Symmetry C8 (3.0 mm × 150.0 mm) at 50°C
    Flow rate: 0.6 mL/min

The LC/MS system used is a Waters Integrity system consisting of:
  ThermaBeam mass detector
  Waters 996 photodiode array detector
  Waters 2690 separations module
  Millennium 2010 data system

The gradient program in this work is less complex than that listed in Table 5.
Use of MS and photodiode array detection provides extensive information
about the sample.

For the multifunctional plastics analysis laboratory, the most versatile,
basic chromatography instrument would be an HPLC system containing a
PDA detector plus an RI detector and a supply of HPLC and GPC columns. This would enable the system to be configured for GPC analyses and
yet still provide enough solvent-delivery flexibility to perform the more
complex operations that are associated with additive analyses. All of the
detection enhancements, viscometry and light scattering for GPC as well as
mass spectrometry for HPLC, could also be accommodated as dictated by
budget and analytical requirements. The only limitation would be the analysis of polyolefins and other polymers that are not soluble at room temperature. Here a dedicated, integrated GPC system is required, as noted in
Secs. II.A and III.D.

III. CHROMATOGRAPHY RESULTS

A. Design Influences on Performance

The primary purpose of GPC is to accurately and reproducibly characterize
polymeric materials, typically a complex mixture and distribution of closely
related components. Therefore it is essential that the GPC system is able to
reliably delineate differences that, although very slight, may be quite significant. For this to occur, the pumping system must deliver the solvent accurately, precisely, and uniformly. Flow fluctuations during an analysis or
from one analysis to the next will seriously impair the quality of the results

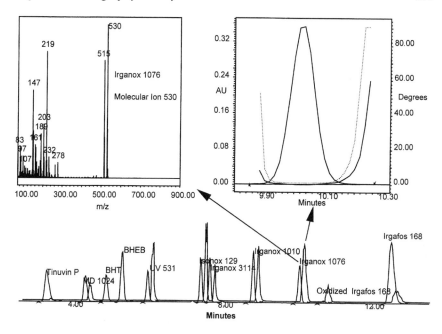

Figure 14 Additive analysis results. The chromatogram is an overlay of the MS and PDA responses. Each response curve contains extraordinary amounts of information. The mass spectrum, upper left, and a spectral measure of peak purity, upper right, have been extracted from the chromatogram. — TMD: total ion chromatogram from 100 to 700 amu. ... PDA: extracted chromatogram at 230 nm.

and dependability of the information. In the extreme, these fluctuations may mask any true differences between samples.

The critical dependence on the flow characteristics arises from the calibration procedure that is the basis for quantitating the GPC chromatograms, Sec. II.B. Recall that calibration involves chromatographing a series of narrow-dispersity standards of known molecular-weight and plotting their log molecular-weight versus elution time. Although the time scale is used, GPC is actually a volumetric effect. The assumption is made that the flow rate remains constant throughout the analysis and from analysis to analysis, enabling time and volume to be used interchangeably. Any variation in the volume of solvent being delivered will cause the particular molecular-weight fraction to elute at a different time and shift the calibration curve. The error is magnified because of the log scale used for the molecular-weight, Fig. 6.

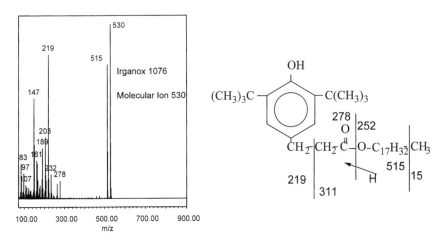

Figure 15 The classical electron impact mass spectra produced enable definitive compound identification, Irganox 1076 in this example. Electron impact mass spectra are interpretable as well as library-searchable.

The impact of flow fluctuation on the accuracy of molecular-weight calculations can be shown by analyzing a known sample at a slightly different flow rate than the flow rate used to develop the calibration curve. Dow 1683 is a well-characterized polystyrene introduced a number of years ago by the research department at Dow Chemical Company (Midland, MI, USA). Many investigators have used it as a reference material during studies of various polymer characterization techniques. The accepted molecular-weight values for this material are Mn = 100,000 and Mw = 250,000.

A sample of Dow 1683 was analyzed at 1.0 mL/min, the same flow rate that was used to construct the calibration curve. Then it was analyzed again at 0.95 mL/min and quantitated with the 1.0-mL/min calibration curve (Table 7).

The use of the log molecular-weight scale versus time (volume) for calibration results in about a 10-fold magnification of any flow error. This general relationship is shown in Fig. 16. Hence the emphasis on using the most precise solvent delivery system possible. It is still good practice to periodically bracket samples with narrow-distribution calibration standards and compare them to the calibration curve to monitor the condition of the system and columns.

A manufacturer of medical devices found that product made from a new batch of resin differed from that fabricated from the resin he had been using. The resin supplier claimed the batches were identical. Comparison of

Table 7 Analysis of a Sample of Dow 1683 (at the same flow rate used to construct the calibration curve)

	Accepted values	Values at 1 mL/min	Values at 0.95 mL/min
Number avg., Mn	100,000	98,457	43,264
Weight avg., Mw	250,000	246,501	108,143

the two batches with an Alliance GPC system revealed slight but reproducible differences, Fig. 17. To ensure that these subtle changes were not due to fluctuations in system operation, each sample was analyzed five times. All 10 molecular-weight distributions are included in the display. The tables show the extraordinary reproducibility of the analyses. The ability to compare the entire distributions by overlaying the composite chromatograms enabled the manufacturer to identify the likely causes of changes in processability and product performance.

In addition to flow precision, flow smoothness is also essential. Flow smoothness results in a very stable, low-noise baseline signal from the RI detector. Detector sensitivity is governed by the ratio of the signal that is produced to the baseline noise. With lower noise it is possible to reduce the scale of operation and still obtain a good signal for quantitation. Reducing

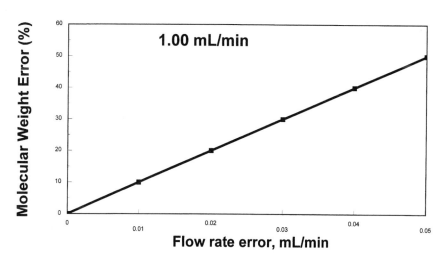

Figure 16 Effect of pump flow rate precision. Each 1% error in GPC flow rate results in a 10% error in molecular-weight value.

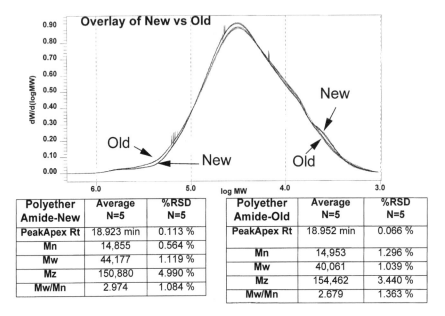

Polyether Amide-New	Average N=5	%RSD N=5
PeakApex Rt	18.923 min	0.113 %
Mn	14,855	0.564 %
Mw	44,177	1.119 %
Mz	150,880	4.990 %
Mw/Mn	2.974	1.084 %

Polyether Amide-Old	Average N=5	%RSD N=5
PeakApex Rt	18.952 min	0.066 %
Mn	14,953	1.296 %
Mw	40,061	1.039 %
Mz	154,462	3.440 %
Mw/Mn	2.679	1.363 %

Figure 17 Alliance narrow-bore GPC (medical, pastic-grade polyether amide). Very subtle differences in the molecular-weight distributions are determined reliably with modern, high-performance instrumentation. Each chromatogram is actually an overlay of five consecutive analyses.

the scale of operation involves using 4.6-mm-ID instead of 7.8-mm-ID columns. Flow rates and injection volumes are reduced proportionately, lowering solvent consumption by about two-thirds. The polyether amide medical-grade resin, Fig. 17, is soluble only at room temperature in a very expensive solvent, 1,1,1,3,3,3-hexafluoro-2-propanol (HFIP). Being able to operate with only one-third the amount of HFIP that would have been necessary with a conventional GPC system was a very significant economic benefit to this laboratory.

An interlaboratory comparison of GPC results provides another demonstration of the exceptional reproducibility provided by the modern high-performance instrumentation. Three polycarbonate resins were independently analyzed by two laboratories, one in the United States and the other in Europe. The conditions each used were similar but not identical (Table 8). This work was also done with 4.6-mm-ID narrow-bore columns and absorbance detection, Sec. II.D, photodiode array in the United States and tunable UV (TUV) in Europe. The most significant difference in the setup between the two labs was the column sets that were used. Polystyrene

Table 8 Conditions for Analyzing Three Polycarbonate
Samples in Two Laboratories

	US Lab	Europe Lab
Calibration	PS Stds. in Me_2Cl_2	PS Stds. in Me_2Cl_2
Columns $4.6 \times 300\,mm$	HR 2, 3, & 4	HR 3 & 4
Injection	$25\,\mu L$	$25\,\mu L$
Concentration	0.15%	0.10%
Flow rate	$350\,\mu L/min$	$350\,\mu L/min$
Detection	PDA @ 239 nm	TUV @ 254 nm

The conditions were chosen independently. Methylene chloride was the solvent
used in both locations.

standards were used for calibration, so the results in Fig. 18 and Table 9 are
relative molecular-weights, Sec. II.B.

Each laboratory performed all analyses in triplicate, with precision of
molecular-weights less than 1% for virtually every value. The exceptional
agreement between labs proves that using a highly reproducible system for
calibration and analysis will ensure consistent results within and between
labs, even if the operating conditions between labs are not identical.

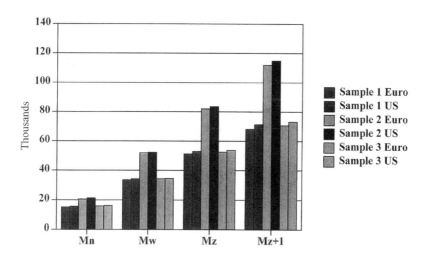

Figure 18 Alliance GPC interlaboratory results (polycarbonates). Reproducibility within labs and accuracy of results, as measured by the agreement between labs, was exceptional, despite slightly different operating
conditions.

Table 9 Alliance GPC Interlab Results
(Polycarbonates)

	Mn	Mw	Mz	Mz + 1
Sample 1 Euro	15,144	33,607	51,142	68,104
Sample 1 US	15,628	34,145	52,980	71,360
Sample 2 Euro	20,482	52,033	81,931	111,986
Sample 2 US	21,257	52,330	83,426	114,851
Sample 3 Euro	15,878	34,300	52,536	70,647
Sample 3 US	16,124	34,398	53,782	73,078

Nevertheless, for best results the operating conditions and column sets should be as similar as possible, and each system must be calibrated under the exact conditions at which it will be used to analyze samples. It should also be noted that all of the results reported by the European lab were very slightly lower than those from the United States. Slight differences in factors such as procedure, data reduction parameters, or the difference in detector absorbance wavelengths might have caused this consistent bias, rather than random variability.

B. Structural Determination

Depending on monomer chemistry and polymerization conditions, polymer chains may be linear, resembling a long, randomly coiled strand of repeating units, or they may be branched, with side chains extending off the main, linear backbone. For some polymers, branching has as much influence on processability and properties as molecular-weight averages, if not more. GPC with multiple detectors, Secs. II.B and II.D, can be used to determine whether a polymer is linear or branched, and if branched, the degree of branching.

When an on-line viscometer is used together with the refractive index detector to generate the intrinsic viscosity [η] in order to build the universal calibration curve, Sec. II.B, the intrinsic viscosity [η] can also be used to determine the presence and degree of branching. This is done by plotting the log of [η] versus log molecular-weight for each slice of the distribution. This plot is called the viscosity law plot, or the Mark-Houwink plot. It is described by the equation

$$[\eta] = KM^{\alpha}$$

in which K is the intercept and α is the slope, also known as the Mark-Houwink constant [9]. If the polymer is linear, the viscosity will increase linearly with molecular-weight and the slope will remain constant. However, if there is any long-chain branching, [η] will not increase linearly with molecular-weight and may even approach a constant value. Corresponding α values will decrease in the branched region of the distribution.

The slope of the Mark-Houwink curve for a branched polymer will still be linear at low molecular-weights when there is no long-chain branching. This linear region can be extrapolated into the high-molecular-weight region to enable determination of the branching index, g', which is defined by the ratio of intrinsic viscosities at a specific slice or molecular-weight point:

$$g' = \frac{[\eta]_{\text{branched}}}{[\eta]_{\text{linear}}}$$

The divergence of the viscosity law plot from linear and subsequent decline in g' for a lightly branched polyethylene is shown in Fig. 19. The work was done with an integrated GPC system that can maintain all zones at the temperature needed to keep the sample in solution. In this presentation the molecular-weight scale is high to low. This is a user preference that is chosen in the software. In addition to the degree of branching indicated by

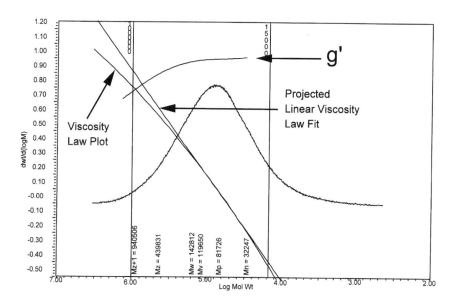

Figure 19 Molecular-weight distribution and branching information for a low-density polyethylene heavy-wall trash bag. The decrease in branching index, g', indicates that the higher-molecular-weight fractions have significant long-chain branching.

g', the software is also capable of calculating how often branches occur along the backbone. This is known as branching frequency, λ, and is determined using the Zimm-Stockmayer equation [10].

Light scattering and viscometry detectors are molecular-weight-dependent detectors. As such, they respond more strongly to higher-molecular-weight fractions than RI detectors. Conversely, the RI detectors are more sensitive to low-molecular-weight regions of the distribution. Light scattering is especially useful for determining the conformation of polymers in solution, the radius of gyration, and whether there is agglomeration or aggregation of polymer chains. Branching information is also provided by light-scattering detectors.

C. Compositional Determination

Analysis of polymer blends and alloys often requires the ability to separate two or more types of polymers or copolymers with varying monomer ratios. Size separation is less useful than other HPLC modes in which the separation is based on differences in chemical composition, Sec. II.F.

Polymer separations by HPLC are usually optimized by changing the solvent composition during the analysis. This is termed solvent programming or gradient elution. Since the refractive index of the solvent changes significantly during the gradient, differential RI detection cannot be used. Absorbance detectors are very popular for this work, providing the polymers being analyzed contain UV absorbing (chromophoric) groups. The evaporative light-scattering detector (ELSD) is a general-purpose detector that is compatible with gradient HPLC of polymers.

In an evaporative light-scattering detector the eluent leaving the column is mixed with a gas and forced through a nozzle to form a mist of uniform droplets that passes into a heated chamber. As the droplets are carried through this heated chamber or "drift tube" by the gas stream, the solvent evaporates, leaving a fine dispersion or cloud of sample particles. The particles travel through a laser light beam, scattering the light. The scattered light is detected in response to the presence of the sample.

One scheme for separating polymer mixtures by HPLC is based on a selective precipitation–redissolution model [11]. It has been used with the evaporative light-scattering detector to monitor the separation of a three-polymer mixture (Fig. 20). Solvent programming was used. The author suggests that this technique can be used to determine chemical composition of copolymers, but no examples are shown.

A subsequent study by another investigator used gradient elution to control the separation of a series of polymethyl methacrylate/butyl acrylate copolymers with varying monomer ratios. Elution times increased linearly

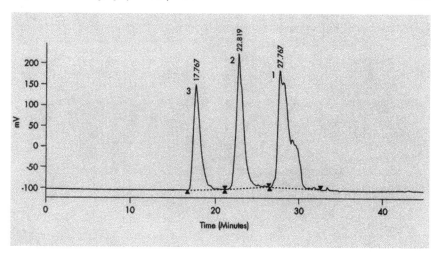

Figure 20   Chromatogram of poly(methyl methacrylate) 3, poly(styrene) 2 and poly(butadiene) 1. Chromatographic conditions: 30-min. linear gradient, 100% methanol to 100% tetrahydrofuran at 1 mL/min. NovaPak Cyano-Propyl 3.9 × 75 mm, 4-μm column at 30°C, evaporative light-scattering detector.

with butyl acrylate content. Monomer ratios were determined with an HPLC/mass spectrometer employing an electron ionization interface [12]. MS of polymers has been practiced for many years, but using the technique on-line to combine it with the separating power of HPLC makes it possible to examine the composition of more complex samples.

The monomer mass spectra were similar to those produced by pyrolysis GC/MS (Fig. 21). The MS detector was calibrated for the amount of butyl acrylate present and was used to determine the amount of this monomer present in unknown polymers. Linear relationships were observed between ion intensity/concentration and ions characteristic of both methyl methacrylate ($m/z = 100$) and butyl acrylate ($m/z = 127$). The peak compositions in Fig. 21 range from methyl methacryate homopolymer, A, to butyl acrylate homopolymer, F, in 20% butyl acrylate increments.

D. Analysis of High-Temperature Soluble Polymers

More polyolefins are produced and used worldwide than any other polymer group. All polyolefins, whether linear or branched, amorphous or crystalline, homopolymer or copolymer, are soluble only at elevated temperatures (Figs. 1 and 19). This imposes extra demands on instruments that are used

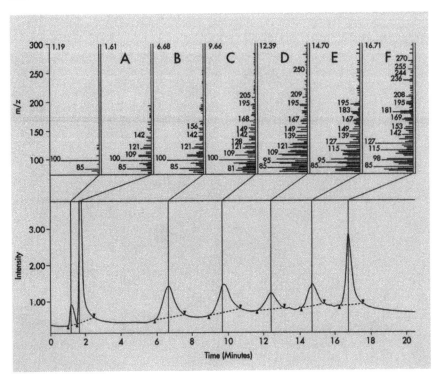

Figure 21 HPLC/MS spectrum index plot of pMMA/BA copolymers. Mass spectra extracted from the total ion chromatogram show the progression from polymethyl methacrylate homopolymer (A) to polybutyl methacrylate homopolymer (F) in 20% butyl acrylate increments.

to determine polyolefin molecular-weight values by GPC. The most common conditions for this work are to dissolve the sample in either TCB or ODCB and operate the instrument with injector, column, and detectors at 135–145°C (Appendixes A and C).

The most dependable approach to controlling temperature throughout the analytical instrument is to integrate all components into a self-contained system such as the Waters Alliance GPC 2000 (Fig. 11). In this way, the modules and connecting tubing can be maintained at the desired temperatures without concern for cold spots that might cause the sample to precipitate and possibly plug the system. Considering that some of the organic solvents used for high-temperature GPC are hazardous, safety features such as flow and pressure monitors, leak sensors, and spill containment can be readily incorporated into an integrated system with greater reliability than in an open, modular configuration. A unique user interface based on an on-board computer produces real-time displays of detector signals and system operating conditions as well as facilitating setup, documentation, reporting, and archiving of results.

These same considerations apply to other important polymer groups that must be dissolved at elevated temperature. They include polyacetals, polyvinylidene fluoride, polyetherketone (PEK), polyetheretherketone (PEEK), polyether sulfone, polyimide, and imide copolymers. Traditionally polyamides and polyesters also were analyzed at elevated temperature, but HFIP will dissolve them at room temperature (Fig. 17).

IV. NATIONAL AND INTERNATIONAL STANDARDS FOR POLYMER LIQUID CHROMATOGRAPHY

ASTM Methods (U.S.)

D 1996-92: Determination of Phenolic Antioxidants and Erucamide Slip Additives in Low Density Polyethylene Using Liquid Chromatography (LC)

D3016-78(1992): Use of Liquid Exclusion Chromatography Terms and Relationships

D5296-92: Molecular Weight Averages and Molecular Weight Distribution of Polystyrene by High Performance Size-Exclusion Chromatography

D5524-94: Determination of Phenolic Antioxidants in High Density Polyethylene Using Liquid Chromatography

D5815-95: Determination of Phenolic Antioxidants and Erucamide Slip Additives in Linear Low-Density Polyethylene Using Liquid Chromatography (LC)

D5910-96: Determination of Free Formaldehyde in Emulsion Polymers by Liquid Chromatography

D6042-96: Determination of Phenolic Antioxidants and Erucamide Slip Additives in Polypropylene Homopolymer Formulations Using Liquid Chromatography (LC)

DIN Method (Germany)

DIN 55672-1: Gelpermeationschromatographie (GPC)

ISO (International Organization for Standardization)

Draft International Standard ISO/DIS 13885: Binders for paints and varnishes-Gel permeation chromatography (GPC) with tetrahydrofuran (THF) as eluent—draft under consideration, 1997

Draft International Standard ISO/DIS 16014: Plastics—Determination of average molar masses and molar mass distribution of polymers using size exclusion chromatography. Part 1: General Principles. Part 2: Measurement at lower temperatures. Part 3: Measurement at higher temperatures—draft under consideration, 1999

JIS (Japan Industrial Standards)

JIS K0124: General Rules for Analytical Methods in High Performance Liquid Chromatography

APPENDIX A: ORGANIC SOLVENTS FOR GPC

Solvent/full name	Boiling point (°C)	Comments
THF (tetrahydrofuran)	66	Highly flammable peroxides may form
Toluene	111	Highly flammable
DMF (N,N'-dimethylformamide)	153	Strong irritant/toxic (0.05 M LiBr added to minimize polar effects)
DMAC (N,N'-dimethylacetamide)	166	Same as DMF
HFIP (1,1,1,3,3,3-hexafluoro-2-propanol)	58	Very expensive, corrosive vapors
TCB (1,2,4-trichlorobenzene	213	Toxic; 1.0 g/4 L of an antioxidant should be added
ODCB (orthodichlorobenzene)	173	Somewhat toxic, also needs antioxidant
NMP (n-methyl pyrrolidone)	202	Strong irritant/toxic (0.05 M LiBr added as for DMF)
CHCl$_3$ (chloroform)	61	Known carcinogen
CH$_2$Cl$_2$ (methylene chloride)	40	Toxic, possible carcinogen
DMSO (dimethyl sulfoxide)	189	Nontoxic

Several other solvents and solvent blends are used occasionally, such as decalin, (for polyolefins), xylene, (for amorphous polypropylene), m-cresol, and trifluoroethanol (for certain nylons and polyesters), etc. These are the solvents used most often as eluents in GPC analysis.

APPENDIX B: SOLVENT SELECTION FOR ROOM-
TEMPERATURE ORGANIC SOLUBLE POLYMERS

Polymer	Solvent	Column temp. (°C)
Acrylonitrile/methylmethacrylate	THF	40
Cellulose acetate	THF	40
Cellulose acetate/butyrate	THF	40
Cellulose acetate/propionate	THF	40
Cellulose triacetate	THF	40
Diallyl phthalate	THF	40
Ethyl cellulose	THF	40
Epoxy resins	THF	40
Phenolic resins	THF	40
Polyglycolic acid	THF	40
Polyester alkyd resins	THF	40
Polymethylmethacrylate and other acrylics and acrylates	THF	40
Polypropyleneglycol	THF	40
Polystyrene	THF	40
Polystyrene/acrylonitrile (SAN) (low ACN)	THF	40
Polysulfone	THF	40
Polyurethane (some)	THF	40
Polyvinylacetate	THF	40
Polyvinylbutyral	THF	40
Polyvinylcloride	THF	40
Polyvinylchloride/acetate	THF	40
Polyvinylidenechloride	THF	40
Polyvinylformal	THF	40
Thermosetting polyesters	THF	40
Rosin acids	THF	40
Polybutadiene	Toluene	75
Polychloroprene (neoprene)	Toluene	75
Polydimethylsiloxane (silicone oils and rubber)	Toluene	75
Polyisobutylene (butyl rubber)	Toluene	75
Polyisoprene (natural rubber, chlorinated rubber)	Toluene	75
Styrene/butadiene rubber (SBR)	Toluene	75
Styrene/isoprene (SI)	Toluene	75

APPENDIX B: CONTINUED

Polymer	Solvent	Column temp. ($^\circ$C)
Acrylonitrile/butadiene styrene (ABS)	DMF[a] with 0.05 M LiBr	85
Acrylic/butadiene/acrylonitrile (ABA)	DMF[a] with 0.05 M LiBr	85
Acrylic/butadiene/acrylonitrile (ABA)	DMF[a] with 0.05 M LiBr	85
ABS/polycarbonate	DMF[a] with 0.05 M LiBr	85
Carboxymethylcellulose	DMF[a] with 0.05 M LiBr	85
Polyacrylonitrile	DMF[a] with 0.05 M LiBr	85
Polybutadiene/acrylonitrile	DMF[a] with 0.05 M LiBr	85
Polystyrene/acrylonitrile (SAN) (high ACN)	DMF[a] with 0.05 M LiBr	85
Polyurethane (most)	DMF[a] with 0.05 M LiBr	85
Melamine/formaldehyde	HFIP with 0.05 M NATFA (trifluoroacetic acid, sodium salt)	35
Nylon (all nylons plus most other polyamides)	HFIP with 0.05 M NATFA	35
Polyethylene terephthalate (PET)	HFIP with 0.05 M NATFA	35
Polybutylene terephthalate (PBT)	HFIP with 0.05 M NATFA	35

[a]In most cases, DMAC may be substituted for DMF, again with the 0.05 M LiBr.

APPENDIX C: SOLVENT SELECTION FOR ELEVATED
TEMPERATURE ORGANIC SOLUBLE POLYMERS

Polymer	Solvent	Column/injector temp. (°C)
Chlorinated polyethylene	TCB[a]	135–145
Ethylene/propylene diene monomer (EPDM)	TCB	135
Polyethylene	TCB	135–145
Polyethylene/ethyl acrylate	TCB	135–145
Polyethylene/vinyl acetate (EVA)	TCB	135
Polyethylene/methacrylic acid	TCB	135–145
Polyphenylene oxide	TCB	135
Poly-4-methyl pentene	TCB	135
Polypropylene	TCB	135–145
Ultra-high-molecular-weight PE (UHMWPE)	TCB	145
Polyamide-imide	NMP with 0.05 M LiBr	100
Polyether-imide	NMP with 0.05 M LiBr	100
Polyether sulfone	NMP with 0.05 M LiBr	100
Polyimide	NMP with 0.05 M LiBr	100
Polyvinylidene fluoride	NMP with 0.05 M LiBr	100
Polyacetals	DMF with 0.05 M LiBr	145
Polyetherketone (PEK)	1 : 1 TCB/Phenol	145
Polyetheretherketone (PEEK)	1 : 1 TCB/Phenol	145
Starch	DMSO	50–100
Cellulose	DMF with 6 M LiCl	85

[a]In most cases, ODCB may be substituted for TCB.

APPENDIX D: SOLVENT SELECTION FOR WATER-SOLUBLE POLYMERS WITH METHACRYLATE GEL COLUMNS

Polymer	Class	Eluent
Polyethylene oxide	Neutral	0.10 M NaNO$_3$
Polyethylene glycol	Neutral	0.10 M NaNO$_3$
Polysaccharides, Pullulans	Neutral	0.10 M NaNO$_3$
Dextrans	Neutral	0.10 M NaNO$_3$
Celluloses (water-soluble)	Neutral	0.10 M NaNO$_3$
Polyvinyl alcohol	Neutral	0.10 M NaNO$_3$
Polyacrylamide	Neutral	0.10 M NaNO$_3$
Polyvinyl pyrrolidone	Neutral, hydrophobic	80% 0.10 M NaNO$_3$/20% acetonitrile
Polyacrylic acid	Anionic	0.10 M NaNO$_3$
Polyalginic acid/alginates	Anionic	0.10 M NaNO$_3$
Hyaluronic acid	Anionic	0.10 M NaNO$_3$
Carrageenan	Anionic	0.10 M NaNO$_3$
Polystyrene sulfonate	Anionic, hydrophobic	80% 0.10 M NaNO$_3$/20% acetonitrile
Lignin sulfonate	Anionic, hydrophobic	80% 0.10 M NaNO$_3$/20% acetonitrile
DEAE dextran	Cationic	0.80 M NaNO$_3$
Polyvinylamine	Cationic	0.80 M NaNO$_3$
Polyepiamine	Cationic	0.10% TEA
n-Acetylglucosamine	Cationic	0.10 M TEA/1% acetic acid
Polyethyleneimine	Cationic, hydrophobic	0.50 M sodium acetate/0.50 M acetic acid
Poly(n-methyl-2-vinyl pyridinium)I salt	Cationic, hydrophobic	0.50 sodium acetate/0.50 M acetic acid
Lysozyme	Cationic, hydrophobic	0.50 acetic acid/0.30 M sodium sulfate
Chitosan	Cationic, hydrophobic	0.50 acetic acid/0.30 M sodium sulfate
Polylysine	Cationic, hydrophobic	5% ammonium biphosphate/ 3% acetonitrile (pH = 4.0)
Peptides	Cationic, hydrophobic	0.10% TFA/40% acetonitrile
Collagen/gelatin	Amphoteric	80 : 20 0.10 M NaNO$_3$/CH$_3$CN

Note that in the many cases where sodium nitrate is shown, many workers have used acetate, sulfate, sodium chloride, etc. Sodium nitrate tends to minimize ionic interferences very consistently for neutral and anionic compounds. These various eluents are used because the methacrylate-based gel packing for aqueous GPC has overall anionic charges, which can cause ion exclusion for anionic samples and ion adsorption for cationic samples if run in water alone.

APPENDIX E: MANUFACTURERS OF GEL PERMEATION CHROMATOGRAPHY AND HPLC SYSTEMS

Chromatography, gel permeation
  Alltech Associates Inc.
  Gilson Inc.
  Hitachi Instruments Inc.
  Polymer Laboratories Inc.
  Shimadzu Scientific Instruments Inc.
  Waters Corp.

Chromatography, LC, Complete Systems
  Alltech Associates Inc.
  Beckman Instruments Inc.
  Bioanalytical Systems Inc.
  Dionex Corp.
  Gilson Inc.
  Hitachi Instruments Inc.
  Jasco Inc.
  Perkin-Elmer Corp.
  Polymer Laboratories Inc.
  Shimadzu Scientific Instruments Inc.
  Thermo Separation Products Inc.
  Varian Instruments
  Waters Corp.

Source: *'97–'98 Lab Guide*, published by ACS Publications, American Chemical Society.

BIBLIOGRAPHY

Application Reviews, published in alternating years by Analytical Chemistry, includes a section on analysis of synthetic polymers and rubbers in which all references to specific analytical methods, including liquid chromatography, are grouped together, e.g., Smith, P. B., et al., Analytical Chemistry, 1997, 69, 101R–103R.

Detection and Data Analysis in Size Exclusion Chromatography, T. Provder, ed., American Chemical Society Symposium Series, 352, 1987.

Handbook of Size Exclusion Chromatography, C.-S. Wu, ed., Marcel Dekker, New York, 1995.

Liquid Chromatography of Polymers and Related Materials II. J. Cazes, X. Delamare, eds., Marcel Dekker, New York, 1981.

Modern Methods of Polymer Characterization, Chaps. 1–5, H. G. Barth, J. W. Mays, eds., Wiley, New York, 1991.

Polymer Characterization, Chaps. 14–18, C. D. Craver, ed., American Chemical Society Advances in Chemistry Series, 203, 1983.

Proceedings, International GPC Symposium '96, R. Nielson, ed., Waters Corp., Milford, MA, 1996.

Proceedings, International GPC Symposium '98, R. Nielson, ed., Waters Corp., Milford, MA, 1999.

REFERENCES

1. Yau, W. W., Kirkland, J. J., Bly, D. D., Modern Size-Exclusion Liquid Chromatography. Wiley, New York, 1979, chap. 12.
2. Fallick, G., Cazes, J., Modern Plastics, 1977, 54, 12, 62–66.
3. Foster, G. N., MacRury, T. B., Hamielec, A. E., Chromatographic Science Series, 1980, Vol. 13, 143–171 (Liquid Chromatography of Polymers and Related Materials II).
4. McCrum, N. G., Buckley, C. P., Bucknall, C. B., Principles of Polymer Engineering. Oxford University Press, New York, 1988, 199–200.
5. Tung, L. H., SPE Conference Proceedings (1958), 959.
6. Balke, S., Hamielec, A., LeClair, B., Pearce, S., Ind. Eng. Chem., Prod. Res., Des., 1969, 8, 54.
7. Benoit, H., Grubisic, Z., Rempp, P., Decker, D., Zillox, J. G., J. Chem. Phys., 1969, 63, 1507.
8. Adams, H. E., in Atgelt, K. H., Segal, L., eds., Gel Permeation Chromatography, Marcel Dekker, New York, 1971, 391.
9. Mark, H., Techniques of Polymer Characterization. Butterworth, London, 1969, chap. 6.
10. Zimm, B. H., Stockmayer, W. H., J. Chem. Phys., 1949, 17, 301.

11. Staal, W. J., Sc.D. thesis, Eindhoven University of Technology, Eindhoven, Netherlands, 1996.
12. Murphy, R. E., Schure, M. R., Foley, J. P., Themabeam Liquid Chromatography Mass Spectrometry Analysis of Polymers for the Calculation of Chemical Dispersity. Presentation at 10th International Symposium on Polymer Analysis & Characterization, Toronto, Canada, 1997.

12

Particle Size Measurement of Plastics and Polymers Using Laser Light Scattering

Philip Plantz

Microtrac, Inc., Largo, Florida, USA

INTRODUCTION

The initial interest in the production of plastics evolved from the desire to create synthetic rubber. The basic building block during these early polymerization reaction attempts was the water-insoluble monomer, butadiene. An emulsion of the monomer was developed by adding a surfactant and homogenizing the mixture. An initiator caused the attachment of the monomer molecules, resulting in larger molecules termed polymers. Today the science of polymerization has grown extraordinarily to include ABS resins, acetal resins, acetate fibers, fiber-reinforced composites, foamed plastics, and many others. These polymers may then be used to form a wide variety of products such as wood glue, carton sealants, paint, paper coatings, water-proofing materials, carpet backing, as well as in the production of more commonly observed products including pipe, bottles, plasticwares, computer cases, desk organizing items, and many others. The importance of these items to the populace cannot be understated. As a result, the quality of the products is of great importance, which, in turn, feeds back to the original control of the polymer characteristics during production. The physical analysis of these suspensions and the resulting plastic precursors can influence upstream processibility, e.g., pellet formation, and, ultimately, the final product characteristics and quality. This chapter will review one aspect of the physical characterization of the polymers and its products—particle size measurement. From the smallest sizes of polymers during the polymerization reaction, to larger sizes of the pellets used in extrusion, the size range generally includes 0.003 to 1000 μm (and possibly larger). The technique to be discussed is light scattering as applied to static (diffraction) and dynamic (Brownian motion) measurements. Several examples of the potential applications will be provided, as well as an overview of the more recently accepted approaches of analyzing the scattering information.

LASER DIFFRACTION PARTICLE SIZE MEASUREMENTS

During the 1970s, one of the early applications of commercial lasers was to measure particle size distributions within the range of approximately 2–300 μm [1]. During the 1980s the technique was extended to as small as 0.1 μm [2] and as large as 1200 μm [3], while modern instrumentation has extended the range to as small as 0.020 μm and as large as 3500 μm [4,5]. In addition, the advent of desktop computer technology and high-speed integrated circuits has advanced the capability to allow for correct light-scattering calculations for transparent particles and the effect of particle size on the relative ability to scatter light [6].

In the simplest situation, a laser is used to provide a well-defined wavelength of light which allows development of a defined pattern of forward-directed scattered light [7,8]. This phenomenon is termed low-angle scattering since it is normally associated with angles of scatter from a few tenths to approximately 40–60°. (The angle is measured from the direction of the incident light). It generally will allow inclusion of the size ranges as small as 0.7 μm depending on the wavelength of the laser. Lasers in the red or infrared spectral region may be used in combination with selected lenses and optical component spacings to provide a light pattern, which, after interaction with particles, can be analyzed. The analysis is performed by measuring the angle of scattering and the intensity pattern of the scattered light. To accumulate the information, an array of silicon detectors is used to transform the scattered light into a series of measured voltages or signals to give a pattern which can be mathematically analyzed to provide a particle size distribution. One the primary equations describing this phenomenon is the Airy formula:

$I(w)$ = scattering intensity as a function of angle

$$= Ek^2 A^4 \left[\frac{J_1(kAw)}{kAw} \right]^2$$

where:

E = light flux per unit area of incident light

$$k = \frac{2\pi}{\lambda}$$

$w = \sin \theta$

θ = angle relative to the incident beam direction

A = radius of particle

J_1 = Bessel function of the first kind

The important aspects of this formula are the wavelength, intensity, and angle of diffraction, and the influence of changes of these values on the size parameter. By using a laser, wavelength becomes a constant. Then, if values for the diffraction angle and intensity are known or can be determined, the size parameter can be calculated. Using appropriate optical components in combination with a silicon detector system, the angle of diffraction and intensity profile for a range of particle sizes can be measured (Fig. 1). The unknown value remaining in the formula is the size value, which can then be calculated using the known or measured values. By measuring the light

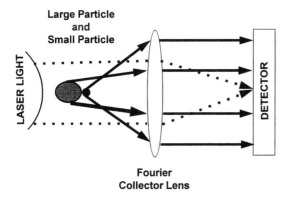

Figure 1 Interaction of light with large and small particles resulting in different angles of diffracted light. A special lens is employed to focus the scattered light onto a silicon detector array.

scattered at wider angles (and possibly shorter wavelengths), smaller-sized particles may be included in the report, down to the submicrometer region or as low as 0.020 μm; however, the issues of light-scatter pattern definition and resolution become apparent at very small sizes, below 0.1 μm. These issues are the subject of extremely in-depth discussions on the scattering of light and will not be addressed in this chapter. Many fine references are available for such study of this complex phenomenon [9].

The preceding discussion applies to a single particle that produces a well-defined diffraction pattern. In the more realistic cases of measuring particle size distributions, many particles of varying size are present, thus demanding extension and more complex analysis than these simple principles afford. One approach to solving the dilemma of this complexity is development of a nonlinear measurement of the scattered [10]. Light scattering is a nonlinear phenomenon, and one approach to its measurement can best be described by applying a nonlinear detection system. One design of such a detector system allows measurement of the angles and the intensities at intervals of the root of 2. Thus, when scattered light illuminates a series of detectors spaced at defined intervals of some root of 2, the requirements for intensity and angular information are satisfied. The detector system also conforms to the nonlinear phenomenon of scattering. Separating and measuring the light pattern by this approach allows for direct evaluation and calculation of particle size distributions from the combined scattering patterns of a size distribution of particles. This design produces a signal response function that is the same for a given volume of particulate [11] across the measurable particle size range. This, in turn, eases the demands of assigning the correct volume contribution by single- or multiple-mode particle size distributions.

The most commonly applied approach to analyzing the diffraction pattern involves application of a process known as an *interative convolution*. In this approach, the resulting pattern of light is used to calculate an initial particle size distribution. The distribution should, in fact, provide the light-scatter signal pattern if the "reverse" computation were to be used. However, due to slight anomolies occurring in the detected light pattern and in the size computation, the forward and reverse computations may not match. Should this occur, further computation of the particle size distribution is necessary in order to obviate the conversion errors. The "new" particle size distribution is evaluated for its ability to provide the original light-scatter pattern. Errors are noted by the software and another computation is performed taking into account the errors. The computational process continues until the forward and reverse computations are within acceptable error. The mathematical process of calculating the particle size distribution from the scattering pattern is termed *deconvolution*, during which the mixed (convolved) scattered light (from the particles passing through the laser beam) is analyzed. The term *iteration* refers to constant recomputation. Combining the terms allows a shortened terminology, *interative deconvolution*, that describes the mathematical process of computing the particle size distribution from the complex light-scatter pattern.

This analytical approach using a single wavelength of light may be extended to very wide scattering angles (ultrafine particles) by using detectors located at very wide angles, typically up to approximately 160°. While this approach is valid, many more detectors are required, which adds to the complexity of design and manufacturing. An alternative approach is to change the direction at which the laser is incident on the particles and reapply detectors initially used for smaller angle (larger particle) measurement [10]. To satisfy this requirement, multiple lasers of the same wavelength can be used. Sequential illumination of the particles by the specially positioned lasers allows light detection and measurement over the entire angular spectrum. It also maintains the mathematical analysis by avoiding the complications of multiple wavelength sources (refer to the Airy formula given above). The system design requires rapid-response, stable lasers as embodied in commercially available laser diode technology. Commercial instruments embodying these design principles are the Microtrac and the Cilas particle size analyzers. Description of the optical design and the components required for measurement are shown in Figs. 2–4.

An advantage of this technology is that it can be easily extended to a larger size measurement range by changing optical components and optical distances. In order to obtain larger particle information, the low-angle scattered light must be separated into defined segments. The issue is addressed by changing the lens or, alternatively, by increasing the distance from the lens to the detector. These design considerations are available in present commercial instrumentation. Another advantage, aside from the measuring range of diffraction instruments, is the ability to measure particles in the dry powder state. This type of measurement is especially useful when the powers are free flowing and not subject to attrition by shear forces that are commonly used to achieve dispersion of dry powder agglomerates. Disposal or recycling of the powder is generally considered to be more favorable than addressing the issues of fluid disposal. Dry powder particles, however, are subject to the effects of compression during storage or agglomeration due to humidity effects [12].

Dynamic Light Scattering

Another technology [13] often used to provide a particle size distribution is dynamic light scattering (DLS). The application of this technology is to

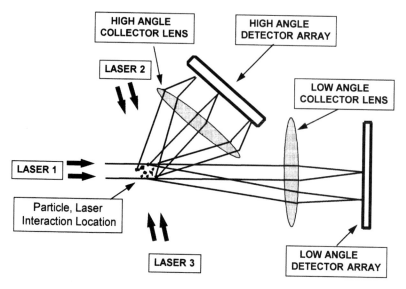

Figure 2 Diagram of the application of multiple lasers that are positioned to change the incident angle of particle illumination in order to reuse detectors applied in low-angle scattering. Laser diodes 1, 2, and 3 are energized sequentially to permit signal production and analysis at angles up to 160°.

those particles that can undergo Brownian motion, which is limited up to approximately 10 μm. However, the practical range is below 1 μm and above 0.001 μm (1 nm). Particles larger than 1 μm can be difficult to maintain in a suspended state, which is a requirement of DLS. Particles smaller than 0.001 μm offer low scattering light flux, which usually prohibits detection of smaller size particles. The approximate measurement capability is, therefore, within the range 0.001–2 μm. Plastics polymers are unique in that the specific gravity is usually close to water and the requirement for suspension offers little or no impact during the course of the measuring period of typically 30–60 s. The measuring time is long compared to static light scattering, but the technology is generally better suited to particle size measurements below 0.1 μm because diffraction techniques exhibit lower scattering intensities at the wider angles necessary to measure in the same, low size range.

The technology under discussion relates to the ability of particles to undergo random motion as a result of the colliding velocity of the suspending fluid molecules. Through this effect, popularly known as Brownian motion, water molecules can impart movement of much larger polymer molecules whose velocity in suspension is a result of the thermal velocity

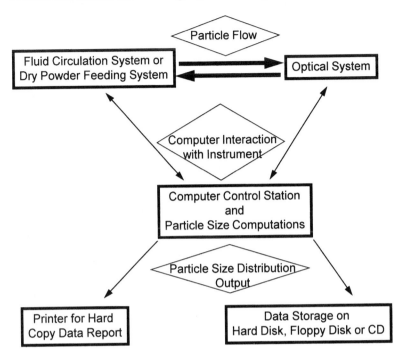

Figure 3 Particle and information flow in a typical laser diffraction particle size analyzer. Fluid or dry powder passes from a sampling system (dry powder passes through the system one time; fluid suspended particles are recirculated). Light is diffracted by the particles and is measured by the optical system under the control of desk-top computer software. Computations of the particle size distribution are performed which are then printed and/or stored on computer storage media.

profile of the water molecules. As temperature increases, the velocity of the fluid molecules increases. This in turn increases the velocity of the particles. As described by kinetic theory [14], at constant temperature, a particle will be moved by fluid molecules and display a velocity dependent on the size of the particle. As the particle collides with water molecules from all directions, the particle is given a variety of movements which may appear to be fully random. Light incident on the moving polymer particle can undergo a change in frequency related to the velocity of the particle. Since a size distribution of particles may be present, a distribution of frequency changes will exist. By monitoring the light scattered by the particles in Brownian motion, information can be obtained that can be used to calculate particle size distributions from a distribution of frequencies.

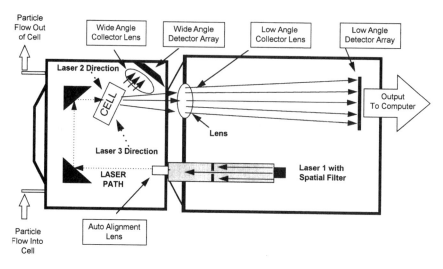

Figure 4 A top-down view of the Microtrac S300 optical system showing the location and direction of the three lasers of wavelength 780 nm. Two lenses and two silicon detector arrays are located to collect and respond to the scattered light pattern from the particles passing through the cell. Flow of the particles is from bottom to the top. The black triangles represent mirrored surfaces used to direct the beam from laser 1 used for low-angle scatter measurements. Following illumination by laser 1, lasers 2 and 3 illuminate in sequence. The sequence is repeated until the analysis time is completed.

While the usual monitoring angles range from approximately 160° to near 0°, the most recently introduced commercial approach to this type of measurement utilizes back-scattered (180°) light to provide the information necessary for size computations (Figs. 5 and 6). In addition to the use of back-scattered light, the technology utilizes a uniquely designed probe to allow laser light to enter the measurement region and collect the backscattered light while simultaneously mixing it with a portion of the non-frequency-shifted laser [15]. This latter heterodyne activity provides for the high signal necessary for measurements utilizing back-scattered light. "Heterodyne" in this sense refers to the mixing of the frequency-shifted scattered light with the non-frequency-shifted light. To analyze this light information, a digital signal processor is used with a fast Fourier transform to calculate a power spectrum to provide a pattern of frequency shifts which are directly related to the size distribution of the particles present. The frequency distribution then can be further analyzed to provide the distribution

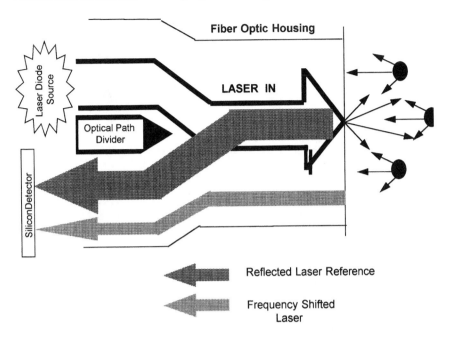

Figure 5 Diagram of the Microtrac Ultrafine Particle Analyzer probe. Light from a laser diode is directed through a fiber-optic waveguide to a transparent window, where it exits to interact with particles exhibiting Brownian movement. A portion of the original light is reflected from the window fluid interface and is directed to a silicon detector. Back-scattered light from the sample is captured and mixed (heterodyne) with the reflected light acting as reference. The heterodyne signal provides high intensity and a reference for mathematical analysis (Controlled Reference Method) of the Doppler-shifted, back-scattered light.

of particle sizes as a function of the signal. This particle size distribution is typically termed an intensity profile, since only the ability of the particles to scatter light is reported. However, depending on the size of the particles and their shape, the ability of the particles to scatter light will vary considerably and the resulting distribution is nontangible in that three-dimensional characteristics are not included. Thus, especially in the size range of 0.1 μm and larger, the amount of particles present may be under–or overreported as a result of the relative particle size and shape. To compensate for the scattering efficiency (ability) of the particles, Mie scattering concepts (which include refractive index considerations) are invoked to "rebalance" the relative amounts of the particle sizes present and to provide a more tangible or

SAMPLE CELL AND PROBE CASE (4" x 6" x 12")

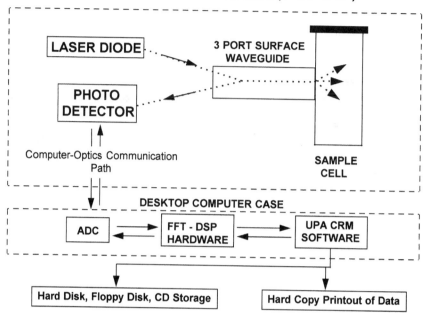

Figure 6 Diagram of the Microtrac UPA light path, data collection, and analysis. As with diffraction, the computer plays a vital role in controlling the measurement, collecting the signals and performing specialized analysis of the raw signals to provide a particle size distribution. Computed particle size distributions are printed and/or communicated to computer data storage media.

realistic display of the distribution. This is generally referred to as the *volume* distribution, representing the *amount* of particulate material in the sample. Two distributions (intensity and volume) can be reported, and each can provide useful, albeit different, information for process and quality control.

APPLICATIONS OF THE TECHNOLOGIES

As noted previously, monomers are used to produce the resins commonly used to manufacture a wide variety of products. The quality of the final product can be directly influenced by a number of process variables, one of which is particle size distribution. This section will address several applications of light-scattering particle size measurements as they pertain to the entire manufacturing scheme leading to the final plastic product. Since the number of monomers, polymers, and resins is immense, limitations of space prevent a review of all possible applications. The examples given are considered by size and distribution, with full recognition that similar applications exist with other chemical analogs involved in plastics manufacturing.

Large Particle Size Applications (200–3000 μm)

Many types of process equipment are used during the manufacture of plastic articles, including injection molds, compression molds, extruders, and rotational molds. A common characteristic of all these processes is that a pellet or powder is used as the starting material. The characteristics of the feed material must meet certain criteria such as melting point, chemical composition, flexural strength, compressive strength, impact resistance, density, chemical resistance, and tensile strength. These provide the characteristics of the resulting particle. Particle size also plays a major role in the processibility of the polymer resin in that flow from the hopper and melting rate are directly influenced by particle size distribution. The forming device processing rate is highly dependent on the melting rate of the pellet or powder. The melting rate is dependent on chemical composition but also on the particle size distribution in that complete melting is required to maintain throughput of processing equipment and the final quality parameters. Particle size distribution affects flow from the hopper which feeds the processing equipment. In many cases, hopper design is influenced by particle size distribution in that narrow distributions of larger particle size tend to flow better than finer particle sizes. In addition, larger particles have much less tendency to agglomerate and plug process orifices. Thus a balance exists between desirable melting rates (which are expectedly higher with smaller particles) and flow conditions (which are usually more easily controlled when the distribution is narrow and the particle size is larger). Particle size measurement information similar to that shown in Fig. 7 can provide a portion of the information required for powder processing quality.

The data in Fig. 7 show a somewhat narrow particle size distribution with an sd value (indicates breadth of the distribution) of 48 μm with a volume mean size (MV) of 650 μm. The 10% value (percent smaller than the size reported) is 591 μm, while the 95% value is 738 μm. Increase of the MV or the 95% value will indicate an increase in larger particles and a broadening of the distribution. While flow may be affected to a small, if not insignificant extent, sufficient increase in the amount of larger particles may disrupt the processing into final product as a result of lowered melting rates.

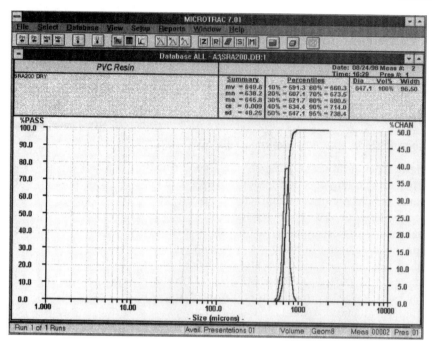

Figure 7 Data presented as a graph of particle size distribution of coarse PVC resin pellet. The distribution is narrow and sufficiently large to permit good process flow and uniform melting.

Correlation of Techniques Used to Measure the Particle Size of Plastics and Resins

Often light scattering replaces sieves or other particle size measuring devices as a means of increasing laboratory analysis rate and providing more timely process data. Often, the particle size data obtained by one technique are different from that obtained by another, which will cause concern to process engineers, laboratory personnel, and plant operators. In general, the data from light-diffraction instrumentation will be reported as being larger than the corresponding sieve measurement [16]. However, as reported in Ref. 16, the particle size distributions from nonidentical techniques are merely shifted, causing the percentile and averaged values to be different from the sieves. The data from the new technique can be mathematically related through correlation. The correlations are generally accepted to follow linear regression analysis, which is available in many common software programs. The results of a correlation study for a PVC slurry are shown in Fig. 8 for

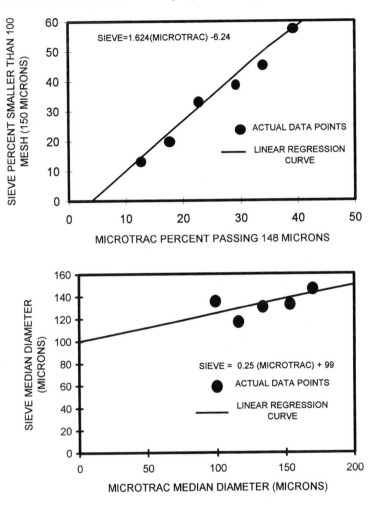

Figure 8 Linear regression curves for the median diameter and the percentage smaller than 100 mesh (148 μm). Actual data are shown overlaid on the regression (correlation) curve. Equations developed from the actual data are presented in the upper left corner. The equation follows the mathematics for a straight line.

dry sieving particle size as a function of diffraction particle size measurement for percent passing 100 mesh (148 μm). The median volume size, which is a general average particle size, can show the effect of an increase or decrease in the amount of coarse particles in the distribution. The percentage smaller than 148 μm is an important "marker" for increase in fines, which can, as discussed previously, have a strong impact on process flow. Both data sets show a strong linear relationship with the sieve data. While the values are different from corresponding sieve data, the data are useful albeit slightly different. Using these diffraction data, a corresponding sieve value can readily be determined graphically or calculated using the equation for a straight line. The diffraction instrumentation can then be used to decrease measuring time, while removing operator influences inherent in the sieve measurement.

Medium Particle Size Applications (0.1–700 μm) Such as High-Impact Polystyrene

Another application of the diffraction instrument is the analysis of high-impact polystyrene (HIPS). Data for such a sample is shown in Fig. 9. One literature citation describes a study in which a sample of HIPS was dissolved in several solvents to test the viability of measuring the rubber particles

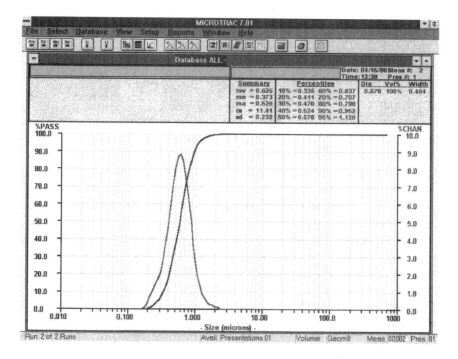

Figure 9 Diffraction data for a sample of rubber particles obtained from dissolving HIPS pellets in MEK. Excessively large rubber particles may cause a change in impact resistance to the final product.

which exist as part of the HIPS pellet. Toluene, MEK, *trans*-cinnamalde-hyde, and isopropanol were tested to determine the ability to dissolve the polystyrene without effects on the rubber particles during preparation or measurement. Toluene is accepted to cause large amounts of swelling of the rubber particles and thus distorts the meaningfulness of the particle size distribution.

Isopropanol suspension required isolation and extraction of the rubber particles prior to suspension because of its inability to dissolve the polystyrene matrix and thus release the rubber particles for measurement. In addition, suspensions in isopropanol, while not causing any measurable swelling, did cause agglomeration of the small particles. Agglomeration in such prepara-tive situations can be random in the size and amount formed. Since lengthy isolation procedures were required prior to measurement, the isopropanol procedure was eliminated. The *trans*-cinnamaldehyde and MEK caused minor swelling which was reproducible, and thus could be considered to be under kinetic control and therefore predictable. Particle size average diameter using the isopropanol resuspension after extraction was 5.4 μm, while the values for direct MEK and *t*-cinnamaldehyde were 5.7 and 5.9, respectively. In contrast, the value for direct suspension of HIPS in toluene resulted in a value of 8.8 μm, indicating considerable swelling of the rubber particles. Because of the relatively lower cost of MEK as compared to *t*-cinnamalde-hyde, the most applicable solvent was considered to be MEK. In addition, size ranking, as determined by diffraction measurement, was identical to those data obtained by phase-contrast microscopy, in which the HIPS pellet is melted to provide a thin film for examination of the rubber particles. This reference method avoided deleterious effects to the rubber particles. Thus, microscopy offers the best solution to avoiding swelling effects but is far too tedious and operator-dependent for routine measurements, while the light-scattering technique offers speed and high reproducibility.

Medium Particle Size Applications (0.1–700 μm) Such as Fillers

Plastic fillers are used to enhance many plastic properties such as increasing the impact strength of polypropylene, PVC, and polystyrene (calcium car-bonate); providing extra slip to improve molding characteristics (thin mica); and decreasing the moisture sensitivity of nylon (kaolin). Each of these can be measured by diffraction technology. Calcium carbonate and mica can be prepared for water suspension measurement using a dilute solution of an anionic surfactant, while calcium carbonate and kaolin can be prepared using a 1% solution of sodium hexametaphosphate. The surfactant and dispersant modify the surface properties of the powders to allow for com-plete dispersion prior to measurement, as well as an opportunity to maintain cleanliness of the windows of the sample cell through which the suspended particles pass to interact with the laser beam to produce the scattering phenomena. Often the use of mild energy is required to separate particles that have attached themselves to each other (agglomerates). The usual choice of energy is ultrasound, which provides high agitation to the particle fluid mixture by a process of cavitation (formation and collapse of pressure waves to produce energy concentration packets). Data using such preparation are shown in Figs. 10 and 11, respectively, for calcium carbonate and glass filler.

Small Particle Size Applications (0.003–6 μm) Such as Polymer Slurries

As explained above, it is often desirable to use other light-scattering technology than diffraction to measure particles in the submicron particle size

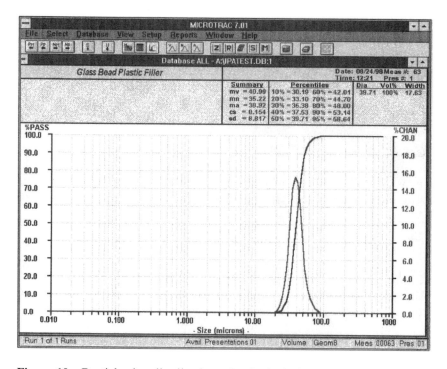

Figure 10 Particle size distribution of spherical glass filler used in a wide variety of plastic products as an adjunct to the resins prior to pellet formation.

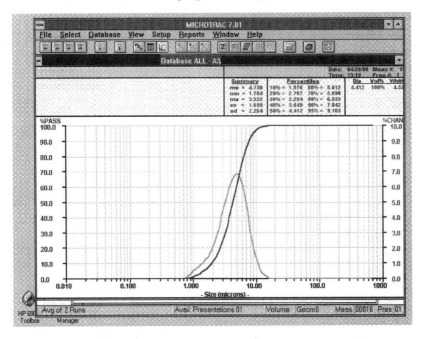

Figure 11 Particle size of calcium carbonate filler. Note the slight "tail" or "skew" to the distribution at the smaller size range of the distribution. Fines can prohibit proper mixing and dispersion prior to or during use in the resin.

range. One of the advantages of DLS as performed by the heterodyne probe method (controlled reference method) is the capability to measure particle suspensions at high concentrations as compared to classical photon correlation spectroscopy (PCS). Using polystyrene as an intermediate material prior to final application, data were developed (Fig. 12) to evaluate the performance of the controlled reference method (CRM). The accompanying graph shows the particle size results of increasing concentration of small size polystyrene microspheres when measured by classical PCS and the controlled reference method. As noted from the graph, concentration as high as 20% produced no deleterious effects on the CRM data, while it is apparent that only a narrow range of concentration can be used for measurement by the PCS instrument. The data in Fig. 13 show the advantage of using the heterodyne (CRM) technique in that a broad distribution of polymers can be measured without advanced knowledge of distribution characteristics, while PCS requires operator intervention to allow selection of a proper mathematical approach to providing data. An additional advantage of CRM as discussed above is the ability to measure high concentrations

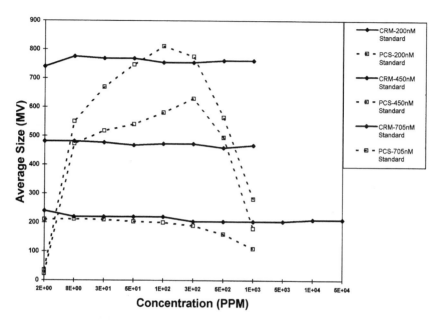

Figure 12   Graphs showing the response of the Ultrafine Particle Analyzer to increasing concentrations of polystyrene of different submicrometer sizes. Similar data for commercial photon correlation spectroscopy show non-linear response, suggesting that specialized sample concentrations are necessary for measurement.

(10%) by the instrumental heterodyne measurement as well as providing opportunities to perform the measurements in the process line.

CONCLUSIONS AND SUMMARY

Spectroscopy provides the laboratory analyst, and the production engineer, the ability to perform particle size measurement over a very broad range of sizes in a timely manner. Spectroscopy in the form of diffraction and dynamic light-scattering technology have been demonstrated to provide information that can be applied to research and process studies as well as to the routine measurement demands of the plant. Examples of particle size measurement capability for the plastics industry from 0.003 to 3000 μm were presented, as well as a short discussion of means to convert the data from the newer technology to the classical sieve measurements. A final comment on the advancing science of spectroscopic particle size measurements was also presented.

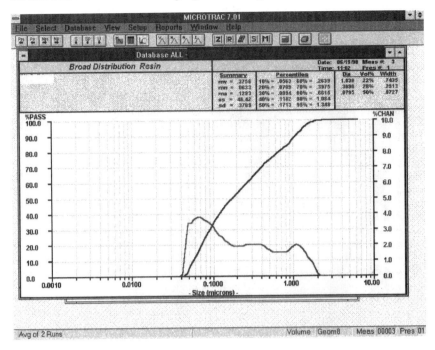

Figure 13 UPA data for a 10% solids sample of a commercial latex used in blow molding. The distribution was obtained without preselection of distribution modes or distribution parameters.

REFERENCES

1. J. Cornillault. Particle size analyzer. Appl. Opt. 11:265–287, 1972.
2. E. C. Muly, H. N. Frock, E. L. Weiss. The application of Fourier imaging systems to fine particles. Seventh Annual Fine Particle Conf., Philadelphia, 1975.
3. E. C. Muly, H. N. Frock. Industrial particle size measurement using light scattering. Opt. Eng. 19:861–869, 1980.
4. P. E. Plantz, M. N. Trainer. Measurement of submicron particles using static and dynamic light scattering. Am. Lab. News, p 20–22, May 1998.
5. P. E. Plantz. Particle size measurement by laser light scattering. In: H. Barth, ed., Modern Methods in Particle Size Measurement. Wiley Interscience, New York, 1981, pp. 88–111.
6. R. Pecora. Dynamic light scattering. In: Applications of Photon Correlation Spectroscopy. Plenum Press, New York, 1985.
7. J. Meyer-Arendt. Introduction to Classical and Modern Optics. 2nd ed. Prentice Hall, Englewood Cliffs, NJ, 1984, pp. 194–196.

8. E. L. Weiss, H. N. Frock. Rapid analysis of particle size distributions by laser light scattering. Powder Technol. 14:287–296, 1976.
9. M. Kerker. The Scattering of Light and Other Electromagnetic Radiation. Academic Press, New York, 1969.
10. P. J. Freud, A. H. Clark, H. N. Frock. Unified scatter technique for full-range particle size measurement. Pittsburg Conf., Atlanta, GA, March 8–12, 1993.
11. M. N. Trainer. Optimization of particle size measurement from 40 nanometers to millimeters using high-angle light scattering. In: T. Provder, ed., Particle Size Distribution III: Assessment and Characterization, ACS Symp. Ser. 693, 1998, pp. 130–147.
12. P. E. Plantz, Concepts on the preparation of particulate for measurement by light scattering particle size analyzer. In: M. Sharma and F. J. Micale, ed., Water-Based Coatings and Printing Technology, 1991, pp. 225–246.
13. R. M. Johnson, W. Brown. An overview of current methods of analysing QLS data. In: S. E. Harding, D. B. Sattelle, and V. A. Bloomfield, eds., Laser Light Scattering in Biochemistry, 1990, pp. 77–91.
14. A. Einstein. Investigation of the Theory of Brownian Movement. A. D. Cowper, transl. Dover, New York, 1956.
15. P. E. Plantz. Ultrafine particle size measurement in the range 0.003 to 6.5 micrometers using the controlled reference method. In: T. Provder, ed., Particle Size Distribution III: Assessment and Characterization, ACS Symp. Ser. 693, 1998, pp. 103–129.
16. H. N. Frock and P. E. Plantz. Correlation Among Particle Sizing Methods. Honeywell Microtrac Applications Note, 1985.
17. R. A. Hall, R. D. Hites, and P. Plantz. Characterization of rubber particle size distribution of high-impact polystyrene using low-angle laser light scattering. J. Appl. Polymer Sci. 27:2885–2890, 1982.

Appendix: ASTM Methods for Analysis of Plastic and Rubber Materials

PLASTICS: ASTM TESTING METHODS

D 785	ROCKWELL HARDNESS
D 789	RELATIVE VISCOSITY, MELT POINT & % POLYAMIDE
D 790	FLEXURAL STRENGTH & MODULUS
D 792	SPECIFIC GRAVITY BY DISPLACEMENT METHOD
D 794	PERMANENT EFFECT OF HEAT ON PLASTICS
D 881	DEVIATION OF LINE OF SIGHT
D 882	TENSILE PROPERTIES OF THIN PLASTIC SHEET
D 952	BOND OR COHESIVE STRENGTH
D 953	BEARING STRENGTH OF PLASTICS
D 1004	INITIAL TEAR RESISTANCE OF PLASTIC FILM
D 1042	DIMENSIONAL CHANGES UNDER ACCEL COND
D 1043	STIFFNESS BY MEANS OF A TORSION TEST
D 1044	RESISTANCE TO PLASTICS BY ABRASION
D 1180	BURSTING STRENGTH OF PLASTIC TUBING
D 1203	VOLATILE LOSS USING ACTIVATED CARBON
D 1204	DIMENSIONAL CHANGES @ ELEVATED TEMP
D 1238	FLOW RATE BY EXTRUSION PLASTOMETER
D 1239	RESISTANCE OF PLASTIC FILM TO EXTRACTION
D 1242	RESISTANCE OF PLASTIC FILM TO ABRASION
D 1243	DILUTE SOLUTION VISCOSITY OF PVC
D 1299	SHRINKAGE AT ELEVATED TEMPERATURES
D 1433	RATE OF BURNING AT 45 DEGREE ANGLE
D 1435	OUTDOOR WEATHERING OF PLASTICS
D 1505	DENSITY BY GRADIENT TECHNIQUE
D 1525	VICAT SOFTENING TEMPERATURE
D 1598	TIME TO FAILURE OF PIPE UNDER PRESSURE
D 1599	SHORT TERM HYDRAULIC FAILURE/PLASTIC PIPE
D 1601	DILUTE SOLUTION VISCOSITY OF POLYETHYLENE
D 1603	DETERMINATION OF CARBON BLACK IN OLEFINS
D 1621	COMPRESSIVE PROPERTIES OF CELLULAR PLASTICS
D 1622	APPARENT DENSITY OF RIGID CELLULAR PLASTICS
D 1623	TENSILE OR ADHESION OF CELLULAR PLASTICS
D 1637	TENSILE HEAT DISTORTION TEMPERATURE
D 1652	EPOXY CONTENT OF EPOXY RESINS
D 1693	ENVIRONMENTAL STRESS CRACKING /PE
D 1694	THREAD ANALYSIS OF PIPE (SEM)

D 1708	MICROTENSILE PROPERTIES OF PLASTICS
D 1709	IMPACT BY FREE FALLING DART
D 1712	RESISTANCE TO SULFIDE STAINING
D 1746	TRANSPARENCY OF PLASTIC SHEETING
D 1784	LONGITUDINAL TENSILE OF PIPE
D 1790	BRITTLENESS TEMPERATURE BY IMPACT
D 1870	ELEVATED TEMPERATURE AGING BY TUBULAR OVEN
D 1894	STATIC AND KINETIC COEFFICIENT OF FRICTION
D 1895	DENSITY/BULK FACTOR/POURABILITY
D 1925	YELLOWNESS INDEX OF PLASTICS
D 1938	TEAR PROPAGATION
D 1939	RESIDUAL STRESS BY GLACIAL ACETIC ACID IMRSN
D 2105	LONGITUDINAL TENSILE OF REINFORCED PIPE
D 2115	OVEN HEAT STABILITY OF PVC COMPOSITIONS
D 2122	DIMENSIONS OF THERMOPLASTIC PIPE & FITTINGS
D 2124	PVC ANALYSIS BY FTIR
D 2126	RESISTANCE TO THERMAL & HUMID AGING
D 2151	STAINING OF PVC BY COMPOUNDING MATERIALS
D 2152	DEGREE OF FUSION BY ACETONE EXTRACTION
D 2222	METHANOL EXTRACT OF VINYL CHLORIDE RESINS
D 2238	ABSORBANCE OF POLYETHYLENE AT 1378 CM-1
D 2240	RUBBER PROPERTY DUROMETER HARDNESS
D 2288	WEIGHT LOSS OF PLASTICIZER ON HEATING
D 2289	TENSILE AT HIGH SPEEDS
D 2290	TENSILE BY SPLIT DISK METHOD
D 2299	DETERMINING RELATIVE STAIN RESISTANCE
D 2383	PLASTICIZER COMPATIBILITY OF PVC
D 2412	EXTERNAL LOAD BY PARALLEL PLATE METHOD
D 2444	TUP IMPACT OF PIPE FITTINGS
D 2445	THERMAL OXIDATIVE STABILITY
D 2463	DROP IMPACT OF BLOW MOLDED CONTAINERS
D 2471	GEL TIME AND PEAK EXOTHERMIC TEMP/RESINS
D 2552	ENVIRONMENTAL STRESS RUPTURE/CONSTANT LOAD
D 2561	ENVIRONMENTAL STRESS CRACK/CONTAINERS
D 2562	CLASSIFYING VISUAL DEFECTS IN THERMOSETS
D 2563	CLASSIFYING VISUAL DEFECTS IN LAMINATES

D 2566	WETTING TENSION OF FILMS
D 2578	PUNCTURE PROPAGATION TEAR RESISTANCE
D 2583	BARCOL HARDNESS
D 2584	IGNITION LOSS OF CURED REINFORCED RESINS
D 2587	ACETONE EXTRACTION AND IGNITION
D 2659	COLUMN CRUSHING THERMOPLASTIC CONTAINERS
D 2683	PE SOCKET FITTING SPECIFICATIONS
D 2684	PERMEABILITY OF CONTAINERS TO REAGENTS
D 2732	LINEAR THERMAL SHRINKAGE OF FILMS & SHEET
D 2734	VOID CONTENT OF REINFORCED PLASTICS
D 2839	USE OF A MELT INDEX STRAND FOR DENSITY
D 2841	SAMPLING OF HOLLOW MICROSPHERES
D 2842	WATER ABSORPTION OF RIGID CELLULAR PLASTICS
D 2856	OPEN CELL CONTENT OF RIGID CELLULAR PLASTICS
D 2857	DILUTED SOLUTION VISCOSITY OF POLYMERS
D 2863	OXYGEN INDEX
D 2911	DIMENSIONS AND TOLERANCES OF PLASTIC BOTTLES
D 2951	RESISTANCE OF PE TO STRESS CRACKING
D 2990	COMPRESSIVE CREEP
D 2991	STRESS RELAXATION
D 3012	THERMAL OXIDATIVE STABILITY—BIAXIAL ROTOR
D 3014	FLAME HT, BURN TIME, WT LOSS IN VERT POSITION
D 3029	TUP IMPACT RESISTANCE
D 3030	VOLATILE MATTER OF VINYL CHLORIDE RESINS
D 3045	HEAT AGING OF PLASTICS WITHOUT LOAD
D 3124	VINYLIDENE UNSATURATION NUMBER BY FTIR
D 3291	COMPATIBILITY OF PLASTICIZERS/COMPRESSION
D 3349	ABSORPTION COEFFICIENT/PE & CARBON BLACK
D 3351	GEL COUNT OF PLASTIC FILMS
D 3354	BLOCKING LOAD OF PLASTIC FILM
D 3367	PLASTICIZER SORPTION OF PVC
D 3417	HEAT OF FUSION BY DSC
D 3418	TRANSITION TEMP OF POLYMERS BY DTA
D 3536	MOLECULAR WEIGHT BY GPC
D 3576	CELL SIZE OF RIGID CELLULAR PLASTICS
D 3591	LOGARITHMIC VISCOSITY OF PVC

D 3593	MOLECULAR WT BY GPC/UNIVERSAL CALIB
D 3594	EA/EEA BY FTIR
D 3713	MEASURING RESPONSE TO SMALL FLAME IGNITION
D 3748	EVALUATING HD RIGID CELLULAR PLASTICS
D 3810	EXTINGUISHING CHARACTERISTICS-VERT POSITION
D 3835	RHEOLOGICAL PROPERTIES/CAPILLARY RHEOMTR
D 3846	IN PLANE SHEAR OF REINFORCED PLASTICS
D 3895	COPPER INDUCED OXIDATIVE INDUCTION BY DSC
D 3914	IN PLANE SHEAR OF RODS
D 3916	TENSILE OF PLASTIC ROD
D 3917	DIMENSIONAL TOLERANCE OF PULTRUDED SHAPES
D 3998	PENDULUM IMPACT OF EXTRUDATES
D 4065	DYNAMIC MECHANICAL PROPERTIES OF PLASTICS
D 4093	PHOTOELECTRIC BIREFRINGENCE
D 4100	GRAVIMETRIC ANALYSIS OF SMOKE/PLASTICS
D 4218	DETERMINATION OF CARBON BLACK CONTENT
D 4272	IMPACT BY DART DROP
D 4274	DETERMINATION OF HYDROXYL # IN POLYOLS
D 4321	DETERMINATION OF PACKAGE YIELD
D 4385	CLASSIFYING VISUAL DEFECTS PULTRUDEDS
D 4475	HORIZONTAL SHEAR OF PLASTIC RODS
D 4476	FLEXURAL PROPERTIES PULTRUDED RODS
D 4508	CHIP IMPACT
D 4591	DETERMINING TEMP TRANSITIONS BY DSC
D 4603	INHERENT VISCOSITY OF PET
D 4660	TDI ISOMER CONTENT BY QUANT FTIR
D 4662	ACID AND ALKALINITY #S OF POLYOLS
D 4664	FREEZING POINT OF TDI MIXTURES
D 4665	ASSAY OF ISOCYANATES
D 4666	AMINE EQUIVALENT OF ISOCYANATES
D 4667	ACIDITY OF TDI
D 4668	QUANTITATION OF Na & K IN POLYOLS
D 4669	SPECIFIC GRAVITY OF POLYOLS
D 4670	DETERMINATION OF SUSPENDED MATTER/POLYOLS
D 4671	DETERMINATION OF UNSATURATION/POLYOLS
D 4672	DETERMINATION OF WATER IN POLYOLS

D 4754	EXTRACTION BY FDA MIGRATION CELL
D 4804	FLAMMABILITY OF NONRIGID SOLID PLASTICS
D 4812	UNNOTCHED IZOD IMPACT
D 4875	QUANTITATION OF PEO IN POLYETHERS BY NMR
D 4876	ACIDITY OF ISOCYANATES
D 4877	DETERMINATION OF COLOR/ISOCYANATES
D 4890	GARDNER/APHA COLOR OF POLYOLS
E 96	WATER VAPOR TRANSMISSION/FISHER CUP
E 252	THICKNESS OF THIN FOIL BY WEIGHING
E 831	LINEAR THERMAL EXPANSION BY TMA

COATINGS: ASTM TESTING PROCEDURES

D 86	DISTILLATION OF PETROLEUM PRODUCTS
D 95	WATER IN PETROLEUM AND BITUMENS
D 153	SPECIFIC GRAVITY OF PIGMENTS
D 185	COARSE PARTICLE SIZE-PIGMENTS, PASTES,PAINTS
D 279	BLEEDING OF PIGMENT
D 280	HYGROSCOPIC MOISTURE IN PIGMENTS
D 281	OIL ABSORPTION OF PIGMENTS BY SPATULA RUB
D 305	SOLVENT EXTRACTABLES BLACK PIGMENTS
D 332	TINT BY VISUAL OBSERVATION
D 344	RELATIVE HIDING POWER BY VISUAL EVALUATION
D 464	SAPONIFICATION NUMBER OF ROSIN
D 465	ACID NUMBER OF ROSIN
D 522	MANDREL BEND
D 562	CONSISTENCY USING STORMER VISCOMETER
D 610	EVALUATING DEGREE OF RUSTING
D 611	ANILINE POINT, MIXED ANILINE POINT
D 658	ABRASION RESISTANCE BY AIR BLAST ABRASIVE
D 659	EVALUATING DEGREE OF CHALKING
D 660	EVALUATING DEGREE OF CHECKING
D 661	EVALUATING DEGREE OF CRACKING
D 662	EVALUATING DEGREE OF EROSION
D 711	NO PICK UP TIME OF TRAFFIC PAINT
D 714	EVALUATING DEGREE OF BLISTERING OF PAINTS
D 772	EVALUATING DEGREE OF FLAKING

D 868	EVALUATING DEGREE OF BLEEDING/TRAFFIC PAINT
D 869	EVALUATING DEGREE OF SETTING OF PAINT
D 870	TESTING OF WATER RESISTANCE/WATER IMMERSION
D 913	EVALUATING RESISTANCE TO WEAR TRAFFIC PAINT
D 968	ABRASION RESISTANCE BY FALLING ABRASIVE
D 969	DEGREE OF BLEEDING OF TRAFFIC PAINT
D 1005	MEASUREMENT OF DRY FILM THICKNESS
D 1078	DISTILLATION RANGE OF VOLATILE ORGANICS
D 1186	FILM THICKNESS OF NONMAGNETIC COATING
D 1198	SOLVENT TOLERANCE OF AMINE RESINS
D 1208	COMMON PROPERTIES OF CERTAIN PIGMENTS
D 1209	APHA COLOR OF CLEAR LIQUIDS
D 1210	FINENESS OF DISPERSION PIGMENT VEHICLE SYSTEM
D 1211	TEMPERATURE CHANGE RESISTANCE WOOD
D 1212	MEASUREMENT OF WET FILM THICKNESS
D 1306	PHTHALIC ANHYDRIDE CONTENT OF ALKYDS
D 1308	EFFECT OF HOUSEHOLD CHEMICALS OF FINISHES
D 1309	SETTING PROPERTIES OF TRAFFIC PAINTS-STORAGE
D 1353	NONVOLATILE MATTER IN VOLATILE SOLVENTS
D 1358	DIENE VALUE OF DEHYDRATED OIL
D 1363	PERMANGANATE TIME/METHANOL $125
D 1364	WATER BY KF METHOD $150
D 1462	ANALYSIS OF REFINED SOYBEAN OIL QUOTE
D 1468	VOLATILE MATTER IN TRICRESYL PHOSPHATE $75
D 1475	DENSITY OF PAINT, VARNISH, LACQUER $50
D 1476	HEPTANE MISCIBILITY OF LACQUER SOLVENTS $100
D 1483	GARDNER COLEMAN OIL ABSORPTION ANALYSIS $100
D 1542	QUALITATIVE DETECTION OF ROSIN IN VARNISHES $200
D 1543	COLOR PERMANENCE OF WHITE ENAMELS $200
D 1544	GARDNER COLOR OF TRANSPARENT LIQUIDS QUOTE
D 1545	BUBBLE VISCOSITY QUOTE
D 1613	ACIDITY IN VOLATILE SOLVENTS
D 1614	ALKALINITY IN ACETONE
D 1617	ESTER VALUE OF SOLVENTS AND THINNERS
D 1639	ACID VALUE OF ORGANIC COATING MATERIALS
D 1643	GAS CHECKING AND DRAFT OF VARNISH FILMS

D 1644	NONVOLATILE CONTENT OF VARNISHES
D 1647	RESISTANCE OF VARNISHES TO WATER AND ALKALI
D 1653	WATER VAPOR PERMEABILITY ORGANIC COATINGS
D 1721	PERMANGANATE TIME OF TRICRESYL PHOSPHATE
D 1722	MISCIBILITY OF WATER-SOLUBLE SOLVENTS
D 1736	EFFLORESCENCE OF INTERIOR WALL PAINTS
D 1848	CLASSIFICATION OF FILM FAILURES–LATEX
D 1926	CARBOXYL CONTENT OF CELLULOSE
D 1951	ASH IN DRYING OILS AND FATTY ACIDS
D 1952	QUANTITATIVE DETERMINATION OF BREAK
D 1954	FOOTS IN RAW LINSEED OIL
D 1955	GEL TIME OF DRYING OILS
D 1957	HYDROXYL VALUE OF FATTY ACIDS AND OILS
D 1958	CHLOROFORM INSOLUBLE MATTER IN OITICICA OIL
D 1959	IODINE VALUE OF DRYING OILS AND FATTY ACIDS
D 1960	LOSS ON HEATING OF DRYING OILS
D 1962	SAPONIFICATION VALUE
D 1963	SPECIFIC GRAVITY OF DRYING OILS
D 1965	UNSAPONIFIABLE MATTER IN DRYING OILS
D 1966	FOOTS IN RAW LINSEED OIL–GRAVIMETRIC METHOD
D 1967	DETERMINATION OF COLOR AFTER HEATING/OILS
D 1980	ACID VALUE/FATTY ACIDS
D 1981	DETERMINATION OF COLOR AFTER HEATING/ACIDS
D 1982	TITER OF FATTY ACIDS
D 2072	H2O IN FATTY QUAT AMMONIUM CHLORIDES
D 2077	ASH IN FATTY QUAT AMMONIUM CHLORIDES
D 2074	INDICATOR METHOD, FATTY AMINES
D 2075	IODINE VALUE OF FATTY AMINES
D 2076	ACID/AMINE #FATTY QUAT AMMONIUM CHLORIDE
D 2078	IODINE # OF FATTY QUAT AMMONIUM
D 2079	NONVOLATILES/FATTY AMMONIUM CHLORIDES
D 2081	PH OF FATTY QUATERNARY AMMONIUM CHLORIDES
D 2086	ACIDITY IN VINYL ACETATE AND ACETALDEHYDE
D 2090	CLARITY/CLEANNESS OF PAINT AND INK LIQUIDS
D 2091	PRINT RESISTANCE OF LACQUERS
D 2197	SCRAPE ADHESION OF ORGANIC COATINGS

D 2352	SULFUR DIOXIDE IN WHITE PIGMENT
D 2369	VOLATILE CONTENT IN COATINGS
D 2370	TENSILE PROPERTIES OF ORGANIC COATINGS
D 2372	SEPARATION OF VEHICLE/SOLVENT
D 2376	SLUMP OF FACE GLAZING/BEDDING COMPOUNDS
D 2379	ACIDITY OF FORMALDEHYDE SOLUTIONS
D 2438	DETERMINATION OF SILICA IN CELLULOSE
D 2454	DETERMINING EFFECT OF OVERBAKE ON COATINGS
D 2485	EVALUATING TEMPERATURE RESISTANT COATINGS
D 2572	ISOCYANATE GROUPS IN URETHANES
D 2621	IR IDENTIFICATION OF VEHICLE SOLIDS
D 2641	CHLORINE IN CELLULOSE
D 2697	VOLUME OF NONVOLATILE MATTER IN COATINGS
D 2698	DETERMINATION OF PIGMENT/CENTRIFUGE
D 2745	TINTING OF WHITE PIGMENTS BY REFLECTANCE
D 2792	SOLVENT AND FUEL RESISTANCE OF TRAFFIC PAINT
D 2793	BLOCK RESISTANCE OF COATINGS ON WOOD
D 2794	IMPACT RESISTANCE
D 2921	QUALITATIVE PRESENCE OF WATER REPELLENTS
D 3002	EVALUATION OF COATINGS FOR PLASTICS
D 3003	PRESSURE MOTTLING AND BLOCKING
D 3021	ANALYSIS OF PHTHALOCYANATE GREEN PIGMENTS
D 3023	STAIN AND REAGENT RESISTANCE OF COATINGS
D 3132	SOLUBILITY RANGE OF RESINS AND POLYMERS
D 3133	CELLULOSE NITRATE QUANT IN ALKYDS BY FTIR
D 3168	QUALITATIVE IDENTIFICATION OF POLYMERS/FTIR
D 3170	CHIPPING RESISTANCE OF COATINGS
D 3260	ACID AND MORTAR RESISTANCE OF COATINGS
D 3272	VACUUM DISTILLATION OF SOLVENTS FOR ANALYSIS
D 3335	LOW CONCENTRATIONS OF PB, CD, AND CO BY AA
D 3359	MEASURING ADHESION BY TAPE TEST
D 3363	FILM HARDNESS BY PENCIL TEST
D 3424	EVALUATING LIGHTFASTNESS
D 3516	ASHING OF CELLULOSE
D 3618	DETECTION OF LEAD IN PAINT & DRIED FILMS
D 3624	LOW CONCENTRATIONS OF HG BY AA

D 3717	LOW CONCENTRATIONS OF SB BY AA
D 3718	LOW CONCENTRATIONS OF CR BY AA
D 3723	PIGMENT CONTENT BY LOW TEMPERATURE ASHING
D 3732	REPORTING CURE TIMES OF UV CURED COATINGS
D 3733	SILICA CONTENT OF SILICONE POLYMERS
D 3912	CHEMICAL RESISTANCE/NUCLEAR PLANT COATINGS
D 3926	PERCENT SOLIDS IN TITANIUM DIOXIDE SLURRIES
D 3960	DETERMINING VOC CONTENT IN COATINGS
D 3971	DICHLOROMETHANE SOLUBLES IN CELLULOSE
D 4017	WATER BY KF METHOD
D 4060	TABER ABRASION
D 4085	METALS IN CELLULOSE BY AA
D 4138	MEASUREMENT OF DRY FILM THICKNESS BY SEM
D 4139	VOLATILE & NONVOLATILE CONTENT OF PIGMENTS
D 4144	ESTIMATING PACKAGE STABILITY FOR UV CURING
D 4145	COATING FLEXIBILITY OF PREPAINTED SHEET
D 4212	VISCOSITY BY DIP-TYPE VISCOSITY CUPS
D 4214	DEGREE OF CHALKING OF EXTERIOR PAINTS
D 4262	PH OF CLEANED OR ETCHED CONCRETE
D 4263	MOISTURE IN CONCRETE/PLASTIC SHEET METHOD
D 4285	INDICATING MOISTURE IN CONCRETE METHOD
D 4301	TOTAL CHLORINE IN EPOXY COMPOUNDS
D 4359	DETERMINING MATERIAL AS LIQUID OR SOLID
D 4370	ACID AND BASE MEQ CONTENT OF ECOAT BATHS
D 4414	MEASUREMENT OF WET FILM THICKNESS BY NOTCH
D 4451	PIGMENT CONTENT BY LOW TEMPERATURE ASHING
D 4518	MEASURING STATIC FRICTION OF COATED SURFACES
D 4563	TIO2 CONTENT BY AA
D 4584	APPARENT PH OF ELECTRODE BATHS
D 4613	APPARENT PH OF WATER IN PHENOL FORMALD
D 4639	VOLATILE CONTENT IN PHENOLIC RESINS
D 4640	STROKE CURE TIME THERMOSET PHENOL FORMALD
D 4706	QUALITATIVE DET. METHYLOL GROUPS/PHENOLICS
D 4713	NONVOLATILE CONTENT OF INKS, RESINS,SOLNS
D 4752	MEK RESISTANCE OF ZINC RICH PRIMERS
D 4758	NONVOLATILE CONTENT OF LATEXES

D 4796	BOND STRENGTH OF TRAFFIC MARKING MATERIALS
D 4946	BLOCKING RESISTANCE OF ARCHITECTURAL PAINTS
D 4948	DETERMINATION OF UPPER LAYER VISCOUS

COATINGS: ASTM SPECIFICATION COMPLIANCE OF RAW MATERIALS AND FABRICATED COMPONENTS

D 29	LAC RESINS
D 34	WHITE PIGMENTS
D 49	RED LEAD
D 50	MANGANESE PIGMENTS
D 79	ZINC OXIDE PIGMENTS
D 81	WHITE LEAD PIGMENTS
D 84	RED LEAD PIGMENTS
D 85	OCHRE PIGMENT
D 126	PB CHROMATE, CR GREEN
D 154	VARNISHES
D 207	DRY BLEACHED LAC
D 209	LAMPBLACK PIGMENT
D 210	BONE BLACK PIGMENT
D 211	CHROME YELLOW, ORANGE
D 212	CHROME GREEN
D 215	WHITE LINSEED OIL PAINT
D 237	ORANGE SHELLAC
D 261	IRON BLUE PIGMENT
D 262	ULTRAMARINE BLUE PIGMENT
D 263	CHROME OXIDE GREEN
D 267	GOLD BRONZE POWDER
D 281	SPATULA OIL ABSORPTION
D 283	COPPER PIGMENTS
D 333	CLEAR, PIGMENTED LACQUER
D 360	SHELLAC VARNISHES
D 365	NITROCELLULOSE SOLUTION
D 444	ZINC YELLOW PIGMENT
D 475	PARA RED TONER PIGMENT
D 476	TITANIUM DIOXIDE PIGMENT
D 480	FLAKED AL POWDERS

D 520	ZINC DUST PIGMENTS
D 521	METALLIC ZINC POWDER
D 561	CARBON BLACK
D 600	LIQUID PAINT DRIERS
D 601	OITICICA OIL
D 602	BARIUM SULFATE PIGMENT
D 603	ALUMINUM SILICATES
D 604	DIATOMACEOUS SILICA
D 605	MAGNESIUM SILICATES
D 607	WET GROUND MICAS
D 656	PURE TOLUIDINE RED TONER
D 715	BARIUM SULFATE PIGMENT
D 717	MAGNESIUM SILICATE PIGM.
D 718	ALUMINUM SILICATE PIGM.
D 719	DIATOMACEOUS SILICA
D 763	RAW AND BURNT UMBER
D 765	RAW AND BURNT SIENNA
D 768	YELLOW IRON OXIDE
D 769	SYNTHETIC FE-OXIDE
D 784	INDIAN LACS
D 802	PINE OIL
D 803	TALL OIL
D 856	PINE TAR, PINE TAR OILS
D 960	RAW CASTOR OIL
D 962	AL POWDER/PASTE
D 970	PARA/TOLUIDINE REDS
D 1064	IRON IN ROSIN
D 1257	HIGH GRAVITY GLYCERIN
D 1258	HIGH GRAVITY GLYCERIN
D 1467	FATTY ACIDS
D 1537	SOYBEAN FATTY ACIDS
D 1538	LINSEED FATTY ACIDS
D 1539	DEHYDRATED CASTOR OILS
D 1540	CHEMICAL RESISTANCE
D 1541	IODINE VALUE/DRYING OILS
D 1585	FATTY ACID CONTENT/ROSIN

D 1640	DRY, CURE TIME
D 1648	EXTERIOR DURABILITY
D 1654	CORROSION RESISTANCE
D 1841	COCONUT FATTY ACIDS
D 1842	CORN FATTY ACIDS
D 1843	COTTONSEED FATTY ACIDS
D 1845	STRONTIUM CHROMATE
D 1984	TALL OIL FATTY ACIDS
D 2571	WOOD FURNITURE LACQUERS
D 2575	POLYMERIZED FATTY ACIDS
D 2931	LATEX FLAT WALLS PAINTS
D 2932	EXTERIOR HOUSE COATINGS
D 3129	LATEX FLAT WALLS PAINTS
D 3256	PHTHALOCYANIN BLUE, GREEN
D 3280	WHITE ZINC PIGMENTS
D 3322	PRIMER SURFACES
D 3323	SOLVENT/FLAT WALL PAINTS
D 3358	WATER BASED FLOOR PAINTS
D 3383	SOLVENT/FLOOR PAINTS
D 3425	INTERIOR SEMIGLOSS
D 3451	POWDER COATINGS
D 3619	ANHYDROUS AL-SILICATES
D 3721	SYNTHETIC RED IRON OXIDE
D 3724	SYNTHETIC BROWN FE-OXIDE
D 3730	ARCHITECTURAL WALL COAT
D 3794	COIL COATINGS
D 4302	ARTISTS' PAINT
D 4540	INT. SEMI-GLOSS/GLOSS
D 4712	INT. WATER BASE PAINTS

RUBBER AND RUBBER PRODUCTS, NATURAL AND SYNTHETIC: ASTM TESTING PROCEDURES

C 542	SPECIFICATION FOR LOCK-STRIP GASKETS
C 564	SPECIFICATION FOR RUBBER GASKETS, PIPE FITTINGS
D 149	DIELECTRIC BREAKDOWN STRENGTH
D 150	DIELECTRIC CONSTANT

D 256	IZOD IMPACT
D 297	CHEMICAL ANALYSIS
D 378	STANDARD TESTS FOR RUBBER BELTING
D 380	STANDARD TESTS FOR RUBBER HOSE
D 395	COMPRESSION SET
D 413	ADHESION TO A FLEXIBLE SUBSTRATE
D 429	ADHESION TO A RIGID SUBSTRATE
D 471	EFFECT OF LIQUIDS
D 518	SURFACE CRACKING
D 573	DETERIORATION IN AIR OVEN
D 575	COMPRESSION
D 746	BRITTLENESS ON IMPACT
D 814	VAPOR TRANSMISSION
D 865	DETERIORATION BY HEATED AIR
D 925	STAINING OF SURFACES-PER COATING
D 1043	TORSIONAL FLEX
D 1084	VISCOSITY OF ADHESIVE
D 1148	HEAT AND UV LIGHT DISCOLORATION
D 1149	SURFACE OZONE CRACKING IN A CHAMBER
D 1171	OZONE CRACKING-TRIANGULAR METHOD
D 1229	COMPRESSION SET AT LOW TEMPERATURES
D 1278	CHEMICAL ANALYSIS OF NATURAL RUBBER
D 1416	CHEMICAL ANALYSIS OF SYNTHETIC RUBBER
D 1460	CHANGE IN LENGTH DURING IMMERSION IN LIQUID
D 1506	ASH OF CARBON BLACK
D 1508	FINES CONTENT OF CARBON BLACK
D 1509	CARBON BLACK HEATING LOSS
D 1510	CARBON BLACK IODINE ABSORPTION NUMBER
D 1511	PELLET SIZE DISTRIBUTION OF CARBON BLACK
D 1512	PH OF CARBON BLACK
D 1513	POUR DENSITY-PELLETED CARBON BLACK
D 1514	SIEVE ANALYSIS
D 1519	MELTING RANGE RUBBER COMPOUNDING MATLS
D 1618	TOLUENE DISCOLORATION
D 1766	SOLUBILITY OF RUBBER CHEMICALS
D 1817	DENSITY WITH IMMERSION LIQUID

D 1870	TUBE OVEN AGING QUOTE
D 1871	ADHESION OF WIRE TO RUBBER
D 2000	CLASSIFICATION AUTOMOTIVE RUBBER PRODUCTS
D 2240	DUROMETER HARDNESS OF RUBBER
D 2527	SPLICE STRENGTH
D 2630	STRAP PEEL ADHESION
D 2702	FTIR IF RUBBER CHEMICALS
D 2703	UV ABSORPTION OF RUBBER CHEMICALS
D 2934	COMPATIBILITY WITH SERVICE FLUIDS
D 3137	HYDROLYTIC STABILITY
D 3265	TINT STRENGTH
D 3677	IDENTIFICATION BY FTIR
D 3849	PRIMARY AGGREGATE DIMENSIONS BY SEM/TEM
D 3900	DETERMINATION OF ETHYLENE UNITS IN EPM/EPDM
D 3959	DISCOLORATION SENSITIVITY TO TOBACCO SMOKE
D 4004	METAL CONTENT BY AA–PER METAL
D 4005	DEGREE OF FUSION OF PVC DISPERSION COATINGS
D 4075	METAL CONTENT BY AA
D 4527	SOLVENT EXTRACTABLES OF CARBON BLACK
D 4569	ACIDITY OF SULFUR
D 4570	PARTICLE SIZE BY SIEVE, FOR SULFUR
D 4571	VOLATILE CONTENT
D 4572	WET SIEVE ANALYSIS FOR SULFUR
D 4573	OIL CONTENT IN SULFUR
D 4574	ASH CONTENT OF SULFUR
D 4578	PERCENT INSOLUBLE SULFUR BY CS2 EXTRACTION
D 4004	METAL CONTENT BY AA-PER METAL
D 4005	DEGREE OF FUSION OF PVC DISPERSION COATINGS
D 4075	METAL CONTENT BY AA
D 4527	SOLVENT EXTRACTABLES OF CARBON BLACK
D 4569	ACIDITY OF SULFUR
D 4570	PARTICLE SIZE BY SIEVE, FOR SULFUR
D 4571	VOLATILE CONTENT
D 4572	WET SIEVE ANALYSIS FOR SULFUR
D 4573	OIL CONTENT IN SULFUR
D 4574	ASH CONTENT OF SULFUR
D 4578	PERCENT INSOLUBLE SULFUR BY CS2 EXTRACTION

RUBBER AND RUBBER PRODUCTS: ASTM SPECIFICATION COMPLICANCE OF RAW MATERIALS AND FABRICATED COMPONENTS

D 378	RUBBER BELTING
D 380	RUBBER HOSE
D 622	AUTOMOTIVE VACUUM LINE
D 751	COATED FABRICS
D 846	RUBBER CEMENTS
D 876	PVC TUBING
D 1330	SHEET GASKETS
D 1414	O-RINGS
D 1417	SYNTHETIC LATEX
D 1565	OPEN CELL FOAM
D 1764	LATEX AUTOMOTIVE PARTS
D 1992	PLASTICIZERS
D 2000	RUBBER AUTOMOTIVE PARTS
D 3453	FOAMS, FURNITURE, BEDDING
D 3489	MICROCELLULAR URETHANE
D 3490	BONDED URETHANE FOAMS
D 3571	LAUNDRY INLET HOSE
D 3572	LAUNDRY OUTLET HOSE
D 3573	SINK SPRAY HOSE
D 3574	SLAB/MOLDED FOAMS
D 3575	CELLULAR OLEFINS
D 3676	RUG UNDERLAY CUSHION
D 3738	RUBBER HOSPITAL SHEETS
D 3851	SHOE SOLING MATERIALS
D 4295	ZINC OXIDE
D 4316	RUBBER WATER BOTTLES
D 4528	SULFUR
D 4677	TITANIUM DIOXIDE
D 4679	RUBBER GLOVES
D 4817	STEARIC ACID
D 4819	CELLULAR OLEFINS
D 4924	WAXES

Index